Dear John,

You might enjoy looking into this volume which describes in the first part my scientific endavor but of course that is mainly outside your field.

Cordially

Mike

The Rudolf Mössbauer Story

Michael Kalvius • Paul Kienle

Editors

The Rudolf Mössbauer Story

His Scientific Work and Its Impact on Science and History

 Springer

Editors
Michael Kalvius
Paul Kienle
Technische Universität München
James-Franck-Str. 1
85748 Garching
Germany
kalvius@ph.tum.de
Paul.Kienle@ph.tum.de

ISBN 978-3-642-17951-8   e-ISBN 978-3-642-17952-5
DOI 10.1007/978-3-642-17952-5
Springer Heidelberg Dordrecht London New York

Library of Congress Control Number: 2011944215

Printed on acid-free paper

Springer is part of Springer Science+Business Media (www.springer.com)

Rudolf Mössbauer

Dedicated in sincere gratitude to our colleague and friend Rudolf Mössbauer.

Foreword

Before the Second World War, Göttingen and Berlin were the German hubs of physics. Munich also had a very good reputation for physics, thanks to Conrad Röntgen and Max von Laue, and this was enhanced by Arnold Sommerfeld's famous school, from which a number of Nobel Laureates emerged. Munich had also had an excellent reputation for chemistry for a long time. In 1875, Adolf von Baeyer was appointed to Munich as the successor to Justus von Liebig, although to the sister university. Nevertheless, it was the appointment of scientists such as Liebig and von Baeyer to Munich which turned Munich into a Mecca for chemistry alongside Heidelberg and later Berlin.

Our university, then still known as the "Technische Hochschule München," excelled in the decades before the Second World War, primarily thanks to its chemistry. Two Nobel Laureates held professorships here – the organic chemists Heinrich Otto Wieland and Hans Fischer. This proud tradition continued with the awarding of Nobel Prizes to the metal-organics chemist Ernst Otto Fischer (1973) and the biochemist Robert Huber (1988).

After 1945, science in Germany stood on the threshold of a new beginning. Munich succeeded in attracting important physicists, including such illustrious names as Werner Heisenberg and Heinz Maier-Leibnitz. Outstanding scientists also helped the physics at the TH München in achieving an international reputation. The Professor for Technical Physics, Walther Meissner, carried out pioneering work in low temperature research and in 1946 – only 1 year after the war ended – he set up a commission of the Bavarian Academy of Sciences and Humanities with associated research institute. This is where "magnetic flux quantization" was observed for the first time. The Walther Meissner Institute for Low Temperature Research has maintained its close links with the TUM's physics department to this day.

Heinz Maier-Leibnitz, who was appointed Meissner's successor in 1952, was a student and colleague of the Nobel Laureates James Franck and Walther Bothe. In the USA, he was able to acquire crucial experience. Maier-Leibnitz introduced a new, promising discipline to the Institute for Technical Physics at our university: Applied nuclear physics. In 1957, he was also able to start up the first research reactor at a German university in Garching. Consequently, his Institute had an excellent research and experimental facility for nuclear physics, neutron physics, and solid state physics at its disposal.

Maier-Leibnitz gave his diploma and doctoral students topics which led to the development of measuring methods. He tried to apply these new methods to a wide range of research fields. Improvisation was required, given the limited equipment at his Institute. The students had to work in cramped laboratory space, share measurement equipment with each other, and even build it themselves in many cases.

One of the most gifted of Maier-Leibnitz' students was Rudolf Mössbauer, who was born in Munich in 1929. Like many of his contemporaries, he had to wait more than a year before he was admitted to the bomb-damaged technical university, only taking up his studies in the winter semester 1949/1950. He spent the time in between working in a laboratory at a company which manufactured optical lenses, Optische Werke Rodenstock, in Munich. His physics teachers at high school had been disappointing – and it was this that motivated him to turn to precisely this discipline at university and to investigate it further.

As Mössbauer later acknowledged, research at Professor Maier-Leibnitz' institute could be conducted "in an atmosphere of great freedom" – and this surely paved the way for his subsequent success. For his doctoral thesis, the young student investigated the "nuclear resonance fluorescence of gamma radiation in iridium 191." His meticulous way of working paid off, because for iridium 191, of all things, the measurement deviations were minimal. The doctoral student did not let himself be misled and did not write off his surprising observations as measurement errors.

Since the research possibilities in Munich were limited, Maier-Leibnitz made it possible for Mössbauer to work at the Max Planck Institute for Medical Research in Heidelberg. In 1957, this is where he made the unexpected discovery that the cross-section increases greatly with decreasing temperature, instead of decreasing as would have been expected. In a further experiment in Heidelberg, he succeeded in measuring the recoilless spectrum of the gamma-ray line directly. After successfully defending his thesis, Mössbauer was awarded his doctorate in January 1958 at the TH München and subsequently worked as a research assistant with his teacher Maier-Leibnitz.

The "Mössbauer effect" named after its discoverer enables researchers to measure the interaction of a nucleus with its environment and the gamma quantum with the gravitational field with great accuracy. It can be used for many different applications not only in nuclear physics and solid state physics, but also in chemistry, biomedicine, geology, mineralogy, and archaeology. The high-precision Mössbauer spectroscopy has even been used to prove predictions of Albert Einstein's General Theory of Relativity in the laboratory. A current example of its application is space research: The two Mars robots "Spirit" and "Opportunity" used this type of spectrometer to investigate the rocks on the red planet. It was thus possible to prove not only the earlier existence of water, but also that the atmosphere once contained much more oxygen than it does today. In 2013, when the Russian "Fobos Grunt" probe lands on Phobos, one of Mars' moons, a Mössbauer spectrometer will again be part of the instrumentation.

In autumn 1958, the newly qualified doctor of physics was full of anticipation as he presented his discovery at the Meeting of German Physicists in Essen, but his German colleagues showed only moderate interest. A year later, however, during a research colloquium at the University of Heidelberg, Mössbauer attracted the attention of the Swiss physicist Felix Böhm, who was teaching in California, and he informed his university, the "California Institute of Technology." Böhm's colleague Richard Feynman sent a wire: "Get this guy!." In Pasadena, CalTech provided Mössbauer with excellent research facilities, and in 1961 the 32-year-old scientist shared the Nobel Prize for Physics with Robert Hofstadter "for his researches concerning the resonance absorption of gamma radiation and his discovery in this connection of the effect which bears his name."

This was the first time that a doctoral thesis which originated from our university had been awarded a Nobel Prize. The smart young physicist was celebrated by the students in Germany like a pop star. The joy was mixed with melancholy on the "brain drain" to the USA, however.

It was almost a miracle that the Nobel Laureate could not only be persuaded to return to Germany, but to his Alma Mater in Munich as well, because a great many German universities and research facilities were competing for the highly talented young scientist. And he had his price. Mössbauer laid down conditions which the bureaucracy of the German Ministries found it hard to deal with, such as converting the classical university faculty into a "department" of professors on equal footing which was managed by an elected board of directors, 20 new chairs and world class research facilities. Mössbauer's demands were vigorously supported by his colleagues in Munich – Heinz Maier-Leibnitz, Nikolaus Riehl, Wilhelm Brenig, and Wolfgang Wild – and the farsighted assistant secretary of state Dietrich Bächler (now an honorary senator of the TUM). Even as late as April 1963, Mössbauer complained in a letter to his teacher Maier-Leibnitz after prolonged, laborious negotiations that the "attitude displayed so far" by the Bavarian Government still required a "drastic change." Finally, the Minister of Culture, Theodor Maunz, and Minister of Finance, Rudolf Erhard, were convinced after all.

The outcome of the negotiations was fantastic: 16 chairs were approved, and four more positions were funded like chairs; there were also hundreds of positions for academic staff. The new physics department in Garching was given an outstandingly well-equipped new building, designed by Professor Angerer, which even today is still setting standards. The press celebrated the "Second Mössbauer Effect."

This investment did actually bear fruit. Klaus von Klitzing, who taught at the TUM from 1980 to 1984 as Professor for Solid State Physics, was awarded the 1985 Nobel Prize for Physics for his discovery of the "quantum Hall effect." Erwin Neher studied physics at the TUM from 1963 to 1966 and was awarded the 1991 Nobel Prize for Physiology or Medicine jointly with Bert Sackmann for their discoveries regarding the function of individual ion channels in cells. Wolfgang Ketterle studied physics at the TUM from 1978 to 1982. In 2001, he shared the Nobel Prize for Physics with Eric A. Cornell and Carl E. Wieman "for the achievement of Bose-Einstein condensation in dilute gases of alkali atoms, and for early fundamental studies of the properties of the condensates."

Rudolf Mössbauer was a flagship of the TUM for decades with his teaching and research work and was deemed to be a brilliant teacher. Every year he travelled to the USA for 3 months to keep in touch with developments there. Countless conferences, working groups, and doctoral students worldwide were concerned with the application possibilities of Mössbauer spectroscopy.

In 1972 Rudolf Mössbauer took on a new challenge: For 5 years he was Director of the Institut Laue-Langevin (ILL) in Grenoble and its high-flux neutron reactor. During this period of time, he turned his attention to low-energy neutrino physics, another promising field of work. Mössbauer became a pioneer in this discipline. The discovery that these elementary particles have a rest mass after all could lead to the correction or even the complete replacement of standard models to explain the world.

Our Alma Mater can consider itself fortunate that Rudolf Mössbauer remained loyal to it for most of his years of active research: For almost four decades he worked here as a student, doctoral student, assistant, and professor. His meteoric rise to become a Noble Laureate at just 32 years of age is still an academic fairytale which enthuses and motivates. It is thanks to scholars like Rudolf Mössbauer that Munich of today is the outstanding German location for physics, and that Munich cannot be overlooked in the world of physics research and finally – that Munich now ranks among the global elite in more disciplines than ever before.

I am delighted that this anthology of renowned scientists and colleagues pays due recognition to the pioneering discovery of Rudolf Mössbauer in its multitude of applications and would like to thank all the authors and editors for their collaboration on this project.

Munich   *Wolfgang A. Herrmann*
August 2011

Preface

In 1958, Rudolf Mössbauer, a student of Heinz Maier-Leibnitz, professor of technical physics at the Technische Hochschule München, published his doctoral thesis in the journal *Zeitschrift für Physik*, **151**, 124 (1958) with the unpretentious title "Kernresonanzfluoreszenz von Gammastrahlung in [191]Ir". It contained "dynamite" in science. A second article with a very similar title "Kernresonanzabsorption von Gammastrahlung in Ir",[191] published in *Naturwissenschaften* **45**, 538, (1958), showed a spectrum of 129 keV γ-rays with the natural line width of 10^{-5} eV, a resolution never observed before. The science community, busy with studying parity break down of the weak interaction, did not care for these exotics, or even thought these results were likely to be wrong. They were checked in 1959 by two US groups. In one of them, a participant betted a dime that Mössbauer was right. It was soon proven that all of the Mössbauer's findings were correct and now physicists were truly electrified. Following the discovery of the resonance absorption of other γ-lines in crystals with "zero energy loss," such as the 14.4 keV γ-transition in [57]Fe, unique uses of what was soon called the "Mössbauer Effect" spread rapidly into many fields of research. The Second International Mössbauer Conference in Paris held in the summer of 1961 was the show case for these fascinating applications and lead directly in the fall of 1961 to the Nobel Prize for the discovery of Rudolf Mössbauer with the citation: *For his researches concerning the resonance absorption of gamma radiation and his discovery in this connection of the effect which bears his name.*

In January 2009, Dr. Claus E. Ascheron from Springer Science and Business Media invited us to edit a book in honor of Rudolf Mössbauer on the occasion of the 50th anniversary in 2011 of the award of the Nobel Prize (1961) for the discovery of the "Mössbauer Effect." We accepted gladly this offer to honor Mössbauer's scientific work and its impact on science and history including his later pioneering work on neutrino oscillations. Mössbauer spectroscopy has provided many new insights in nearly all fields of natural sciences and into technological problems as well. One other unique feature is that, except for the recent applications using synchrotron radiation, the standard experimental setup is simple and measurements can be carried out in a standard laboratory. Its educational value is very high: it introduces the experimenter, especially students, into the basic phenomena of nuclear and solid state physics together with applications of quantum theory. Funding agencies like the International Atomic Energy Agency soon became aware of this aspect of the Mössbauer effect and provided special grants to set up Mössbauer spectrometers in countries with an emerging scientific community. This important development was especially pleasing to Rudolf Mössbauer to whom scientific education was a prime issue.

We decided to present in this book a mixture of contributions reflecting the discovery of the effect and its early history followed by reports on unique applications in various areas of science and views on future developments. For the historical part, we invited contemporary witnesses using a last opportunity to get their views on record. The authors of the scientific contributions dealing with important applications of the Mössbauer spectroscopy have highest scientific reputation and most of them are still working in the field.

After a reprint of Rudolf Mössbauer's Nobel lecture, the first book chapter, entitled *Rudolf Mössbauer in Munich*, makes the reader familiar with the fascinating discovery story of a diploma and a follow-up doctoral thesis on the temperature dependence of nuclear resonance absorption of γ-rays. Remarkably, it started in the after-war time when nuclear physics was still forbidden in Germany. In careful experiments, performed at the Max-Planck-Institut für Medizinische Forschung in Heidelberg – the Munich institute of Maier-Leibnitz was too crowded with students and equipment was scarce – Mössbauer found that by decreasing the temperature of source and absorber to that of liquid nitrogen the resonance absorption effect *increased*, instead of the expected decrease. He explained this unexpected discovery, using basically a paper published by Lamb in 1939 on the resonance absorption of *neutrons* in nuclei bound in a crystal, as an increasing fraction of "zero recoil energy loss" events in the emission and absorption of γ-rays by a nucleus bound in a crystal when its temperature was lowered. In a follow-up experiment, Mössbauer directly showed by introducing a small Doppler shift between the source and the absorber that the measured resonance absorption cross section as function of the energy introduced by the Doppler shift is given by the folding of the emission line of Lorentz shape with natural line width with the corresponding resonance absorption cross section. A subchapter *Dawn of the Mössbauer Spectroscopy in Munich* reports on technical developments for recording Mössbauer spectra and unique studies of isomer shifts, electric quadrupole, and magnetic dipole splitting of rotational transitions in deformed rare-earth nuclei, important for determining the exceptional properties of rotating super-fluid deformed nuclei. In a short personal report on a failed visit of the Louvre of one of us (PK) with Maier-Leibnitz on the occasion of the Second Mössbauer Conference in Paris in 1961, the tale is narrated why the Nobel Prize award for the discovery of the Mössbauer effect was given to Rudolf Mössbauer alone. In a final paragraph, the story how Mössbauer changed his field of research and got involved in the new adventure "Neutrino Physics" is told.

Following a reprint of Rudolf Mössbauer's own account of the discovery of his effect, the early days of the Mössbauer Effect and the exciting discovery of the 14.4 keV Mössbauer line of ^{57}Fe, which allowed many new applications until today, are recalled by authors who acted at the forefront of fundamental discoveries during these pioneering times. Early developments of the theory of the Mössbauer Effect are communicated by a famous pioneer who paved the way for understanding "zero energy loss" transitions of γ-rays in a solid and who may well be credited for introducing the term "Mössbauer Effect" into the world of science. The section is concluded by original contributions on the CalTech years of Rudolf Mössbauer and by the review on "The World beyond Iron" written by one of us (MK).

This historical Part is then complemented by comprehensive reviews of high-lights about the impact of Mössbauer spectroscopy in the fields of nuclear physics, applications of isomer-shift measurements in solid-state chemistry, and studies of internal magnetic fields in solids. A review of studies of dynamic processes in condensed matter, extraordinary biological and medical applications, and unique results of extraterrestrial Mössbauer spectroscopy follows. Scientific highlights continue with investigations of fundamental relativistic phenomena, studies of phase coherence in nuclear resonant scattering and by a chapter on performing Mössbauer-Effect studies using synchrotron radiation, which points into a bright future of ultrahigh-resolution, time-resolved synchrotron radiation spectroscopy. In particular, unique aspects of nuclear resonance scattering of synchrotron radiation as a probe of atomic vibrations are described.

The third Part reports on future developments of the Mössbauer spectroscopy. In a contribution *Dreams with Synchrotron Radiation*, the author describes challenges and opportunities for Mössbauer spectroscopy in the next 50 years showing that its future is bright indeed. The most challenging performance requirement for the Mössbauer studies of non-equilibrium systems is, first, to collect information, preferably with a single X-ray pulse from the radiation source when the system dwells in the non-equilibrium (excited) state and second, to probe mesoscopic structures with a spatial resolution of nanometers. A second contribution, *Mössbauer Effect with Electron Antineutrinos*, discusses in detail the proposal to use the mono-energetic neutrinos from bound-state ß-decay of ^{3}H to ^{3}He to perform an ultra-high resolution neutrino resonance absorption experiment. There could be a very interesting fundamental physics program ahead if such an experiment would work, but the authors concluded that this is rather unlikely at present.

The last Part, termed "Epilogue" contains two contributions. The first, entitled simply *Neutrinos*, refers to the second Love of Rudolf Mössbauer, the study of the physics of neutrino oscillations. As pointed out Rudolf Mössbauer and Felix Böhm initiated more than 30 years ago first experiments to study the mass and the mixing matrix of neutrinos by observing neutrino oscillations, following a proposal of Bruno Pontecorvo. Nowadays, such experiments created a so-called Neutrino Industry. Two Nobel Prizes went already into this exciting field which is thoroughly reviewed in this chapter; more are expected to come. The second and final contribution to this book, *The Second Mössbauer Effect*, uses the title of a headline of the "Spiegel" magazine, announcing the foundation of the Physik-Department at the Technische Hochschule München on the occasion of Mössbauer's return from CALTEC in 1964. It is a short review honoring the engagement of Maier-Leibnitz and his former student Mössbauer in successfully reforming the structure of German universities several years before the even more famous "Spiegel" article *Unter den Talaren der Muff von tausend Jahren* appeared, a trigger of the 1968 student revolution in Germany.

The selection of themes treated in this book certainly is short of reflecting all the aspects of Mössbauer spectroscopy. As editors we had to make choices, but we hope the reader gets an idea of Rudolf Mössbauer's outstanding contribution to modern science. One special other aspect of the Nobel Prize to Rudolf Mössbauer should be pointed out. It signaled to the world that Germany after a dark past and the destructions of World War Two had returned into the world of science. All of us are particularly grateful to Rudolf Mössbauer for his long-lasting efforts in that direction.

We appreciate the excellent collaboration with the authors of the contributions, the advices we got from various members of the "Mössbauer Community" and are especially grateful for the help in editing of the book by Dr. Josef Homolka and the steady support of Dr. Claus E. Ascheron and his staff from Springer Media.

This book was intended for appreciation of the Fiftieth Anniversary of the presentation of the Nobel Prize to Rudolf Mössbauer in 1961 and planned to be delivered to him December 10, 2011 in Munich. Sad enough he passed away during printing September 14, 2011 in the arms of his beloved wife, Christel Mössbauer. We express our sincere condolence to her.

Garching   *G. Michael Kalvius*
August 2011   *Paul Kienle*

Contents

Part I Discovery and Early Period of the Mössbauer Effect

**1 Recoilless Nuclear Resonance Absorption
of Gamma Radiation** 3
Rudolf L. Mössbauer
References.. 17

2 Rudolf Mössbauer in Munich 19
G.M. Kalvius and P. Kienle
2.1 Years of Study at the Technische Hochschule München 19
2.2 Diploma Thesis of Mössbauer
with Prof. H. Maier-Leibnitz 20
2.3 Doctor Thesis of Mössbauer with Prof. H. Maier-
Leibnitz Performed with Experiments
at the Max-Planck Institute for Medical Research
in Heidelberg 22
2.4 The Dawn of Mössbauer Spectroscopy in Munich 25
2.5 The Second International Mössbauer Conference
in Paris (1961) 31
2.6 The "Second Mössbauer Effect"................................... 32
2.7 Back to Munich from Grenoble with a New
Scientific Challenge 33
2.8 Outlook 34
References.. 34

3 The Discovery of the Mössbauer Effect 37
Rudolf L. Mössbauer
3.1 Introduction... 37
3.2 The Discovery of the Mössbauer Effect........................... 37
References.. 47

4   **The Early Period of the Mössbauer Effect and the**
     Beginning of the Iron Age   49
     John P. Schiffer
     References   68

5   **The Early Iron Age of the Mössbauer Era**   69
     Stanley S. Hanna
     References   80

6   **The Early Developments of the Theory**
     of the Mössbauer Effect   83
     Harry J. Lipkin
     6.1   Introduction   83
     6.2   Prehistory   83
     6.3   The Iridium Age   84
     6.4   How I Got into the Mössbauer Business   86
     6.5   The Iron Age   88
     6.6   Why Mössbauer Couldn't Use ^{57}Fe   89
     6.7   The Generalized Mössbauer Effect   90
     6.8   A Connection with Mössbauer's Later Work
            on Neutrino Oscillations   91
     6.9   A Curious Recollection from the Past   92
     References   92

7   **The CalTech Years of Rudolf Mössbauer**   93
     Richard L. Cohen
     7.1   Introduction   93
     7.2   Recollections of Prof. Felix H. Boehm   93
     7.3   Richard Cohen Gets Involved   94
     7.4   The Next Wave   104
     7.5   Comments By Gunther K. Wertheim   107
     7.6   The Sunset of the CalTech Period   108
     7.7   Epilogue   109
     References   109

8   **The World Beyond Iron**   111
     G. Michael Kalvius
     8.1   Introduction   111
     8.2   The More Commonly Used Mössbauer Elements
            Other Than Iron   112
            8.2.1   Elements of the 5th Main Group (Sn to I)   114
            8.2.2   Alkalies   119
            8.2.3 d-Transition Elements   120
            8.2.4 f-Transition Elements   123
            8.2.5   High Resolution Mössbauer Resonances   134
     References   139

Part II Highlights of Applications of the Mössbauer Effect

9 Nuclear Physics Applications of the Mössbauer Effect 145
Walter F. Henning
9.1 Introduction: Some Historical Remarks 145
9.2 The Nuclear Landscape and the Mössbauer Effect 147
9.3 Mössbauer Studies of Nuclear Properties 149
    9.3.1 Measurement of Nuclear Lifetimes 151
    9.3.2 Isomer Shifts in Nuclei 151
    9.3.3 Isomer Shifts and Nuclear Models 154
    9.3.4 Hyperfine Splitting and Nuclear Moments 160
    9.3.5 Hyperfine Anomaly 164
    9.3.6 Tests of Time Reversal Invariance 167
    9.3.7 Dispersion Phenomena in Mössbauer Spectra 168
9.4 Conclusion and Perspectives 169
References 173

10 Isomer Shifts in Solid State Chemistry 175
F.E. Wagner and L. Stievano
10.1 Introduction 175
10.2 History and Basic Concepts 176
10.3 Theory of the Isomer Shift 180
10.4 The Measurement of Isomer Shifts 182
10.5 The Calibration Problem 183
10.6 Isomer Shifts in d Transition Elements 188
10.7 Isomer Shifts in Main Group Elements 192
10.8 Outlook 195
References 195

11 The Internal Magnetic Fields Acting on Nuclei in Solids 199
Israel Nowik
11.1 Introduction 199
11.2 Internal Fields in Magnetic Solids 200
    11.2.1 The Magnetic Hyperfine Interactions in Free
             Atoms or Ions 201
    11.2.2 The Magnetic Hyperfine Interactions in nd
             (n = 3, 4, 5), and nf (n = 4, 5) Elements in Solids 201
    11.2.3 Influence of External Magnetic Fields 203
    11.2.4 Static Internal Fields in Magnetically Ordered Systems .. 204
    11.2.5 Dynamics of Internal Fields in Magnetically
             Ordered Systems 209
    11.2.6 Internal Magnetic Fields at Phase Transitions 210
    11.2.7 Internal Fields in Unusual Spin Structures 211
    11.2.8 Transferred Hyperfine Fields 216
    11.2.9 Internal Fields in Thin Layers 217

11.3   Summary.. 217
References.. 218

12   Time-Dependent Effects in Mössbauer Spectra.......................... 221
F. Jochen Litterst
12.1   Introduction.. 221
12.2   General Considerations 222
12.3   Fluctuations of Ionic Charge State and Symmetry 225
12.4   Spin Relaxation.. 229
        12.4.1   Relaxation Spectra in Paramagnets
                  and Magnetically Ordered States 229
        12.4.2   Superparamagnetic Relaxation............................ 233
        12.4.3   Examples of Non-white Noise Relaxation 235
12.5   Diffusion.. 237
12.6   Conclusion... 240
References.. 241

13   Mössbauer Spectroscopy of Biological Systems......................... 243
Eckard Münck and Emile L. Bominaar
13.1   Introduction... 243
13.2   Spin Hamiltonian Analysis 244
13.3   Sulfite Reductase: The Discovery
        of Coupled Chromophores... 246
13.4   Biomimetic Chemistry: Insights into the Dependence
        of Reactivity on Structure... 249
13.5   Fe$_3$S$_4$ Clusters, a New Role for Iron–sulfur Clusters 253
13.6   Core Distortion in an All-ferrous Fe$_4$S$_4$ Cluster 256
13.7   An Example of Whole-cell Mössbauer Spectra...................... 259
References.. 260

**14   Relativistic Phenomena Investigated
       by the Mössbauer Effect** 263
W. Potzel
14.1   Introduction... 263
14.2   Time Dilatation and Second-Order Doppler Shift 264
        14.2.1   Thermal Motion of a Mössbauer Nucleus
                  in a Lattice 264
        14.2.2   Zero-Point Motion of a Mössbauer Nucleus
                  in a Lattice 268
14.3   Experiments with High-Speed Centrifuges 269
        14.3.1   Ether Drift and Relativity Theory........................ 269
        14.3.2   Ether-Drift Experiments................................. 272
        14.3.3   Dependence of Clock Rates on the Velocity
                  Relative to Distant Matter 274
        14.3.4   Distinction Between the Special Theory
                  of Relativity and Covariant Ether Theories................ 274

14.4   Gravitational Redshift 276

     14.4.1   Importance of Gravitational Redshift Experiments 277

     14.4.2   Gravitational Redshift Implies Curved Spacetime 278

     14.4.3   Changes of Length and Time
           in a Gravitational Field 278

14.5   Gravitational Redshift Experiments Using
       the Mössbauer Effect 280

     14.5.1   GRS Experiments with the 14.4 keV
           Transition in ^{57}Fe 280

     14.5.2   GRS Experiments with the 93.3 keV
           Transition in ^{67}Zn 282

14.6   Relativity and Anisotropy of Inertia 286

14.7   Selected Modern Tests of the General Theory
       of Relativity 286

14.8   Conclusions 288

References 289

15 Extraterrestrial Mössbauer Spectroscopy 293
     Göstar Klingelhöfer

15.1   Introduction 293

15.2   Comparative Planetology 294

     15.2.1   The Planet Venus 295

     15.2.2   The Planet Mars 297

15.3   Laboratory Studies of Extraterrestrial Materials: Lunar
       Samples from the Apollo Era 297

     15.3.1   Lunar Samples from the Apollo Era 298

15.4   In Situ Exploration of the Martian Surface
       by Mössbauer Spectroscopy: The Mars Exploration
       Rover Mission 2003 300

     15.4.1   The Instrument MIMOS II 301

     15.4.2   Gusev Crater Landing Site, Mars 304

     15.4.3   Meridiani Planum Landing Site 306

     15.4.4   Victoria Crater 309

     15.4.5   Meteorites on Mars 309

15.5   Recent Developments 311

15.6   Summary and Outlook 311

References 313

16 Coherent Nuclear Resonance Fluorescence 317
     G.V. Smirnov

16.1   Introduction 317

16.2   When and How the Radiative Channel Starts Playing
       a Key Role 320

     16.2.1   Absorption at Nuclear γ-Resonance
           from the Point of View of the Optical Theory 320

     16.2.2   Visualization of the Coherent Forward Scattering 321

       16.2.3   Nuclear Resonant Bragg Scattering:
              Peculiar Features.. 326
       16.2.4   Suppression Effect.. 331
       16.2.5   Speed Up Effect 334
       16.2.6   On Synchrotron Mössbauer Source........................ 335
   References.. 337

**17 Nuclear Resonance Scattering of Synchrotron Radiation
   as a Unique Electronic, Structural, and Thermodynamic Probe......** 339
   E. Ercan Alp, Wolfgang Sturhahn, Thomas S. Toellner,
   Jiyong Zhao, and Bogdan M. Leu
   17.1   Introduction.. 339
   17.2   Synchrotron Radiation and Mössbauer Spectroscopy 340
   17.3   Nuclear Resonance and X-ray Scattering 342
   17.4   Synchrotron Mössbauer Spectroscopy 344
   17.5   Mössbauer Spectroscopy as a Tool for Lattice
        Dynamics in the Synchrotron Era 346
   17.6   Why Another Vibrational Spectroscopy? 348
   17.7   Geophysical Applications of NRIXS............................... 350
   17.8   Biological Applications of NRIXS 351
   17.9   Materials Science Applications of SMS and NRIXS 351
   References.. 353

Part III   Future Developments of the Mössbauer effect

18 Dreams with Synchrotron Radiation 359
   G.K. Shenoy
   18.1   Introduction.. 359
   18.2   Future Synchrotron Radiation Sources............................ 360
        18.2.1   Storage Rings 360
        18.2.2   X-ray Free Electron Laser (XFEL) Sources............... 361
        18.2.3   Return of Classical Mössbauer Energy
              Spectroscopy at Synchrotron Sources 363
   18.3   Potential Applications of New Generation Sources 365
   18.4   Exotic Mössbauer Resonances and Cosmology 367
   18.5   Entanglement and coherent control of nuclear excitons 370
   18.6   Conclusions.. 371
   References.. 371

19 Mössbauer Effect with Electron Antineutrinos 373
   W. Potzel and F.E. Wagner
   19.1   Introduction.. 373
   19.2   Bound-State β-Decay and Its Resonant Character 374
        19.2.1   Usual β-Decay... 374
        19.2.2   Bound-State β-Decay 374
        19.2.3 ^{3}H–^{3}He System 375

19.3   Mössbauer Antineutrinos 375
        19.3.1   Resonance Cross Section 375
        19.3.2   Phononless Transitions 376
        19.3.3   Linewidth 378
19.4   Fundamental Difficulties: Alternative Systems 383
19.5   Principle of Experimental Setup 384
19.6   Interesting Experiments 385
19.7   Conclusions 387
References 388

Part IV   Epilogue

20   Neutrinos 393
      F. von Feilitzsch and L. Oberauer
      20.1   Neutrino Physics 393
              20.1.1   Phenomenology of Neutrino Oscillations 393
      20.2   Oscillation Experiments at the ILL and Gösgen 395
              20.2.1   The ILL Experiment 395
              20.2.2   The Gösgen experiment 397
              20.2.3   The Neutrino Decay experiment at Bugey 398
      20.3   Solar Neutrino Experiments 400
      20.4   The GALLEX and GNO experiment 404
      20.5   Observation of Neutrino Flavor Transitions 405
      20.6   The Proof of Neutrino Oscillations 407
      20.7   The Future of Neutrino Physics 412
      References 415

21   The Second Mössbauer Effect 417
      Paul Kienle
      21.1   Excellence Clusters 422
      21.2   Sonderforschungsbereiche Transregios 422
      References 423

Index 425

Contributors

E. Ercan Alp Advanced Photon Source, Argonne National Laboratory, Argonne, IL 60439, USA, Eea@aps.anl.gov

Emile L. Bominaar Department of Chemistry, Carnegie Mellon University, 4400 Fifth Avenue, Pittsburgh, PA 15213, USA

Richard L. Cohen 43 Caribe Isle, Novato, CA 94949, USA, campcohen@aol.com

Fritz von Feilitzsch Physics Department, Technical University Munich, 85747 Garching, Germany, franz.vfeilitzsch@ph.tum.de

Stanley S. Hanna Physics Department, Stanford University, Stanford, CA 94305, USA

Walter F. Henning Physics Department, Argonne National Laboratory, Argonne, IL 60439, USA, wfhenning@anl.gov

G. Michael Kalvius Physics Department, Technical University Munich, 85747 Garching, Germany, kalvius@ph.tum.de

Paul Kienle Physics Department, Technical University Munich, 85747 Garching, Germany, paul.kienle@ph.tum.de

Göstar Klingelhöfer Institute for Inorganic and Analytic Chemistry, University of Mainz, 55099 Mainz, Germany, klingel@mail.uni-mainz.de

Bogdan M. Leu Advanced Photon Source, Argonne National Laboratory, Argonne, IL 60439, USA

Harry J. Lipkin Department of Particle Physics, Weizmann Institute of Science, 76100 Rehovot, Israel, Harry.lipkin@weizmann.ac.il

F. Jochen Litterst IPKM, Technical University Braunschweig, 38106 Braunschweig, Germany, j.litterst@tu-bs.de

Rudolf L. Mössbauer Physics Department, Technical University Munich, 85747 Garching, Germany

Eckard Münck Department of Chemistry, Carnegie Mellon University, 4400 Fifth Avenue, Pittsburgh, PA 15213, USA, emunck@cmu.edu

Israel Nowik The Racah Institute of Physics, The Hebrew University, 91904 Jerusalem, Israel, nowik@vms.huji.ac.il

Lothar Oberauer Physics Department, Technical University Munich, 85747 Garching, Germany, lothar.oberauer@ph.tum.de

Walter Potzel Physics Department, Technical University Munich, 85747 Garching, Germany, walter.potzel@ph.tum.de

John P. Schiffer Physics Division, Argonne National Laboratory, Argonne, IL 60439, USA, schiffer@anl.gov

Gopal K. Shenoy Advanced Photon Source, Argonne National Laboratory, Argonne, IL 60439, USA, gks@aps.anl.gov

Genadii V. Smirnov Russian Research Center, "Kurchatov Institute", 123182 Moscow, Russia, g.smirnov@gmx.net

Lorenzo Stievano Université Montpellier 2, Case 1502, 34095 Montpellier cedex 5, France, lorenzo.stievano@univ-montp2.fr

Wolfgang Sturhahn Advanced Photon Source, Argonne National Laboratory, Argonne, IL 60439, USA

Thomas S. Toellner Advanced Photon Source, Argonne National Laboratory, Argonne, IL 60439, USA

Friedrich E. Wagner Physics Department, Technical University Munich, 85747 Garching, Germany, friedrich.wagner@ph.tum.de

Jiyong Zhao Advanced Photon Source, Argonne National Laboratory, Argonne, IL 60439, USA

Part I
Discovery and Early Period
of the Mössbauer Effect

Rudolf Mössbauer started studying physics at the Technische Hochschule München in 1949. In 1958, he graduated as student of Heinz Maier–Leibnitz with a doctoral thesis on nuclear resonance fluorescence of γ-rays in ^{191}Ir. He discovered nuclear resonance absorption of 129 keV γ-rays "without recoil" in a crystal exhibiting the natural line width of 10^{-5} eV thus leading to an energy resolution of about 10^{-10}, never achieved before. Following this pioneering thesis work, many other "Mössbauer-lines" were discovered, such as the 14.4 keV γ-line of ^{57}Fe with high recoil free fraction at room temperature and even narrower line width of about 10^{-7} eV. This led to fascinating applications in many fields of sciences and to the Nobel Prize for Rudolf Mössbauer in 1961 *For his researches concerning the resonance absorption of gamma radiation and his discovery in this connection of the effect which bears his name.*

Chapter 1
Recoilless Nuclear Resonance Absorption of Gamma Radiation

Rudolf L. Mössbauer

It is a high distinction to be permitted to address you on the subject of recoilless nuclear resonance absorption of gamma radiation. The methods used in this special branch of experimental physics have recently found acceptance in many areas of science. I take the liberty to confine myself essentially to the work which I was able to carry out in the years 1955–1958 at the Max Planck Institute in Heidelberg, and which finally led to establishment of the field of recoilless nuclear resonance absorption. Many investigators shared in the preparations of the basis for the research we are concerned with in this lecture. As early as the middle of the last century Stokes observed, in the case of fluorite, the phenomenon now known as fluorescence – namely, that solids, liquids, and gases under certain conditions partially absorb incident electromagnetic radiation which immediately is reradiated. A special case is the so-called resonance fluorescence, a phenomenon in which the re-emitted and the incident radiation both are of the same wavelength. The resonance fluorescence of the yellow D lines of sodium in sodium vapour is a particularly notable and exhaustively studied example. In this optical type of resonance fluorescence, light sources are used in which the atoms undergo transitions from excited states to their ground states (Fig. 1.1). The light quanta emitted in these transitions ($A \rightarrow B$) are used to initiate the inverse process of resonance absorption in the atoms of an absorber which are identical with the radiating atoms. The atoms of the absorber undergo a transition here from the ground state (B) to the excited state (A), from which they again return to the ground state, after a certain time delay, by emission of fluorescent light.

As early as 1929, Kuhn [1] had expressed the opinion that the resonance absorption of gamma rays should constitute the nuclear physics analogue to this optical resonance fluorescence. Here, a radioactive source should replace the optical light source. The gamma rays emitted by this source should be able to initiate

Nobel Lecture © The Nobel Foundation 1961, reprinted by permission.

R.L. Mössbauer (✉)
Physics Department, Technical University Munich, 85747 Garching, Germany

M. Kalvius and P. Kienle (eds.), *The Rudolf Mössbauer Story*,
DOI 10.1007/978-3-642-17952-5_1, © Springer-Verlag Berlin Heidelberg 2012

Fig. 1.1 Scheme of
resonance absorption

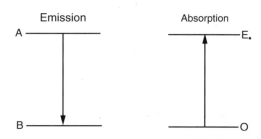

the inverse process of nuclear resonance absorption in an absorber composed of nuclei of the same type as those decaying in the source. Again, the scheme of Fig. 1.1 would hold here, but the radiative transitions would now take place between nuclear states. Nevertheless, all attempts in the next two decades to find this nuclear resonance absorption proved fruitless. Before I can approach the subject of my talk, it is appropriate to consider the reasons why the discovery of nuclear resonance absorption was so long delayed.

For simplicity, we shall first consider a nuclear transition of a free nucleus at rest. The gamma quantum emitted in the transition A B imparts a recoil momentum $\vec{p}$ to the emitting nucleus and consequently a kinetic energy ΔE, which is given by

$$\Delta E = \frac{\vec{p}^{\,2}}{2M} = \frac{E_0^{\,2}}{2Mc^2}, \tag{1.1}$$

where M is the mass of the nucleus and c is the velocity of light. The energy liberated in this nuclear transition is divided, in accordance with the law of conservation of momentum, the larger part being carried away by the emitted quantum, the other part going to the emitting nucleus in the form of recoil energy. This recoil-energy loss of the quantum has the consequence that the emission line does not appear at the position of the transition energy E_0 but is displaced to lower energy by an amount ΔE (Fig. 1.2). The absorption line, on the contrary, is displaced to a higher energy by the same amount ΔE, because in order for the process of resonance absorption to occur, a quantum must provide, in addition to the transition energy E_0, the amount of energy ΔE which is taken up by the absorbing nucleus in the form of a recoil kinetic energy. Typical values for the line shifts ΔE lie in the range from 10^{-2} to 10^2 eV; they are therefore very small in comparison with the energies of the gamma quanta, which frequently are of the order of magnitude of millions of electron volts.

Since there is an uncertainty in the energies of the individual excited states of the nuclei, the lines associated with transitions between an excited state and the ground state have a certain minimum width. This so-called natural width Γ is, according to the Heisenberg uncertainty principle, connected with the lifetime τ of an excited nuclear state by the relation $\Gamma\tau = h$. The usual values for the lifetimes τ of the low-lying excited nuclear states lie in the range from 10^{-7} to 10^{-11} s, corresponding to an interval of 10^{-8} to 10^{-4} eV for the natural line widths appearing in ground-state transitions. Such extraordinarily sharp lines exhibiting the natural line

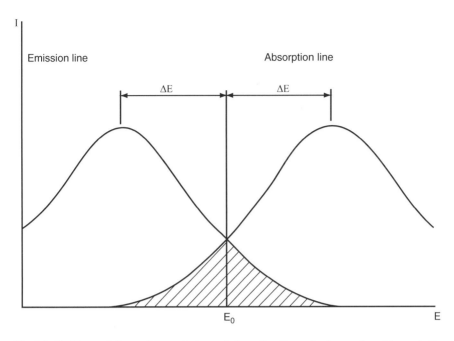

Fig. 1.2 Position and shape of the emission and absorption lines of a free nucleus (shown in the case of the I29-keV transition in ^{191}Ir for $T = 300°$K)

width are normally not observed. Rather, a series of side effects always exist, which lead to considerable broadening such that the line widths ordinarily associated with low-energy gamma transitions exceed the natural minimum width by many orders of magnitude. The most important broadening mechanism is the thermal motion of the nuclei in the source and in the absorber; this leads to Doppler shifts in the energy of the gamma quanta and therefore produces corresponding line broadenings. Even a temperature drop to absolute zero does not extinguish this thermal broadening, since the existence of zeropoint energy at absolute zero frequently produces at least in solids, line widths of the same order of magnitude as the ones existing at room temperature.

Usually the line shifts ΔE are large in comparison to the thermal line widths, and they are also always very large in comparison to the natural line widths associated with the low-energy nuclear transitions we are concerned with here. As a consequence, the energy of an emitted quantum is usually too small for the inverse process of resonance absorption to be carried out, or in other words, the probability of occurrence of nuclear resonance absorption is so small that the process escapes detection. Therefore, the long and unsuccessful search for nuclear resonance absorption is to be blamed on the high recoil-energy losses of the gamma quanta.

On the other hand, entirely different conditions hold for optical resonance absorption. There, because of the much lower energies of the light quanta, the recoil-energy losses produced by light quanta are small in comparison with the line widths. Emission and absorption lines therefore overlap in an ideal manner, the resonance condition is satisfied, and the optical effect is, at least in principle, easily observable.

The unsatisfactory situation with respect to nuclear resonance absorption first changed in 1951, when Moon [2] succeeded in demonstrating the effect for the first time, by an ingenious experiment. The fundamental idea of his experiment was that of compensating for the recoil-energy losses of the gamma quanta: the radioactive source used in the experiment was moved at a suitably high velocity toward the absorber or scatterer. The displacement of the emission line toward higher energies achieved in this way through the Doppler effect produced a measurable nuclear fluorescence effect.

After the existence of nuclear resonance fluorescence had been experimentally proved, a number of methods were developed which made it possible to observe nuclear resonance absorption in various nuclei. In all these methods for achieving measurable nuclear resonance effects, the recoil-energy loss associated with gamma emission or absorption was compensated for in one way or another by the Doppler effect.

At this point, let me speak of my esteemed teacher, Heinz Maier-Leibnitz, who in 1953 directed my attention to this newly advancing field of nuclear resonance fluorescence, and who stimulated me to turn to this area of research. He was, also, the one who made it possible for me to conduct my research throughout the years 1955–1958 at the Heidelberg Max Planck Institute, in an undisturbed and fruitful atmosphere – research which finally led to the discovery of recoilless nuclear resonance absorption. I want to express my warmest thanks to my esteemed teacher for his efforts on my behalf.

The method which I shall now proceed to discuss differs fundamentally from the methods described above in that it attacks the problem of recoil-energy loss at its root in a manner which, in general, insures the complete elimination of this energy loss. The basic feature of this method is that the nuclei in the source and absorber are bound in crystals. The experiments described in the following paragraphs exclusively employed radioactive sources which emitted the 129-keV gamma line leading to the ground state in ^{191}Ir.

The first experiment [3] aimed at measuring the lifetime of the 129-keV state in ^{191}Ir by utilizing methods of nuclear resonance absorption known at that time. The experimental set-up used for this purpose is shown schematically in Fig. 1.3. A method first employed by Malmfors [4] appeared to be especially suitable for the planned measurement. In this method, a broadening of the emission or absorption line, leading to a corresponding increase in the degree of overlap of the two lines, is achieved by increasing the temperature. If the relative shift of the emission and the absorption lines resulting from the recoil-energy losses is only of the order of magnitude of the line widths, a temperature increase leads, under favourable conditions, to a measurable nuclear absorption effect. In the case of the 129-keV transition in ^{191}Ir, there is considerable overlap of the two lines even at room temperature as a consequence of the small energies of the quanta and the small line

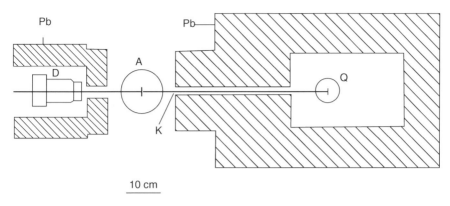

10 cm

Fig. 1.3 Experimental arrangement. A absorber-cryostat; Q cryostat with source; D scintillation detector; K lead collimator

shifts resulting therefrom (see Fig. 1.2). In this case, not only an increase but also a decrease in temperature can result in a measurable change in the nuclear absorption.

My decision between these two possibilities was made in favour of a temperature decrease. It was motivated essentially by the consideration that at low temperature, effects of chemical binding would be more likely than at elevated temperatures. This hypothesis was vindicated in an unexpected way during the course of the experiments. The simultaneous cooling of the source and the absorber with liquid air led to inexplicable results, for which I first blamed effects associated in some way with the cooling of the absorber. In order to eliminate these unwanted side effects, I finally left the absorber at room temperature and cooled only the source. In very tedious experiments, which demanded extremely stable apparatus, a small decrease in the absorption with respect to the value at room temperature was in fact obtained - a result consistent with my expectations. The evaluation of these measurements finally led to the determination of the lifetime sought for.

In a second series of experiments I attempted to explain the side effects which had appeared in the simultaneous cooling of the source and the absorber during the earlier experiments. The result of this attempt was striking: instead of the decrease expected, a strong increase in the absorption clearly manifested itself when the absorber was cooled. This result was in complete contradiction to the theoretical expectation. The observed temperature dependence of this absorption is shown in Fig. 1.4.

In considering the possible sources of the anomalous resonance effect, I now began to subject the hypothesis of the existing theory to a critical examination.

The views originally held as to the shape and energy of the emission and absorption lines were based on the assumption that the emitting and absorbing nuclei can be treated as free particles. It was therefore natural to modify this assumption, taking into account the fact that source and absorber were each used in crystalline form. Therefore, I first attempted to explain the observed anomalous resonance absorption by assuming that the recoil momentum was not transferred

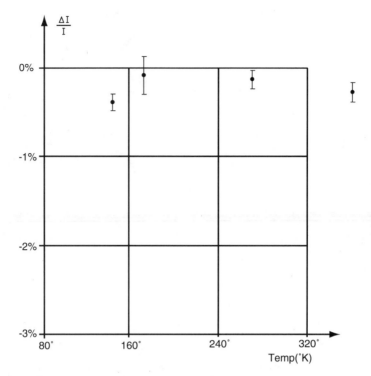

Fig. 1.4 Temperature dependence of the absorption. Relative intensity change change $\Delta I/I$ in comparison with that of a non-resonant absorber

to the single nucleus. It should rather be transferred to an assembly of nuclei or atoms which include nearest or nextnearest neighbours surrounding the nucleus under consideration. After the failure of this and other attempted explanations, based on a purely classical point of view, I turned my attention to a quantum-mechanical treatment of the problem.

Let me here introduce some concepts from crystal physics, by means of which I will develop the reasoning that finally led to a solution of the problem. All the internal motions of the particles forming the crystal lattice can be described in terms of a superposition of a large number of characteristic vibrations whose distribution is called the frequency spectrum of the crystal. The nature of the binding of the particles forming the crystal determines the structure of the lattice vibrational spectrum, this spectrum often being very complicated. Nevertheless, the substitution, for the true vibrational spectrum of the lattice, of a much simpler frequency distribution often suffices for qualitative considerations.

This first, still purely classical picture of the internal motions in a crystal corresponds, in the quantum–mechanical description, to a system of uncoupled harmonic oscillators with quantized energy states. While the vibrational spectrum describes only the distribution of the fundamental frequencies of these oscillators, the temperature determines the so-called occupation numbers of these crystal oscillators; these occupation numbers simply tell which of the possible energy states the particular oscillators occupy.

The recoil energy appearing in the emission or absorption of a quantum by a nucleus bound in a crystal is taken up by the crystal partly in the form of translational energy and partly in the form of internal energy. The resultant increase in translational energy is always negligible because of the enormous mass of the crystal as a whole in comparison to the mass of a single nucleus. An increase in the internal energy leads to changes in the occupation numbers of the individual crystal oscillators. Because of the quantization of the oscillator energies, the crystal can absorb the recoil energy only in discrete amounts. The nuclear transitions of the bound nuclei are normally accompanied by simultaneous transitions of the crystal oscillators. Thus, for example, a gamma quantum can be emitted and simultaneously one of the crystal oscillators can undergo a transition to a neighbouring energy state. Likewise, a gamma emission process can be accompanied by simultaneous transitions of two crystal oscillators, and so forth. As a consequence of the quantization of the oscillators, there also exists in principle the possibility that the gamma transition takes place with none of the crystal oscillators changing their states.

The problem now was to compute the probabilities of the various processes. Significant was the calculation of the probability of nuclear transitions leaving the lattice state unchanged – that is, transitions in which no recoil energy is transferred to the lattice in the form of internal energy. Similar problems had already been solved much earlier. The coherent scattering of X-rays from crystals had been known for decades; in that case a momentum transfer to the reflecting lattice takes place without simultaneous transfer of internal energy to the crystal. And for a long time, the analogous problem of the elastic scattering of slow neutrons from crystals had been thoroughly studied, both experimentally and theoretically. Lamb [5] had, as early as 1939, developed a theory for the resonance capture of slow neutrons in crystals. This theoretical work was somewhat premature, inasmuch as Lamb assumed considerably smaller values for the widths of the neutron lines than were observed in later experiments. For this reason, this extraordinarily beautiful work was of no practical significance in the area of application originally intended. It remained only to apply this Lambs theory to the analogous problem of the resonance absorption of gamma radiation. This indeed enabled me to show that under the conditions chosen in the experiments described above, there exists a high probability of nuclear transitions with no simultaneous change of the lattice state. Since these nuclear transitions are not associated with any energy losses caused by recoil phenomena, I shall in the following discussion characterize these particular transitions as recoilless transitions, and the lines associated with such transitions, as recoilless lines. In such a recoilless emission process, the entire excitation energy is therefore transferred to the emitted quantum, and the corresponding situation holds for the recoilless absorption. Here the notation recoilless relates only to recoil energy transferred in a nuclear transition, and not to the transferred momentum. The value of this transferred momentum is determined by the energy of the gamma quantum and is essentially a constant, independent of any change in the internal state of motion of the crystal. This momentum is, therefore, transferred to the lattice in all emission or absorption processes, even in the recoilless processes. It is always taken up by the crystal as a whole, and therefore the corresponding translational velocity is negligibly small.

What are the conditions under which the recoilless nuclear resonance absorption can be observed? In answering this question, I wish to develop here, without presenting the mathematical formulation of the theory, a detailed picture of the processes which take place in radiative transitions in nuclei bound in crystals. The recoil-energy loss that occurs in a nuclear transition of a free nucleus is given by (1.1). It can be shown that in a transition of a nucleus bound in a crystal, (1.1) is no longer valid for the individual process but holds in the means over many processes; that is, instead of (1.1), we now have

$$\overline{\Delta E} = \frac{E_0^{\,2}}{2Mc^2}. \tag{1.2}$$

Let us now consider a very simple model, in which we describe the vibrational state of the crystal by a single frequency ω the so-called Einstein frequency. It is instructive to consider the two limiting cases, in which the mean recoil energy is either large or small in comparison to the transition energy of the Einstein oscillator:

$$\overline{\Delta E} > \hbar\omega(\text{case}1)$$

$$\overline{\Delta E} < \hbar\omega(\text{case}2)$$

In case 1, many oscillator transitions are required to take up the energy contribution $\overline{\Delta E}$ in the lattice. The nuclear processes will therefore in general be accompanied by simultaneous transitions of many oscillators. The probability of a nuclear transition taking place without any oscillator transition – that is, the probability of a recoilless process – is correspondingly small. The situation is entirely different in case 2. Here it is immediately evident that a nuclear transition which is accompanied by an oscillator transition occurs relatively seldom, leading under these circumstances to a high probability for recoilless processes. The probabilities, under these conditions, for the occurrence of gamma emission processes accompanied by $0, 1, 2, \ldots, n$ oscillator transitions are shown in Fig. 1.5 for these two cases.

For qualitative considerations, this simple picture can well be applied to the case of the real crystal. The frequency spectrum of the real crystal exhibits an increasingly high oscillator density at high frequencies. It is sufficient, for this simplified consideration, to replace the Einstein frequency by the upper frequency of the vibrational spectrum of the real crystal. This limiting frequency ω_g, is related, approximately, to the characteristic temperature θ of the crystal by the equation $\hbar\omega_g = k\theta$.

The essential condition for a high probability of recoilless nuclear transitions now has the form

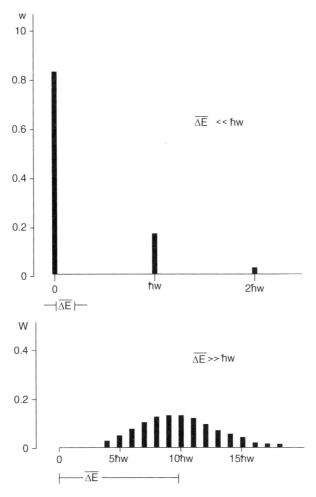

Fig. 1.5 Relative probability for a gamma transition associated with simultaneous transitions of $0, 1, 2, \ldots, n$ crystal oscillators, for two different values of $\overline{\Delta E}$ for the temperature $T = 0°\mathrm{K}$

$$\frac{E_0^2}{2Mc^2} < k\theta \tag{1.3}$$

The condition given here is quite restrictive because it limits the observation of recoilless resonance absorption to nuclear transitions of relatively low energy; the upper limit lies at about 150 keV.

If $\Delta E = E_0^2/2Mc^2$ is small in comparison to the upper energy limit of the frequency spectrum, the percentage of recoilless processes that occur is high even at room temperature. However, if $\overline{\Delta E}$ is about equal to the upper energy limit of the frequency spectrum, then the probability of transitions of the crystal oscillators must be correspondingly reduced by the use of low temperatures in order to arrive at measurable effects.

After the appearance of the observed strong resonance absorption in [191]Ir had been traced back to the phenomenon of recoilless nuclear resonance absorption, the next step was to compute the probability of the effect in a general form. This probability, also known as the Debye-Waller factor, in analogy with the terminology used in X-ray scattering, is, as I have already pointed out, strongly dependent on the temperature and the energy of the nuclear transition. This dependence is illustrated in Fig. 1.6 by two examples. The shape and location of the emission and absorption lines, as shown in Fig. 1.2 for the case of a free nucleus, are modified considerably by the influence of the chemical binding. While the centres of the lines as given by (1.2) are retained, each of them shows a complicated structure which reflects different single and multiple oscillator transitions, hence the notion line is applied here in a more generalized sense.

The most interesting prediction of the theory is the appearance of a strong line with the natural line width appearing in the structure of both lines at the position of the transition energy E_0. This line represents the recoilless processes. The strong prominence of these lines with the natural line width within the total line structure is not so surprising when one considers that all the recoilless processes in the emission and absorption spectra appear within an energy range of the order of magnitude of the natural line width. The gamma transitions associated with the oscillator transitions appear, on the other hand, in an energy range of the order of magnitude

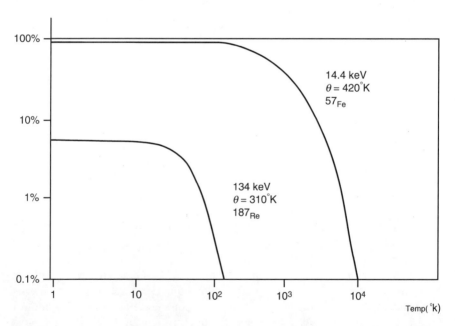

Fig. 1.6 Fractions of recoil-free nuclear transitions (Debye-Waller factors) in [57]Fe and [187]Re, shown as functions of the temperature

of the Debye temperature of the crystal; this energy range is always broader than the natural line width by many orders of magnitude.

All these considerations provided a very plausible explanation of the origin of the observed resonance absorption in ^{191}Ir at low temperatures. The agreement between the experimentally observed temperature dependence of the absorption (Fig. 1.4) and the theoretically computed dependence was satisfactory when one considered that in the first calculation the Debye approximation was used for the frequency spectrum because the actual frequency spectra of the crystals used were not known. Notwithstanding this qualitative agreement between experiment and theory, the situation still appeared to be somewhat unsatisfactory. The theory predicted that under the prescribed conditions lines of natural width should appear in the structure of the emission and absorption lines at the value of the transition energy E_0.

While the assumption of the existence of these lines made it possible to explain the experiments thus far carried out, the keystone was evidently still missing – namely, direct experimental proof of the existence of these lines and, especially, a demonstration that their widths were indeed the natural line widths. It was necessary to find a detector which had the necessary energy resolution to measure the profiles of these extremely sharp lines. The use of conventional detectors was excluded from the start. For example, the scintillation detectors frequently used for gamma-ray measurements in the relevant energy region have resolutions of the order of magnitude of 10^4 eV, while the natural line widths of the observed lines were only about 10^{-6} eV. The possibility of using the atomic nuclei themselves as detectors was suggested as a way out of this situation. As was shown above, an essential prediction of the theory was that the recoilless lines in both the emission and the absorption spectra should appear at the same position – that is, at the value of the transition energy.

Both lines would, therefore, overlap completely, this being the reason for the strong absorption effect. If one could succeed in partially removing the perfect overlap by a relative displacement of the lines, the absorption effect of the recoilless lines should disappear correspondingly.

In the experiment carried out to test the validity of this prediction, the idea was to accomplish the relative shift of the lines by means of the Doppler effect, so that a relative velocity would be imparted to the nuclei of the source with respect to the nuclei of the absorber. Here, we had a sort of reversal of the experiment carried out by Moon. Whereas in that experiment the resonance condition destroyed by the recoil-energy losses was regained by the application of an appropriate relative velocity, here the resonance condition fulfilled in the experiment was to be destroyed through the application of a relative velocity. And yet there was an essential difference between this and Moon's experiment. There, the width of the lines that were displaced relative to one another was determined by the thermal motion of the nuclei in the source and absorber; here, the line widths were sharper by four orders of magnitude. This made it possible to shift them by applying velocities smaller by four orders of magnitude. The indicated velocities were in the region of centimeters per second.

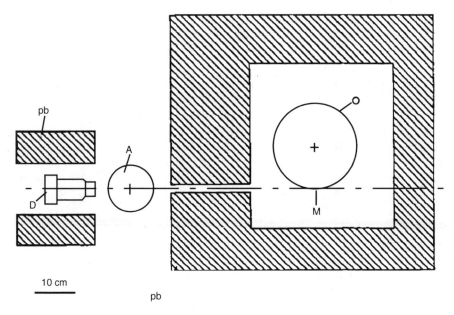

Fig. 1.7 Experimental arrangement. A absorber-cryostat; Q rotating cryostat with source; D scintillation detector

Figure 1.7 shows the experimental arrangement [6, 7]. For simplicity, I decided to move the source by means of a turn table. Only the part of the rotational motion marked by the heavy line in Fig. 1.7 was used for the measurement – namely, that part in which the source was moving relative to the absorber with approximately constant velocity. The intensity at the detector was measured as a function of the relative velocity between the source and the absorber. Since the preparation of the conical-gear assembly necessary for adjusting the various velocities caused a disagreeable delay in this experiment which was so exciting for me, I took advantage of the existence in Germany of a highly developed industry for the production of mechanical toys. A day spent in the Heidelberg toy shops contributed materially to the acceleration of the work.

Figure 1.8 shows the result of this experiment, a result which was just what had been expected. As the figure demonstrates, a maximum resonance absorption was actually present at zero relative velocity as a result of the complete superposition of the recoilless emission and absorption lines; therefore, minimal radiation intensity passing through the absorber was observed in the detector. With increasing relative velocity the emission line was shifted to higher or lower energies, the resonance absorption decreased, and the observed intensity correspondingly increased. The necessary relative velocities were manifestly only of the order of centimeters per second. Since the experiment consisted essentially of producing a shift of an emission line of width Γ relative to an absorption line of width Γ, the observed line possessed a width which, with a small correction, was equal to $2\,\Gamma$. It was especially

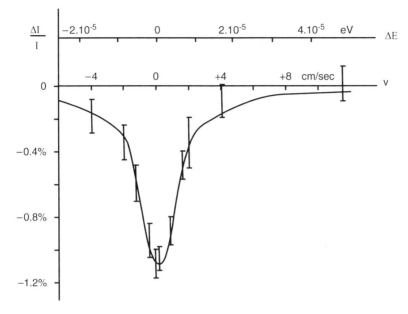

Fig. 1.8 Relative intensity ratio $\Delta I / I$ of gamma radiation measured behind the resonant iridium absorber, in comparison with intensities measured behind a nonresonant absorber

satisfying that the line width thus obtained agreed with the width determined in the first experiment [3] under much more difficult conditions. While absorption effects of the order of 1% were observed in the second experiment, an effect of the order of a hundredth of 1% had been achieved in the earlier work. Thus, direct proof of the existence of recoilless absorption was achieved.

The significance of the new method was immediately apparent, although not all of its consequences were immediately realized. With 4.6×10^{-6} eV the line width observed in the 129 keV gamma transition of ^{191}Ir was already smaller than the usual thermal line widths by many orders of magnitude. Let us define the energy resolving power $E/\delta E$ by the ratio of the energy E_0 of the nuclear transition and the natural width Γ of the line; that is, $E/\delta E = E_0/\Gamma$ Then we obtain in the present case $E/\delta E = 2.8 \times 10^{10}$. The actual energy resolution achieved was already much greater in this experiment, since energy shifts amounting to a small fraction of the natural line width were actually observed. Since then, it has become possible, by utilizing other nuclear isotopes, to improve the obtainable energy resolution once more by several orders of magnitude. It is this property of the recoilless nuclear resonance absorption – namely, that it is possible by this means to measure extraordinarily small energy differences between two systems – which gave the method its significance and opened up a broad field of possible applications.

Thus, the extraordinary sharpness of the recoilless gamma lines brought direct investigation of the hyperfine structure of nuclear transitions within the range of possibility. As a rule, atomic nuclei possess electric and magnetic moments in their various excited states. The interaction of these moments with internal or external fields leads to a splitting of nuclear levels into a number of states that are very close to one another. This hyperfine structure normally remains hidden in gamma lines, since the thermal width of a gamma line is always very large in comparison to the spacing of the hyperfine levels of the nuclear state. When the distances between the individual hyperfine components are larger than the natural widths of the gamma lines, as is frequently the case, and when the conditions necessary for the observation of recoilless absorption are fulfilled, this method makes possible direct measurements of the hyperfine structure of both participating nuclear states. Instead of a single line with the natural line width, a whole set of lines now appears in the emission spectrum, corresponding to recoilless transitions between the hyperfine levels of the nuclear excited and ground states. The same situation holds for the absorption spectrum. Studies of this type yield predictions on the magnetic dipole and electric quadrupole moments of the nuclear states involved, as well as predictions on the magnetic fields and gradients of the electric field prevailing at the site of the nucleus. The special promise of such measurements lies primarily in the possibility of obtaining information on the hyperfine splitting of excited nuclear states. In fact, several moments of excited nuclear states have been determined in this way in various laboratories. However, in our laboratory and in laboratories of many others now working in this region, this method is used mainly for studying the internal fields existing at the site of the nucleus; such studies have already led to a series of interesting results.

In addition to measurement of the fields located in crystals at nuclear sites and to measurement of the moments of excited nuclear states, studies of a number of important effects have been made during the past 2 years in a large number of laboratories. The observation of these effects was made possible by means of even sharper nuclear transitions, especially that of the 14.4 keV transition in ^{57}Fe.

Particular mention should be made here of the beautiful measurements of the energy shift of radiation quanta in the gravitational field of the earth [8], the observation of the second-order Doppler effect, and the measurements of the isomeric shift. However, discussion of these and of a whole series of other effects which have been studied by means of the method of recoilless nuclear resonance lies outside the framework of this address.

The interpretation of the formalism which underlies the quantitative description of recoilless nuclear resonance absorption has led to extremely active discussions. The question particularly raised has been whether the effect can be explained classically and, in this connection, whether the momentum transfer takes place as a continuous process during the lifetime t of the excited nuclear state or as a spontaneous process in the sense of quantum electrodynamics. The question raised here is closely connected with the problems encountered in the dualistic description of radiation as a wave process or as a stream of free particles. Notwithstanding the fact that many details of the recoilless resonance absorption can be described by classical models, I should characterize this effect as a specifically quantum-mechanical one. In particular, it can be shown by the mathematical formulation of the theory that the momentum transfer takes place spontaneously. This can be demonstrated experimentally, not by an individual process but, rather, by the measurement of certain integral quantities, as, for example, the Debye-Waller factor.

In conclusion, let me now say a few words on the limits of the methods described. An upper limit for the usable natural line widths has certainly been given when the widths approach the width of the vibrational spectrum of the crystal oscillators.

In this particular case, it becomes impossible to distinguish clearly between nuclear processes which are simultaneously accompanied by oscillator transitions and those which are not. However, this would require nuclear lifetimes of less than 10^{-13} s, which do not occur in the nuclear levels available for this type of experiment. Therefore, this upper limit is both uninteresting and unrealistic.

More difficult is the question of a lower limit for the available line widths. To go beyond the limit so far attained of about 10^{-8} eV for the natural line widths is quite possible in principle. There are, however, a number of factors which have delayed extension of the method into the region of higher sensitivity. On the one hand, there are very few nuclei of stable isotopes whose first excited states possess the desired lifetime of greater than 10^{-7} s. On the other hand, all the previously mentioned side effects of such sharp lines play a dominant role; small disturbances of the crystal symmetry and small contaminations quickly lead to individual shifts in the nuclear states, and these shifts, as a group, produce a very considerable broadening of the extremely sharp lines. In this way, the resonance condition is so far violated that the lines are not observed. However, there is a well-founded view that still more sharply defined nuclear transitions will be available before long.

These should lead to multiplication of the possibilities for applying the method of recoilless absorption. We may therefore hope that this young branch of physics stands only at its threshold, and that it will be developed in the future, not only to extend the application of existing knowledge but to make possible new advances in the exciting world of unknown phenomena and effects.

References

1. W. Kuhn, Phil. Mag. **8**, 625 (1929)
2. P.B. Moon, Proc. Phys. Soc. (London) **64**, 76 (1951)
3. R.L. Mössbauer, Z. Physik **151**, 124 (1958)
4. K.G. Malmfors, Arkiv Fysik **6**, 49 (1953)
5. W.E. Lamb Jr., Phys. Rev. **55**, 190 (1939)
6. R.L. Mössbauer, Naturwiss. **45**, 538 (1958)
7. R.L. Mössbauer, Z. Naturforsch. **14a**, 211 (1959)
8. R.V. Pound, G.A. Rebka Jr., Phys. Rev. Lett. **4**, 337 (1960). A complete bibliography of all pertinent researches is included in H. Frauenfelder, *The Mössbauer Effect, Frontiers in Physics*, Benjamin, New York, 1962.

Chapter 2
Rudolf Mössbauer in Munich

G.M. Kalvius and P. Kienle

2.1 Years of Study at the Technische Hochschule München

Mössbauer and one of the authors (PK) started in 1949 studying physics at the Technische Hochschule München (THM), which was still under reconstruction from the war damages. It offered two directions for studying physics: "Physik A" and "Physik B." I took courses in "Physik A," which meant Technical Physics; Mössbauer studied "Physik B," which was General Physics. Actually, the lectures of both directions were not too different up to the forth semester, followed by a "pre-diploma" examination, which Mössbauer passed in 1952. I as "Physik A" student had besides the various physics, chemistry, and mathematics courses, in addition lectures in Technical Electricity, Technical Mechanics, Technical Thermodynamics, and later Measurement Engineering offered by very famous professors, such as W.O. Schumann, L. Föppl, W. Nußelt, and H. Piloty. Our physics teachers were G. Joos (Experimental physics), G. Hettner (Theoretical Physics), and W. Meissner (Technical Physics); in mathematics, we enjoyed lectures by J. Lense and R. Sauer, and interesting chemistry lectures by W. Hieber. Thus we received a high-class classical education, but quantum mechanics was not a compulsory subject. Mössbauer complained about this deficiency when he realized that the effect he found was a quantum mechanical phenomenon. Quantum mechanics was offered as an optional subject by Prof. Fick and Prof. Haug. Mössbauer just missed to take these advanced lectures, although he was highly talented in mathematics and received even a tutoring position in the mathematics institute of Prof. R. Sauer, while I worked in engineering projects and had extensive industrial training.

In 1952, Prof. Walther Meissner retired and Prof. Heinz Maier-Leibnitz, a nuclear physicist, coming from the famous Max-Planck Institute for Medical Research in Heidelberg, became his successor and Director of the as well famous "Laboratorium

G.M. Kalvius (✉) · P. Kienle
Physics Department, Technical University Munich, 85747 Garching, Germany
e-mail: kalvius@ph.tum.de

M. Kalvius and P. Kienle (eds.), *The Rudolf Mössbauer Story*,
DOI 10.1007/978-3-642-17952-5_2, © Springer-Verlag Berlin Heidelberg 2012

für Technische Physik." Maier-Leibnitz was a doctoral student of James Franck, and after obtaining his doctoral degree, an assistant and collaborator of Walther Bothe working on basic and applied projects in nuclear physics. The Laboratorium für Technische Physik so far had been devoted to research in low temperature physics, especially superconductivity and technical thermodynamics ever since K. von Linde held its first chair. Maier-Leibnitz (for short ML, as he was called among his students) turned it into a very successful Laboratory for applied nuclear physics emphasising a novel direction called "Nukleare Festkörperphysik" (meaning solid state physics using nuclear methods). It became one of the sustained growth centres of the THM and a palladium for creative education of many young scientists by original research. We all went to ML with the hope to get an interesting research project for a Diploma thesis. ML took up everybody who asked him for a thesis and all of them got a project of original research, each one carefully selected so that they covered methodically interconnected areas of science. So, we learned to collaborate very early in our science education. Prof. Joos who held the chair for Experimental Physics was extremely selective. He only took students, he thought were the very best ones. They had to participate in advanced seminars and pass rigorous tests. Mössbauer prepared himself for a Diploma thesis with Prof. Joos and passed all hurdles, but finally he did not take it.

2.2   Diploma Thesis of Mössbauer
with Prof. H. Maier-Leibnitz

Mössbauer had in addition applied to Maier-Leibnitz for a Diploma project and decided to join this laboratory in 1953 with a very interesting thesis, which ML gave to his supposedly very best student. It concerned the development of a precision method for measuring the absorption of γ-rays. This study was then to be taken as a basis for a later to follow doctoral thesis studying the nuclear resonance absorption of γ-rays in experiments of the type K.G. Malmfors did at this time (in scattering geometry) by producing a partial overlap of the Doppler shifted and broadened emission- and absorption lines through heating the γ-ray source [1]. For these difficult absorption experiments with small changes of the order of 10^{-4}, ML proposed to build proportional counters filled with pure methane [2] which showed a so-called "long and flat counting plateau" as function of the applied voltage. Such counters were also used by L. Köster for precise lifetime measurements of long-lived radioactive nuclei in short measuring times. Mössbauer succeeded in constructing and operating this type of detectors which had very excellent stability but low counting efficiency. Even with a bundle of fifteen "gold plated" counters, the efficiency amounted to only 5% for the γ- rays he had in mind for the nuclear resonance absorption experiment. As a real test of their performance he succeeded

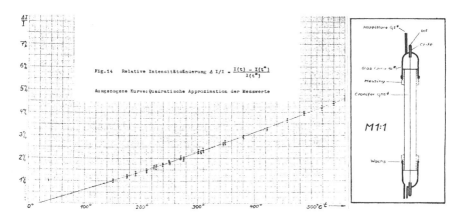

Fig. 2.1 *Left*: temperature dependence of the change of 1.25 MeVγ-ray intensity through a 40 mm Cu disk normalized to the intensity at 21°C. The curve fitted to the data points is not a simple straight line, but the calculated function taking into account not only the change in density, but also the influence of temperature on the absorption cross-section due to weakening of the chemical bond. *Right*: drawing (reduced, no longer 1:1) of the proportional counters used. The tube length was 96 mm, and the inner diameter 12 mm. The counter wall was 0.1 mm thick brass gold plated (5 μ) on the inside. Counting gas flow was maintained through the capillaries. From the Diploma thesis of Rudolf Mössbauer

in measuring the temperature dependence of the electronic absorption of γ-rays of ^{60}Co (energy range of 1.25 MeV) in a copper foil as a function of thermal expansion This is a tiny effect, caused by the temperature dependence of the density of the copper metal. Figure 2.1 (left) shows the measured change $\Delta I/I$ (in %) of the absorption of the ^{60}Co γ-rays through a 40 mm thick Cu foil as a function of temperature (normalized at 21°C). Note the many measuring points, all reaching an accuracy of about 10^{-3}. Figure 2.1 (right) shows a sketch of the proportional counter Mössbauer used. In order to increase the γ-detection sensitivity, the 0.1 mm thick brass wall was plated inside with a 5 μm gold layer.

After having finished his Diploma thesis in 1954, Mössbauer was, according to ML, excellently prepared for measuring the temperature dependence of the nuclear resonance absorption of γ-rays. ML proposed to use for such an experiment a low energy γ-transition of a heavy nucleus, thus reducing the recoil shift of the emission and absorption lines in order to search for possible chemical effects. These had been studied within classical theory by Steinwedel and Jensen already in 1947 [3], but they did not predict the quantum effect Mössbauer found in 1958. As the Laboratorium für Technische Physik was crowded with students and had only a small financial budget, ML found a better suited laboratory for his student Mössbauer to work on his doctorate, sending him to the Institute for Physics in the Max-Planck Institute for Medical Research in Heidelberg, the laboratory ML came from.

2.3 Doctor Thesis of Mössbauer with Prof. H. Maier-Leibnitz Performed with Experiments at the Max-Planck Institute for Medical Research in Heidelberg

Already in December 1953, while still working on his diploma thesis, the 129 keVγ-transition in ^{191}Ir populated by the β^--decay of ^{191}Os was selected by Mössbauer for his nuclear resonance absorption experiment. To get knowledgeable in this matter, he used the reference library of the Deutsche Museum in Munich. It was the only source for nuclear physics literature accessible at this time in Munich. All of us spent days and weeks in this library, because one could not take out books or journals. Also, one had to copy by hand all paragraphs of interest, since Xerox machines did not yet exist.

The selection of the transition was made for the following reasons:

1. The transition energy was considered low enough to lead, together with the large mass of ^{191}Ir, to small recoil energies, so that an overlap of the emission and absorption lines could well be achieved via the thermal broadening method of Malmfors.
2. The ^{191}Os radioisotope decays with a half-life of 15.4 days by β^--decay to ^{191}Ir and populates strongly the 129 keV ground state transition, the basis for the nuclear resonance absorption experiment since ^{191}Ir is a stable isotope with a natural abundance of 37.3%. The ^{191}Os radioactive mother nuclei can conveniently be produced by thermal neutron capture in ^{190}Os and were offered at that time in the famous "Harwell" catalogue, the only accessible source for this radioisotope. In 1953, Germany had no reactor and was even not allowed to do nuclear research until 1955.
3. The lifetime of the 129 keV M1 transition was unknown, so there was hope that its determination could be the result of Mössbauer's thesis work.

ML provided his student with a reprint of the famous Lamb paper [4] on resonance neutron capture of nuclei in crystals, which Mössbauer later used to explain the effect he observed. In addition, he had with him the Steinwedel–Jensen paper [3] containing the proof that narrow resonant lines should not be observable, in contrast to his later discovery.

So the start of the project in Heidelberg was loaded with problems, but Mössbauer was free to do what he wanted to do, because his supervisor could visit him only from time to time. In contrast to our poor Munich University Laboratory, the Max-Planck Institute in Heidelberg was rich, so Mössbauer could buy all the electronics he needed, whereas we used to build it ourselves. This he considered an advantage for a precision measurement down to the order of 10^{-4}, albeit he had already achieved this goal in Munich with homemade electronics and stable proportional counters. At ML's first visit at Heidelberg, a surprise awaited him. His self-willed student had decided following other researchers in Heidelberg to get rid of the proportional counters he had developed and successfully applied for measuring the temperature dependence of γ-ray absorption by a copper foil in

Munich. He declared his truly irritated professor that he is going to use modern NaI(Tl) scintillation detectors, because of their high efficiency of nearly 100% for the detection of the 129 keVγ-line. ML was shocked and tried to explain his student that the success of the experiment does not depend on high detection efficiency, but on the stability of the detector during the time consuming absorption measurements. Mössbauer refused this valuable advice of his supervisor and embarked on using scintillation detectors. One of his colleagues in Heidelberg recalled [5]: "It so happened that in June 1955 a charming and very determined doctoral student from Munich took up a little room in the... Institute at Heidelberg. He brought with him a bundle of counting tubes, gold plated inside... which, however, never came to great use.... Scintillation counters had reached a state of development that it appeared to Mössbauer their use would be of advantage. The main problem [now] was the fight to stabilize the counting system." Indeed, this was a lot of trouble because the gain of the photomultipliers he used to detect the scintillation light was unstable. As a main requirement of the experiment, he had to make use of an efficient electronic gain stabilization system for his photomultipliers. Rudolf Mössbauer believes even today that without scintillation counters he would have never observed his effect [6].

Truly important for the success of the experiment was a decision concerning the temperature variation in the resonance absorption experiment for changing the overlap of the Doppler-broadened and shifted emission and absorption lines. Originally Mössbauer went up with the temperature as Malmfors had done, and as ML had proposed, in order to increase the overlap of the emission and absorption lines for observing an increase of the resonant absorption according to classical expectations. But then Mössbauer realized that going down with the temperature to the boiling point of liquid nitrogen would double the effect he had measured already by increasing the temperature. This was the decisive move, which he planned to do just for improving the accuracy of his data. The measurement showed indeed the expected magnitude of $\Delta I / I = (2.7 \pm 0.7) \times 10^{-4}$, *but with the wrong sign.* The absorption increased, yet was expected to decrease at the lower temperature because of the smaller overlap of the emission and absorption lines. Mössbauer checked and rechecked the accuracy of this data until he was firmly convinced to be right. He had found a new effect, which was later termed the "Mössbauer Effect." Now the question arose how to explain this novelty.

Jensen (Professor of theoretical physics at Heidelberg and later Nobel Laureate) could not help Mössbauer in explaining this strange effect, as it was not contained in the Jensen–Steinwedel paper, which was, as was shown later, the classical approximation. So Mössbauer returned to Munich and studied Lamb's paper on resonance absorption of neutrons by the (n, γ) absorption process in a crystal. It showed narrow recoil-free absorption lines, which, however, could not be observed, because the natural line widths of the neutron resonances were too large. The narrow lines predicted theoretically by Lamb were thus unobservable in neutron capture. But Mössbauer realized that the natural line width of the γ-transition was only 10^{-5} eV compared to 10^{-2} eV width of the neutron resonances. Therefore, a narrow resonance absorption line having the natural width may well appear just at the transition energy, the signature of recoil-free emission and absorption in a crystal. In the following time, Mössbauer worked out theoretically the appearance of recoil-free emission and absorption lines and found two Lorentz shaped lines of natural width, both located at the resonance energy, later known as Mössbauer lines. These lines overlap automatically. In this way he was able to explain on the basis of the work of Lamb the signal he had observed in his experiment as an increase of the recoil-free resonance absorption which he found to raise with lower temperature. He so discovered "new physics" revealing old principles, well described by quantum mechanics. With this result, he finished his doctoral thesis and published it in Zeitschrift für Physik, **151**, 124 (1958) with the title "Kernresonanzfluoreszenz von Gammastrahlung in ^{191}Ir."

I (PK) remember this paper very well because I was during this time for 1 1/2 years at the Brookhaven National Laboratory learning "Health Physics" since we had in the meantime a swimming pool reactor in operation at Garching near Munich and I was on the way to return to Munich in order to take care of radiation safety at the FRM together with Martin Oberhofer. At this time, Maurice Goldhaber came back from the Second United Nations International Conference on the Peaceful Uses of Atomic Energy, held in Geneva, September 1958, where he had met ML and brought the exciting news that a student of ML had discovered nuclear resonance fluorescence of γ-rays in solids at low temperatures. Some American colleagues asked me to translate the paper written in German, but I hesitated to do so, because the paper was difficult to understand and to translate and I was busy preparing my return to Munich. Actually, I was tempted to try myself another transition with lower energy for getting a higher recoil-free fraction. But I got tied up with the organization of Health Physics at the FRM and my attempt to get a ^{57}Co source for a resonance experiment with the famous 14.4 keVγ-line in ^{57}Fe from the cyclotron in Heidelberg failed.

Mössbauer in the mean time realized that he had not yet shown directly the narrow lines of recoil-free resonance absorption he had found. He suddenly remembered vividly the technique of Moon to study resonance fluorescence by changing the overlap of emission and absorption line with an externally generated Doppler shift. Since the γ-radiation used by Moon (411 keV transition in ^{198}Hg) was thermally broadened and recoil shifted, large Doppler velocities (up to 800 m/s) were needed and produced by an ultracentrifuge. Mössbauer realized that with his narrow, recoil free line, he could achieve the same result and the definite proof of his discovery with velocities of only a few cm/s. Being frightened that competition would do this crucial experiment before him, he received immediate permission of ML to go back to Heidelberg for doing the Doppler shift experiment using a slowly moving source on a turntable, and a stationary absorber, both cooled by liquid nitrogen. Fortunately, his detection equipment still existed. The experiment was easier than the original one, because in measuring the folding of two Lorentz lines, he observed at maximum overlap of the lines an absorption effect of 1%. In the hope to hold back the competition, he published this fantastic result again in German with the non-spectacular title "Kernresonanzabsorption von Gammastrahlung in Ir^{191}" in Naturwissenschaften **45**, 538, (1958), an even less internationally known journal. This was the true birth of Mössbauer spectroscopy, as it became to great use the world over.

2.4   The Dawn of Mössbauer Spectroscopy in Munich

When Rudolf Mössbauer returned from Heidelberg to Munich for his doctoral examination, he had just received the preprints of his publication in Zeitschrift für Physik. Seeing his work in print for the first time makes any young scientist excited and proud. Mössbauer was no exception. It was known at the time to the nuclear fluorescence community that one of its prominent members, Franz R. Metzger from the Franklin Institute in Pennsylvania, was preparing an extensive review on the subject. So Mössbauer quickly mailed him a copy of his article, only to realize shortly after, when the scientific opportunities of his effect became apparent to him, that this had been an enormous error. He felt he had created a most powerful competition. He needed not to have worried. In the review [7] one only finds a footnote (!) in the chapter on "Recoil Energy Losses" reading: "The complicated structure of the absorption line in the case of stronger binding was also discussed by Lamb (1939) and has recently been studied experimentally by Mössbauer (1958)." This is just one of the several instances where the potential of Lamb's and Mössbauer's work was not seen. Still, after the shock of informing Metzger, Rudolf Mössbauer had been more careful in publishing his Doppler shift result, as mentioned. But now it did not work as expected, as the reader can see in the contributions by John Schiffer and Stanley Hanna in this volume.

Around the time Rudolf Mössbauer got his PhD the other author of this story (GMK) had finished a few month earlier his Diploma thesis with ML on another subject of "Nuklearer Festkörperphysik," dealing with three quantum annihilation of positrons in plastic materials. Although proportional counters were the vocation of ML, he agreed, perhaps after the incident with Mössbauer, that one better uses NaI(Tl) scintillation detectors for performing coincidence measurements with the 511 keV annihilation radiation. So I was allowed to buy two NaI crystals and two photomultipliers. This was a huge expenditure for the Institute at the time being and certainly very generous of ML. The detector as well as all electronics we built ourselves. Yet, I got only two detectors, hence I had to measure the three-quantum rate as the reduction of a normalized two-quantum rate. I mention this as another example that we all learned with ML to perform precision counting experiments looking for tiny rate differences.

I then started my Doctoral thesis also on Malmfors type nuclear resonance fluorescence measurements with a rather energetic (> 1 MeV)γ-ray with the aim of lifetime measurements. But, shortly after Mössbauer had performed the Doppler shift experiment, ML came into my office, dropped Mössbauer's thesis and a preprint of the Doppler paper on my desk and said: "Have a look at that. I think this is more exciting than measuring standard resonance fluorescence. Perhaps you would consider switching the theme of your thesis" (Typical ML, he did not say "You better change..."). Although at my time quantum mechanics had become part of the curriculum in theoretical physics, we still had only rudimentary training. I struggled through Mössbauer's paper, understood perhaps 50%, mainly because I was familiar with momentum transfer and recoil energies. Yet, this was enough to decide immediately to switch my thesis and to enter the new exciting field.

By now, Rudolf Mossbauer had fully settled in Munich and began experimental work in his field. Soon a small group of "Mössbauer Physicists" was established, including the master himself. Some worked directly with Mössbauer, others more with ML and Paul Kienle. But we were one group; if for nothing else than a simple practical reason: measurements were made "point by point," meaning one measured briefly ($\sim$10 min) but repetitively at different velocities within the desired velocity range. A fixed scheme for the data taking at different velocities was established (like a railroad timetable) and had to be followed rigorously to keep disturbing influences like the decay of source activity, etc. to a minimum. Consequently at least two persons had to be present around the clock, one reading the scalers (they were binary) and the other setting the next velocity by changing gear. To have the necessary manpower, all of us participated in each experiment. We also had a weekly discussion session, "the Mössbauer Seminar." One (admittedly curious) question was how to call the effect we measured and discussed. With young Mössbauer present, it was a bit awesome to speak of the Mössbauer resonance. "Recoil free nuclear resonance absorption" was not only cumbersome but also much disfavored by Mössbauer because, as he liked to point out, it is incorrect: The recoil is still there, albeit with zero recoil energy loss. Furthermore, Mössbauer insisted that the non-Doppler broadening was the more important feature. In fact, in his first paper at that time, he spoke of "Nuclear resonance absorption of non-Doppler broadened Gamma rays" [8]. Also too complicated a name. We settled on the "M-effect." Most amusingly, Mössbauer used this term many years later when he wrote his account of the discovery of his effect [6]. The term "Mössbauer Effect" was established by H. Lipkin in his famous article "Some simple features of the Mössbauer Effect" [9] and it stuck. By the way, it took the "Mössbauer Seminar" several weeks to understand the "simple features."

An often asked question is why Munich not immediately embarked on measuring ^{57}Fe. As Mössbauer and PK mention, ^{57}Co is a source conveniently produced with accelerated deuterons using a ^{56}Fe(d, n)^{57}Co reaction and could have been produced in Germany only using the Heidelberg cyclotron. Its scientific person in charge, however, refused using it as a radioisotope production facility and turned our requests down. ML realized somewhat later, that a cyclotron uses small iron plates for accurately adjusting the magnetic field. The beam hits these "shim plates" on occasion and thus they should contain some ^{57}Co activity. He sent me (GMK) to Heidelberg and I had to persuade Prof. Schmidt-Rohr the director of the cyclotron laboratory to hand me a piece of such a plate. I got it, it was weakly active, I wrapped it in a bit of lead foil, put it in my little suitcase, and went back by train to Munich. Of course it turned out to be impossible to separate this bit of activity from all the iron and unfortunately nothing came out of it. Purchasing activity was not in the budget of the laboratory and, in particular, we were sort of obliged to use reactor generated activities since the laboratory had just gotten a swimming pool reactor (the famous "Garchinger Atom-Ei") featuring in particular irradiation facilities for the production of radioisotopes for research. In addition, it was of course convenient especially for short living sources. Furthermore, Mössbauer was suspicious, that the line width of ^{57}Fe was too narrow to be observed, suffering from severe broadenings by imperfections in the source matrix. This was not unfounded, but, as it turned out, the width of the iron line is just at the limit not to suffer prominently from such broadenings. Last but not least, being the former Institute of Walther Meissner, we had excellent low temperature facilities, which allowed measuring the higher energy transitions with comparative ease. Liquid helium was available, but precious. In the beginning, all low temperature measurements were carried out with liquid hydrogen. The lowest temperature was 14 K when pumping on the hydrogen. Precooling was initially only possible with liquid air. All this sounds scary in retrospect, but we never had the slightest accident, mainly due to the outstanding low temperature physicist Werner Wiedemann who had joined the group and who taught us very efficiently the basics of low temperature technology (we built also the cryostats ourselves). The lure of a room temperature Mössbauer resonance was there, nevertheless. After detailed discussions with ML, GMK chose as doctoral thesis to study the resonance of the 8.42 keVγ-ray in ^{169}Tm for which a convenient reactor generated source was available. It also meant returning to the beloved proportional counters.

Like all early "Mössbauer Physicists," we were trained in nuclear physics without deeper understanding of solid state physics. So, we first concentrated on measuring nuclear parameters like the lifetime (where unaccounted line broadenings would have been destructive), the electromagnetic moments of excited states and changes of nuclear radii in γ-transitions from observing isomer shifts. Yet, it was obvious that a hard learning process was in front of us (including Mössbauer) to understand the intricacies of solid state physics and chemical bonding. Another question often asked is whether the Munich Mössbauer group realized the many applications of the Mössbauer effect. Already in May 1958, Maier-Leibnitz produced a list of 15 different possible applications of the Mössbauer effect ranging from lifetime measurements, hyperfine splitting, and isomer shifts (which he called the nuclear volume effect and referred to as influence of bonding on the γ-ray energy) to various polarization studies combined with angular distribution experiments. Only gravitational studies were missing. Angular distribution of nuclear radiation was a highly active subject at the time, especially the PAC (perturbed angular correlation) studies, another technique of the "Nukleare Festkörper-physik" [10]. The original of ML's note, which GMK found one morning on his desk, unfortunately got lost. But photographic copies were distributed to the "Mössbauer Group" and one of them (in the possession of PK) survived. It is shown in Fig. 2.2. ML's handwriting is not easy to read and of course the text is in German. Thus we added a list of the subjects mentioned in English.

It is to be noted that the landmark paper by Kistner and Sunyar [11] establishing the isomer shift and the electric quadrupole splitting appeared only in 1960 and selective absorption was pursued only in the mid seventies! Maier-Leibnitz himself was most interested in the optical properties of Mössbauer radiation. For example, he initiated a study of total reflection of the 8.42 keV Mössbauer radiation in ^{169}Tm [12].

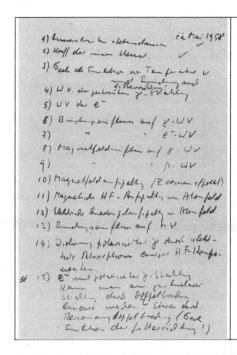

	1) Line widths and life times.
	2) Coefficient of the internal conversion
	3) σ_{coh} as function of temperature and bonding and lattice direction
	4) Ang. Distr. of the scattered γ-radiation
	5) Ang. Distr. of e⁻ (conv. electr.)
	6) Influence of bonding on the γ-ang. distr.
	7) Influence of bonding on the e⁻-ang. distr.
	8) Influence of magnetic field on the γ-ang. distr.
	9) Influence of magnetic field on the β-ang. distr.
	10) Magnetic splitting (Zeeman effect)
	11) Magnetic hyperfine splitting in the atomic field
	12) Electric Quadrupole splitting in the atomic field
	13) Influence of bonding on hv
	14) Separation of polarized γ by selective absorption of some hyperfine components
	15) e⁻ with polarized γ-radiation. Is it possible to create from circular radiation linear polarized one by double refraction? For example by resonant double refraction (σ_{coh} as function of lattice direction!)

Fig. 2.2 Note of Maier-Leibnitz (11 Mai 1958) listing possibilities of applications of the Mössbauer effect

One problem, misjudged in its severity by all early Mössbauer spectroscopists, was the velocity drive for the Doppler shift spectroscopy. The situation is well described by L. Grodzins, another pioneer of Mössbauer studies. In a somewhat free quote he said [13]: "When you first consider the problem of making a drive, you think this is trivial and can be solved with $100. You get your source and scatterers, but your drive has 50% vibrations, which you would like to reduce to 10%. Then you really have to reduce them to 1%. The final solution is simple. You place the source on the chest of a graduate student, having found that his chest beats up and down with constant acceleration." Originally the drives were all mechanical. The turntable-type drive used by Mössbauer (see Fig. 7 in [6]) in establishing the narrow recoil free lines worked fine, but was inefficient. Later, in Munich, Mössbauer designed a rather sturdy constant velocity drive with a stationary source cryostat and an absorber cryostat mounted on a movable trolley (Fig. 2.3) driven by a gear and chain drive. A change of velocity meant changing the toothed wheels in the gear box. To keep vibrations to a minimum, all moving parts were bathed in oil and so was soon the whole laboratory. Insiders came only in their oldest clothes. ML (being in more formal attire) entered only when it was absolutely needed. Once he visited us with Prof. Christoph Schmelzer from Heidelberg, but they stayed at a safe distance from our oily monster in which they were not interested at all. They talked only about how to construct a high duty factor linear accelerator to accelerate all ions

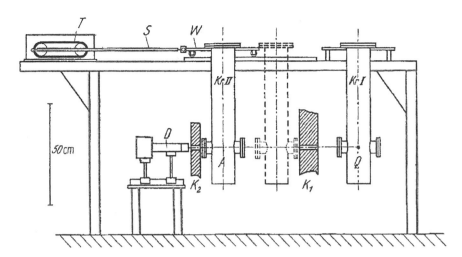

Fig. 2.3 Mössbauer's velocity drive in Munich. KrI = stationary cryostat containing the radioactive source Q; KrII = movable cryostat with a resonant absorber A mounted on a trolley W; T = motor, gearbox, and chain drive; S = connecting rod; K_1 and K_2 = lead collimators; and D = scintillation detector

up to uranium to energies above their respective Coulomb barriers. This was the first time I (PK) heard a discussion about what later became the famous UNILAC, built by Schmelzer and his group in the early seventies at GSI in Darmstadt.

Taking the whole Mössbauer spectrum with a multichannel analyzer (MCA) was the next advance, after the unreliable tube systems were replaced by transistorized MCAs. But price wise, they seemed out of reach to us. GMK was lucky: a guest group working at the reactor lent him for a few days their MCA (the famous fully transistorized TMC). The first day I measured with it, ML coming through my lab (which was not oiled), looked suspiciously at the MCA box asking: "What is that?" I explained how I took the whole spectrum storing it in the MCA. ML said: "Mr. Kalvius, you are wasting time which you will spend in repairing the system. But that's your choice." The next day ML came through again and asked: "Is the monster still working?" I showed him the complete Mössbauer spectrum on the screen. ML looked at me very seriously, and then declared: "I will buy two for the Mössbauer Group." This must have eaten up a good part of the Laboratory's annual budget but it is another example of the flexibility of ML, once you had convinced him. Then Egbert Kankeleit came up with two most important advances in the Mössbauer effect measuring techniques. He constructed the feedback controlled double loudspeaker drive based on easily available commercial operational amplifiers and with it, the digital time wise channel advance controlled by a precision quartz oscillator. This worked because the velocity sweep generated by the "Kankeleit drive" (as it became to be known later) was exactly known and stable to a few parts in 10^{-3}. That eliminated the necessity to take an additional baseline control spectrum. The early availability of Kankeleit's system was a great advantage for the

Munich group in its effort to produce internationally competitive results. ML was so taken with these developments that he proudly remarked on them at the second Mössbauer conference [14]. Kankeleit published his inventions only with much delay [15]. It became and still is the standard method of Mössbauer spectroscopy.

Mössbauer pursued in Munich first the measurements of nuclear life times. Due to their nuclear physics background, other members of the group concentrated on the determination of magnetic dipole moments, electric quadrupole moments, and isomer shift especially of rotational excitations of rare earth nuclei. A special case, leading to an unexpected isomer shift, was the $K = 1/2$ band of ^{169}Tm, the radius of which turned out to become smaller in the rotational state than in its ground state [see the contribution by W. Henning]. These were very interesting nuclear structure results for advanced nuclear models, with two of the early ones leading to Nobel Prices, i.e., H.J. Jensen and M. Göppert-Mayer and A. Bohr and B. Mottelson.

Studying rare earths from the nuclear point of view naturally initiated investigations of their atomic, solid state and chemical properties. Also, since now money was more easily available from newly founded government agencies, we could buy ^{57}Co activity and use the iron resonance. We benefited from the expertise in proportional counters. A lucky circumstance was that in the chemistry department of the University of Munich famous work on metal–organic (also on Fe containing) compounds under the leadership of Prof. E.O. Fischer (who in 1973 received the Nobel Prize in chemistry for this work) was carried out. Via one of his former student, Ursula Wagner (at the time U. Zahn), we became engaged in studies of the quadrupole interaction and isomer shifts of those novel and fascinating materials [16]. Another development was the measurements of Mössbauer resonance scattering by detecting the resonance production of conversion electrons using a double focusing beta spectrometer [17].

Rudolf Mössbauer left us in 1960 to accept an offer as a research fellow at the California Institute of Technology (CalTech) continuing his research of Doppler-free resonance spectroscopy. He carried, so to speak, with him the ^{169}Tm resonance [see contribution by R.L. Cohen] and in a sense Pasadena and Munich became competitors. Yet, the field was wide and both laboratories had interesting results under different aspects [18, 19]. Already during Mössbauers stay in Munich the "Nobel" smell was in the air. Shortly after Mössbauer had left, a guest scientist from Stockholm arrived in Munich. Officially, he carried out an experiment at the nuclear reactor in Garching. Cleverly, he managed to drive with one of us graduate students in and out of Munich. Then he started the inquiry, how, in our view, the situation was in the laboratory at the time Rudolf Mössbauer discovered the effect. The real reason for this fellow's stay in Munich was clear. In 1961, Rudolf Mössbauer was awarded the Nobel Prize of Physics for his doctoral thesis, jointly with Prof. R. Hofstadter from Stanford. (An amusing side view: Hofstadter was the inventor of the modern NaI(Tl) scintillation detector which Mössbauer esteemed so highly. The Nobel Prize was awarded to Hofstadter, however, for his work on elastic electron scattering off

nuclei). Rudolf Mössbauer was then also appointed as Professor of Physics at Cal-
Tech and enjoyed the discussion with the famous theoretician colleagues very much.

2.5 The Second International Mössbauer Conference in Paris (1961)

The second Mössbauer Conference [20] took place in summer 1961 in Paris. Actually it was the first conference on the subject with broad international attendance.
It is called the second conference since in June 1960 a rather informal discussion
meeting on the Mössbauer effect had taken place at the University of Illinois. It was
mostly attended by US scientists who had just embarked in the field and proceedings
were not published [see contribution by S.S. Hanna].

ML and I (PK) attended this conference and stayed together in a charming small
hotel in the Rue Saint Jacques in Saint Germain des Prés. It was one of the nicest
experiences for me attending with ML this conference. I was invited to report on our
first results on Mössbauer transitions of rare earth nuclei, especially on the 8.42 keV
resonance of ^{169}Tm studied in collaboration with Mike Kalvius, my coeditor of this
book, and our first ^{57}Fe-Mössbauer spectra of metal–organic compounds, showing
quadrupole splitting. I enjoyed listening to many interesting applications of the
Mössbauer-effect and could meet and discuss for the first time with very famous
people.

One day the Conference participants were invited to visit the Louvre. I went there
with ML and he promised to show me the most famous paintings. At the entrance of
the museum, ML was expected by a gentleman, who did not introduce himself, but
asked ML, whom he obviously knew very well, for a short interview. ML excused
himself, but asked me to wait for him a short while at the entrance. He did not return
for more than an hour and I still waited. He came back finally all alone, very pale
and asked me to call a taxi and bring him back to the hotel; he had never done that
before or later. He did not say anything at all only in front of his room he proposed:
"Let's meet as usual for breakfast." So we met at breakfast and he excused himself
for letting me wait so long at the entrance of the Louvre and explained: "This Waller
wanted to talk me into receiving the Nobel Price 1961, but I told him that my student
found the nuclear resonance effect all by himself." I was dazzled at that time, but
I now understand that I was a silent witness of an event in the history of science:
"This Waller" was the famous Prof. Waller, from the Nobel-Foundation and known
to me from the "Debye–Waller factor" which determined the recoil-free fraction in
a Mössbauer experiment. At this point I became speechless. But I thought then, and
still think, ML became very famous due to this remarkable attitude concerning the
discovery of his student especially in the science community. What was not correct,
I still think, is that the Nobel Prize Committee did not award ML the "Prize" later
for his innovative work in neutron optics, while Brockhouse and Shull received the
Nobel Prize for neutron-spectroscopy and -scattering in 1994!

2.6 The "Second Mössbauer Effect"

"The Second Mössbauer Effect" was the headline of the "Spiegel" (a German weekly journal like the "Times") for announcing the foundation of the "Physik-Department" at the Technische Hochschule München (THM), as it was still called at this time. Originally a project drawn up by Maier-Leibnitz and his colleagues, Mössbauer made its realization a "sine, qua non" condition for his return. More details on this historical development in German university structure can be found in the contribution by P. Kienle.

After the main parts of his claims were indeed accepted by the Bavarian government, Rudolf Mössbauer returned to the THM in 1964. Included in his demands was a new department building in Garching with well-equipped laboratory space and nice offices into which we moved in 1970. In 1965, I (PK) returned to the THM from the "Institut für Strahlen- und Kernphysik" at the Technische Hochschule Darmstadt at which I had started a Mössbauer effect group, which was then taken over by E. Kankeleit who returned from Caltech. In the new Department building in Garching, Mössbauer and I had our offices close together with only our secretaries' offices between us. This led to a very close collaboration of our groups in the field of application of the Mössbauer effect.

Mössbauer became interested mainly in two special subjects. One was the precise difference between the Debye–Waller factor (X-ray scattering) and the Lamb–Mössbauer factor (recoilless fraction in resonant γ-radiation) together with the differences in the reaction time between resonant and non-resonant scattering [21,22]. The other subject was concerned with the coherent properties of Mössbauer radiation [23] (as foreseen by ML in his note – Fig. 2.2). One of the unusual effects in this connection is the so-called "speed up" in coherent decay modes of Mössbauer radiation in perfect single crystals. Theorists like Yu. Kagan, G.T. Tremmel, and J.P. Hannon were common guests to discuss the intricacies of those problems [for more details see the contribution by S.G. Smirnov]. Another point capturing Mössbauer's interest was the so-called "phase problem" in structure determinations by X-rays. In conventional X-ray diffraction, the phase information on the scattered radiation is lost, preventing the direct transformation of the diffraction pattern into the location of the scattering centers (e.g., the atoms) in real space. Creating the pattern via resonant scattering of Mössbauer radiation, the phase information could be obtained, a feature especially of importance in the study of (Fe containing) bio-molecules [24]. The problem of realizing this approach was the low intensity of ^{57}Co sources compared to modern X-ray tubes. Mössbauer, however, clearly foresaw that properly filtered synchrotron radiation would offer a solution [25]. So he initiated a group concerned with the filtering process together with general Mössbauer spectroscopy of biological substances headed by F. Parak.

I (GMK) had left in 1963 for the US, working mostly at Argonne National Laboratory, an old "Hochburg" of Mössbauer studies. Being in summer 1969 as a guest in Stanford collaborating with S.S. Hanna, I got an invitation by Rudolf Mössbauer to come to Pasadena, where he annually spent several months. In the evening, we had dinner together and there out of the blue sky he told me that he had succeeded in getting double professorships for some institutes of the Physics Department and whether I would like to share with him the responsibility for the Institute E15 in Munich. Surely I would, and we celebrated this by going to the "in" nightclub "The Pink Panther" at the strip. A brief story showing that Mössbauer really meant sharing: when I arrived in Munich, he proudly told me that, despite the extreme shortage of rooms (the new department building was not yet finished), he had managed to find me a nice office. When he showed me the place, there were four students working on high-pressure experiments, who proudly had finally found an empty room to build up their equipment. "Oh well" said Mössbauer "why should I be better off than my students, you share my office with me." We did so until Rudolf Mössbauer accepted in 1972 the offer of the directorship of the Institute Max von Laue-Paul Langevin (the German-French and then also British High-Flux Reactor) in Grenoble, France, for a period of 5 years as the successor of Maier-Leibnitz. During his time in Grenoble, Mössbauer was mainly involved in administration, because the finance situation of the ILL was very critical, and conventional neutron physics did not interest him much. The Mössbauer group in Munich was taken care of by PK and GMK. However, PK turned his interest to heavy ion physics with experiments at the Accelerator Laboratory in Munich and at the GSI in Darmstadt so it was agreed to melt our Mössbauer groups together.

2.7   Back to Munich from Grenoble with a New Scientific Challenge

When Mössbauer returned from Grenoble to Munich, he visited me (PK) and was quite excited. "Paul, I will get into new physics." "What will you do?" I asked him curiously. "I want to search for neutrino oscillations." "Please, explain?" So he explained to me that neutrino oscillation is a quantum mechanical phenomenon predicted by Pontecorvo, whereby a neutrino created with a *specific* lepton flavor (electron, muon or tau) can later be measured to have a *different* flavor. The probability of measuring a particular flavor for a neutrino varies periodically as it propagates in space and time. The observation of this phenomenon implies that the neutrino has a non-zero mass, which is not part of the original Standard model of Particle Physics. "Very exciting, but what neutrino source do you want to use?" "The ILL high flux neutron source of course." "Very interesting, and with whom do you want to perform this experiment?" "This is exactly why I come to you" replied Mössbauer "I like to ask you whether you have for me an excellent, young post-doctoral student who could construct and perform such an experiment right away." I hesitated for a moment and then said "Yes, I have one, Franz von Feilitzsch, but he has still, do take his Dr. examination." "No problem, we will fix this, but do you have a position for him?" "Yes, I do." "So I take him with the position, but do you also have some money for us to start the project right away?" This was really a bit too much for me. He finally got the extra support he needed and von Feilitzsch joined his neutrino oscillation project and two other former students from my group followed him (V. Zacek and G. Wild, now G. Zacek). Together with Felix Böhm they performed the very first neutrino oscillation experiment at the ILL with a follow up experiment at the Swiss power reactor station in Gösgen [26]. The details of that story can be found in the contribution by F. von Feilitzsch and L. Oberauer. Mössbauer had by now left nearly altogether the field he had created, with the exception of the above-mentioned coherent effects which he continued to pursue with U. van Bürck and several guests.

About 25 years later the KamLAND collaboration in Kamioka, Japan, found indeed these oscillations using the combined antineutrino flux of about 40 Japanese power reactors [27]. But this exciting chapter of physics is by far not finished. We have recently observed in a storage ring for heavy ions in Darmstadt two-body Electron Capture decays of hydrogen like ions with one electron in the K-shell showing time modulation of the exponential decay with periods of a few seconds [28]. We propose as origin of the observed modulation a quantum beat like effect produced by massive flavor mixed neutrinos [29]. Their quadratic mass differences and the structure of the neutrino flavor mixing matrix can be deduced from the precisely measured modulation period of the decay and the modulation amplitude [30]. So Rudolf's and my (PK) scientific interest come very close again after nearly 50 years of research.

2.8   Outlook

A final word about the scientist Rudolf Mössbauer. He was hard working following up with great determination any project he had set his mind on. He would freely discuss his ideas with colleagues, assistants, and graduate students, taking criticism from anyone. He thoroughly liked university atmosphere. He several times declined to take a directorship at a Max Planck Institute. He took university life serious, giving lectures as everyone, including the introductory physics course. His lectures were brilliant and tremendously liked by the students. If time permitted he liked to take part personally in experiments.

References

1. K.G. Malmfors, Arkiv f. Fysik **6**, 49 (1953)
2. L. Koester, H. Maier-Leibnitz, Sitzungsberichte d. Heidelberger Akademie der Wissenschaften, Math. Kl., Jg. (1951)
3. H. Steinwedel, J.H. Jensen, Z. Naturforsch. **A2**, 125 (1947)
4. W.E. Lamb Jr., Phys. Rev. **55**, 190 (1939)
5. T. Mayer Kuckuk, quoted (in German) in *Deutsche Nobelpreisträger*, ed. by A. Herman, Karl Moos, Partner (München, 1987)

6. R.L. Mössbauer, Hyperfine Interact. **126**, 1 (2000) see also reprint in this volume.
7. F.R. Metzger, in *Progress in nuclear physics*, vol. 7, ed. by O.R. Frisch (Pergamon, New York, 1959)
8. R.L. Mössbauer, W.H. Wiedemann, Z. Physik **159**, 33 (1960)
9. H.J. Lipkin, Ann. Phys. **9**, 332 (1960)
10. H. Frauenfelder, R.M. Steffen, in *Alpha-, Beta- and Gamma-Ray Spectroscopy*, vol. II, ed. by K. Siegbahn (North Holland, 1965)
11. O.C. Kistner, A.W. Sunyar, Phys. Rev. Lett. **4**, 412 (1960)
12. F.E. Wagner, Z. Physik **210**, 361 (1968)
13. L. Grodzins in [20], p.52
14. H. Maier-Leibnitz in [20], p.57
15. E. Kankeleit, Rev. Sci. Instrum. **35**, 194 (1964)
16. U. Zahn, P. Kienle, H. Eicher, Z. Physik **166**, 220 (1962)
17. E. Kankeleit, Z. Physik **164**, 442 (1962)
18. R. Cohen, R.P. Hausinger, R.L. Mössbauer *Recoilless Resonant Absorption in Nuclear Transitions of the Rare Earths* in [20], p.172ff
19. P. Kienle, G.M. Kalvius, F. Stanek, F.E. Wagner, H. Eicher, W. Wiedemann, *Hyperfine Splitting of Gamma rays from Rare Earth Nuclides* in [20], p. 185ff
20. *The Mössbauer Effect (Proc. 2nd Int. Conf. on the Mössbauer Effect, Paris, France)*, ed. by D.M.J. Compton, A.H. Schoen (Wiley, New York, 1962)
21. R.L. Mössbauer, J. Phys. (Paris), Coll. **37**(Suppl. 11) (1976)
22. Yu. M. Kagan, J.S. Lyubutin (ed.), *Applications of the Mössbauer Effect*, vol. 1 (Gordon and Breach, 1985), p. 1
23. R.L. Mössbauer, J. Phys. (Paris), Coll. **37**, Suppl., 5 (1976)
24. R.L. Mössbauer, Naturwissenschaften **60**, 493 (1973)
25. G.T. Trammel, J.P. Hannon, S.L. Ruby, P.A. Flinn, R.L. Mössbauer, F.G. Parak, AIP Conf. Proc. **38**, 46 (1977)
26. L. Oberauer, F. Von Feilitzsch, R.L. Mössbauer, Phys. Lett. B **198**, 113 (1987)
27. S. Abe et al., Phys. Rev. Lett. **100**, 221803 (2008)
28. Y.N. Litvinov et al., Phys. Lett. B **664**, 664 (2008)
29. A.N. Ivanov, P. Kienle, Phys. Rev. Lett. **103**, 062502 (2009)
30. P. Kienle, Progress in Particle and Nuclear Physics, **64**, 434 (2010)

Chapter 3
The Discovery of the Mössbauer Effect

Rudolf L. Mössbauer

A historical outline is given of the discovery of recoilless nuclear resonance absorption of γ-radiation, also called the Mössbauer (M-) effect.

3.1 Introduction

I shall initiate this talk with two remarks:

1. More than 20 years ago, following my stay at Grenoble as the director of the Institute Laue-Langevin (ILL) and the associated European High Flux Reactor, I have decided to change my field of research: this decision was based on the circumstance that hundreds of laboratories were engaged worldwide in applications of the M-effect, with thousands of papers appearing every year. I, therefore, wanted to do something else. None of the thousands of neutron experiments performed at the ILL was of sufficient interest to me, but a reactor is also a substantial source of neutrinos, and these attracted my attention. I, therefore, decided to study neutrino physics.
2. Today, I shall speak on the discovery of the M-effect, i.e., on very old material. You will, therefore, forgive me that I shall use for brevity reasons the word "M-effect" instead of my preferred old notation "recoilless nuclear resonance absorption of γ-radiation."

3.2 The Discovery of the Mössbauer Effect

γ-Resonance spectroscopy, synonymous with recoilless nuclear resonance absorption of γ-radiation or in short with the M-effect, forms part of the general field of resonance fluorescence. In fact, my story starts early this century, when the

R.L. Mössbauer (✉)
Physics Department, Technical University Munich, 85747 Garching, Germany

M. Kalvius and P. Kienle (eds.), *The Rudolf Mössbauer Story*,
DOI 10.1007/978-3-642-17952-5_3, © Springer-Verlag Berlin Heidelberg 2012

Fig. 3.1 Experimental setup for measuring resonance fluorescence (resonant scattering) or resonant absorption

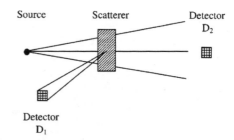

Fig. 3.2 Atomic or nuclear energy states

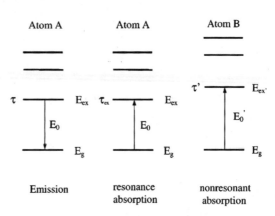

English physicist Robert Wood succeeded in demonstrating the existence of optical resonance fluorescence, by observing the yellow lines resonantly scattered from sodium vapor [1]. Figure 3.1 shows schematically the arrangement for studying both atomic and nuclear resonance fluorescence. Figure 3.2 gives an explanation of the optical phenomenon in terms of an atomic level scheme. Atom A in a light source emits a spectral line, passing from an excited state with energy E_{ex} and lifetime τ_{ex} to a stable ground state with energy E_g. The widths of the lines are even for gaseous atoms typically not the natural line widths Γ, but experience broadenings due to the linear Doppler effect or due to collisions. The energy $E_0 = E_{ex} - E_g$ released in the transition will be emitted in the form of electromagnetic radiation. The quanta emitted by the source would produce in a scatterer the inverse process of resonance absorption, if the scatterer is composed of atoms of the same type A than those decaying in the source, hence the name resonance absorption. The atoms excited in the scatterer will decay back to their ground state, giving rise to an electromagnetic radiation in all directions. As shown in Fig. 3.1, a detector D_1 will register this fluorescent radiation, while a detector D_2 would instead notice a corresponding reduction in the transmitted intensity, a phenomenon called absorption. The optical phenomenon was indeed observed, but not the apparently equivalent nuclear resonance phenomenon of γ-radiation, replacing the atomic levels of Fig. 3.2 by nuclear levels, i.e., replacing the optical transitions by nuclear γ-transitions. It was as early as 1929, when Werner Kuhn pointed out the distinction between the optical and the nuclear phenomenon [2]: the energy E_0 produced in

Fig. 3.3 Relative displacement of emission and absorption lines due to the recoil energies taken up by the emitting and absorbing nuclei, assuming nuclei in gaseous form

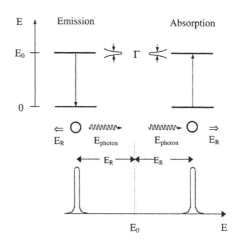

the transition from an excited to the ground state is used by both methods for the dominant emission of a photon with energy E_{photon}, yet a small amount is also used in the form of kinetic energy E_R for the recoiling particle, $E_{\text{photon}} = E_0 - E_R$, where $E_R = p_{\text{photon}}^2/2m_{\text{atom}} \approx E_{\text{photon}}^2/2m_{\text{atom}}c^2$. In a similar fashion, absorption requires the energy E_0 of the transition and in addition the energy E_R of the recoiling particle, i.e., $E_{\text{photon}} = E_0 + E_R$. By consequence, there results in an energy deficit $2 \times E_R$ between emission and absorption lines, as indicated in Fig. 3.3. The recoil energy E_R is proportional to E_{photon}^2. It is for this reason, that even in gases recoil effects are small compared to the Doppler level widths for optical transitions, but they play a major role for γ-transitions. After the paper of Kuhn, nothing essential happened for the next 20 years. It was only in 1951, when Moon [3] at Birmingham realized the first observation of nuclear resonance fluorescence, by placing a source of 411 keV γ-radiation, which is emitted in the decay of ^{198}Hg to ^{198}Au, on the tip of an ultracentrifuge. Applying the enormous speed of 670 m/s to the source, he used the linear Doppler effect to overcome the energy deficit between source and absorber. The scattering cross-section increased in his experiment by a factor of 10^4 and in this way he managed to observe the resonance phenomenon.

Another method was introduced by the Norwegian physicist Malmfors [4], who tried successfully a thermal method for observing nuclear resonance fluorescence, increasing the overlap between emission and absorption lines through heating procedures. Moon at Birmingham and Metzger at the Franklin Institute of Philadelphia performed in the following years dozens of experiments involving nuclear resonance fluorescence, mainly employing ultracentrifuges, but also using recoils of preceding transitions and preceding nuclear reactions. All these experiments provided, in particular, the lifetimes of some 50 excited nuclear states. A review of these experiments is given by Malmfors [5].

My own research started in 1953. My thesis supervisor was Professor Maier-Leibnitz, who provided me with one subject for both my master thesis and my doctorate thesis, in suggesting the application of the thermal method of Malmfors to another isotope. I worked on my masters thesis in Munich, during the period of summer 1953 until March 1955. During this thesis, following the suggestions of my thesis advisor, I performed:

1. The building of 12 proportional counters. This counter bundle exhibited a total efficiency for the selected isotope of some 5%, which on the one hand was a world record, while on the other hand it turned out to be entirely insufficient for the later purposes.
2. The selection in December 1953 of the 129 keV transition emitted in the decay of ^{191}Os to ^{191}Ir. This selection was made in the library of the Deutsches Museum in Munich, which at that time was the only reasonable library in Munich. The selection of this isotope was made because:

   (a) The energy of the transition appeared sufficiently low for a thermal experiment.
   (b) This radioisotope was available in the "Harwell" catalogue. Germany at this time had no reactor of its own and the Harwell catalogue provided by my thesis supervisor was Germany's only source for isotope importation.
   (c) The lifetime of the 129 keV state was unknown and its determination might be the new result of my thesis.

From my thesis supervisor, I had also four preprints at my disposal:

- Number 1 and 2 were preprints concerning the work of Moon.
- Preprint number 3 was the famous paper by Lamb [6] on neutron capture γ-rays, which I used later on for the theoretical explanation of the results of my measurements.
- Preprint number 4 was a theoretical paper by Steinwedel and Jensen [7], which contained on the basis of the paper by Lamb [6] the mathematical proof for the impossibility, to obtain the narrow lines, which I later found.

In 1955, after finishing my master's thesis, I received an offer from Professor Maier-Leibnitz to go to the Institute, where he had come from, the Max Planck Institute for Medical Research at Heidelberg, containing amongst its three sub-Institutes the Institute for Physics directed by Prof. W. Bothe, who had received the Nobel Prize already in 1954 for his discovery of the coincidence method and unfortunately died already half a year later. I accepted this offer to Heidelberg within less than a second, because it was a dream for me. I also know by now, that I could never have succeeded in Munich. So I went to Heidelberg in May 1955 and started my doctorate thesis there. The work in Heidelberg involved several major changes:

1. Following the example of other researchers at Heidelberg, I immediately replaced my Munich proportional counters by NaJ counters, which for the 129 keV radiation offered a nearly 100% efficiency. This decision turned out to be crucial for the following work.
2. In contrast to Munich, the Max Planck Institute had the money for buying electronic components, though otherwise, the complete electronics had to be built by oneself. This turned out to be very fortunate, facilitating an accuracy

of measurements down to the order of 10^{-4}, which typically is not possible nowadays, though tube equipment is replaced by transistorized equipment. The reason is simple, because I was able to constantly supervise my only channel, while nowadays one is dealing with hundreds or thousands of correlated channels, which, of course, cannot be supervised in detail.

3. There existed still a military government in this postwar period in Germany. With the Nazis fortunately gone, this military government looked for new tasks and found them in, for instance, administering spectroscopically pure materials, which I needed for my reactor irradiations. Following instructions from this military government, we had to write to the Max Planck Institute in Göttingen as the only place, which was allowed to communicate with the British. I, finally, gave up and wrote directly to England, which was strictly against the law. Fortunately, an English lady was helping me to bypass regulations and manufacture the radioactive sources I actually needed. At that time it was not so natural for an English lady to help a poor German graduate student, but without her help I would not have succeeded.

4. The crucial thing in my work was a decision concerning temperature changes. I had the opportunity in going up in temperature, as had previously been done by Malmfors and was also suggested by my thesis supervisor. I realized, however, that going down in temperature to the boiling point of liquid nitrogen should give about the same difference effect as going up in temperature. Fortunately, liquid nitrogen was available in our Institute at Heidelberg. I decided to lower the temperature, because it appeared to me much simpler to build a cryostate rather than a furnace. This decision was crucial for the later discovery. The cryostate was built without any vacuum, consisting of a Dewar with an insert, which permitted to alternately bring a resonant absorber (iridium) and a non-resonant absorber (platinum) into the γ-ray beam. This revolving system got stuck, but by wrapping the Dewar with tissue (developing snow all around it), the system worked perfectly well.

I have built at Heidelberg two experimental arrangements, aiming at intensity differences of $\Delta I / I \approx 0.1\%$ and 0.01%, respectively. Only the latter equipment worked, with a final result of $\Delta I / I = (2.7 \pm 0.7) \times 10^{-4}$. This corresponded roughly to the expectation, yet the sign appeared wrong: decreasing the temperature should give a reduced overlap of emission and absorption lines, resulting in an increase in the transmitted intensity. The observation yielded the opposite result. This situation of the lines is illustrated in Fig. 3.4 for the case of freely recoiling nuclei experiencing only Doppler shifts. The real situation, of course, was different, but unknown at that time.

At first, I searched for a dirt efffect, but this was in vain. I then approached Professor Jensen for help, who let me alone, but suggested, if anything strange happens, it must be in Lamb's paper. Jensen, of course, on the basis of Lamb's paper had written a paper himself, but I must simply confess that I was too stupid to understand the quantum mechanical arguments presented in it. I should note in this context, that at that time quantum mechanics was not required for obtaining a

Fig. 3.4 Position of the emission and absorption lines at room temperature for the 129 keV transition in ^{191}Ir, indicating the common region of overlap and assuming nuclei subject only to Doppler broadening

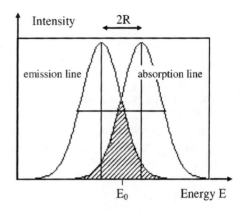

129 keV resonance line in ^{191}Ir

master's degree at Munich. By contrast, partially because of Jensen's presence, the situation in Heidelberg had been quite different. It had been my mediocre knowledge of quantum mechanics which made it quite impossible for me to understand the contents of Jensen's paper. This behavior is an example of the fact, that young people often are more apt to attack problems with unconventional approaches, which would not be touched by older, more experienced and more knowledgeable persons. It is quite interesting to regard nowadays the reason why Jensen's paper had been wrong. The authors made the classical assumption, that the line would be narrowed down to the natural width Γ only if the phonon spectrum was associated with only one frequency (Einstein model). A Debye spectrum increasing with ω^2, which incidentally always holds in any case at low temperatures, would according to the classical arguments always exhibit line widths way above the natural line width Γ, because such a spectrum would form a continuum, containing any number of small frequencies. It had been the delusion of Jensen and Steinwedel to assume that the recoilless lines could be approached in a classical steady way, which our experiment has shown to be wrong. The experiment has indeed verified the M-effect as a typical quantum mechanical phenomenon, with the recoilless lines reached in an unsteady manner, shooting steeply out above a phonon background. This situation was later on also explained by Dosch and Stech [8].

After my approach to Jensen, I went for help to my thesis supervisor in Munich. His advice was to take the train and to visit the various institutes for theoretical physics in Germany, starting with Professor Leibfried in Aachen, who then was the pope for phonon physics in Germany. I accepted, but I would first give one more try myself. I had been fascinated by two of the figures given in Lamb's paper, which are reproduced in Fig. 3.5. Both figures exhibited a peak near the resonance $E = E_0$, but I did not understand the reason for it. In case I narrowed the resonance by several orders of magnitude, maybe this peak would grow even more. Lamb had performed his work in 1939, at which time very little had been known about the widths of

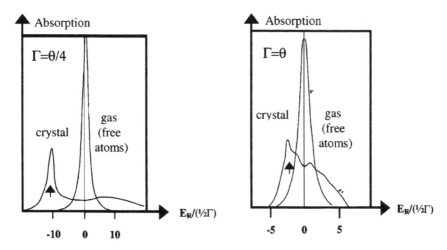

Case of moderate binding. For the cases of very strong binding the curves for the crystal would be identical to the gaseous case, except that they would show up at the position of the arrows.

Neutrons: Recoil energy of the nucleus M: $E_R = (m/M)E_n$

Gamma-radiation: Recoil energy of the nucleus M: $E_R = \dfrac{\bar{p}_\gamma^2}{2M} \approx \dfrac{E_\gamma^2}{2Mc^2}$

m, E_n = mass and energy of incident neutron
M = mass of recoiling nucleus (atom)
$\bar{p}_\gamma, E_\gamma$ = momentum and energy of emitted gamma quantum

Fig. 3.5 Figures 3.2 and 3.3 from the paper by Lamb [6], showing the neutron resonance absorption curves in cold solid silver for nuclei with level widths Γ of 4θ (*right side*) and $\theta/4$ (*left side*). The curves applying for a gas (free atoms) are also shown. The abscissa measures the distance from resonance in units of $1/2\Gamma$. For increasing widths, the impact of phonons upon the γ-lines would become less pronounced. If the binding would be very strong, then the curves for the solid would have the same form as for the gas, except being centered at $E0 - ER$

the neutron resonances, which later on turned out to be generally much wider than envisaged in his work. This is the reason why Lamb's work has never gained any importance in the field of neutron physics. Lamb had indeed performed his work for the capture of neutrons (n, γ) which now are heavily used for obtaining information on nuclear levels by looking at the energies of the emitted γ-quanta. Lamb at his time was more interested in the process of the capture of neutrons, which exhibit a recoil phenomenon with the absorbing nuclei. In the neutron case, the recoil energy E_R is given by $E_R = (m/M)E_n$, where m, E_n, and M are the masses and energy of the incident neutron and the mass of the hit nucleus. Could this also be applied to γ-resonance absorption? Knowing for historical reasons a lot of mathematics, I picked up my slide-rule (we did not have any PCs these days), and tried a series expansion right at resonance. I clearly obtained a divergence. I then used the five digit logarithm table left over from my own high school and tried again, again with the result of a divergence. So I went to the library of the Deutsches Museum in Munich, where they had a 12 digit logarithm table. With this, I obtained two steeply increasing points. Now the behavior was understandable: I had replaced the neutron capture energy by the energy E_γ of a γ-quantum appearing right at the resonance energy, completely avoiding any recoil energy E_R by putting the relevant isotope into a crystal. The observed γ-quantum exhibited then only the natural width given by the uncertainty principle, $\Gamma \approx 10^{-5}$eV, instead of the neutron width $\Gamma \approx \theta \approx 10^{-2}$eV as obtained by Lamb. There thus appeared a narrow resonance line showing up right at the resonance position, occurring as a recoilless transition. This situation might be explained likely by a person throwing a stone from a boat. The majority of the energy is submitted to the stone, but a small amount goes into the kinetic energy of the recoiling boat. During summer time, the boat will simply pick up this recoil energy. If, however, the person throws the stone during winter time, with the boat frozen into the lake, then practically all energy is going into the stone thrown and only a negligible amount is submitted to the boat. The entire lake will, thus, take up the recoil and this procedure occurs as recoilless process.

In the following period, I successfully performed a theoretical treatment in closed form, first for the absorption line and, knowing by then some quantum mechanics, also for the emission line and their fold. Two incredibly narrow lines emerged right at the resonance position, nowadays called the recoilless lines or the M-lines and the sign observed in my thesis, which had been opposite to expectation, could be explained: with decreasing temperature, the extraordinarily narrow lines grew, increasing their overlap, thus increasing the absorption and by consequence decreasing the transmission. With this result, I finished my doctorate thesis [9]. The result is demonstrated in Fig. 3.6 both for a free recoiling nucleus, such as existing in a gas at sufficiently low pressure, and for a nucleus embedded in a crystal, where one obtains γ-lines both in the emission and absorption spectra, i.e., γ-lines associated with zero-phonon lines, one-phonon lines and, finally, multiple-phonon lines. The thesis work thus could explain the presence of the sharp lines, but I had no idea how to measure them. Crucial for my work, of course, were only the zero-phonon or the equivalent recoilless lines, involving always the momentum transfer, but lacking an associated energy transfer.

I returned then from Heidelberg to Munich, where I thought I would work on the first atomic reactor station we just had obtained in Germany. So I prepared myself for neutron work. Three months later, the printed version of my first publication appeared. One still reads this first publication of oneself, while later publications are of no interest. During this reading, it occurred to me that I had not performed the main experiment: it should be possible to measure the sharp resonance lines by using the linear Doppler effect. I was so excited that I dashed across the hallway into the office of Professor Maier-Leibnitz without even knocking on the door and cried: I take the next train back to Heidelberg, because I forgot about the major experiment. Explaining my idea, he immediately agreed with me. Back in Heidelberg, where my old equipment still existed, I really panicked: I had sent two preprints in German to my main competitors, Professors Moon at Birmingham and Professor Metzger

I. Free recoiling nucleus

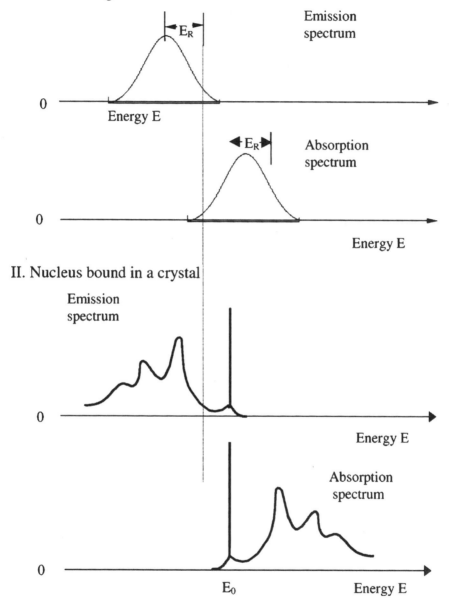

Fig. 3.6 Emission and absorption lines of resonant γ-radiation. The *upper part* applies to a nucleus free to recoil and subject only to the ordinary Doppler effect. The *lower part* applies schematically to a nucleus bound in a crystal, where the various atoms oscillate around their equilibrium positions and one obtains zero-phonon, one-phonon, and multi-phonon peaks

at Philadelphia. Both were really the popes in Doppler-shift experiments, having performed them over many years. With Moon I felt reasonably sure, because I thought he would not know any German. I did not know, however, that the German emigrant Peierls was sitting next door to him, who, of course, knew perfectly German and immediately jumped on him. Peierls, fortunately for me, thought I was wrong. So Moon did not do any experiments. Metzger in Philadelphia, who was of Swiss origin, knew German. He had even written to me, wondering about my error calculation, which carried a Boltzmann factor involving 1/4 kT instead of 1/2 kT, because I was dealing with the overlap of two lines. I thought that Metzger was simply trying to gain time; in reality he was doing the Doppler experiment. Every morning I went trembling to the library of the Max Planck Institute, regarding the newly incoming journals. It later turned out, that I could have been quiet, with none of my competitors having thought about the Doppler shift experiment. The Doppler shift experiment itself was rather easy, involving an effect of order 1%, while the previous measurement of the cross-section gave only an intensity change of order 0.01%. In setting up the equipment, I required a turntable, as illustrated in Fig. 3.7. A first effort in our mechanical shop showed, that it took about three days for a skilled guy to finish one cone-shaped wheel, of which I needed an entire set. Thus, I went to the mechanical toy shops of Heidelberg and bought up all their teeth wheels. The turntable did not work very smoothly with the wheels not fitting together, but fortunately that did not matter. The first curve obtained in such a Doppler shift experiment is shown in Fig. 3.8. Because I was moving narrow lines with natural widths against each other, the necessary speeds were some 10^6 times smaller than in the experiments of my competitors. After this experiment, I knew precisely what I had at my hands. I immediately went to our Institute, which had the only cyclotron then available in Germany and asked for the production of ^{57}Fe. Unfortunately, I was turned down, perhaps because I had asked for too much activity. I also went to my former thesis supervisor to ask for his advice concerning the publication. In this postwar period in Germany, I wanted to quietly cream off the results of my work, publishing my first Doppler experiment in a journal as obscure

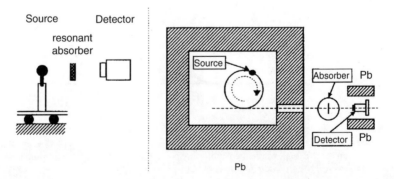

Fig. 3.7 Principle and actual realization of the observation of recoilless nuclear resonant absorption of γ-radiation, using Doppler shifts of the emitting relative to the absorbing nuclei

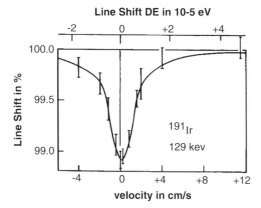

Fig. 3.8 First observation of recoilless nuclear resonance absorption of γ-radiation [10] in the case of the 129 keV transition in ^{191}Ir. Maximum resonance absorption causing minimum transmitted intensity, occurring whenever emission and absorption lines are located right at the resonance energy $E0$. This perfect resonance is destroyed, if source and absorber are moving relative to each other, moving an emission line of natural width Γ over an absorption line of identical width

as possible. Unfortunately, I got the advice to publish in the "Naturwissenschaften" [10]. Though still published in German, I got within one week requests for 260 preprints, which clearly showed that my case was lost. I conclude with two stories: At Los Alamos they threw a bet that I would be wrong. The entire investment was 1 nickel. I feel that they should have used at least a dime instead. My second story goes as follows: in 1959 I gave a colloquium at Heidelberg and Felix Böhm, spending a sabbatical year there, was in the audience. He was a Swiss-American residing at the California Institute of Technology. He asked me for a preprint, sent it to Jesse DuMond, who turned it over to the two main theoreticians then at Caltech, Bob Christie, and Dick Feynman. Both gentleman met in the evening and decided to tell each other in the following morning what they thought about the paper, which was still written in German. Feynmans remarks were: the whole thing is crazy, but I cannot find any error in the mathematics. So a famous cable arrived at Heidelberg, containing only three words: get the guy. Signed Dick Feynman. That was the basis of my invitation to Caltech, but here I exceed already the scope of my topic and, therefore, will finish.

References

1. R.W. Wood, Am. Acad.**11**, 396 (1904)
2. W. Kuhn, Philos. Magn.**8**, 625 (1929).

 3. P.B. Moon, Proc. Phys. Soc.**64**, 76 (1951)
 4. K.G. Malmfors, Ark. Fys. 6, 49 (1952)
 5. K.G. Malmfors, Alpha-, Beta-, Gamma-Ray Spectrosc.**2**, 1281 (1965)
 6. W.E. Lamb Jr., Phys. Rev.**55**, 190 (1939)
 7. H. Steinwedel, J.H. Jensen, Z. Naturforsch. A**2**, 125 (1947)
 8. H.G. Dosch, B. Stech, Semper Apertus**3**, 417 (1986)
 9. R.L. Mössbauer, Z. Phys.**151**, 124 (1958)
10. R.L. Mössbauer, Naturwissenschaften**45**, 538 (1958); Z. Naturforsch. A**14**, 211 (1959)

Chapter 4
The Early Period of the Mössbauer Effect and the Beginning of the Iron Age

John P. Schiffer

As the first papers of Mössbauer became known in late 1958 and early 1959, they first generated skepticism. Then, as the observations were reproduced at several laboratories, skepticism was followed by curiosity. The discovery of the huge recoil-free fraction in ^{57}Fe in late 1959 generated enormous excitement and intense activity, and by early 1960 the door was opening to large classes of hitherto inaccessible measurements. In the following article, I give a personal account of my involvement with the Mössbauer effect as a 29-year old nuclear physicist in 1959–1960, which includes the discovery of ^{57}Fe as the ideal nucleus for this effect, and the activities that sprung from this in the following months. What is presented is mostly in the form of photographs and old letters, recalling some of the events of that period and is largely connected with my own work. I surely had some correspondence with Mössbauer, we certainly sent him preprints of our papers and there is a reference to a private communication from him in our first paper. But I have not found any copies of this correspondence. I believe that I did not meet Rudolf Mössbauer until later, probably in 1961 in the US.

It was in the winter of 1959 that I was invited to give a seminar at Los Alamos, on work I had been doing on nuclear reactions. I remember that at lunch, a Los Alamos physicist, John Marshall, asked whether I had seen this result in a German paper in Naturwissenschaften, claiming to see nuclear resonance absorption of gamma-rays at low temperatures. I had not seen or heard of this work. Apparently somebody reported an effect when a source was cooled to low temperatures. "Everybody knew" that to achieve resonant absorption from a nuclear transition, one had to overcome the energy lost into nuclear recoil that is imposed by momentum conservation. One either had to heat a radioactive source to a temperature sufficient

J.P. Schiffer (✉)
Physics Division, Argonne National Laboratory, Argonne, IL 60439, USA
e-mail: schiffer@anl.gov

M. Kalvius and P. Kienle (eds.), *The Rudolf Mössbauer Story*,
DOI 10.1007/978-3-642-17952-5__4, © Springer-Verlag Berlin Heidelberg 2012

for a very small fraction of the nuclei to have the right thermal velocity to compensate for the recoil energy, or else, mount a source on a centrifuge to provide the appropriate velocity. It seemed incredible that there could be an effect at low temperatures.

When I got back to Argonne, I looked up and read the two papers [1, 2] by somebody called Mössbauer and consulted with Luise Meyer-Schützmeister, a colleague who was born and educated in Germany and had been working with gamma-rays. (I knew German – but wanted to make sure that I understood all the fine points in the description of the measurements). The results seemed very strange. A gamma ray from an excited state of [191]Ir was reported to show enhanced absorption (by about a percent) in an iridium metal absorber, when the absorber was cooled. Stranger yet, this effect was reported to disappear as a function of velocity between source and absorber in the few cm/s range, which was consistent with the Doppler shift from the natural line width corresponding to the lifetime of the state. There was no apparent sign of the recoil, nor of the thermal broadening that one would have expected in a solid. We agreed that this result was likely wrong – but that if it turned out to be correct it could be very exciting. Luise and I, together with my longer term collaborator, Linwood Lee, decided that we must attempt to repeat the measurement, either to show that it was wrong, or to confirm it. (In my career, I have felt compelled to do such measurements a number of times, because I strongly believe that the scientific process requires that strange and unexpected results be followed up – either to confirm or to refute. In fact, this was the only time that I found one of these "strange" results to turn out to be correct).

Together with Luise, my colleague Lin Lee and Dietrich Vincent, a visitor at Argonne that year, we set out to design and build a suitable apparatus. We tried to slightly improve on Mössbauer's design by having three absorbers mounted on a rotating cryostat and having a timing device to count only when the source and absorber were aligned within appropriate limits. This we did quickly, as the date (April 4, 1959) on some lantern slides, shown in Fig. 4.1, indicates. We completed the measurement and confirmed the result.

We did not think that reproducing a published result was worth another paper and felt we needed more to publish. From the Mössbauer paper and an incomplete understanding of the reference to Lamb on neutron scattering that was given there, we concluded that apparently this was an effect that took place only in materials with high Debye temperatures and that a nucleus with heavy mass was essential; so we went on to another nucleus, an isotope of tungsten, and also saw a comparable effect. With this additional result we thought we had enough that was new and submitted a Letter to the newly formed journal Physical Review Letters. At that point, we discovered that a group at Los Alamos had also done an experiment that had confirmed the effect and our two papers were published side by side. Note that in those days the delay between receipt and publication in PRL was a matter of weeks (submitted on August 3 and out in the September 1st issue) as shown in Fig. 4.2. The first results must have been obtained in March; the measurements on tungsten must have been in about May or possibly June. While the results were fascinating,

Fig. 4.1 Apparatus and data (transparency dated April 4, 1959) of the first confirmation of the Mössbauer result on Ir at Argonne

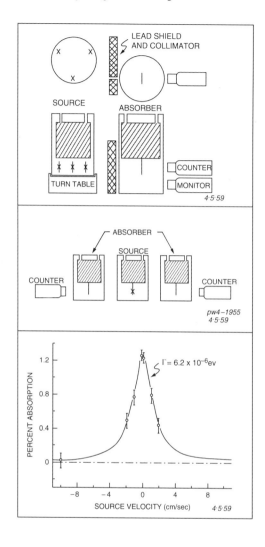

the effect was on the order of a percent or so in both cases, and applications seemed limited. We had many discussions with theorists at Argonne (mostly Mort Hamermesh, but also Dave Inglis and Maria Mayer) and the question of hyperfine effects came up, which had also been noted by Mössbauer.

But I had been awarded a Guggenheim fellowship and had planned to spend a year in England. I left in June 1959 to go to Harwell (a leading laboratory at that time in nuclear physics, that was comparable to Argonne in many ways but had superior accelerators). My plans were to work on nuclear reactions with the new tandem accelerator at the A.E.R.E., Harwell. Of course I discussed the Mössbauer effect with my English colleagues when I arrived – they had not yet heard of it.

VOLUME 3, NUMBER 5   PHYSICAL REVIEW LETTERS   SEPTEMBER 1, 1959

NUCLEAR RESONANCE ABSORPTION OF GAMMA RAYS AT LOW TEMPERATURES[*]

L. L. Lee, Jr., L. Meyer-Schutzmeister, J. P. Schiffer, and D. Vincent
Argonne National Laboratory, Lemont, Illinois
(Received August 3, 1959)

Recent experiments by Mössbauer[1] have shown that nuclear resonance absorption of gamma rays can take place when both source and absorber are placed at liquid nitrogen temperature. This effect can be ascribed qualitatively to the fact that at low temperatures some of the emitting and absorbing atoms are not able to recoil in the crystal lattice and thus the recoil momentum is taken up by the entire crystal. This means that the emitting gamma ray loses virtually no energy to the recoiling system and that no energy is lost on recoil in absorption, so that nuclear resonance absorption can readily take place. In subsequent papers[2] Mössbauer reported that the resonance absorption could be destroyed by introducing a Doppler shift a few times as large as the width

used in his measurements was Os^{191} which decays to Ir^{191} with a 16-day half-life. This beta decay populates a 4.9-sec isomeric state which cascades through the 129-kev excited state of this nucleus. The experiment was designed to measure the lifetime of this excited state. A schematic diagram of our experimental arrangement is shown in Fig. 1. The sources and absorber were mounted on the bottom of liquid nitrogen containers in such a way that filling the containers did not change the amount of absorber the detected gamma rays had to pass through. A lead collimator was placed so that a source could be seen by the detectors only while its velocity along the line of sight differed from the mean by less than 5%. By the use of three sources, the

P. P. Craig, J. G. Dash, A. D. McGuire, D. Nagle, and R. R. Reiswig
Los Alamos Scientific Laboratory, University of California, Los Alamos, New Mexico
(Received August 3, 1959)

Conditions for resonant absorption of nuclear gamma rays are difficult to obtain, since the natural widths of nuclear excited states are typically small compared to the energy lost to nuclear recoil by the emitted gamma ray.[1] If E_0 is the energy of the nuclear excited state and the nucleus is free to recoil, the gamma-ray energy will be $E_0 - R$, where R is the recoil energy of the nucleus. Moreover, a gamma ray can be resonantly absorbed by a free nucleus at rest only if the gamma-ray energy equals $E_0 + R$, since the free absorbing nucleus must also recoil with energy R. Earlier systems for observation of resonance absorption all involved supplying the $2R$ energy deficit in one way or another; e.g., by Doppler-shifting the emitted gamma ray by ultracentrifuging the source.

Recently Mössbauer[2] has demonstrated a different technique for observing nuclear resonant absorption, based on the fact that in crystals at

tion events, the energy deficit $2R$ is small compared to a typical nuclear energy level width, and the absorption cross section very large. Mössbauer demonstrated the existence of nuclear resonance absorption of the 129-kev gamma-ray of Ir^{191} in metallic Ir (38.5% Ir^{191}) with source and absorber at a temperature of 88°K. By mounting the source on a turntable, he measured the variation in transmission as a function of the Doppler shift of the emitted gamma rays, and thereby obtained the width of the first excited level of Ir^{191}.

This Letter reports our repetition of Mössbauer's measurements, the extension of the temperature range to 1.5°K, measurements on various foil thicknesses, and a demonstration of semiquantitative agreement of the results with theoretical calculations by Visscher[3] based upon a theory of Lamb.[4]

By neutron bombardment of metallic osmium we prepared 16-day Os^{191}, which decays to an

Fig. 4.2 The first two confirmations of the Mössbauer effect

Somebody suggested that I talk to Walter Marshall, a young and very bright theorist just back from Berkeley. He had worked on understanding magnetic fields in ferromagnetic atoms and was an expert on neutron scattering, very much aware of Lamb's work. Walter listened to me for a few minutes and immediately knew what the effect must be. He explained it to me and introduced me to the Debye-Waller factor [3, 4] that is critical in the description. This factor was first derived by Debye in 1913 in connection with trying to explain how the diffraction of X-rays by

Debye-Waller factor:

$$f \equiv e^{-\frac{\langle r^2 \rangle}{3\lambda^2}}$$

f is the recoil-free fraction,

$<r^2>$ is the mean-square displacement of the emitting
       atoms in the lattice,

λ is the wavelength of the emitted or absorbed photon.

4. *Interferenz von Röntgenstrahlen und Wärmebewegung;
von P. Debye.*

In einigen Notizen habe ich versucht, Belege dafür bei-
zubringen[1]), daß die Wärmebewegung der Kristallatome einen
wesentlichen Einfluß hat auf die von Friedrich-Knip-
ping-Laue entdeckten und von Laue[2]) schon theoretisch be-
handelten Interferenzen von Röntgenstrahlen. Ebenso wie das
bei der ursprünglichen Einsteinschen Theorie der spezifischen
Wärme der Fall war, wurden bei dieser Berechnung alle Atome
als voneina-   Utrecht, 29. September. 1913   erselben als
(Eingegangen 10. Oktober 1913.)

Zur Frage der Einwirkung der Wärmebewegung
auf die Interferenz von Röntgenstrahlen.

Von Ivar Waller in Göttingen.

(Eingegangen am 18. Juni 1923.)

Die ersten Berechnungen der Einwirkung der Wärmebewegung
der Kristallatome auf die Interferenz von Röntgenstrahlen hat be-
kanntlich Debye ausgeführt. Eine ausführliche Behandlung des
Problems hat er in seiner Abhandlung „Interferenz von Röntgenstrahlen und Wärmebewegung"[1]) gegeben. Als Ausgangspunkt seiner
Berechnungen benutzt Debye Vorstellungen, welche der Laueschen
Theorie der Interferenz von Röntgenstrahlen in einem starren Atom-
gitter zugrunde liegen. Debye vervollständigt diese Theorie durch
die Berücksichtigung der Wärmebewegung. Dabei wird mit einer

crystals, which had been observed by von Laue and the Braggs, could be consistent with the thermal motion of atoms. This was well before quantum mechanics. The beginnings of the two papers are reproduced in Fig. 4.3.

The physics behind the recoil-free emission of gamma rays was exactly the same as that behind the coherent scattering of X-rays from crystals that allowed the observation of X-ray diffraction and had been "well understood" in that connection. But the simple connection, the fact that the same relationship must also apply to photons emitted by an atomic nucleus and the scattering of photons of similar energies had not been made! With Marshall's explanation it was immediately clear

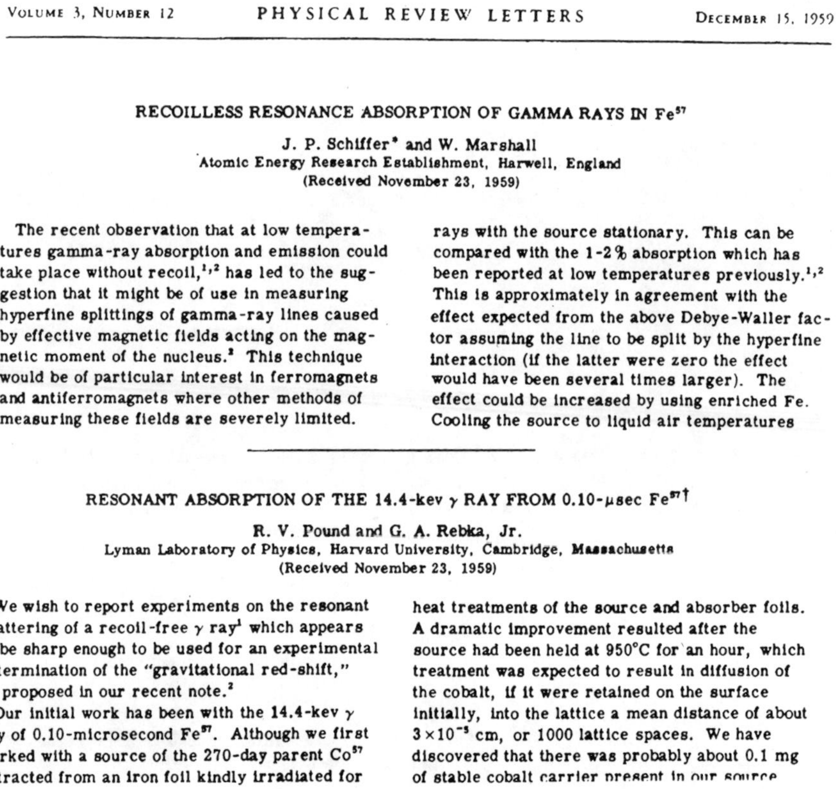

VOLUME 3, NUMBER 12   PHYSICAL REVIEW LETTERS   DECEMBER 15, 1959

RECOILLESS RESONANCE ABSORPTION OF GAMMA RAYS IN Fe[57]

J. P. Schiffer[*] and W. Marshall
Atomic Energy Research Establishment, Harwell, England
(Received November 23, 1959)

The recent observation that at low temperatures gamma-ray absorption and emission could take place without recoil,[1,2] has led to the suggestion that it might be of use in measuring hyperfine splittings of gamma-ray lines caused by effective magnetic fields acting on the magnetic moment of the nucleus.[2] This technique would be of particular interest in ferromagnets and antiferromagnets where other methods of measuring these fields are severely limited.

rays with the source stationary. This can be compared with the 1-2% absorption which has been reported at low temperatures previously.[1,2] This is approximately in agreement with the effect expected from the above Debye-Waller factor assuming the line to be split by the hyperfine interaction (if the latter were zero the effect would have been several times larger). The effect could be increased by using enriched Fe. Cooling the source to liquid air temperatures

RESONANT ABSORPTION OF THE 14.4-kev γ RAY FROM 0.10-μsec Fe[57][†]

R. V. Pound and G. A. Rebka, Jr.
Lyman Laboratory of Physics, Harvard University, Cambridge, Massachusetts
(Received November 23, 1959)

We wish to report experiments on the resonant scattering of a recoil-free γ ray[1] which appears to be sharp enough to be used for an experimental determination of the "gravitational red-shift," as proposed in our recent note.[2]

Our initial work has been with the 14.4-kev γ ray of 0.10-microsecond Fe[57]. Although we first worked with a source of the 270-day parent Co[57] extracted from an iron foil kindly irradiated for

heat treatments of the source and absorber foils. A dramatic improvement resulted after the source had been held at 950°C for an hour, which treatment was expected to result in diffusion of the cobalt, if it were retained on the surface initially, into the lattice a mean distance of about 3×10^{-5} cm, or 1000 lattice spaces. We have discovered that there was probably about 0.1 mg of stable cobalt carrier present in our source

Fig. 4.4 Two parallel Letters in PRL with first reports on the Mössbauer effect in ^{57}Fe

that the energy of the gamma-rays was much more important than the mass of the nuclei or the Debye temperature. After talking to Marshall, I immediately went to the library to search through the compilations of nuclear data – and very quickly found the 14-keV first-excited state in ^{57}Fe (a 2% isotope of iron) as an ideal candidate. I talked to my colleagues at Harwell and very quickly, made a ^{57}Co source on the small Van de Graaff accelerator at Harwell using the ^{56}Fe(d,n)^{57}Co reaction, mounted it on a loudspeaker to give a Doppler shift, and within a couple of weeks, we had observed a huge effect – more than an order of magnitude larger than in Ir or W, a line with a lifetime that promised a very sharp line, and an effect that was evident at room temperature, with natural iron in which the mass 57 isotope is only 2%, and the iron was magnetic, so the Zeeman splitting of the line was likely to be large. Anybody with even a very modest accelerator and a piece of natural iron could produce ^{57}Co sources.

THE   DEBYE-WALLER   FACTOR   IN   THE   MÖSSBAUER   EFFECT

The purpose of this note is to point out that in the Mössbauer effect
the fraction of recoiless gamma-rays can be calculated exactly within the
framework of the Debye theory of solids by the Debye-Waller factor.   This
factor is well known in X-ray crystallography and is given for instance in
Compton and Allison, X-rays in Theory and Experiment, p. 487.

(1) $f = e^{-2W}$

where $W = \dfrac{3}{2} \dfrac{E_\gamma^2}{Mc^2 k\theta} \left[\dfrac{1}{4} + \left(\dfrac{T}{\theta}\right)^2 \int_0^{\frac{\theta}{T}} \dfrac{x\,dx}{e^x - 1} \right]$

Fig. 4.5 Beginning of a mimeographed note by Marshall and the author, prepared and distributed
in the fall of 1959 to acquaint the relevant community about the simple theory of the Debye-Waller
factor

The sharp line and the magnitude of the effect meant that this result was much
more than a curiosity and could have many applications. Marshall and I wrote up our
results and sent it off to Physical Review Letters as shown in Fig. 4.4. Also, since
I knew that many of my colleagues had not understood the elementary physics,
we wrote a short note explaining the Debye-Waller factor to other physicists,
mimeographed and distributed it to all who might be interested, shown in Fig. 4.5.

With the lifetime of the state in question providing a very sharp line, we discussed
further experiments and the potential of studying the hyperfine structure and the
prospects of using this as a tool in solid-state physics. But I was not so familiar
with the significant problems in magnetism; I did not know the Harwell system
(where to find tools and equipment, how the shops worked, personal relationships,
and all that is needed to start new experiments. And Harwell experimentalists I
knew were mostly engaged in other projects). Marshall was enthusiastic, but he
was a theorist. I did, however, write to my colleague Stan Hanna at Argonne,
because I knew that the equipment and interest existed there and also because
Stan had been trying to measure the internal fields in Fe by positron annihilation
for years. Some of the correspondence is shown in Fig. 4.6. But [57]Fe with its
huge absorption dip and sharp line provided an exquisite precision tool, unique
before the availability of lasers, and I tried to think what could be done with
it. It did not seem that within nuclear structure physics there would be many
applications.

At Harwell, meanwhile, I did get together with Ted Cranshaw, a cosmic-
ray physicist, and we discussed the possibility of using [57]Fe for measuring the

Argonne National Laboratory
OPERATED BY THE UNIVERSITY OF CHICAGO
BOX 299 LEMONT, ILL.

TELEGRAM VUE LS LEMONT, ILL.   CLEARWATER 7-7711   TELETYPE TWX LEMONT, ILL. 1710

November 12, 1959

Dr. John Schiffer
Nuclear Physics Division
Atomic Energy Research Establishment
Harwell, Didcot, Berks
England

Dear John:

Thank you for your recent letter with the exciting news it contained. We have of course all been following your work to some extent by means of your letters to Mort and others. We are all excited about it and in fact a group of us (Perlow, Vincent, Littlejohn, Preston and Heberle with consultation with Mort) have just now seen the effect you describe in Fe^{57}. With a very crude set-up at R. T. we see ~ 50% effect using an Fe^{57} absorber. Ideas and suggestions- many half-baked no doubt - are flying around thick and fast and have been for some time now. Before plunging ahead, however, we wanted to let you know what we are doing - now that we have seen something - and to assure you that our aim is to supplement your work and to aid you in any way that we can.

Argonne National Laboratory
OPERATED BY THE UNIVERSITY OF CHICAGO
BOX 299 LEMONT, ILL.

TELEGRAM VUE LS LEMONT, ILL.   CLEARWATER 7-7711   TELETYPE TWX LEMONT, ILL. 1710

November 25, 1959

Dr. John Schiffer
Nuclear Physics Division
Atomic Energy Research Establishment
Harwell, Didcot, Berks
England

Dear John:

Just a note to keep you informed on things here. Lin is going to give the paper at Cleveland since Luise doesn't yet feel quite up to it. In view of your letters to Luise and to me and especially since Pound has exposed Fe^{57} to print and since we know people like Frauenfelder are working on it (he heard about your work through Lipkin) -- I think Lin should mention your work on Fe^{57} briefly, if he is agreeable. Since the word is out, his mentioning it can only help. This will not reach a very large audience of course and I personally feel that you should send off a brief note on Fe^{57} right away (with no reference to Pound). If you don't, someone else will I feel sure.

With very best wishes. Yours,

S̶t̶a̶n̶

S. S. Hanna

Fig. 4.6 Two letters that were written by Stanley Hanna at Argonne to the author, in response to previous letters (that did not survive) in which I told the Argonne group about having found ^{57}Fe. The second letter refers to the APS meeting in November, where our work on iridium was to be presented by Lin Lee (since Luise Meyer-Schützmeister was expecting her second child). Jesse DuMond of Caltech was at that meeting and commented on the work, remarking that Mössbauer had accepted a position at Caltech

gravitational red shift. This was one of the three gravitational effects from general relativity predicted by Einstein some 55 years earlier, but it had never been observed, and a measurement seemed just feasible with this sharp resonance. Such a measurement required a vertical drop of several tens of meters – and this existed in a water tower at Harwell. But it would require a much stronger source, a larger absorber, and detector.

Peierls was the eminent theorist in England at the time. He was at the University of Birmingham, and invited me to give a colloquium. It was also in Birmingham that Moon and his coworkers had studied nuclear resonance absorption by attempting to compensate the recoil by spinning the source very fast in a centrifuge. Peierls, who was one of the founders of solid-state physics, told me about the calculations he had carried out years earlier with a student for the experimental work of Moon to include the effects of thermal broadening in a solid environment on the gamma-emission. When Peierls' student came up with the calculation reflecting the Debye spectrum in the solid, Peierls told both the student and Moon that they should ignore the unphysical delta-function peak at zero momentum transfer. This was characteristic of Peierls – that he was bemused by his own mistakes and relished telling about them.

The late R.V. Pound, well known for his work with Purcell on nuclear magnetic resonance, apparently recognized the importance and the physics of Mössbauer's result, and had understood the effect much better than the rest of us – except perhaps for Marshall. Pound had come up with ^{57}Fe and ^{67}Zn as possible candidates independently and wrote a Physical Review Letter in which he proposed to measure the gravitational red shift with the effect. We found out about this when we heard of an article in the New York Times, shown in Fig. 4.7, that Pound at Harvard announced plans to measure the red shift in a similar experiment. While he had apparently not yet observed the Mössbauer effect in ^{57}Fe experimentally, he did have a press conference to discuss the proposal to use it for a measurement of the gravitational red shift, before his paper in PRL was published. We were quite upset about this – newspaper stories and proposals staking out a claim, before the effect had even been observed. In fact, the publicity prior to publication prompted an editorial from Goudsmit in Physical Review Letters, shown in Fig. 4.8, saying that prior publication in the press was not acceptable. (The editorial was not entirely effective, and prepublication press conferences are not unknown to this day.) A second Letter by Pound and Rebka, with the measurement of the large effect in ^{57}Fe, was published alongside ours a month after the one with the red-shift proposal, as was shown in Fig. 4.4. The ^{57}Fe story and the possibility of measuring the red shift ("testing General Relativity") eventually caused quite a stir in the English as well as the US press as is shown in Fig. 4.9.

As we started to plan the gravitational red shift measurement at Harwell, I had a conversation with Heinz London (both he and his brother Fritz were well-known low-temperature physicists, one at Harwell and the other at Duke in the United States) and he asked me whether while we were at it, had we thought of also measuring the effect in an accelerated system. By then I began to know the Harwell system and I knew that there was a neutron chopper group, similar to the one at

Fig. 4.7 Article in the New York Times in which Robert Pound announced plans for a measurement of the gravitational red shift

Argonne. I talked to Peter Egelstaff who was in charge of the chopper. He referred me to a young postdoc from New Zealand, Hal Hay and, together with Cranshaw, we undertook to look for the accelerated shift in a fast vacuum rotor that was a by-product of the chopper work.

I had met Harry Lipkin at Argonne, while he was on a sabbatical at Urbana from the Weizmann Institute, and he was quite interested in this new effect. He visited me at Harwell in transit and we exchanged some letters afterward, that perhaps help give the flavor of those days, and are shown in Fig. 4.10.

Volume 4, Number 1   PHYSICAL REVIEW LETTERS   January 1, 1960

EDITORIAL

Publicity

Newspapers and magazines are printing more stories about scientific advances than ever before. The general public is slowly learning that an interest in the progress of science is essential in our modern society. A large portion of the financial support for scientific research comes from public funds. It is, therefore, the duty of the scientist to cooperate with the press and to make sure that the news articles convey the proper meaning of the discoveries and are free from exaggerated claims of importance and usefulness.

Since many of the advances in physics are first published in PHYSICAL REVIEW LETTERS, we send complimentary copies to several science writers. Moreover, we alert them in advance about potentially newsworthy papers, usually through the excellent services of Mr. Eugene Kone, Director of Public Relations for the American Institute of Physics, who has also greatly improved the handling of press conferences and releases at our meetings.

We recommend that our authors check with us or Mr. Kone on matters of press relations. As a matter of courtesy to fellow physicists, it is customary for authors to see to it that releases to the public do not occur before the article appears in the scientific journal. Scientific discoveries are not the proper subject for newspaper scoops and all media of mass communication should have equal opportunity for simultaneous access to the information. In the future, we may reject papers whose main contents have been published previously in the daily press. While we welcome the intensified interest of the layman in physics research we recognize that formerly crackpots often made the front page with their spectacular stories, and this still happens occasionally. We are sure that our authors do not wish to be confused with these pseudo-scientists in the minds of the public. This can be avoided by using the right publicity channels which will give these stories an authoritative stamp of reliability and the proper dignity. In addition, careful planning and preparation of news material will achieve increased accuracy in the interpretation of science.

Fig. 4.8 Editorial by Sam Goudsmit in Physical Review Letters, reacting to the New York Times article that preceded the PRL publication date. The highlighting in red was added

Fig. 4.9 Pictures accompanying an article in the London Times about our experiment

We had started with a measurement of both effects with what we could pull together, and by the time of the annual meeting of the American Physical Society in New York, at the end of January 1960, we had some results and I flew back to present a postdeadline paper, which followed an invited paper by Pound (who had no results at that point). It was clear that Pound did not take kindly to the competition, which was perhaps understandable, but I remember being somewhat taken aback by his distant attitude. I knew of Pound's pioneering work on nuclear magnetic resonance and was in awe of him; it was somewhat of a shock.

Our first results from both the gravitational and the accelerated red shift were published side by side as shown in Fig. 4.11. I show a letter from Hans Frauenfelder at Urbana (whom I had not met up to that time) congratulating us on the results, shown in Fig. 4.12. Our uncertainties were still large and we wanted to do better, with a larger source, absorber, and detector.

Shortly after this, I received a letter from a Cambridge undergraduate who had read about our work in the Times – he had a question about an effect that was not mentioned – the temperature shift. I took the letter (Fig. 4.13) to Marshall and he immediately saw the point and felt very embarrassed about having missed this. I remember his gesture, slapping his forehead and saying that this young man should have his job. We contacted the undergraduate and Marshall arranged for a shiny laboratory limousine to go to Cambridge and bring him to Harwell. This was Bryan Josephson – and his idea was indeed correct. We persuaded him to write this up for Physical Review Letters, and I believe that this was his first scientific publication.

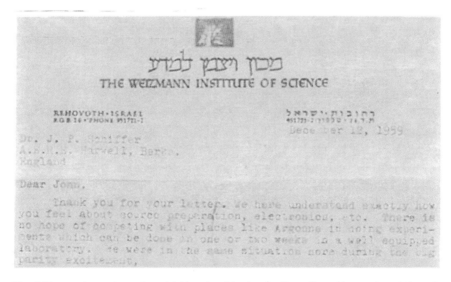

Fig. 4.10 Two letters from Harry Lipkin who visited me at Harwell, on his way from Urbana in the United States to Israel. Note the mention of Peierls in the first letter – I had not yet met Peierls at this point and Harry had carried the news about ^{57}Fe from Harwell to Birmingham. The second letter is sympathizing with my complaints to him at the time, about having found something new and exciting, and not being able to follow it up at a strange laboratory as effectively as I could have at Argonne

Volume 4, Number 4   P H Y S I C A L   R E V I E W   L E T T E R S   February 15, 1960

MEASUREMENT OF THE RED SHIFT IN AN ACCELERATED SYSTEM USING THE MÖSSBAUER EFFECT IN Fe[57]

H. J. Hay, J. P. Schiffer,* T. E. Cranshaw, and P. A. Egelstaff
Atomic Energy Research Establishment, Harwell, England
(Received January 27, 1960)

In an adjoining paper[1] an experiment is described in which the change of frequency in a photon passing between two points of different gravitational potential has been measured. Einstein's principle of equivalence states that a gravitational field is locally indistinguishable from an accelerated system. It therefore seemed desirable to measure the shift in the energy of 14-kev gamma rays from Fe[57] in an accelerated system. In order to do this we have plated a Co[57] source on to the surface of a 0.8-cm diameter iron cylinder. This cylinder was rigidly mounted between two Dural plates which also held a cylindrical shell of Lucite, 13.28 cm in diameter and 0.31 cm thick, concentric with the iron cylinder. An iron foil 3.5 mg/cm² thick and enriched in Fe[57] to 50% was glued to the inside surface of the Lucite. This assembly was mounted in a neutron chopper drive unit[2] and rotated at angular velocities up

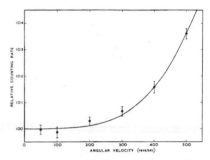

FIG. 2. Comparison of the calculated curve with experimental points. The statistical errors of each point are indicated. The curve was calculated from the parameters given in the text.

MEASUREMENT OF THE GRAVITATIONAL RED SHIFT USING THE MÖSSBAUER EFFECT IN Fe[57]

T. E. Cranshaw, J. P. Schiffer,* and A. B. Whitehead
Atomic Energy Research Establishment, Harwell, England
(Received January 27, 1960)

The change in the frequency of spectral lines with gravitational potential, generally referred to as the gravitational red shift, was first predicted by A. Einstein in 1907.[1] The effect can be calculated from the time dilatation in a gravitational potential which follows from the principle of equivalence. From the point of view of a single coordinate system two atomic systems at different gravitational potentials will have different total energies. The spacings of their energy levels, both atomic and nuclear, will be different in proportion to their total energies. The photons are then regarded as not changing their energy and the expected red shift results only from the difference in the gravitational potential energies of the emitting and absorbing systems. Astronomical observations, though somewhat ambiguous, have tended to confirm this effect.[2] The recent discovery by Mössbauer[3] of recoilless nuclear resonance absorption of gamma rays as a precise resonance process has suggested to several groups[4-6] the possibility of using this effect to measure the gravitational red shift. More specifically the discovery that Fe[57] could absorb 14-kev gamma rays in a resonance whose width is approximately 6.4×10^{-13} of the gamma-ray energy[4,6,7] has made this experi-

ment a practical possibility.

We have performed this experiment using a total difference in height of 12.5 meters. A source of Co[57] of approximately 30 millicuries was electrodeposited on the surface of an iron disk which was then heated for five hours at 700°C in a hydrogen atmosphere, then for an equal length of time in a vacuum. This disk was mounted on a transducer device which consisted of a coil mounted between the poles of an electromagnet and supported on an elastic spider. The gamma rays passed through an evacuated tube with 0.005-in. Mylar windows on both ends. Antiscattering baffles were mounted inside the tube. The detector consisted of a proportional counter with a five-inch diameter window also covered by 0.005-in. Mylar. The counter was filled with krypton gas to approximately $\frac{1}{4}$ atmosphere pressure. Krypton is especially favorable because its absorption edge is just below the energy of the 14-kev gamma rays.[8] A five-inch diameter foil containing 4 mg/cm² of Fe enriched in Fe[57] to 24.1% was directly above the counter.

The transducer was driven sinusoidally at 50 cps and counts were recorded in two scalers for alternate halves of the cycle. Two other scalers were switched simultaneously and recorded tim-

Fig. 4.11 Our adjacent papers on the first results on gravitational and rotationally accelerated red shifts

UNIVERSITY OF ILLINOIS
DEPARTMENT OF PHYSICS
URBANA
February 17, 1960

Dr. J. P. Schiffer
Atomic Energy Research Establishment
Harwell, Didcot, Berkshire
England

Dear Dr. Schiffer:

I have just seen your two beautiful letters in the
Physical Review Letters and I would like to congratulate you and
your group very much. You certainly did a very good job very
quickly and I hope that you will follow up these measurements
with equally beautiful continuations.

Just as a side remark, I would like to add that we were
all very glad to see that you and not Pound succeeded first in
doing this experiment.

I hope I will get to know you personally after you are back
at Argonne after your sabbatical leave. Please give my best regards
also to Dr. Bretscher.

Very sincerely yours,

Hans Frauenfelder

HF:ek

Fig. 4.12 Letter from Hans Frauenfelder at Urbana congratulating us on the red-shift measurements

The rest of my year at Harwell was spent in accumulating data for the gravitational red shift. Meanwhile, the iron (literally) was in the fire: with the information I sent Stan Hanna, he and Gil Perlow (Fig. 4.14) devised a way to use a precision lathe to provide vibration-free motion, and pinned down (after some false starts elsewhere) the magnetic hyperfine structure of ^{57}Fe and showed that the internal field of an Fe atom was opposite to the external field. Also, Mort Hamermesh at Argonne, a theorist who followed this work closely, suggested a measurement of the decay spectrum as a function of time when the radiation was filtered by the Mössbauer effect. It was done by Holland and Lynch, who had specialized in fast timing measurements – and they found the nonexponential behavior predicted by Hamermesh. Utilizing

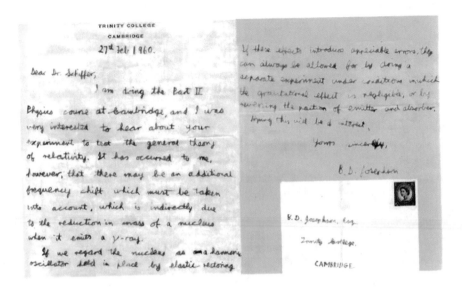

Fig. 4.13 Letter from B.D. Josephson, (future Nobel Prize winner, then an undergraduate at Cambridge), asking about the temperature shift, which we had ignored. The stamped, self-addressed envelope that he enclosed, is also shown

the very large Mössbauer effect and the very sharp line width in ^{57}Fe a number of other laboratories pursued other aspects, including Urbana, Los Alamos, Bell Labs, Brookhaven (where the quadrupole hfs and the chemical shift were established) and the number of papers using the effect increased enormously, as shown in Fig. 4.15.

The Mössbauer effect was getting very hot and the rate of publications was increasing rapidly in the first few months of 1960, as is illustrated in the Table 4.1. I worked on getting a substantially expanded red-shift experiment in my remaining time at Harwell. By the time I came back to Argonne at the end of 1960 I had to make a decision, either to switch fields and work on the Mössbauer effect as a tool in condensed-matter physics, or to make use of the new tandem accelerator that meanwhile was almost completed at Argonne. For better or worse, I chose the latter and stuck with nuclear physics. I briefly considered doing a larger red-shift measurement in a mine in northern Michigan, but after looking down a large vertical mine shaft with a drop of 1,600 m I decided that this was not for me. But at Argonne there remained a very active group working on the Mössbauer effect in the following decade, with Gil Perlow and later Stan Ruby in the Physics Division (Stan Hanna went to Stanford in 1961), and with a large group in the Solid State Division including Mike Kalvius, Gopal Shenoy, and Bobby Dunlap. Similarly, at Harwell further work was done by Ted Cranshaw and Charles Johnson and collaborators.

Fig. 4.14 Some of the author's collaborators in this period are shown. On top, left Luise Meyer-Schützmeister from Argonne (1915–1981); on the top right, Walter C. Marshall at Harwell (later Lord Marshall) (1932–1996), and below Stan Hanna and Gil Perlow in 1960, with the apparatus they used to determine the hyperfine structure of ^{57}Fe

C O N T E N T S

Physical Review Letters - 15 April 1960
(Volume 4, Issue 8)

EDITORIAL... 395

Attempts to Detect Resonance Scattering in Zn^{67}; The Effect of Zero-Point Vibrations
....R. V. Pound and G. A. Rebka, Jr.   397

Search for the Anisotropy of Inertia Using the Mössbauer Effect in Fe^{57}
..C. W. Sherwin, H. Frauenfelder, E. L. Garwin, E. Lüscher, S. Margulies, and R. N. Peacock   399

High-Frequency Studies on Superconducting Tin.........M. S. Dresselhaus and G. Dresselhaus   401

Measurement of Local Fields at Impurity Fe^{57} Atoms Using the Mössbauer Effect
....G. K. Wertheim   403

Recoilless Rayleigh Scattering in Solids.......................C. Tzara and R. Barloutaud   405

Study of the Intermediate State in Superconductors Using Cerium Phosphate Glass
....Warren DeSorbo   406

Sputtering Thresholds and Displacement Energies ..Robley V. Stuart and Gottfried K. Wehner   409

Proposal for an Electron Spin Resonance Experiment of S-State Ions Under High Hydrostatic
Pressure..Hiroshi Watanabe   410

Evidence for Quadrupole Interaction of Fe^{57m}, and Influence of Chemical Binding on Nuclear
Gamma-Ray Energy.................................O. C. Kistner and A. W. Sunyar   412

Structure of Nuclear Matter...A. W. Overhauser   415

Fig. 4.15 Table of contents of the April 15, 1960 issue of Physical Review Letters, with the articles related to the Mossbauer effect highlighted in *yellow*

Table 4.1 Mössbauer chronology from 1958 to April 1960

1958:
R.L. Mössbauer
Kernresonanzabsorption von γ Strahlung in Ir^{191},
Kernresonanzfluorescence von γ Strahlung in Ir^{191}.
August 1959:
L.L. Lee, L. Meyer-Schutzmeister et al.
Nuclear resonance absorption of gamma rays at low temperatures.
P.P. Craig, J.G. Dash et al.
Nuclear resonance absorption of gamma rays in Ir^{191}.

(continued)

Table 4.1 (continued)

November–December 1959:

R.V. Pound and G.A. Rebka
  Gravitational red-shift in nuclear resonance.
J.P. Schiffer and W. Marshall
  Recoilless resonant absorption of gamma rays in Fe^{57}.
R.V. Pound and G.A. Rebka
  Resonant absorption of the 14.4-keV γ-ray of 0.1 μs Fe^{57}.

January 1960:

S.S. Hanna, J. Heberle et al.
  Observation of the Mössbauer effect in Fe^{57}.
G. DePasquali, H. Frauenfelder et al.
  Nuclear resonance absorption and nuclear Zeeman effect in Fe^{57}.
G.J. Perlow, S.S. Hanna et al.
  Polarization of nuclear resonance radiation in ferromagnetic Fe^{57}.

February 1960:

S.S. Hanna, J. Heberle et al.
  Polarization spectra and hyperfine structure in Fe^{57}.
T.E. Cranshaw, J.P. Schiffer et al.
  Measurement of gravitational red shift using the Mössbauer effect in Fe^{57}.
H.J. Hay, J.P. Schiffer et al.
  Measurement of the red shift in an accelerated system using the Mössbauer effect in Fe^{57}.
G. Cocconi
  Upper limit on the anisotropy of inertia from the Mössbauer effect.
R.E. Holland and F. Lynch
  Time spectra of filtered resonance radiation of ^{57}Fe.

March–April 1960:

D. Nagle, P.P. Craig et al.
  Nuclear resonance fluorescence in Au^{197}.
R.V. Pound and G.A. Rebka
  Variation with temperature of the energy of recoil-free gamma rays from solids.
R.V. Pound and G.A. Rebka
  Apparent weight of photons.
B.D. Josephson
  Temperature-dependent shift of γ-rays emitted by a solid.
R.V. Pound and G.A. Rebka
  Attempts to detect resonance scattering in Zn^{67}: the effects of zero-point vibrations.
C.W. Sherwin, H. Frauenfelder et al.
  Search for the anisotropy of inertia using the Mössbauer effect in Fe^{57}.
G.K. Wertheim
  Measurement of local fields at impurity Fe^{57} atoms using the Mössbauer effect.
O.C. Kistner and A.W. Sunyar
  Evidence for quadrupole interaction of Fe^{57m}, and influence of chemical binding on nuclear gamma-ray energy.

References

1. R.L. Mössbauer, Kernresonanzfluorescence von γ Strahlung in Ir^{191}. Z. Phys. **151**, 124 (1958)
2. R.L. Mössbauer, Naturwissenschaften **45**, 538 (1958)
3. P. Debye, Interferenz von Röntgenstrahlen und Wärmebewegung. Ann. d. Phys. **348**, 49 (1913)
4. I. Waller, Zur Frage der Einwirkung der Wärmebewegung auf die Interferenz von Röntgenstrahlen. Z. Phys. **17**, 398 (1923)

Chapter 5
The Early Iron Age of the Mössbauer Era

Stanley S. Hanna

*This account of the early days of Mössbauer spectroscopy in the United States was delivered by Stanley S. Hanna at the International Conference on the Mössbauer Effect 1989 in Vancouver, BC, Canada. It is one of a series of invited talks discussing the history and some newer developments of Mössbauer studies. They all appeared in Hyperfine Interactions **90** (1990). Stanley's narrative gives a vivid account of the struggle to understand the hyperfine spectrum of iron, which nowadays is often just an experiment a physics major has to carry out in the physics lab course. With the permission of the author, one of the editors (GMK) has made a few alterations and abridgments to adjust this text to the present volume. GMK came to Argonne Nat'l. Lab. at a much later time than the one described in this article. But he got to know personally most of the actors of the wild time recounted here, and also was told their personal experiences. GMK also had the good fortune to work with Stanley Hanna and his (then) graduate student Gene Sprouse at Stanford.*

The discovery of the Mössbauer effect in [191]Ir [1] ushered in the Iridium Age as reckoned on the Mössbauer calendar created by H. Lipkin [2] shown in Table 5.1. At first, life did not change much in this new age. But after some investigators established the correctness of the new effect [3, 4] and others began to understand its true meaning [5], interesting new possibilities seemed to lie in the future, and life became more exciting. However, on the horizon there were stunning developments that would doom the Iridium Age. A new nucleus had just been studied [6, 7] that had properties (see Fig. 5.1) far superior to [191]Ir [8–10] for realizing the full potential of the Mössbauer effect. This new nucleus was [57]Fe, a 2.14% abundant isotope of iron. It was the autumn of 1959 on the Gregorian calendar that saw the dawning of this new Ironium Age. The early period of this age is the subject of this paper. As an active participant in this period, the following is a personalized account of the

S.S. Hanna (✉)
Physics Department, Stanford University, Stanford, CA 94305, USA

M. Kalvius and P. Kienle (eds.), *The Rudolf Mössbauer Story*,
DOI 10.1007/978-3-642-17952-5__5, © Springer-Verlag Berlin Heidelberg 2012

Table 5.1 The Mössbauer calendar

Period	Date (Gregorian calendar)	Remarks
Prehistoric	Before 1958	Might have been discovered, but was not
Early iridium age	1958	Discovered, but not noticed
Middle iridium age	1958–1959	Noticed, but not believed
Late iridium age	1959	Believed, but considered not interesting
Early ironium age	1959	Wow!! Mössianic power was transferred from iridium to iron
Middle ironium age	1960	Ushered in by Allerton Meeting. Shows no sign of abating

Fig. 5.1 Properties of ^{57}Fe as known in 1959. After [7]

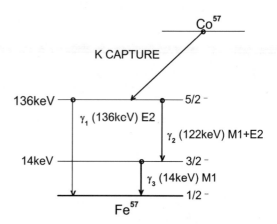

exciting, frenetic, but wonderful time during which so much was accomplished in such a short time.

At Argonne National Laboratory, where one of the confirmations of the Iridium Age had taken place [3], the new possibilities afforded by ^{57}Fe were a favorite topic of conversation. What could be done with this sensational new effect? Were we on the threshold of exciting new discoveries? One of the intriguing ideas that came up was the possibility of measuring the shift due to the centripetal acceleration in a high-speed rotor. One day after lunch, six of us – Juergen Heberle, Carol Littlejohn, Gil Perlow, Dick Preston, Dieter Vincent and myself – got together and decided to give it a try. We received the immediate and enthusiastic support of our director, Mort Hamermesh (who also made valuable contributions to our investigations), and of the Argonne Laboratory. A special red shift account was set up and all the needed facilities of the Laboratory were placed at our disposal. Almost immediately, Mina Rea Perlow joined our group and within a very short time ^{57}Co activity had been produced in the Argonne cyclotron. Gil and Mina Rea soon produced suitable sources of ^{57}Co diffused into iron metal. Frank Karasek went to work rolling very thin iron foils and each day, or so it seemed, he produced a new record for thinness. The detector group fabricated very thin NaI crystals to detect the low energy γ-rays and we were soon ready to try to see the effect in ^{57}Fe.

We needed a suitable "drive" and considered an electronic device, but we felt we did not have time to design and construct one. It was suggested that a lathe could be used, so we went to the foreman of our machine shop and asked for his best lathe. At first he was reluctant to consign his best instrument to a bunch of physicists, but when we mentioned the words red shift, he was happy to comply. We set the experiment up using the drive mechanism of the lathe carriage to produce the needed slow, smooth motion. We worked at night so as to eliminate vibrations from all the other machinery in the shop. As I recall, we were successful in seeing a Mössbauer "dip" in our first full-fledged attempt.

Meanwhile, we were busy fabricating and testing small air-jet-driven rotors made of aluminum. With the help of many people, we soon produced a successful high-speed model that would rotate at almost the breaking point of aluminum (we believed!). We ran the experiment several times and undoubtedly observed the red shift, but we were not satisfied that we had reduced the effect of vibration enough to report the experiment. This kind of measurement was somewhat later successfully carried out at Harwell [11] using a neutron chopper, and at Birmingham with an ultracentrifuge [12].

However, during these attempts it became clear to us that much more interesting physics was waiting to be discovered. In an effort to increase the size and sensitivity of the effect, we tried several different chemical forms of source and absorber. We found the intriguing result that only with identical source and absorber (in our case metal) did we observe maximum absorption at zero velocity (Fig. 5.2, left). We also reported there is evidence that the width of the transmission dip is influenced by the environment of the emitter or the absorber [13]. Yet, we did not perceive clearly at the time that these observations predicted the presence of hyperfine interactions. But shortly after we submitted the letter [13], the first report of a Mössbauer spectrum

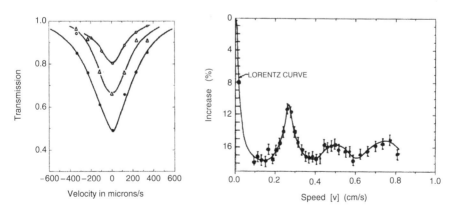

Fig. 5.2 *Left*: First ^{57}Fe "dips" observed at Argonne. The different curves refer to different absorber thicknesses (from *top* to *bottom*: 0.2, 0.9, 2.5 mg ^{57}Fe cm^{-2}) The resonant absorption cross-section was derived from these measurements. After [13]. *Right*: First spectrum of ^{57}Fe showing hyperfine splitting. The source is ^{57}Co in Fe, the absorber Fe metal. The spectrum is folded about zero velocity. The spectrum was terminated too soon to show two additional lines. After [14]

of ^{57}Fe appeared [14] which clearly showed a spectrum split into seven lines. Since the lines appeared symmetrically about zero velocity, they appear as four lines when plotted against the speed |v|. The spectrum is shown in Fig. 5.2, right. When this report appeared, we were already busy following up another provocative line of thought. Very little was known to us then about fields at nuclei in magnetic media. But it was argued that in iron, the nucleus was probably in some kind of magnetic field and the nuclear transitions should be polarized according to the "classical Zeeman effect." Gil devised a simple experimental setup to test this idea. We obtained some "five-and-dime" magnets and by constructing a simple apparatus arranged to measure the Mössbauer absorption when the direction of magnetism in the metallic absorber was set at different angles relative to that in the metallic source (Fig. 5.3). We reasoned that the absorption should be greatest when the magnetizations (i.e., assumed polarizations) were aligned and least when they were perpendicular. One evening we were ready to try this experiment and were rewarded by a beautiful sine curve for the Mössbauer absorption measured [15] as the angle between the polarizations was varied from 0 to 180°. This occurred a few days before the holiday season in December and contributed much to the festive spirit.

On the basis of these two experiments [14, 15], we were now convinced that some form of magnetic hyperfine interactions is present in the Mössbauer spectrum of magnetic iron. Despite the fact that the four-line folded spectrum of [14] had been confirmed by the Illinois group [16] we set out to also reproduce this result. In those days, a velocity spectrum was obtained in a laborious fashion, point by point, as the velocity of the lathe carriage was increased in small steps. By now, the lathe had been moved to its own room (in the basement, to reduce vibration). I remember well plotting the evolving spectrum as each point from adding up the 14.4 keV counts. We decided to beautify the four line spectrum by continuing on to higher velocities. To our astonishment, another line appeared and then still another. We continued on,

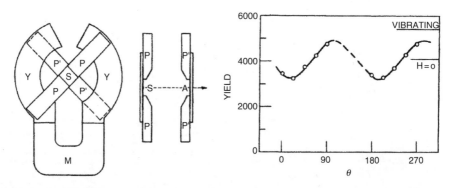

Fig. 5.3 *Left*: Apparatus for detecting polarization of resonance radiation in ^{57}Fe. On one side of the alnico magnet M, the yoke structure Y is fastened. The pole pieces P and P′ bridge the circular yoke gap on either side of M. Source S and absorber A bridge the gaps of P and P′, respectively. P and P′ can be rotated relative to each other. *Right*: Transmission curve of resonant radiation as function of angle Θ between the magnetizations of S and A. The yield obtained with a vibrating source (resonance destroyed) and with zero field in source and absorber are indicated. After [15]

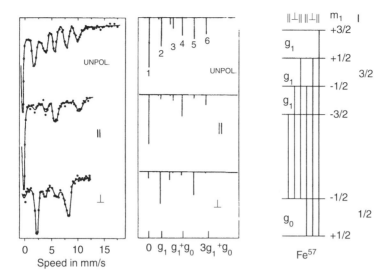

Fig. 5.4 The nuclear Zeeman effect in ^{57}Fe. *Left*: Spectra (^{57}Co in Fe *vs.* Fe metal) with, from top to bottom: unpolarized source and absorber, source and absorber polarized parallel, and source and absorber polarized perpendicular. *Center*: Bar diagram showing positions and intensities of the hyperfine lines of the spectra on the left for different polarization. *Right*: Zeeman levels and hyperfine transitions with their polarizations for ^{57}Fe. After [17]

but no other lines appeared. By this time, the lathe velocity had increased to such a point that we were afraid that vibration might have obscured another weak line. Nevertheless, we were so excited by the six lines (see the top spectrum on the left in Fig. 5.4), in contrast to the four reported earlier, that we fired off a Letter to Physical Review Letters. By return mail, the Letter came back with the admonition that we simply could not publish every new step. I hope some day to frame this rejected paper!

By this time, however, we had taken the next step. The key lay in combining our static measurement with the polarized source and absorber [15] with the dynamic measurements we had made on the lathe [13]. So, our five-and-dime magnets were mounted on the lathe and complete polarized spectra were taken, first with source and absorber polarizations aligned and then with the polarizations perpendicular. This particular night it was snowing outside the basement window where the lathe was set up, and as the snow piled higher and higher, one line after another appeared in the polarized spectra. We were delighted to see that the line intensities varied dramatically with the relative orientation of the polarizations. Each line was enhanced in one orientation and suppressed in the other, or vice versa [17] (see the lower two spectra in Fig. 5.4, left).

It now appeared that all that remained was to solve the puzzle presented by these spectra. A favorite game was to draw the presumed set of polarized Zeeman lines for a magnetically split 3/2 excited state and a split 1/2 ground state on one sheet of paper for the source and an identical set of lines on another sheet for the absorber, and then to pass one over the other to arrive at the predicted polarization spectra. We knew, of course, what the relative line intensities and polarizations should be for each Zeeman component. The big unknown was the relative spacings of the excited state and the ground state splittings. But no matter what we did, no solution emerged. I remember well sitting one Sunday after supper and wondering what could possibly be wrong when a light flashed across my mind. What would happen if I inverted the excited state Zeeman multiplet relative to that of the ground state? In other words, there was no reason to assume that the signs of the magnetic moments of the excited and ground states were the same. In a short while I tried this scheme out, and after a few adjustments of the relative splittings had achieved a remarkable agreement with the observed polarized spectra. The Zeeman levels of ^{57}Fe are shown in Fig. 5.4, right. It occurred to us that the correctness of our scheme could be tested in a simple way. The solution predicted that the line appearing at speed $|v| = 4\,\mathrm{mm\,s}^{-1}$ should be a doublet, with one member appearing in parallel polarizations and the other member in perpendicular polarizations (see center part of Fig. 5.4). Careful measurements were immediately carried out and gave the confirming result shown in Fig. 5.5 Fortunately, the ground-state moment of ^{57}Fe had just been reported by an ENDOR measurement [18]. From their value, we immediately obtained the excited state moment. In addition, we could deduce the hyperfine field at the nucleus. In a day or two, a paper was ready for Physical Review Letters. One seeming hurdle remained. The deduced value of the hyperfine field, 330,000 G, seemed ridiculously large in those early days before such large nuclear hyperfine fields in solids became routine. Could anything possibly be wrong? It was with considerable trepidation that the Letter [17] was dropped into the mailbox.

About this time, we became distracted from the excitement of the hyperfine chase. Gil kept insisting that the Mössbauer effect should exhibit interesting phenomena if the time evolution of the radiation was perturbed in some way -in

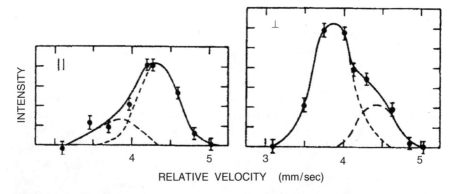

Fig. 5.5 The "line" appearing at $|v| = 4\,\mathrm{mm\,s}^{-1}$ with source and absorber parallel and then perpendicular, obtained to confirm the interpretation of the hyperfine level scheme shown in Fig. 5.4, right

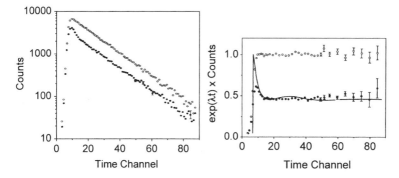

Fig. 5.6 Time Filtering of Resonance Radiation. *Left*: Time spectrum of the 14 keV radiation of ^{57}Fe (source ^{57}Co in Fe metal) through a Fe metal absorber. *Open symbols* through a vibrating absorber (resonance absorption "off"), *filled symbols* through a stationary absorber (resonance absorption "on"). *Right*: Same data as left but with the exponential decay $e^{-\lambda t}$ taken out. The *solid line* is the theoretical fit for a thin absorber. One time channel corresponds to 8.5 ns, the mean life $1/\lambda$ of the 14 keV state to 17 time channels. After [19, 20]

this case by passing through a Mössbauer absorber. Again we were fortunate in that Bob Holland and Frank Lynch had a setup working for the measurement of the time delay of nuclear radiations, and I had recently made nuclear lifetime measurements with Bob and Frank. We set the apparatus up to measure the time-delay spectrum in the successive $\gamma - \gamma$ decay in ^{57}Fe with and without Mössbauer absorption. Thus, the "time-effect" of Mössbauer radiation was established [19] (see Fig. 5.6). Subsequently, Mort Hamermesh gave a nice theoretical treatment of the observed effects [20].

There still remained some loose ends to tie up on the properties of the magnetic hyperfine field at the nucleus. Our experiments had definitely shown that the field was aligned with the direction of magnetization. But was it parallel or antiparallel? It seemed a simple experiment could settle this. We looked around the basement of the physics building and found a large unused electromagnet (as opposed to our five-and-dime toys). It was argued that if the source was placed in the large field of this magnet, the observed hyperfine splitting would increase if the external and internal fields were parallel, but would decrease if they were antiparallel. The experiment was soon carried out and the internal hyperfine field was shown to be antiparallel to the magnetization [21] (see Fig 5.7). By still another happy coincidence, this turned out to be a hot theoretical question at the time. Walter Marshall visited us just before we carried out our experiment and gave us his prediction that the field would be parallel [22, 23]. This seeming contradiction to the actual result in no way detracted from his calculation that showed that the nuclear hyperfine field was the result of cancellations of very large terms, so that the sign of the result depended on very small refinements in the calculation. Thus, in the end the nuclear hyperfine field in iron was remarkable by its smallness rather than its largeness as we had feared earlier. More recently, Das [24] has given a refined calculation from first principles that reproduce both the magnitude and the sign of the field in iron quite well.

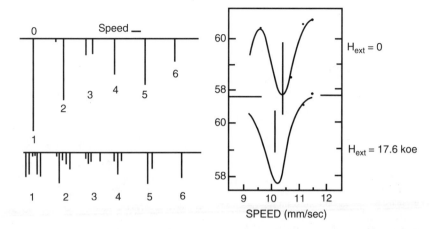

Fig. 5.7 Experiment to determine direction of the hyperfine field. *Left*: predicted effect on the spectrum (^{57}Co in Fe *vs.* Fe metal)if the external and internal fields are parallel. *Right*: observed opposite effect on line 6. After [21]

There were still surprises to come in Mössbauer research. While all the hyperfine phenomena discussed above were being discovered, Pound and Rebka in the United States. [10, 14], and Schiffer et al. [8, 9, 25] in England were vigorously pursuing programs to measure the red shift produced by the gravitational potential. One day, a letter came to the latter group in which the writer pointed out with some diffidence that the Mössbauer effect should be temperature dependent. Thus, in any attempt to measure the effect of the gravitational potential by placing the source and absorber at different heights, a difference in temperature could mask the sought-after red shift. This communication was about to be consigned to the waste basket, along with the others decrying attempts to tamper with relativity, when the light of truth flashed across Walter Marshall's mind, leading him to exclaim that the author of this letter should be the head of theory at Harwell rather than himself. The author turned out to be Brian Josephson, who would go on to future fame in other areas, but who was then at Cambridge studying for his senior exams. It was with considerable difficulty that the Harwell group persuaded Brian to come to Harwell (in a "shiny black limousine") [8,9,22,23] and then to publish a note on his important contribution [26]. In the meantime, the Harvard team had already become aware of the temperature effect [27] and soon produced a published result on the gravitational red shift [28].

At about this time, the first three-line folded spectrum of ^{57}Fe appeared [29] (see Fig. 5.8, top), which unfolded becomes the standard six-line spectrum so familiar today. This spectrum was produced by a magnetically split (six-line) iron source and a nonmagnetic (single line) ferrocyanide absorber. Essentially simultaneously with this report came the landmark paper of Kistner and Sunyar [30] in which they observed the six-line spectrum from Fe_2O_3 (see Fig. 5.8, bottom) and interpreted the overall shift and perturbation of the six lines as being due to the interactions of

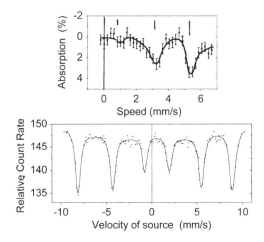

Fig. 5.8 *Top*: First publication of a "six-line" spectrum, appearing here as three lines because of folding at $v = 0$. The source is Fe metal at 300 K, the absorber $Na_4Fe(CN)_6$ at 80 K. After [29]. *Bottom*: Six-line spectrum of ^{57}Fe in Fe_2O_3. The source is ^{57}Co in stainless steel (single emission line). The spectrum is no longer symmetric around $v = 0$ and cannot be folded. The shift from zero velocity of the center of the spectrum was attributed to arise from the difference in electronic charge distribution at the nucleus between source and absorber material (later named isomer shift). The difference in separation between lines 1 and 2 at negative velocities and lines 5 and 6 at positive velocities is attributed to the shift of nuclear Zeeman levels (Fe_2O_3 is a ferrimagnet) caused by the additional interaction between the electric field gradient generated by the atomic charges and the nuclear quadrupole moments in the absorber (quadrupole splitting). After [30]

the nuclear charge distribution and the quadrupole moment, respectively, with the electric fields of the atomic electrons at the nucleus.

This experiment, together with the earlier observation of the magnetic interaction, completed the observation of the basic interactions of the nucleus with its electronic environment, as expressed through its static nuclear moments – the electric monopole, the magnetic dipole, and the electric quadrupole. (Owing to symmetry considerations, a nucleus does not possess a static magnetic monopole or quadrupole moment nor an electric dipole moment, and the allowed higher moments are too small to be readily observed.)

There still were many exciting possibilities to investigate. The nucleus ^{119}Sn appeared to be a promising Mössbauer case [31], but coming on the heels of the ^{57}Fe investigations, it seemed to be less exciting since it was nonmagnetic. Perhaps this was a case where we could test a hunch of Gil Perlow's that a nonmagnetic atom combined with a magnetic one would become endowed with magnetic properties. I remember searching through the book of alloys and hitting upon the magnetic alloys Mn_2Sn and Mn_4Sn which we could use as absorbers. We located some old samples of neutron-irradiated tin containing metastable ^{119m}Sn to use as a nonmagnetic source. (Such activity can inevitably be found stored around a reactor laboratory such as Argonne.) The measurements were carried out on our lathe and both compounds revealed a dominant doublet structure. In the case of Mn_2Sn, the

Fig. 5.9 The nuclear Zeeman effect in ^{119}Sn. *Left*: Spectrum of ^{119}Sn in Mn$_2$Sn. *Right*: Zeeman level diagram derived from the spectrum. From [32]

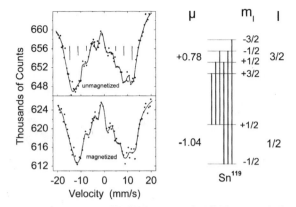

doubt splitting was wide enough that, under good statistics, the expected six-line spectrum (unsplit source and split absorber) was revealed [32] (see Fig. 5.9). Before publishing this result, we carried out polarization measurements and found the line intensities to change properly, although very slightly, according to expectation. Nevertheless, some of our friends suggested that we were in fact seeing only a large quadrupole splitting. Our interpretation was soon amply confirmed and opened the way for extensive studies of the so-called transferred hyperfine interaction in alloys and compounds.

It seemed to us that one more important test of the nuclear hyperfine field remained to be done. Although we had shown convincingly that the field was correlated in direction with the magnetism (albeit of opposite sign), there remained the question of the magnitude of the field -was it indeed proportional to the magnitude of the magnetism. It appeared that a simple test would be to measure the overall Zeeman splitting in ^{57}Fe as a function of temperature and compare it to the well-known dependence of the magnetism on temperature in magnetic iron. We were in the midst of carrying out this experiment at the time of the Allerton Conference (see Fig. 5.10) and so reported our results there [33]. In the spirit of the conference, no actual data were reproduced in the proceedings. To quote the note takers: the results up to 700 K were "as one expects from the known variation in the saturation magnetization."

We have chosen to terminate the Early Ironium Age, and thus our talk, with the Allerton Conference. We do not imply that there were no more exciting and wonderful things to do. Rather, we feel the conference summarized the basic features and potential of the Mössbauer effect, and laid the foundation for a long and enduring field of research.

It is with deep gratitude that I acknowledge the many and important contributions of my colleagues in the Argonne investigations described in this talk. Special thanks are due to John Schiffer, who not only was our colleague but in effect an overseas collaborator. Many great and wondrous events occurred during this time and the associations in our group were very lively and pleasant.

TN 60-698

AD

M Ö S S B A U E R   E F F E C T

Recoilless Emission and Absorption of Gamma Rays

University of Illinois

Allerton House

June 6 and 7, 1960

Edited by

Hans Frauenfelder and Harry Lustig

Contract AF 18(603)-49

These discussions were held as an On-Site Conference of the
Advisory Committee to the Directorate of Solid State Sciences,
Air Force Office of Scientific Research, Washington, D. C.

Fig. 5.10 Cover page of the Proceedings of the Allerton Conference held at the Allerton House,
June 6 and 7, 1960. Also referred to as the "First Conference on the Mössbauer Effect."

Epilogue

I would like to close this brief and personal account of this exciting period in the history of physics by one further historical observation. During the summer of 1961, a well-known member of the Nobel Prize Committee was visiting Argonne. We had many long discussions on the fundamental importance of Rudolf Mössbauer's discovery. I remember well his final question: "Would the Mössbauer effect really become a widespread and important tool for all of science?" I expressed my opinion that it would. I think this book is only one of many striking confirmations of this prediction.

References

1. R.L. Mössbauer, Z. Phys. **151**, 124 (1958)
2. H.J. Lipkin, reproduced in *The Mössbauer Effect*, ed. by H. Frauenfelder (Benjamin, New York, 1963)
3. L.L. Lee Jr., L. Meyer-Schützmeister, J.P. Schiffer, D. Vincent, Phys. Rev. Lett. **3**, 223 (1959)
4. P.P. Craig, J.G. Dash, A.D. McGuire, D. Nagle, R.R. Reiswig, Phys. Rev. Lett. **3**, 221 (1959)
5. H.J. Lipkin, Hyperfine Interact. **72**, 3 (1992)
6. H.R. Lemmer, O.J.A. Segaert, M.A. Grace, Proc. Phys. Soc. London **A68**, 701 (1955)
7. A.T.G. Ferguson, M.A. Grace, J.O. Newton, Nucl. Phys. **17**, 9 (1960)
8. J.P. Schiffer, private communication
9. J.P. Schiffer, W. Marshall, Phys. Rev. Lett. **3**, 556 (1959)
10. R.V. Pound, G.A. Rebka Jr., Phys. Rev. Lett. **3**, 439 (1959)
11. H.J. Hay, J.P. Schiffer, T.E. Cranshaw, P.A. Egelstaff, Phys. Rev. Lett. **4**, 165 (1960)
12. D.C. Champeney, P.B. Moon, Proc. Phys. Soc. (London) **77**, 350 (1961)
13. S.S. Hanna, J. Heberle, C. Littlejohn, G.J. Perlow, R.S. Preston, D.H. Vincent, Phys. Rev. Lett. **4**, 28 (1960)
14. R.V. Pound, G.A. Rebka Jr., Phys. Rev. Lett. **3**, 554 (1959)
15. G.J. Perlow, S.S. Hanna, M. Hamermesh, C. Littlejohn, D.H. Vincent, R.S. Preston, J. Heberle, Phys. Rev. Lett. **4**, 74 (1960)
16. G. DePasquali, H. Frauenfelder, S. Margulies, R.N. Peacock, Phys. Rev. Lett. **4**, 71 (1960)
17. S.S. Hanna, J. Heberle, C. Littlejohn, G.J. Perlow, R.S. Preston, D.H. Vincent, Phys. Rev. Lett. **4**, 177 (1960)
18. G.W. Ludwig, D.H. Woodbury, Phys. Rev. **117**, 1286 (1960)
19. R.E. Holland, F.J. Lynch, G.J. Perlow, S.S. Hanna, Phys. Rev. Lett. **4**, 181 (1960)
20. F.J. Lynch, R.E. Holland, M. Hamermesh, Phys. Rev. **120**, 513 (1960)
21. S.S. Hanna, J. Heberle, G.J. Perlow, R.S. Preston, D.H. Vincent, Phys. Rev. Lett. **4**, 513 (1960)
22. W. Marshall, private communication
23. W. Marshall, Phys. Rev. **110**, 1280 (1958)
24. T.P. Das, in *Variations on Nuclear Themes Honoring Stanley S. Hanna, Stanford University*, ed. by C.M. Class, L. Cohen (World Scientific, Singapore, 1994)
25. T.E. Cranshaw, J.P. Schiffer, A.B. Whitehead, Phys. Rev. Lett. **4**, 163 (1960)
26. B.D. Josephson, Phys. Rev. Lett. **4**, 341 (1960)
27. R.V. Pound, G.A. Rebka Jr., Phys. Rev. Lett. **4**, 274 (1960)
28. R.V. Pound, G.A. Rebka Jr., Phys. Rev. Lett. **4**, 337 (1960)
29. S.L. Ruby, L.M. Epstein, K.H. Sun, Rev. Sci. Instrum. **31**, 580 (1960)
30. O.C. Kistner, A.W. Sunyar, Phys. Rev. Lett. **4**, 412 (1960)

31. R. Barloutaud, J.L. Picou, C. Tzara, Comp. Rend. **250**, 2705 (1960)
32. S.S. Hanna, L. Meyer-Schützmeister, R.S. Preston, D.H. Vincent, Phys. Rev. **120**, 221 I (1960)
33. S.S. Hanna, in *1st Conference on the Mössbauer Effect*, ed. by H. Frauenfelder, H. Lustig. Air Force Office of Scientific Research Report AF-TN 60-698 (University of Illinois, IL, 1960). pp. 39–40

Chapter 6
The Early Developments of the Theory of the Mössbauer Effect

Harry J. Lipkin

6.1 Introduction

I was at the University of Illinois at Urbana in the academic year 1958–1959 when I first heard about the Mössbauer effect. My contacts with Bardeen and his theory group taught me everything I needed to know about solid state physics to understand the Mössbauer effect. I also learned at Urbana from Fred Seitz that my old friend Kundan Singwi had done pioneering work in neutron scattering which was very relevant to the Mössbauer effect. I had met Kundan and his wife Helga in 1953 when we were both postdocs learning about nuclear energy at Saclay, lived in the same pension operated by the French Atomic Energy Commission, and had dinner together every evening. It was a pleasure to renew our contacts after their arrival at Argonne in 1959 when we were both involved in the Mössbauer effect, and during an extended period while he was at Argonne and we visited every summer. I shall miss both John and Kundan.

6.2 Prehistory

Bruria Kaufman was a graduate student at Columbia at the time when Willis Lamb had written his now famous paper on neutron absorption in crystals [1]. In the paper, he had made an approximation of using only the leading term in (1/N) where N is the number of atoms in the crystal in order to obtain the approximate expression $\exp(\langle -k^2 x^2 \rangle)$ for the Debye-Waller factor. Bruria wrote a letter to Lamb pointing out that this approximation was not necessary; the problem could be solved exactly

H.J. Lipkin (✉)
Department of Particle Physics, Weizmann Institute of Science, 76100 Rehovot, Israel
e-mail: Harry.lipkin@weizmann.ac.il

M. Kalvius and P. Kienle (eds.), *The Rudolf Mössbauer Story*,
DOI 10.1007/978-3-642-17952-5_6, © Springer-Verlag Berlin Heidelberg 2012

for a harmonic crystal by using mathematical techniques developed by Ott for X-ray scattering [2].

Lamb then had thanked Bruria but did not think it was worth publishing this correction. The paper had already been submitted for publication, and terms of higher order in $(1/N)$ were certainly negligible. Besides, this paper wasn't all that important at the time [10]

Bruria had noted the connection between Lamb's work on neutron absorption and the long history of coherent X-ray scattering. This was completely missed by Lamb, not mentioned in his paper and not noticed again until after the accidental experimental discovery of the Mössbauer effect. All the physics needed for the discovery of the Mössbauer effect had been known many years earlier. What was missing was the connection between the absorption of particles and waves by individual atoms. If Bruria's work had been published at the time it might have provided the insight needed to change drastically the future direction of nuclear resonance scattering research and led to a much earlier discovery of the Mössbauer effect.

Lamb had discussed the absorption of a neutron, a particle, by an atom in a crystal. Ott had considered the scattering of electromagnetic waves by the atoms in a crystal. Although these seem at first to be very different Bruria showed that was a connection. Deeper thinking here would have seen that the connection was the wave-particle duality, first noted by Einstein in his treatment in 1905 of the photoelectric effect. The electromagnetic waves striking the crystal was really a shower of particles called photons. Bruria's result showed that when a photon is absorbed by an atom in a crystal the transfer of energy and momentum to the crystal is exactly the same as when a neutron is absorbed by an atom in the crystal.

One of the problems that impeded the understanding of the Mössbauer effect was the necessity of a nucleus to recoil to conserve momentum and therefore change its kinetic energy. But it was already well known that when a photon was scattered in X-ray diffraction experiments the momentum of the photon was changed without energy change,

6.3   The Iridium Age

Rudolf Mössbauer's first paper was followed by an avalanche of papers many of which simply reproduced Mössbauer's experiment, obtained exactly the same results without adding anything new and were published in refereed journals as original results. This unique phenomenon reflected the general feeling in the nuclear physics community that this experiment must be wrong and that it was necessary to correct this error by doing the experiment right.

When I first saw Mössbauer's paper, I consulted a real solid state expert – Fred Seitz. Fred's immediate response was "Who is this fellow Mössbauer? Does anybody know him? Is he reliable? Give me a few days to think about it." Several days later he told me "I've looked at this Mössbauer paper and it is perfectly all right. But I must admit that when I first saw it I thought that it was completely crazy."

Table 6.1 Mössbauer scales

Nanovolts 10^{-9} eV	Millivolts 10^{-3} eV	Kilovolts 10^3 eV	Gigavolts 10^9 eV
Natural line width	Free recoil energy	Gamma ray energy	Nuclear mass
Γ	R	E_γ	M
....	$E_\gamma \cdot \frac{E_\gamma}{M}$	....	....
Other relevant scales (temperatures):			
....	kT (Room)	....	....
....	$k\theta$ (Debye)	....	....

This sums up the situation in physics at the time of Mössbauer's first experiment. It was perfectly all right and the theory needed to understand it was well known and published. But the nuclear physicists did not know nor understand the relevant solid state physics, while the solid state physicists who could understand it were put off because their initial response based on intuition was wrong. The energy and momentum scales of the experiment were unfamiliar and misled their intuition. The energy of a 129 KeV γ ray is so large compared to lattice energies that an experiment on this energy scale cannot possibly leave the lattice undisturbed. What was not clear was that this energy simply leaves the crystal and that a recoil momentum of 129 KeV imparts a kinetic energy of only 0.05 eV to a nucleus with a mass of 190 GeV.

The reason for the confusion is seen in Table 6.1 which shows that the energy scales involved in the understanding of the Mössbauer effect vary by 18 orders of magnitude between nanovolts and gigavolts.

All the physics needed to understand the Mössbauer effect had been published long ago by Lamb [1, 2] and others. That photons could be scattered by atoms in a crystal without energy loss due to recoil was basic to all work in X-ray diffraction and crystallography. All the quantitative calculations including the definition and evaluation of the Debye-Waller factor were well known. But nobody interpreted this as a probability that a photon could be scattered by an atom in a crystal without energy loss due to recoil. The X-ray physicists worked entirely with the wave picture of radiation and never thought about photons. The Debye-Waller factor written as $\exp(\langle -k^2 x^2\rangle)$ clearly described the loss of intensity of coherent radiation because the atoms were not fixed at their equilibrium positions and their motion introduced random phases into the scattered wave.

Nobody noted that scattering the X-rays involved a momentum transfer and that coherence would be destroyed if there was any energy loss in the momentum transfer process. They did not see that the Debye-Waller factor also could be interpreted as the probability that the scattering would be elastic and not change the quantum state of the crystal. They did not see that localizing the atoms to the extent needed to get phase coherence in the wave picture meant introducing an uncertainty in the momentum of the atom which was large enough to absorb the momentum transfer without energy transfer. The uncertainty in the momentum of the atom simply means that there is continuous momentum interchange between the atom and the rest of the crystal so that the total momentum is of course conserved. Thus there can be an appreciable probability that a momentum transfer which is small compared to the momentum oscillations between atom and crystal can be absorbed by the entire crystal without changing the internal energy of the crystal.

There was a clear communications barrier between nuclear and solid state physicists who spoke different languages and did not understand one another. Maurice Goldhaber summed up the situation by saying that this would teach many old things to new people.

6.4 How I Got into the Mössbauer Business

I was fortunate to be in the right place at the right time and with the right background to learn enough of both languages to understand the physics and the significance of the Mössbauer effect at an early stage. I believe that my article [3] was the first to use the name Mössbauer effect. At that time, others either did not believe in the effect or felt that it was not important enough to be called by its discoverer's name.

I was spending a year at the University of Illinois at Urbana, following up my own experimental research program in beta ray polarization measurements after the discovery of parity nonconservation. I directed Hans Frauenfelder's group of students and postdocs in their polarization experiments which paralleled those of my group at Weizmann, while Hans was on sabbatical at CERN. We both knew that the parity game was closing down. The exciting controversial days of exploring the weak interaction in beta decay were over. The new data confirmed completely the universal V–A interaction of Marshak and Sudarshan and the two component neutrino with left-handed neutrinos and right-handed antineutrinos. We needed a new experimental program and Hans had met Mössbauer and was very impressed. He sent us a reprint and suggested that we look into it. This led to my consultation with Fred Seitz and to the beginning of the Urbana Mössbauer program which was just getting started when I left and was continued by Hans with great success.

At the same time I was expanding my theoretical work on collective motion in nuclei to the general many-body problem where ideas from solid state physics were percolating into nuclear physics, mainly as a result of the then new BCS theory of superconductivity. In addition to directing Hans's experimental group I attended all of the seminars of John Bardeen's group at the exciting time when BCS was still controversial and there were many VIP physicists insisting that BCS was nonsense because it was not gauge invariant, etc. I learned a great deal from John and from his group of postdocs about physics and how to look through both the forest and the trees to find the physics.

BCS was among the first bridges between nuclear and solid state physicists and I was there seeing it being built. I become an interpreter between two groups speaking different languages. One example was the quasispin algebra developed independently by Arthur Kerman in nuclear physics and Phil Anderson in superconductivity to describe pairing correlations in many-fermion systems. Neither knew about the work of the other and I was for a time the only one who was sufficiently connected to both the nuclear and solid state grapevines to know about both. I introduced each to the work of the other.

As soon as I heard about the Mössbauer effect and understood it I was able to use my own background in many-body physics and what I had learned at Urbana to write a series of articles explaining the Mössbauer effect in simple terms and deriving a number of sum rules [3] and other results [4, 5] which remain pedagogically useful today for teaching basic principles of quantum mechanics to graduate students [6]

Shortly after I heard about the Mössbauer effect from Hans I gave a colloquium talk at Columbia on our parity experiments and had lunch with C.S. Wu and Rudi Peierls who was then visiting Columbia. When I mentioned that we were considering a Mössbauer experiment Rudi said that a group at Los Alamos was doing an experiment. I then asked members of our Urbana group who were attending the Washington APS meeting to contact the Los Alamos people and find out what they were doing. They reported that no one from the Los Alamos Mössbauer group was at the meeting, but that they heard that a Mössbauer experiment was being done at Argonne. I found this very amusing, because I had been in contact with David Inglis and Maria Mayer at Argonne, and had been invited to spend the following summer at a Nuclear Physics workshop that they were organizing at Argonne. They had been working on the theory of the Mössbauer effect, but somehow the subject had never come up in our previous discussions.

Even more amusing was my first meeting with my old friend Kundan Singwi that summer in the Argonne cafeteria. I told him enthusiastically about the Mössbauer effect pointing out that his experience with neutron scattering in crystals was just what was needed for exciting investigations into the theory of the effect. The head of the Argonne Solid State Division who was sitting with us then said that this was all very interesting and asked whether the experiment had ever been done. He was surprised to hear that it had been done by John Schiffer in the building next door at Argonne. There was truly a communications barrier between nuclear and solid state physicists.

During the summer of 1959, the few experimentalists working on the Mössbauer effect were wondering how to find other isotopes which would give a larger effect and how to embed them in compounds and crystals where the Debye-Waller factor or Mössbauer fraction would be increased. Meanwhile the physics community, now convinced by subsequent experiments that the effect was really there, insisted that it was an unimportant curiosity. It was a nice exercise in quantum mechanics but would not lead to anything new and useful. Nuclear physicists claimed that it would teach us nothing new about nuclear structure. Solid state physicists claimed that its use as a tool in solid state physics could not compete with neutron scattering and nuclear magnetic resonance, which would get the same information more easily.

6.5   The Iron Age

A few months later, after the simultaneous discovery in many places that ^{57}Fe was an ideal source, the field exploded. There was a strong effect with a large Debye-Waller factor. It could be inserted into magnetic materials and enable the measurement of nuclear hyperfine splittings for the first time, as well as studying hyperfine interactions in many materials where NMR did not work. It could be inserted into important biological molecules like hemoglobin, thereby opening a new field in biophysics. It provided a source with a line width many orders of magnitude smaller than previously available sources, thus enabling precision measurements of quantities like the gravitational red shift.

One may well ask why it took several months for the active experimenters to find ^{57}Fe. Here again we see the communications barrier. One of the expressions in the literature for the Debye-Waller factor indicated that it could be increased by using a material with a high Debye temperature. A look through the tables of Debye temperatures immediately led to beryllium, and there were proposals to embed Mössbauer's iridium source in beryllium to enhance the Debye-Waller factor.

But this use of the Debye temperature is based on an erroneous understanding of the basic physics. The standard expression for the Debye-Waller factor as a function of the Debye temperature does not hold for an impurity source with a radically different mass from the lattice host atoms. The general expression $\exp((-k^2x^2))$ shows that the important physics is the localization of the source atom by strong binding which makes its position fluctuations small on the scale of the wave length defined by the momentum transfer k. The expression in terms of the free recoil energy R and the Debye temperature are based on the relations

$$k^2 = 2MR, \qquad (6.1)$$

where we set $\hbar = c = 1$, while for the ground state wave function of a harmonic oscillator with frequency ω

$$\langle x^2 \rangle = \frac{1}{(2M\omega)}. \qquad (6.2)$$

Thus, the mass factors cancel in the product

$$k^2 \langle x^2 \rangle = \frac{R}{\omega} \qquad (6.3)$$

For a source in a homogeneous harmonic crystal, the quantity ω is replaced by a characteristic frequency of the lattice. For the Debye model of the crystal, this characteristic frequency is proportional to the Debye temperature θ. Evaluation of the LHS of (6.3) for the ground state of the lattice which describes the system at zero temperature gives the well known result

$$k^2 \langle x^2 \rangle = \frac{3R}{2K\theta}, \tag{6.4}$$

where K is Boltzmann's constant.

But for a source which is an impurity in the lattice there are two relevant masses, that of the impurity and that of the host atoms. Thus, the mass factors in (6.1) and (6.2) are different and do not cancel in the product. The result is that the high characteristic frequency of beryllium is due to the low mass of the beryllium atom rather than to strong binding. The high Debye temperature is thus irrelevant for the Mössbauer effect of an impurity embedded in beryllium and the suggestions to use it were erroneous. The general state of confusion on this issue can be seen in the panel discussion which took place at the Second International Mössbauer conference [7]

Similar effects with two masses occur in diatomic lattices, where the two constituent atoms have very different masses. This effect was later calculated in detail and gave relations between the Mössbauer fraction and the specific heat of such lattices [8]

6.6 Why Mössbauer Couldn't Use ^{57}Fe

Rudolf Mössbauer himself claims that he had thought of using ^{57}Fe while still in Germany, but was unable to get it. ^{57}Fe could be produced only in cyclotrons, and the authorities running the relevant cyclotron in Germany told Rudolf that they had more important things to do than make sources for graduate students. By the time that Rudolf was settled at Cal Tech, others had discovered ^{57}Fe and the work was in full swing. I remember hearing a number of stories like this from Rudolf when I visited him in Pasadena in November 1962, and he was debating whether to remain at Cal Tech or return to Germany. At Cal Tech he was free to carry out his research and received encouragement and support from experts in all related fields. Feynman and Gell-Mann were always ready to discuss theoretical questions. Solid state physicists, low temperature physicists, metallurgists, and chemists were all very friendly and cooperative in helping him solve the "interdisciplinary" problems arising in proper preparation of sources and absorbers and in interpreting results. In Germany, he received no such support; each expert closely guarded his "secrets" and Rudolf would have to discover all the relevant techniques by himself. Furthermore at Cal Tech he could rise to the highest positions, especially after his Nobel Prize, and still do full time research without large teaching loads or administrative responsibilities. In Germany this was impossible. One would have to become *The Herr Professor* of the department or the director of an institute to advance in status and salary.

But Heisenberg, who was attempting to convince Rudolf to return, responded with arguments very familiar to me as what we in Israel call "Zionism." "You are needed. You should give up your comfortable life in America, return to your homeland and contribute to its rebirth and rebuilding. Of course your criticisms are correct. These things must change. But only people like you can change them. Come." Rudolf responded to this challenge and attempted to induce needed reforms in Germany, like creating a department where there could be many professors with full privileges and no administrative responsibilities. Many years later Rudolf told me that he was amused by my calling him a "Zionist," but admitted that this description was essentially correct, and that unfortunately his achievements had fallen short of his expecutations. He is not alone. Many of us Zionists in Israel feel the same way about our own achievements.

6.7 The Generalized Mössbauer Effect

Once the basic physics underlying the Mössbauer effect was understood, it became evident that this physics of momentum transfer to bound systems occurs everywhere in physics and can be described by the same formalism. This is discussed in detail with many examples from different areas of physics in my Quantum Mechanics book [6] where it can give the student an introduction into exciting frontier physics at a very early stage before he has mastered a great deal of formalism. The analog of the Debye-Waller factor appears everywhere and is called by various names like form factor or structure factor. It is just the probability that momentum can be transferred to a constituent in a bound system with no change in the internal wave function of the bound system and the recoil momentum taken up by the whole system.

$$F = \sum_i P_i \mid \langle i \mid e^{i\vec{k}\cdot\vec{x}} \mid i \rangle \mid^2, \tag{6.5}$$

where $\vec{k}$ denotes the momentum transfer, $\vec{x}$ is the co-ordinate of the active constituent and P_i denotes the probability that the system is initially in the state $|i\rangle$. This can be rewritten

$$F = \sum_i P_i \mid \int \rho_i(\vec{x})e^{i\vec{k}\cdot\vec{x}}d^3x \mid^2 \approx e^{-k^2\langle x^2\rangle}, \tag{6.6}$$

where $\rho_i(\vec{x})$ denotes the probability density for the variable $\vec{x}$ in the state $|i\rangle$, $\langle x^2\rangle$ is averaged over the entire distribution of initial states and the approximate equality holds if the density $\rho_i(\vec{x})$ is a gaussian.

Measuring this factor as a function of the momentum transfer or a scattering angle measures the fourier transform of the probability density for the active constituent. In x-ray crystallography it gives the fourier transform of the electron density. In elastic electron scattering on nuclei it is the probability of elastic as opposed to inelastic scattering and gives the fourier transform of the electric charge distribution in the nucleus.

The probability of momentum transfer without energy loss due to recoil is large when the Debye-Waller factor is small; i.e., when the fluctuations in the position of the active constituent $\sqrt{\langle x^2 \rangle}$ is small in comparison with the wave length $1/k$ defined by the momentum transfer. Such effects occur even in macroscopic experiments. An amusing example occurs in an experiment measuring the tiny mass difference, 3×10^{-12} MeV, between the two neutral kaon states, K_{long} and K_{short}. A K_{long} beam is passed through two slabs of material and the conversion $K_{\text{long}} \rightarrow K_{\text{short}}$ is observed as a function of the macroscopic distance d between the two slabs. by detecting the decay $K_{\text{short}} \rightarrow 2\pi$ in the emergent beam. The measurement depends upon the coherence and interference between the components which underwent conversion in the two slabs. However, the conversion, because of the mass difference, involves a momentum transfer, and the kinetic energy loss in the momentum transfer to the particular slab where the conversion took place would destroy the coherence. But both slabs are bound by gravitational and frictional forces to some apparatus like a table, which is in turn bound to the earth. There is a Mössbauer effect in which the whole earth takes up the recoil and not the individual slab. The probability that this occurs is given again by the Debye-Waller factor $\exp(\langle -k^2 x^2 \rangle)$ which is immediately seen to be close to unity. All practical experiments are automatically designed to make $\sqrt{\langle x^2 \rangle}$ small in comparison with the wave length $1/k$. The oscillations can be observed only if each slab is "bound strongly enough" to the rest of the experimental apparatus so that the fluctuations in its position are small compared to the wave length that one wishes to measure which is just $1/k$. This can be called Lipkin's Mössbauer principle for macroscopic interference experiments: "If you can measure it, then you can measure it. Don't worry about recoil."

6.8 A Connection with Mössbauer's Later Work on Neutrino Oscillations

Neutrinos experiments observe a ν entering a detector, changing the charge of a nucleon and emitting a charged lepton. If energy and momentum are conserved, the ν energy and momentum are determined by energies and momenta of the detector nucleon and lepton. Knowing the ν energy and momentum gives its mass and there are no oscillations. Mössbauer physics and Dicke Superradiance are needed to understand neutrino oscillations. The ν mass is not observable because the momentum change in the detector nucleon is not observable. A "mising momentum" is absorbed by the whole detector like the photon momentum by a whole crystal in the Mössbauer effect [9]. This ν interaction with a detector at rest in the laboratory is overlooked in treatments based on relativistic field theory, The nucleon wave function in a ν detector must vanish outside the detector for all times. This condition shows oscillations are produced by interference between ν states with same energy and different momenta. The detector nucleon is described in quantum mechanics by a wave function which must have coherence and interference between components with different momenta at each energy to cancel out the probability of finding the nucleon outside the detector at all times. Components of an incident ν with the same energy and different momenta can both be absorbed by the same detector nucleon eigenstate, leaving no trace identifying the momentum of the ν that produced the transition. The contributions to the final state amplitude via these different paths are therefore coherent. Interference between them can be observed producing oscillations. Neutrino detection is a "two-slit" or "which-path" experiment in momentum space. Oscillations are transitions between Dicke superradiant and subradiant states.

6.9   A Curious Recollection from the Past

I conclude with an amusing incident that occurred when Willis Lamb visited Israel while traveling around the world. When I met him at the airport upon his arrival from India one of his first questions was whether Bruria Kaufman was at the Weizmann Institute. When he met Bruria, he immediately told her that she had made an important contribution to the theory of the Mössbauer effect and pulled out of his briefcase a letter that she had written to him in 1939. Bruria's response was "What is the Mössbauer effect?" Lamb then explained that although this paper wasn't all that important at the time, it was now crucial with all this attention paid to the Mössbauer effect. It would be useful for pedagogical purposes to publish the exact treatment. Bruria and I then collaborated in writing up the exact treatment for publication [10] with my contribution being to explain the Mössbauer effect to Bruria and introduce the new notations of creation and destruction operators for oscillator quanta which made the treatment much simpler than Ott's use of properties of Hermite polynomials.

References

1. W.E. Lamb Jr., Phys. Rev. **55**, 190 (1939)
2. H. Ott, Ann. Physik **23**, 169 (1935)
3. H.J. Lipkin, Ann. Phys. **18**, 182 (1962)
4. H.J. Lipkin, Ann. Phys. **23**, 287 (1963)
5. H.J. Lipkin, Ann. Phys. **26**, 115 (1964)
6. H.J. Lipkin, *Quantum Mechanics* (North-Holland, Amsterdam, 1973), pp. 33–110
7. M. Hamermesh et al., in *The Mössbauer Effect*, ed. by D.M.J. Compton, A.H. Schoen. Proceedings of the Second International Conference on the Mössbauer Effect held at Saclay, France (1961) (Wiley, New York, 1962), p. 19
8. H.J. Lipkin, Y. Disatnik, D. Fainstain, Phys. Rev. **A139**, 292 (1965)
9. H.J. Lipkin, arXiv:1003.4023[hep-ph]
10. B. Kaufman, H.J. Lipkin, Ann. Phys. **18**, 294 (1962)

Chapter 7
The CalTech Years of Rudolf Mössbauer

Richard L. Cohen

7.1 Introduction

In the summer of 1959, I was starting my third year as a graduate student in Physics at CalTech. I had passed all my course work and the qualifying oral exam, and was ready to begin my thesis research. I joined the group of Jesse DuMond and Felix Boehm, called Physics 34, which dealt mainly with X-rays, spectroscopy, and radioactive decay. The CalTech group that used a Van de Graaf generator as the main research tool was separate, and had a much higher profile.

7.2 Recollections of Prof. Felix H. Boehm[1]

Interviewed Feb 2, 2010, in Altadena CA, by Richard Cohen (The figures refer to prints of letters exchanged between Prof Boehm at Heidelberg and Prof. DuMond at CalTech.)

RLC: Please tell me how you first heard about Rudolf Mössbauer and this resonance he had discovered:

FB: *I spent the academic year 1958–1959 as a guest professor in Heidelberg. I also saw Hans Jensen and we had an excellent time. Things were still post-war-like in many areas of experimental physics. In spring 1959, I met Rudolf. I don't remember*

---

[1]Prof. Boehm was a Professor of Physics at CalTech, and the effective head of the Physics 34 research group.

R.L. Cohen (✉)
43 Caribe Isle, Novato CA 94949, USA
e-mail: campcohen@aol.com

M. Kalvius and P. Kienle (eds.), *The Rudolf Mössbauer Story*,
DOI 10.1007/978-3-642-17952-5_7, © Springer-Verlag Berlin Heidelberg 2012

the exact occasion, but it was in Heidelberg. He did work in Heidelberg, but he was a student of Maier-Leibnitz in Munich. He told me about his work, and I heard from various other scientists, notably Jentschke, about these fantastic results.

He was invited to give a colloquium in Heidelberg, which he did very well. It was very clear and very nuanced. He was doing the experiments at Heidelberg, at that point on ^{191}Ir. He was commuting between Munich and Heidelberg, at least a year or more. His doctoral degree was in 1958, as I remember, and shortly before, he sent in two papers to Zeitschrift für Physik, and Zeitschrift für Naturforschung. Results in these papers were not read too widely, so the word didn't get out.

His presentation and his work struck me as being fantastically interesting, and I asked him in July 1959 if he would like to join me and Jesse DuMond at CalTech, and on July 13 1959 I wrote him a letter (Fig. 7.1) inviting him to come to CalTech and work in our research group for a 1–2 year term.

At the same time, I wrote a letter (Fig. 7.2) to Jesse DuMond describing the results. Jesse was excited, (Fig. 7.3) and immediately took it to Feynman, who agreed that there could be a recoil-free event (Fig. 7.4) for a single crystal. Jesse still had some reservations, but we completed the arrangements, (Figs. 7.5 and 7.6) and in about February 1960 Rudolf Mössbauer arrived in the United States. He arrived with Elizabeth and Peter, and we picked them all up at the airport. We looked for a place for them to live, and finally found a nice place in Altadena.

He went to work immediately. His previous experiments had been done with mechanical linear motion constant velocity drives, and since we had Herb Henrikson, a very skilled and creative mechanical engineer in the group, we decided to continue the tradition with a precision-ground rotating cam driven by a synchronous AC motor and gearbox.

7.3   Richard Cohen Gets Involved

CalTech was a relatively small university. In experimental physics, it offered a choice of high energy, low energy nuclear, low temperature physics, and cosmic ray research. (Contemplate briefly the additional fields which would be required today to claim even a minimal Ph. D. program!)

I chose the low energy nuclear option, partly because I had worked for two summers at Brookhaven, and found the work interesting and relevant to the everyday physical universe around us. I joined the group of Jesse DuMond and Felix Boehm (Physics 34), which dealt mainly with X-rays, spectroscopy, and radioactive decay. The CalTech group that used a Van de Graaf generator as the main research tool was separate, and had a much higher profile.

News of Mössbauer's discovery preceded his arrival at CalTech. Late in the fall of 1959, there was a Physics Colloquium at CalTech, at which Mössbauer's startling experiments and results were presented. I was really excited, especially because the

den 13. Juli 1959

Prof.Boe/Kl

Herrn
Dr. R. Mössbauer
Labor für Technische Physik
Technische Hochschule München

M ü n c h e n

Lieber Herr Mössbauer,

ich möchte Sie gerne anfragen, ob Sie Lust hätten, für ein oder
zwei Jahre zu uns in unsere Forschungsgruppe am California Institute
of Technology nach Pasadena zu kommen. Über die Physik in Pasadena
will ich Ihnen gerne detaillierte Auskunft geben, falls Ihnen der
Gedanke, nach USA zu kommen, attraktiv erscheint. - Ich möchte Ihnen
hier nur sagen, daß wir in unserem Laboratorium, das ich zusammen
mit Professor DuMond leite, über ziemlich gute apparative Ausrüstung,
(mehrere Beta-Spektrometer, Kristall-γ-Spektrometer, viel moderne Elek-
tronik) verfügen. Sie hätten die Möglichkeit, sich gleich einem inter-
essanten Problem der Nieder-Energie Kernphysik zuzuwenden. Falls Sie
gerne Ihre schönen Experimente über elastische Linien fortsetzen
möchten, so könnten Sie das bei uns ohne Verzug tun und von den be-
nachbarten modernen Tieftemperatur-Laboratorien von Professor Pellam
(Collins Verflüssiger, adiabatische Demagnetisierungseinrichtung für
0.001°K) Gebrauch machen.

Berichten Sie mir doch bitte kurz über Ihre Reaktion. Falls Sie
vor dem 24. Juli, dem Tag meiner Abreise nach USA, nach Heidelberg
kommen könnten, hätten wir Gelegenheit, über Pasadena zu plaudern.

Mit den besten Grüßen

Ihr

Fig. 7.1 Letter of F. Boehm to Rudolf Mössbauer inviting him to CalTech

Professor J.W.M.DuMond   July 12 59
Physics 34
California Institutof Technology
Pasadena,Calif. USA

Dear Jesse:
....

Immediately after receiving your letter yesterday I wrote
to a very able young man in Munich whose acquaintance
I recently made. His name is Dr.R.Mössbauer. Mössbauer has
recently made very striking experimental work by being able
to show that nuclear fluorescence excitation can occur
elastically in certain conditions i.e. in certain
crystals near the Curie temperture. In one case (Ir^{191})
he was able to find under these conditions that the nuclear
recoil is taken up entirely by the crystal giving rise to
a resonance peak whose width is only just the natural line
width. There is no Doppler broadening. By moving source
or scatterer with velocities of the order of 1 cm/sec

Mössbauer was able to get in or out of resonance. He plans
now to study the m-splitting of a level in a magnetic
field and hopes to be able to separate the magnetic
substates by finding 2I+1 resonances at differen; source
scatterer velocities. Mössbauer is a nice and clever person,
about 28 or 30. His latest work is published in Zeitschrift
für Naturforschung, 14a, 211 (1959).
....   —   It is funny that the Americans
seem to show little readyness and enthusism to come to work with
us as compared to the Europeans who have the greatest respect
and esteem of our work in particular and of Calt Tech in general.
I have had so many compliments of your and our work and at
least a dozen or more people have asked me seriously if they
might be able to come to work with us. Hauser and Mössbauer
are the two top men among those inquiries.(or latent inquiries
respectiveley).

Please excuse the hastely written letter. To keep you informed
as fast as possible I am not waiting for the secretary to do
a more careful typing. — With our best wishes

                                        sincerely

Fig. 7.2 Excerpts from a letter written by F. Boehm to J.W.M. DuMond. The *dots mark* omitted
passages which refer to internal considerations of the group. Ulrich Hauser was a Postdoc at
Heidelberg. He worked with R. Mössbauer at CalTech and returned later to Heidelberg where
he still pursued Mössbauer spectroscopy for some time

July 15, 1959

Professor Felix Boehm
Department of Physics
University of Heidelberg
Philosophenweg 12
Heidelberg, W. Germany

Dear Felix,

Your letter of July 12 has just arrived and although I have just mailed a letter to you yesterday I think a few words in answer are in order.
....

Your description of Mössbauer's work is very impressive and indeed exciting, He sounds like a very capable man provided you feel he is reliable. One thing that worries me just a little is that I find the result he has obtained very hard to understand. I find it hard to see how the nucleus could be so rigidly coupled to the crystal lattice as a whole that the momentum of the absorbed proton (which is absorbed in an extremely short time interval) could fail to cause a recoil Doppler effect essentially the same as for a free nucleus. Also how about the recoil of the originally emitting source nucleus? Is this also in a similar crystal? Very young men have been known to get startling and dramatic results which turned out later to be misinterpretations of the observations or even downright frauds. The pressure to "ring the bell" is very hard on a young man and perhaps worse in Europe than in this country. I am not accusing Dr. Mössbauer at all, but merely warning that you should be careful not to be misled. If he is right and actually can investigate the magnetic substates of nuclear levels by the method your outline, he has the makings of a Nobel Prize winner in him, I would say, and we certainly should encourage him to join us. I have consulted with Dr. Robt. Christy about the physics of Dr. Mössbauer's effect and have read him my above paragraph with which he says he is in complete agreement. He will investigate Mössbauer's paper.
.... Sincerely and hastily,

Jesse

P.S. Bob Christy, Dick Feynman and I have read Mössbauer's paper and also looked at W.E. Lamb Jr's paper on which the theory is based and we are all three deeply impressed and much excited. It looks as though what Mössbauer reports is indeed possible. Dick F. will let me know tomorrow. He promises to think read "about it tonight". Is "the crystal" in Mössbauer's experiment simply the parent Osmium lattice?

A.W.T,

Fig. 7.3 Response of J.W.M. DuMont to the letter of F. Boehm (Fig. 7.2)

nuclear physics techniques might be used to study the physics of solids. Judging from the audience size and level of attention, I was not the only attendee who was fascinated by this discovery!

As I exited the auditorium I was walking next to Felix Boehm, and I was bubbling over with enthusiasm. He said (approximately) "You're looking for a thesis topic. Is that something you might be interested in?" I allowed that I might be interested. "Well," he continued, "Mössbauer will be coming here in a couple of months as a

July 17, 1959

Professor Felix Boehm
Department of Physics
University of Heidelberg
Philosophenweg 12
Heidelberg, W. Germany

Dear Felix,

I have received the following report from Dick Feynman who along with Robt. Christy and myself has read Mössbauer's paper. He says:- "I checked all the formulas in Mössbauer's paper. They are all correct. There should be a line with zero shift (sizeable as long as T$<$θ and the recoil energy E $<$ kθ where θ is the Debye temperature; Size for low T $\propto e^{-3E/2k\theta}$). Let's try to get the guy." The italics are mine, not Feynman's .

One thing which made me skeptical about him, at the time I wrote my first letter about Mössbauer to you, was that you (mistakenly) wrote Curie temperature instead of Debye and I just could'nt see for the life of me what Curie temperature could have to do with this phenomenon. After consultation with Feynman and Christy, and after reading Mössbauer's paper all is now reasonably clear to me. In the first place not all resonance fluorescence scattering of gamma-rays, even at these low temperatures, is of this "unmodified" or "elastic" type, only a small fraction which, however, is extremely sharply defined as to energy! The phenomenon is almost exactly equivalent to "Bragg scattering" or "unmodified scattering" of x-rays save that in the latter case atomic electrons are the scattering agents while in Mössbauer's case the scattering is from nuclei. That there should be a least energy and momentum that single nuclei in the lattice can accept seems entirely reasonable. The most exciting thing about Mössbauer's effect is the enormous resolving power, which clearly is of the order of one part in 3 x 10^{10}! This is of the same order as the accuracy of the caesium clock (though resolving power is of course not quite the same thing as accuracy).

Feynman, Christy and myself are strongly in favor of urging you to make every effort to get Mössbauer to join us here. I have communicated this information to Dr. Robt. Bacher. He is reluctant to make an immediate move to offer Mössbauer a post without first having you talk to Mössbauer, ask him if the proposal of a year in Pasadena as post doctoral research fellow appeals to him, and hearing directly from you his reaction and your recommendation. Because I know your stay in Heidelberg is drawing within less than a week of its close I have cabled you today substantially what I say here and this letter is merely a confirmation. Try your best to get Mössbauer if in your judgement he is as good as his article makes him look. Feynman and Christy are enthusiastically in favor. *Use the L.D. phone if necessary we will gladly reimburse you.*

With cordial best wishes,

Jesse

Fig. 7.4 Excerpts from a follow up letter by J.W.M. DuMond to F. Boehm

postdoc. He'll need some help to set up and get his experiments running. Would you like to work with him for your thesis?" "That would be great" was all that I could say, and the deal was done.

At the time, it did not occur to me that in that 30 seconds I had set my career on its track for the next 25 years, and even 50 years later it would be a strong influence on my life, because of all the friends I had made!

The Physics 34 soil was very fertile for Rudolf Mössbauer and his experiments. 1960 was at the very beginning of availability of semiconductor radiation detectors and electronics, and they were expensive, temperamental, and unreliable. The group had long experience with proportional counters and scintillation detectors, and a large supply of glow-discharge ring counters. Microprocessors, and the skills and

Prof. Robt. Bacher   c   Jesse W. M. DuMond   July 17, 1959

Mössbauer as a Possible Candidate for Postdoctoral Research
Fellow in DuMond and Boehm's Group.

The enclosed carbon copy of a letter which I have just written to
Felix Boehm is almost self-explanatory.
....
      You undoubtedly know that (1.) If one tries to excite a characteristic
nuclear gamma-ray transition by placing the nuclide in a flux of gamma-rays
from the same nuclear species, the gamma-ray emitted by transition between
the same identical pair of levels fails to produce any effect because the
two Doppler shifts, at the emitting nucleus and also at the absorbing
nucleus (because of the momentum of the photon), shift the latter off
resonance by a great deal more than the width of the gamma-ray transition.
Thermal agitation of both source and receiving nucleii is one way of fur-
nishing sufficient velocity so that a few resonance fluorescence transitions
can take place but the distribution of velocities renders the effect, not
only weak but ill-defined.

      What Mössbauer has demonstrated experimentally (see Zeits. f. Natur-
forschung 14a, 211 (1959)) is that, if the crystal lattice in which the
emitter and absorber nuclei are situated is at or below the Debye temperature,
it is possible to have a type of resonance fluorescence scattering which is
extremely sharply defined because the momentum of the photon is then, for a
small class of cases, not accepted by the absorbing nucleus individually but
is transmitted to the whole crystal lattice of which the nuclei are a part.
The effect is closely analogus to"unmodified"or "elastic" scattering of x-rays.

      The striking and exciting characteristic of Mössbauer's finding is its
enormous resolving power. By rotating a disc with the absorber on its
periphery, a velocity of 1 cm/sec along the direction of propagation is
sufficient to shift the radiation from resonance completely out of resonance!
Since the velocity of light is 3 x 10^{10} cm/sec this is a resolving power, as
regards resonance fluorescence, of 1 part in 3 x 10^{10}! Mössbauer plans to
study the 2I+ 1 fine structure levels due to magnetic splitting of nuclear
gamma-ray transitions in this way which will be a new way of determining spins.

      Both Profs. Feynman and Christy who have examined Mössbauer's paper are
strongly in favor of making every effort to encourage Mössbauer to join our
group and, needless to say, so am I. In accord with our telephone conversation of today I am cabling Felix Boehm to this effect and hope you will see
your way clear to making Mössbauer an offer of appointment provided Felix
Boehm's inquiries justify such a move.

                          Jesse W. M. DuMond

Fig. 7.5 Letter (slightly shortened) of J.W.M. DuMond to Prof. R. Bacher (chairman of the physics department)

infrastructure to use them, were well into the future. Commercial transistorized
multichannel analyzers were just being developed, with ferrite core memories.

Another valuable resource at CalTech was Richard Feynman, the eminent theorist
with a very intuitive way of understanding problems in new areas. Feynman was
very influential in attracting Rudolf Mössbauer to CalTech, and then in establishing
a position for him (see the series of letters quoted above). Mössbauer frequently
consulted with Feynman, especially on the physics underlying the "Recoil-free-
fraction" that was at the core of the discovery .

Professor Jesse W.M. DuMond   21 July 1959

Physics Department   Prof.Boe/Kl
California Institute of Technology

Pasadena, California (USA)

Dear Jesse,

    Thank you for your good letters of July 14, 16, and 17.

    Yesterday evening I talked to Dr. Mössbauer quite explicitly
here and I am happy to report to you that he is willing to come to
Pasadena to work with us. Since he has a few obligations in Munich
the earliest date for his arrival would be Febr. 1. I told him that
I will recommend his appointment as a research fellow to you and
to Dr. Bacher. Enclosed is a short curriculum vitae which he prepared for our information. Mössbauer is besides being quite clever
in physics a very nice guy and I am looking forward to have him in
our group.

+99.9%   It is now virtually certain that Dr. Hauser will be granted a
          fellowship from the German Government. This means that all we have to
* this difference   pay Hauser from our Ph 34 budget would be the difference between the
          amount Dr. Bacher has offered and the fellowship amounting to ~100 $
          a month. I have urged Mössbauer to apply for such a fellowship
          (which besides the one year grant will also pay for travel expense for
          him and his family to US) and there is a good chance that he will get it.

          ....

    (I am sorry of creating confusion by talking about Curie temperature
meaning Debye temperature. I wrote that letter late at night.)

    With cordial best wishes

                              sincerely,

Fig. 7.6 Part of a Letter of F.Boehm to J.W.M. DuMond with the acceptance of Rudolf Mössbauer
to come to CalTech

    People still didn't have much of an idea of how to do the experiments, and what
we would find. It was beginning to be clear that the lower the gamma ray energy,
the larger the recoil-free fraction would be, especially at room temperature, but
the harder it was to detect the gamma rays. For many isotopes and excited states,
there were no convenient precursors. And it was tempting to look for long-lived
excited states to get the highest resolution, but the absorption cross-section was
smaller.

    After discussion within the group, it was decided to first try the 8.4 keV excited
state of Thulium (^{169}Tm), whose precursor could easily be generated by reactor
irradiation of ^{168}Er to make ^{169}Er, which beta decayed to make the excited 8.4 keV
Tm state .

    Additionally, many of the rare-earth elements were interesting since the 4f
electron states that defined the basic properties of the rare earths had been widely
studied, and had very interesting physics and chemistry. Nuclear structure theorists
were also interested, because the recently developed "Nilsson model" was especially
appropriate for nuclides of the rare earth elements, and the Mössbauer studies of

hyperfine interactions could measure the moments of these nuclides[2] and further verify the validity of the Nilsson model .

Once again, luck played a role. As the grad student with the most seniority, I chose the ^{169}Tm system to work on, and proceeded rapidly. The next student in line got the 6 keV resonance in ^{181}Ta, which turned out to be very difficult to work with, despite a very sharp resonance and a very large recoil-free fraction.

And so we set to work. We developed a good basic design for a gas proportional counter, using argon and other gases, with an integral beryllium window to admit the low-energy gamma rays into the gas volume. We developed cryostats and ovens to reach temperatures from 60 K to 1,230 K. We had standard Nixie/Pixie scalers, and linked them to an automatic printer. At first the printed data were reduced manually, but they were later converted to punched cards, which went to a main frame CalTech computer. It actually plotted the data for us. I generated a controller, based on motorized timers and relays, which controlled Herb Henrikson's motorized cam linear motion device, magnetic clutches for velocity control, and the scalers and printer. When it was time to move to the next data point, the equipment sounded like the Glockenspiel at the Marienplatz in Munich. One visionary visitor even commented that I would be a good candidate for Bell Laboratories, judging from the relay-driven controller.

But what would we be looking for? Mössbauer's early experiments at Heidelberg provided much data on the dependence of the recoil-free-fraction on temperature, host lattice, and recoil energy. But there was no detailed analysis that included realistic lattice dynamics, although intuitive theorists were making good approximations, using Debye temperatures and simplified models.

It was clear that hyperfine interactions would be a significant area of study. Both theoretical calculations and magnetic resonance experiments showed that hyperfine splittings were often larger than the resonance width (determined by the nuclear excited state lifetime). Also, perturbed angular correlation (PAC) measurements showed that excited nuclei tended to behave as would be expected under the influence of magnetic fields.

The reactor irradiation obviously damaged the erbium oxide heavily . On the advice of a local radiochemist, we dissolved the irradiated source material in acid, precipitated it as the oxalate, and heated it in an oxygen environment to restore a normal erbium oxide structure, similar to thulium oxide.

And so we started out to measure spectra for combinations of source and absorber materials, and observed how the spectra varied with temperature. And we did see lots of changes! Clearly, the resonance strength, determined by the recoil free fraction, was behaving appropriately, decreasing at higher temperatures. But the splittings of the spectra, obviously resulting from hyperfine interactions, also decreased at higher temperature. The basic spectrum at room temperature, using an erbium oxide source (with reactor-generated ^{169}Er in Erbium oxide) and a thulium oxide absorber , was a triplet: a strong central line, symmetrically flanked by two

---

[2] See book Chap 9.

side peaks. At high temperatures ($\sim$700 K) for the oxide source, the spectrum became a simple symmetric doublet; at $\sim$1,200 K source temperature (absorber at 300 K), the spectrum again became a symmetric triplet, like that observed with both source and absorber at $\sim$300 K.

Our use of the ^{169}Tm resonance proved, fortuitously, to have several features that helped us to understand the many effects we were seeing. First, the 8.4 keV transition we were using went from a spin 3/2 excited state to a spin 1/2 ground state in the source, and conversely from the 1/2 ground state to the 3/2 state in the absorber. We expected little or no magnetic hyperfine interaction, since the environment was only paramagnetic, and there was no applied magnetic field. We thus expected that the main source of hyperfine interactions would be from the interaction of the electric field gradient at the nucleus with the nuclear quadrupole moment of the excited (3/2) state. The 3/2 state would be symmetrically split into two components, equally displaced from the unsplit 3/2 state, and the 1/2 ground state would be unsplit. So the source should emit a symmetric doublet, and the absorber cross section would also be a symmetric doublet, with the same energy splitting as the source.[3]

We had a real "eureka moment" when we realized that this model produced exactly the puzzling 3-peak structure we were seeing from the experiments! With no Doppler shift between the source and absorber, we were seeing the maximum absorption, because the photons from both the lower and upper level of the doublet would find a corresponding absorption peak in the absorber Tm nuclei. If we moved the source toward or away from the absorber just enough to produce a Doppler shift of the full energy of the splitting, there is a weaker absorption, as each emitted photon can be absorbed by only one of the two absorption lines.

We were very fortunate here, because if the nuclear spin values had been larger, both the emission spectra and the absorption spectra would have been split into many different hyperfine lines. The actual measured spectra would have been much weaker, and much more difficult to measure and analyze.

We had originally chosen the oxide materials because they were the easiest to get, and we didn't at that time know what would be better. Once we thought we understood the spectra, we immediately brainstormed to test our theory. We guessed that metals probably had lower internal electric field gradients than insulators, and that the crystal structure of Thulium metal had higher symmetry than the oxide. So I borrowed a vacuum system and evaporated a thin layer of Tm metal on to a beryllium disc, and we put it in a spectrometer that had an oxide source. Within a few hours we could see a well-defined doublet spectrum from the oxide, as we anticipated, but this time without the complications of the splitting in the spectrum of the Tm absorber. We had another successful experiment![4]

---

[3]Assuming that the splitting in the erbium oxide source was the size of that in the Tm oxide absorber, and averaging over all directions.

[4]Rudolf Mössbauer reported this first set of our results at the meeting in Paris 1961 [1] including some of the spectra.

The temperature dependence of the quadrupole splitting of the oxide source was our next puzzle. The observed splitting (in velocity units) went from $\sim$7 cm s^{-1} at 100 K on a gentle curve down to about 0 cm s^{-1} at $\sim$600 K, and then increased to $\sim$3 cm s^{-1} at about 1,300 K, as if the curve were being reflected by the 0 cm s^{-1} splitting.

We were pleased that the low energy of the Tm 8.4 keV radiation allowed us to make measurements at such high temperature. If we had only measured to 600 K, which was already far above where most Mössbauer experiments were carried out, we would have missed this anomaly. After a few days of contemplating this odd curve, we realized that the reflection point was probably an artifact of how we had plotted the data, and that there was no reason to think that the splitting increased above 600 K, but that it had no polarity, and simply "went through zero." This looked less anomalous, but we had no idea of the possible origin of this result.

And then 1 day, we had a visit from Rudolf Sternheimer, from Germany, who came to meet Rudolf Mössbauer and some others in Physics 34. Rudi was a theoretician who was interested in hyperfine interactions in solids, and he spent an afternoon with us looking at our results. He explained how the net field gradient that gave rise to the hyperfine structure we observed came (mainly) from the effects of the nearby ions on the electrons in the 4f shells of the Tm ions. The best way to look at this was to consider the charge distribution (from the ions and valence electrons around a Tm ion), and calculate the effects on the 4f electrons in their orbits. You could then consider the field gradients produced by the nearby ions as being multiplied by the distortions of the 4f shells. This multiplier effect could be several times the direct field gradient from the nearby ions, and tended to be negative.

But there was an additional complication, because the thermal excitation of the 4f electrons competes with the effects of the electrostatic distortions from the surrounding ions. This additional interaction allowed Rudi to explain, in a qualitative way, the temperature dependence that was so puzzling to us: if you assume that the shielding effect at low temperatures is, e.g., -5, the field gradient from the 4f electrons is the dominant component of the field gradient, being -5 times the field gradient produced by the surrounding ions. The total field gradient at the nucleus would be $1 + (-5) = -4$ units. But if the sample temperature is increased, the thermal excitation of the 4f electrons decreases the shielding effect. The multiplier might go from -5 at the lowest temperatures to -0.5 at the highest temperatures (1,300 K). The total field gradient would then be $1 + (-0.5) = 0.5$ units. At an intermediate temperature (e.g., 600 K), the multiplier could be (~ -1), and the total field gradient would be $1 + (\sim -1) = \sim(0)$ and there would be no net hyperfine splitting, exactly as we observed. The behavior described is now widely known as "Sternheimer shielding."

It was encouraging for us all to see that a real theorist had made serious efforts to study the configuration we had been measuring on, and found our results understandable!

We had expanded Mössbauer's original experiments to a new isotope, extended the measured resonance conditions from cryogenic temperatures to almost white heat, observed a number of new phenomena, and had reasonable explanations for them, that bridged the nuclear and solid-state physics worlds. It was clearly time for me to start writing my thesis.

All of the work described above took from February 1960, when Rudolf Mössbauer arrived at CalTech, until 1961. It took approximately 3 months for me to write my thesis, make the figures, etc. During this period, I also visited Bell Laboratories in New Jersey for an employment interview, which was also successful, and I committed to come and work there after my Ph.D. requirements were completed. My thesis oral was then set for Feb. 29, 1962, and I passed smoothly. We had the ritual celebratory beer blast that night. On March 1, the next morning, a moving crew arrived to pack our limited possessions and move them to New Jersey. On March 2, my wife Molly, 8 months pregnant, flew east with me, and I reported to work in Bell Laboratories on March 5.

7.4 The Next Wave

While I concentrated on the thulium experiments at CalTech, several additional graduate students and PostDoc fellows (notably E. Kankeleit from Munich) joined the Physics 34 group, and started working on the other possible resonances that we had originally considered. Research with the ^{169}Tm isotope continued as John Poindexter's thesis and Milton Clauser's thesis, with much more sophisticated choice of materials, and detailed theoretical analysis [2, 3]. Eventually, there were six theses completed under Rudolf Mössbauer's supervision (see Table 7.1).

With our improved understanding of the physics underlying Mössbauer's discovery, we thought we would be seeing much larger effects, especially at low temperatures. But in real experiments, resonance absorption effects were reduced by other radiation that could not be separated from the gamma ray we wanted, other elements in the absorber samples were opaque to the gamma rays, hyperfine

Table 7.1 CalTech theses based on Mössbauer spectroscopy[a]

1.	Richard Lewis Cohen	Mössbauer effect in thulium-169	1962
2.	John Marlan Poindexter	Electronic shielding by closed shells in thulium compounds	1964
3.	Arnold Vincent Lesikar	Magnetic hyperfine interactions in Sm-149 in ferromagnetic Sm Al-2	1965
4.	Frank Snively	Mössbauer effect investigations of the magnetic properties of single crystal thulium ethylsulphate	1965
5.	Milton J. Clauser	Mössbauer effect investigations of the hyperfine interactions and relaxation phenomena in salts of thulium	1966
6.	David G. Agresti	Nuclear hyperfine interactions in W-182, W-183 and PT-195 as studied by the Mössbauer effect	1967

[a]Information provided by Prof. Felix H. Boehm

splitting divided the strength of the resonance, or source strengths were limited. So experimental runs tended to be long- days, or even a week if the source had a long enough half life. There were strong incentives to eliminate the point-by-point data acquisition that we had started with.

By 1962, with solid-state multichannel analyzers(MCAs) readily available, the precise loudspeaker (voice coil) electronic drives and use of the MCA as a time-scanning device had mostly eliminated the early electromechanical systems, with many benefits.[5]

Mössbauer's original experiments with Iridium in Heidelberg yielded extremely small effects. He was careful to assure that voltage and temperature fluctuations, and "dirt effects" such as condensation on cryostat windows, were not giving imprecise data. In the ultimate attempt to stabilize the vacuum tube electronics, he actually soldered the most critical tubes (i.e., hot cathode vacuum tubes) into their sockets! He was also proud that he had eschewed the advice of many more experienced nuclear physicists to "Go to higher energy gamma rays, to get rid of these dirt effects."

In October of 1961, we appreciated the wisdom of Rudolf Mössbauer's approach: an early morning phone call from Stockholm informed him that he had been awarded the Nobel Prize for his discovery. News spread rapidly, and there was a very low-key champagne party in the Physics 34 offices.

There were a number of unusual factors in this award:

1. All the work was done by Rudolf Mössbauer himself – he had consultations from the physics community, but did not have an active collaborator for his initial discoveries. Also he gave the complete and correct theoretical explanation of his discovery in the original paper.
2. It was actually the work done for his Ph.D. that was awarded the prize. Quite unusual.
3. The award was made only about 2.5 years after his initial publication of his results, and the onset of awareness of his findings by experts in this area of physics.
4. When the prize was awarded to him, the publicity resulted in a wave of new research using his discovery – not just nuclear physicists, but chemists, geologists, and later, even biologists, archaeologists, and interplanetary exploration!
5. He was only 32 years old at the time of the award and remains the youngest Nobel laureate in physics (closely followed later by B. Josephson).
6. Although not the first prize in sciences to be won by a German since World War II, it was the first for work done completely after the war. (E.g. the 1958 physics award went to Max Born and Walter Bothe for work done in the 1920s. Also the 1950 and 1953 prizes in chemistry were given for pre-war work). It was seen as a symbol of Germany's return to the first rank of countries with major scientific competence.

---

[5]Note by the editors: The development steps of Mössbauer spectroscopy at CalTech were very close to those in Munich (see book Chap. 1).

What was so special about this discovery that it spread so rapidly and influenced hundreds of scientists to drop what they were doing and pick up Mössbauer Spectroscopy as a principal tool for their work?

I'll give you an example from my own experience. While I was working on my thesis research at CalTech, I was asked to host a visiting class of science students from a local high school, with a "show and tell." I assembled a demonstration Mössbauer experiment from a small cam drive that was no longer in use. I used it to move a small radioactive ($^{57}Co \rightarrow {}^{57*}Fe$) source, using a matching piece of iron foil as an absorber. The foil was taped to a small portable Geiger counter used as a detector. Why a Geiger counter? This was the height of the Cold War – the time of fallout shelters and (almost) nuclear holocaust . Most people (especially an interested group of science students) knew that Geiger counters measured radiation, and they were simple and widely available. With this source and absorber, the resonance absorption effect was high, about 20%. With the source and absorber stationary, the counter clicked away. With the drive motor turned on, the Doppler shift destroyed the resonance absorption, and the clicking of the counter speeded up significantly. These students all went home knowing that they had seen a table-top demonstration of this year's Nobel Prize, and that the Doppler effect works not only for train whistles, but for electromagnetic radiation as well. This was not a simulation, not a computerized emulation of an experiment, not a video purporting to show a scientist twiddling knobs. It was real! And large numbers of experimental physicists who had originally gone into physics research with table-top experiments in mind, saw this as a possible path to a more congenial life style. You were doing research in a bright new area, and if you were clever, you had reasonable hope of designing an experiment that would give you a publishable result in a few days.

I cite this event to show how easy it was (provided you could get a radioactive source) to set up a basic experiment that would actually demonstrate the discovery that had won the prize. In any city that had a reactor and a university with a physics department, a few people could collaborate and generate their own research program that linked them directly to the Nobel Prize that was so exciting. These experiments also provided good training for students.

All this interest from countries that did not normally participate in international physics conferences led to a series of international conferences based entirely on Mössbauer spectroscopy. The first of these was at Cornell University in 1963.[6] The conference language was, as usual, "broken English," but if you walked around, you could hear many active discussions, not just in German, French, Swedish, etc., but Polish, Hungarian, Russian, Hebrew, and multiple Indian languages as well. Mössbauer Spectroscopy has since then always been a strongly international discipline.

---

[6]Strictly speaking, it was not the first Mössbauer conference. It was titled "M^{III}" [4]. But it was the first of the series of Mössbauer conferences with broad international attendance which exist till the present . The two previous meetings were a workshop-type gathering at U.of Illinois, Urbana in June 1960 (see book Chap. 4) and a Conference in Paris, Sept. 1961 (see book Chap. 1), i.e. before the wide international wave of Mössbauer studies began.

7.5   Comments By Gunther K. Wertheim[7]

Gunther Wertheim was the eminent scientist in Mössbauer spectroscopy at the Bell Telephone Laboratories in Murray Hill, N.J. He is one of early pioneers of promoting the applications of the Mössbauer effect, especially in magnetism. He is the author of a monograph on Mössbauer spectroscopy [5] that was the basic reference for (estimated) thousands of graduate students.

GKW: *I recall attending the meeting of a small group in Vienna convened by the Atomic Energy Commission to advise it on applications of Mössbauer spectroscopy. I knew most of the attendees at least by name, but had not encountered the Soviet representative, Vitali Goldanskii. He was urbane, fluent in accented English, and came prepared with typical Russian presents for all the members of the group. As the meeting progressed it became quite obvious that, aside from scientific applications, no one had come up with a practical application of this esoteric phenomenon. Then Goldanskii proposed that we should consider including its utility for "geochemical exploration in space" in our report. It seemed akin to science fiction to most of us. Was he just reminding us of the Soviet's preeminence in rocket propulsion? So finally someone asked him directly whether he really wanted something so futuristic in our report to the AEC. His reply was, "certainly, it helps to get money." At that point we all realized that, East or West, things were the same the world over. I should add that the most recent Mars rovers were actually equipped with a Mössbauer spectrometer to determine the nature of the iron compounds that make it the red planet.[8] So, he was actually a visionary, just 40 years ahead of the time.*

One day John Galt, then a director, came into my office ask me to sign and date an entry in his notebook proposing an application of the Mössbauer Effect. He was doing what we were all supposed to do in order to establish priority for new ideas and patents. His idea was to use the gamma rays as a high resolution communication medium that could accommodate millions and millions of channels. I immediately saw and voiced fundamental obstacles to the realization of his proposal, but he did not care. So, I inscribed the required "Read and understood, my name and date" on his notebook page. Note, that it did not constitute agreement or approval of the proposition. Looking back on that incident I now realize that he, unlike the rest of us, had not lost sight of the fact that our mission at Bell Labs was "communication." He was, in fact, providing a rationale for my work that would be understood by AT&T.

I had begun to do experiments with the low energy gamma ray of ^{57}Fe, which was produced by the decay of the fairly long-lived ^{57}Co parent that could be obtained commercially. One problem was that the gamma ray was split into six components by the magnetism of iron. That made for undesirably complex spectra. So, I went to Vince Jaccarino, one of our experts in magnetism, and asked for help. He suggested

---

[7]Retired Head of the Bell Laboratories Crystal Physics Research Department.
[8]See book Chap. 15.

diamagnetic potassium ferrocyanide or a non-magnetic stainless steel. He must have figured (correctly) that I was one of those nuclear physics types who did not know anything about magnetism. The metallurgy department rolled out a thin layer of high-chromium stainless steel that worked quite well. Later on we actually had some potassium ferrocyanide prepared with separated ^{57}Fe isotope that worked even better.

I quickly did an experiment with ^{57}Co in Fe, Co and Ni, which showed different magnetic splittings and submitted the result to Physical Review Letters, emphasizing the use of the stainless steel, unsplit absorber in the title. It came back with the suggestion from Sam Goudsmit, the editor, that I should shift the emphasis to the physics of the hyperfine splitting in the different host metals. I did and it was published. In retrospect, I marvel that one could do interesting science while being so fundamentally ignorant. And yet, that is the experience that I had every time I switched to an entirely new field of endeavor.

7.6   The Sunset of the CalTech Period

From the moment that they arrived in Pasadena, Rudolf Mössbauer and his (first) wife Elisabeth enjoyed the informal, open, and unstructured relationships they encountered at CalTech. They often compared the US environment favorably with the more formal and traditional atmosphere that dominated the German (and, in fact, almost all European) university systems.

From the time they arrived in February 1960, there was a rapidly increasing schedule of visitors to the Physics 34 offices, to meet Rudolf Mössbauer and see what was going on. In return, Mössbauer received and accepted many invitations to visit and give colloquia at many other universities and laboratories, and he was "on the road" a significant fraction of the time. He got to see that although CalTech, with its small size and southern California location was a somewhat extreme case, most US universities maintained informal relationships among the faculty, students, and other academic employees. He remained affable and approachable, a strong contrast to the "German Professor" model.

From the announcement of the Nobel Prize award in October 1961, this stream of visitors, and the public appearances, turned into a torrent. The Nobel Prizes get broad general publicity, and the contacts were no longer primarily scientists, but included many lay press and international reporters, and he became a visible public figure. In 1960–1961, he was awarded three additional prestigious international prizes, attracting even more visibility, and he was promoted from Research Fellow to Full Professor at CalTech.

During this period, it was common for bright young Ph.D.'s from European universities to come to the US for periods of a few years. The opportunities for advancement were greater in the US, and these young scientists could build up a portfolio of research results, make contacts, and in many cases get faculty appointments. This was a period of rapid growth in the US science and technology community as a side effect of the Cold War. A faculty position in the US was perceived as a useful negotiating chip for those who wanted to return to their home countries with university appointments.

Rudolf Mössbauer joined this group of expatriates with mixed emotions. He and his family very much enjoyed the California environment- the weather, the beaches and the mountains, the friendly people and informal relationships. They had a lot of good will toward CalTech because it had been a supportive and generous host by bringing him to the US before his expertise was so widely appreciated, and he had become close friends with some of the other faculty members. But he also was hoping to modernize university structure and replace the traditional "One Professor" approach with a departmental structure more like that in the US. He made a modernization of the faculty structure a precondition of his return to a German university. He had done both undergraduate and graduate work at the Technische Universität München. He negotiated through the Bavarian Ministry of Culture, Education and Research (the managers of the Technische Universität München) not only for a full professorship for himself, but also for a reorganization of the faculty. This type of change has been called "The Second Mössbauer effect" (Details can be found in book Chap. 21). Early in 1965, it was announced that he and his family would be returning to Munich, terminating the CalTech phase of their lives. Mössbauer returned, however, to CalTech regularly during the university summer recess for many years.

7.7 Epilogue

In mid 1964, I got a letter at Bell Laboratories from Rudolf Mössbauer. It was an invitation for me to come to Munich as a "Wissenschaftlicher Assistent" for a year to set up the laboratory for him and manage and advise the graduate students. I, Molly, and Bell Laboratories thought this would be a good opportunity, and in September of 1964, Molly and I – this time with our 2.5 year old son- set off on the M.S. Bremen for a year of research and cultural enrichment.

References

1. R. Cohen, U. Hauser, R.L. Mössbauer, in *Recoilless Resonant Absorption in Nuclear Transitions of the Rare Earths*, ed. by D.M.J. Compton, A.H. Schoen. The Mössbauer Effect (Wiley, New York, 1962), p. 172
2. R.G. Barnes, E. Kankeleit, R.L. Mössbauer, J.M. Poindexter, Electronic shielding of the crystalline field in thulium ethyl sulfate. Phys. Rev. Lett. **11**, 253 (1963)
3. M.J. Clauser, E. Kankeleit, R.L. Mössbauer, Pseudoquadrupole shift of gamma resonance spectra. Phys. Rev. Lett. **17**, 5 (1966)
4. A.J. Bearden (ed.), *Third international conference on the Mössbauer effect*. Rev. Modern Phys. **36**, 333–503 (1964)
5. G.K. Wertheim, *Mössbauer Effect: Principles and Applications* (Academic, New York, 1964)

Chapter 8
The World Beyond Iron

G. Michael Kalvius

8.1 Introduction

The mere fact that the Mössbauer effect was discovered with the 129.4 keV transition in ^{191}Ir demonstrates immediately the availability of Mössbauer isotopes other than ^{57}Fe. Nevertheless, the ^{57}Fe resonance remains the soul of Mössbauer spectroscopy. It combines a number of favorable properties: a source with convenient half-life (270 days), a large recoil-free fraction which allows measurements well above room temperature, and an energy resolution of $\sim 10^{-8}$ eV which is one to two orders of magnitude smaller than the typical hyperfine interaction energies. Yet, the energy resolution is not high enough to lead to substantial line broadenings by the unavoidable small distortions in the crystalline lattice of a real solid. The low natural abundance (2,2%) of the ^{57}Fe is compensated by the large resonance cross-section and isotopic enrichment is only needed for materials containing iron in very low concentration or for extremely small samples. ^{57}Fe was in fact not the second Mössbauer transition to be used after ^{191}Ir. In establishing the correctness of the, not immediately believed, result of Mössbauer, the group at Argonne National Laboratory [1] measured not only the recoil-free resonance absorption in ^{191}Ir, but also that of the 100 keV transition in ^{182}W. This historically number two resonance has later mainly be used for the establishment of nuclear parameters.

Mössbauer nuclear resonance fluorescence has been observed in 106 γ-transitions. They are distributed over 86 isotopes, meaning that in some nuclei more than one suitable transition exists. Mössbauer active isotopes are found in 45 elements, since a number of elements contain several usable isotopes. The record is held by Gd with six Mössbauer isotopes and eight transitions.

G.M. Kalvius (✉)
Physics Department, Technical University Munich, 85747 Garching, Germany
e-mail: kalvius@ph.tum.de

M. Kalvius and P. Kienle (eds.), *The Rudolf Mössbauer Story*,
DOI 10.1007/978-3-642-17952-5__8, © Springer-Verlag Berlin Heidelberg 2012

To be a Mössbauer candidate, a γ-ray transition needs to fulfill two basic requirements. First, the transition must lead directly into the ground state of a stable, or at least a long lived, nucleus. Second, a sizable portion of the γ-ray must be emitted and absorbed free of thermal Doppler broadening and recoil energy loss, which is possible only, if the atom containing the Mössbauer isotope is bound in a solid, not necessarily a crystalline solid. The fraction of non-shifted and non-broadened γ-ray emission is given by $f = \exp\left[-(E_\gamma/\hbar c)^2 <u^2>\right]$ with $<u^2>$ being the mean square displacement of the atom due to lattice vibrations, and E_γ is the energy of the resonant gamma radiation. One calls f the "recoil free fraction" or the "Lamb-Mössbauer factor" or (slightly incorrect) the "Debye-Waller factor" or simply the f-factor. The size of $<u^2>$ depends on the strength of the bond in the solid, usually parametrized by the Debye temperature Θ. Naturally $<u^2>$ becomes smaller with reduced temperature, yet zero point vibrations prevent $<u^2>= 0$ for $T\to 0$ and $<u^2>$ reaches a constant value in the low temperature limit, making it impossible to compensate for a high E_γ by cooling to very low temperatures. The γ-ray with the highest energy used for Mössbauer measurements is the 186.9 kev transition in ^{190}Os [2]. Last, but not least, a convenient radioactive mother activity (not too short a half life, easy to produce, or even better, available commercially as ready to use source) is important for a wide spread of a specific Mössbauer isotope.

Figure 8.1 displays a periodic table marking the elements with at least one γ-transition exhibiting a sizable Mössbauer effect. Shown is the most suitable transition for the most commonly used isotope together with a star rating of its application in general scientific research. Notoriously absent are Mössbauer candidates in the light elements. What a dream, if one could perform Mössbauer spectroscopy with carbon or oxygen. The reason for their absence lies in the systematics of nuclear structure, which prevents in light nuclei the formation of low energy states just above the ground state. From the point of view of modern solid state physics (e.g., strongly correlated electron systems) the absence of Ce is unfortunate. Systematic studies are possible in the d- and f-transition series, providing novel insights into magnetic properties, especially the dynamics of magnetic moments. Another set of elements for comparative measurements exist from tin to iodine, delivering a wealth of information on chemical bond structures.

8.2 The More Commonly Used Mössbauer Elements Other Than Iron

Only 18 Mössbauer elements (having a rating of two or more stars) are practicable for general research. Their relative usage is shown in Fig. 8.2. The 27 nonlisted Mössbauer elements account for less than 2% of records on Mössbauer studies. The total number of more than 10^5 records demonstrates impressively the impact of

H																	He
Li	Be											B	C	N	O	F	Ne
Na	Mg											Al	Si	P	S	Cl	Ar
K 40 0,01 29.4 *	Ca	Sc	Ti	V	Cr	Mn	Fe 57 2.2 14.4 ****	Co	Ni 61 1.2 67.4 **	Cu	Zn 67 4.1 93 **	Ga	Ge 73 7.8 13.3 *	As	Se	Br	Kr 83 11.6 9.3 *
Rb	Sr	Y	Zr	Nb	Mo	Tc 99 -- 140.5 o	Ru 99 12.7 89.4 **	Rh	Pd	Ag 107 51 93 ?	Cd	In	Sn 119 8.5 23.9 ****	Sb 121 57.3 37.2 ***	Te 125 7 35.5 ***	I 129 -- 27.8 ***	Xe 129 26.4 39.6 *
Cs 133 100 81 **	Ba 133 -- 12.3 o	(Ln)	Hf 178 27.1 93.2 *	Ta 181 100 6.25 **	W 182 26.4 100 *	Re 187 62.5 134.2 o	Os 189 16.1 69.6 *	Ir 193 62.7 73 **	Pt 195 33.8 98.8 *	Au 197 100 77.3 ***	Hg 199 13.2 158 o	Tl	Pb	Bi	Po	At	Rn
Fr	Ra	(An)															

	La	Ce	Pr	Nd	Pm	Sm	Eu	Gd	Tb	Dy	Ho	Er	Tm	Yb	Lu
Lanthanides (Ln)	139 99.9 165.8 o		141 100 145 *	145 8.3 72.5 *	145 -- 61.4 o	149 13.8 22.5 *	151 48 21.6 ***	155 14.7 86.5 ***	159 100 58 o	161 19 25.7 **	165 100 94.7 o	166 33.4 80.6 **	169 100 8.4 **	170 3 84.3 **	175 97.4 113.8 o
	Ac	Th	Pa	U	Np	Pu	Am	Cm	Bk	Cf	Es	Fm	Md	No	Lr
Actinides (An)		232 -- 49.4 o	231 -- 84.2 *	238 99.3 44.9 *	237 -- 59.5 ***	240 -- 42.8 *	243 84 *								

Fig. 8.1 Mössbauer periodic table. Listed is the mass number of the most commonly used isotope, its natural abundance (absent in case of radioactive nuclei), and the γ-ray energy in keV. The star rating means: **** = most widely used, *** = very useful for all types of research, ** = useful for selected applications, * = only of limited use, o = the Mössbauer effect has been established and was used mainly to determine nuclear parameters

Rudolf Mössbauer's discovery on modern science. In our look at the (Mössbauer) world beyond iron, we shall only give a brief overview of the main applications of the various isotopes, restricting ourselves to classical Mössbauer studies (i.e., excluding studies using synchrotron radiation) and without a detailed discussion of the results obtained. These can be found in the references listed and, in several cases, in the special chapters of part 2 of this book. Particularly for tin which remains the second most important resonance, some historical background is also presented. Basic technical aspects of working with these useful Mössbauer isotopes are available in [4]. Reviews often cited here are [5,6].

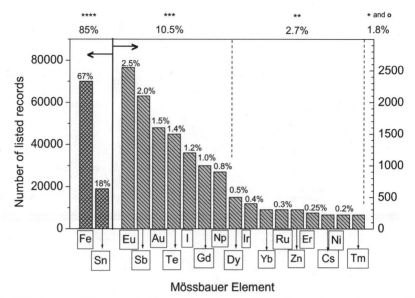

Fig. 8.2 Number of listed records for the commonly used 18 Mössbauer elements. For each of the not listed 27 elements the number of records is less than 100, mostly below 50. The percentage of records for each of the commonly used elements is given above the bar. On top, the percentage of records for each "star" category is listed. The total number of records as of mid 2009 is 105,000. Numbers based on the survey published in the Mössbauer Effect data index [3]

8.2.1   Elements of the 5th Main Group (Sn to I)

The major interest of Mössbauer studies in this series is chemical information which is obtained from isomer shift and quadrupole interaction data.

8.2.1.1 ^{119}Sn

This resonance was discovered only a few months after the storm of papers on ^{57}Fe broke loose at the turn 1959/60. First measurements were reported by a French group [7], interestingly enough in scattering geometry. They were quickly followed by more extensive studies at the University of Manchester (e.g. [8]). Early data were also reported by a Russian group [9]. Alan Boyle, then at Manchester recalls [10]: *The first I heard of Mössbauer was sometime (probably around August) in 1959. Sam Devons [the Professor of Physics and director of the Physics Laboratory] often used to do a walk-about around the labs to keep track of what was happening. He would often broach some idea that he had had and you were expected to research it for discussion at the weekly meeting of our small Van de Graaf group. On this occasion he asked me "How would you like to observe the Nuclear Zeeman effect?" He had recently returned from a meeting in Birmingham and Mössbauer's paper*

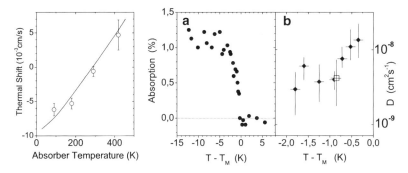

Fig. 8.3 *Left*: Thermal red shift in [119]Sn. Source and absorber are Sn metal. Source is kept at 300K. The *solid line* is the prediction by Josephson. *Right*: Temperature dependences of (**a**) the resonant absorption and (**b**) the diffusion coefficient based on a jump model (*filled symbols* = from absorption, *open symbol* = from linewidth) for tin metal near the melting point (T_M). Data from [8, 11]

had been mentioned by Moon. I believe it was Sam's realization that it would enable the examination of nuclear hyperfine effects. I got to work, and having managed to work through Mössbauer's paper, began a search for suitable transitions. The bible of the time was the Reviews of Modern Physics list of known nuclear energy levels. I started at the high end and worked down until I found [119]Sn – this looked really good and it never crossed my mind that there may be an even lower atomic weight example. As it happens it was probably for the best that we were unaware of the potential of [57]Fe. Early on there was a paper by Barloutaud of Saclay and I went to Paris to see him – I suppose mainly to check out the competition. As it turned out their group quickly lost interest. We soon became aware of the work at Harwell on [57]Fe and over the following years had many contacts with Walter Marshall and the others there. Henry [Hall] came to Manchester from Cambridge and brought to us B. Josephson's ideas on the thermal red shift[1] which led us to our first experiment [8] (see Fig. 8.3, left).

The high recoil free fraction of the 23.9 keV γ-ray allows measurements well above room temperature. Due to the low melting point of tin metal ($\sim 232°C$), it was possible to carry out the first Mössbauer study of the temperature dependence of the f-factor in a metal up to its melting point [11]. Considerable anharmonicity in lattice vibrations was seen and, at the last few degrees below the melting point, a rapid decrease of absorption together with line broadening caused by diffusional motion of the tin atoms (see Fig. 8.3, right). This study coincided with the seminal paper by Singwi and Sjölander on the effect of atomic motion on Mössbauer γ-rays [12].

The 23.9 keV state is populated by the isomeric γ-ray transition from 245 d, [119m]Sn. The fact that the Mössbauer transition is fed from a radioactive Sn state is most advantageous for so called "source experiments" e.g., ppm doping of [119m]Sn

---

[1] See book Chap. 14 for explanation and thermal red shift data taken with [57]Fe.

in metallic matrices. Chemically, the activity introduced is already tin and the low energy isomeric γ-ray (66 keV) causes minimal radiation damage, avoiding problems termed "Aftereffects of Nuclear Transformations" which may occur in cases of β- or, even worse, α-decay sources [13]. Those effects are of interest in the realm of radiation chemistry, but are a disadvantage when studying for dilute impurities the hyperfine interactions or the vibrational couplings to the phonon spectrum of the host [14].

The energy resolution of the ^{119}Sn resonance is about five times worse than that of ^{57}Fe. Isomer shifts are well, quadrupole splittings often only resolved. The latter is also the case for magnetic splittings since hyperfine fields at the Sn nucleus are usually small, as will be discussed below.

The atomic electron configuration of tin is $^3P_0$ ($5s^25p^2$), the 4th shell being completely filled except for the $4f$ states. One finds formally divalent and tetravalent tin compounds. In the stannous compounds (Sn^{2+}), which ideally should have the configuration $5s^25p^0$ considerable $s - p$ hybridization (covalency) is often present. Tetravalent tin forms a huge variety of organometallic compounds where the sp^3 hybridization dominates. The isomer shift (IS) is largely determined by the effective number of s-electrons, while the p-electrons are the main source for an electric field gradient (efg). Correlations between IS and quadrupole splittings have been pursued extensively to gain information on the covalency and spatial distribution of the bonding electron states. Those studies comprise the main application of the Sn Mössbauer resonance [15]. A recent survey [16] lists over 600 tin compounds where IS and quadrupole splittings have been determined. Volume dependences of IS have been investigated in compounds and alloys. As shown in Fig. 8.4, right, the high pressure study of the semiconductor $SnMg_2$ revealed a non linear pressure dependence of IS caused by the decrease of the energy gap under reduced volume. This behavior is in contrast to the behavior of tin metal where the linear increase of IS with pressure arises exclusively from the compression of the outer $4s$ electron shell [17].

Tin, not being a transition element, does not possess a hyperfine field at the nucleus created by open orbitals of its own electron shell. This allowed the easy measurement of the nuclear Zeeman effect in an applied field to fix the ratio of excited state to ground state magnetic moment [11]. The Mössbauer transition in ^{119}Sn has the 1/2–3/2 spin sequence, is of pure M1 character and ground and excited state dipole moments are of opposite sign. A 6-line Zeeman pattern similar to that of ^{57}Fe results.

In the absence of external fields, hyperfine fields at the tin nucleus are only present in compounds containing a magnetic ion like $SnMn_2$[2] or in alloys with a transition metal (eg. Sn in Fe). The hyperfine field in such cases is a so called "transferred hyperfine field" from the neighboring magnetic ions.[3] In metals, one expects the main contribution to come from the polarization of conduction electrons

---

[2]The spectrum is shown in book Chap. 5.

[3]Transferred hyperfine fields are discussed in more detail in book Chap. 11.

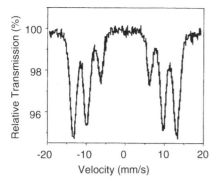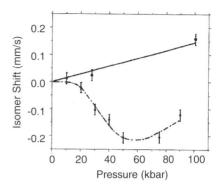

Fig. 8.4 *Left*: Well resolved raw (no fit) ^{119}Sn magnetic spectrum to elucidate shape of the Zeeman pattern. The sample is $Ca_{0.25}Y_{2.75}Sn_{0.25}Fe_{4.75}O_{12}$ at 77 K. After [19]. *Right*: Change of the isomer shift in tin metal (solid line) and the semiconductor $SnMg_2$ (*dash-dot curve*). A positive shift means a decrease of γ-ray energy. After [17]

by the moments on the magnetic ions. This polarization is transmitted to the *s*-electron shells of the Sn atom leading to a Fermi contact field. Yet, other contributions like the overlap of wave functions of *d*-electrons of the magnetic atom with the 4*s*-electrons of tin can play a role as well [18]. In the tin doped yttrium-iron-garnets substantial transferred fields are generated by the polarization of the tin electron shell via the indirect exchange interaction. The probably best resolved magnetic spectrum has been observed [19] in $Ca_{0.25}Y_{2.75}Sn_{0.25}Fe_{4.75}O_{12}$ (Fig. 8.4, left). Although the field transfer mechanism is of interest, magnetic studies make up only a minor portion of measurements with ^{119}Sn, a clear contrast to ^{57}Fe.

8.2.1.2 ^{121}Sb

The electronic structures of antimony and tin compounds are closely related. The higher resonance energy (37 keV) limits the accessible temperature range. In contrast to ^{119}Sn, source experiments are not so common, because the formation of the resonant γ-ray is preceded by an energetic β^--decay from 76 y ^{121m}Sn which can give rise to aftereffects. Like in tin, IS are well resolved. The spin sequence of the 37 keV transition is 7/2–5/2 and its pure M1 character leads, as far as quadrupole interactions are concerned, to a never fully resolved 8 line pattern. Still, the relevant parameters can be determined in most cases. Since the resonant transition occurs between levels with spin $I > 3/2$, one can also obtain the sign and the asymmetry parameter of the efg, in contrast to tin. Correlations between IS and quadrupole interactions in Sb^+ and Sb^{3+} compounds are used to elucidate bonding properties [20]. Hyperfine fields in Sb are also of the transferred type, but in magnetic compounds like MnSb large enough for satisfactory resolution of the Zeeman pattern. The parameters of the 37 keV transition in ^{121}Sb are similar to those of the 22 keV transition in ^{151}Eu, and so are their hyperfine spectra. Typical ^{151}Eu spectra are shown in Fig. 8.7. Overall, magnetic studies play only a minor role in ^{121}Sb. The total number of records for ^{121}Sb is about 10% of those for ^{119}Sn, while the recent list [21] of Sb compounds studied with respect to IS and quadrupole splitting lists 300 entries which is half of the 600 entries for Sn, emphasizing that the strength of ^{121}Sb studies lies in chemical applications.

8.2.1.3 ^{125}Te

The features of the 35.5 keV, 3/2–1/2, M1 Mössbauer transition in ^{125}Te are basically similar to ^{119}Sn, except that the energy resolution is poorer and of the same magnitude as the hyperfine interactions in the majority of Te compounds. Despite the limited resolution, a sizable amount of Te compounds have been investigated with respect to properties of chemical bond [22]. The nature of transferred hyperfine fields at the Te nucleus have been studied for Te impurities in metallic hosts (e.g., Te in Fe [23]) and in magnetic compounds like $MnTe_2$ [24]. The ^{125}I parent activity decays by electron capture (EC) which is particularly powerful in disrupting the electron shell of the daughter atom. Aftereffect studies in $Na^{125}IO_3$ and related compounds [25] lead to complex spectra pointing toward a redistribution of bond structure after the radioactive decay process.

8.2.1.4 $^{127/129}$I

The 57.6 keV transition in ^{127}I and the 27.7 keV transition in ^{129}I are both used for Mössbauer studies. The main difference between them is the resolving power. It is very high in ^{129}I, allowing precision measurements of hyperfine interactions, but about ten times worse in ^{127}I. To demonstrate the excellent quality of ^{129}I hyperfine spectra the quadrupole split spectrum of molecular iodine is depicted in Fig. 8.5. Magnitude and sign of the efg as well as the asymmetry parameter can be extracted accurately. Regarding quadrupole interactions, the ^{129}I Mössbauer resonance is in competition with NQR, but it delivers in addition the valuable IS data. The main complication in using ^{129}I in absorber materials is its weak β^--activity (1.7×10^7 years) which does not influence the Mössbauer measurements, but constitutes a health hazard. ^{127}I is the only stable isotope.

The ^{129}I resonance is perhaps the most favorable candidate for the study of chemical bond structure due to the possibility of obtaining precise values for IS and efg, including the asymmetry parameter η. Also, the bonding in iodine compounds is easier to handle theoretically than that of its neighbors tin, antimony, and tellurium [27, 28]. However, the results on iodine can be expanded rather straightforwardly to analogous Te compounds [20]. The quantitative interpretation of iodine spectral parameters is based in first approximation on a pure valence bond picture involving only the $5p$ orbitals ($s = 0$). Yet, the measured value for η in I_2 (see Fig. 8.5) indicates that intermolecular bonding (overlap effects) is present. That overlap

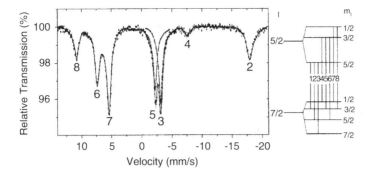

Fig. 8.5 *Left*: ^{129}I spectrum of I_2 at 100 K showing highly resolved asymmetric quadrupole interaction. The very weak line 1 is outside the velocity scan. The measured parameters are IS $= 0.82$ mm s^{-1} (against the Zn^{129m}Te source), $e^2qQ = -63.7$ mm s^{-1}, and $\eta = 0.16$. *Right*: Hyperfine levels under quadrupole splitting and allowed transitions. After [26]

effects need to be considered in iodine bonds, is further emphasized by results on iodine molecules insulated in an inert matrix (e.g., frozen rare gas) [29].

Hyperfine interactions have been measured in various magnetic compounds, partially in standard absorber experiments, partially as ^{129m}Te source experiments. Iodine impurity studies in magnetic hosts have especially been performed by implanting ^{129m}Te into the host with an isotope separator [30]. The question of interest is always the nature of the transfer mechanism of the magnetic field from surrounding moments. A fully satisfactory theory covering conducting and nonconducting materials is still not available.

8.2.2 Alkalies

Two Mössbauer isotopes, ^{40}K and ^{133}Cs, exist. The 29.4 keV transition in ^{40}K (4–3, E1) is difficult to handle. No parent radioactivity exists, the measurements need to be done "on line" using a nuclear reaction like (n,γ). A number of potassium compounds have been studied in that manner [31], but neither isomer shifts nor hyperfine splittings could be observed. From the temperature dependence of the f-factor the Debye temperatures for K metal ($\sim$90 K) and KF ($\sim$230 K) were derived. Unfortunately, the lightest Mössbauer isotope defies applications.

The situation is more favorable for the 81 keV (7/2–5/2, M1) transition in ^{133}Cs. Isomer shifts are readily observed, quadrupole interactions partially resolved. Available data are discussed in terms of population of the lone 6s state outside the xenon core [32].

Caesium is easily intercalated into graphite forming the lamellar compounds C_8Cs and $C_{24}Cs$. Large quadrupole splittings were observed, as expected. From the IS it followed that Cs in $C_{24}Cs$ is fully and in C_8Cs $\sim$50% ionized. The recoil free fraction changed by a factor of 20 with the direction of observation. The analysis lead to values of the mean square displacements of the Cs atoms of $< x^2 >= 35.8 \times 10^{-4}\text{Å}^2$ in the direction perpendicular to the graphite c-axis (i.e. parallel to the sheets), while the mean square displacement of the Cs atoms parallel to the c-axis is only $< z^2 >= 18.9 \times 10^{-4}\text{Å}^2$. The Cs, situated between the sheets, rattles loosely parallel to them, but is tightly constrained by the carbon planes in the other direction [33]. In pure graphite, the opposite is true. One finds $<z^2>= 37.6 \times 10^{-4}\text{Å}^2$ and $<x^2>= 12.5 \times 10^{-4}\text{Å}^2$ which arises fron the very high stability of atomic bonding within a sheet, whereas the sheets are bound weakly by van der Waals forces [34].

The implantation of the β^--parent 5.3 d, ^{133}Xe into iron with an isotope separator [30] lead to a complex magnetic spectrum, explained as the overlay of three Zeeman patterns with markedly different hyperfine fields. Their relative population is temperature dependent. These results were explained by assuming that vacancies are generated in the implantation process and then trapped by the impurity Xe atoms. Three different vacancy formations (linear, triangular, tetrahedral) are created. The large atomic size of Cs is thought to be the cause that a unique magnetic spectrum is not created.

8.2.3   d-Transition Elements

Most important is ^{57}Fe, but it is outside this text. The main part of Mössbauer work in the d-transition elements are isomer shift and magnetic studies which we outline only briefly. More details are available in Chaps. 10, 11, and 12 of this book and also in [35].

8.2.3.1 ^{61}Ni

Nickel, together with Co (for which no Mössbauer isotope exists), is an important ingredient of magnetic alloys, especially permanent magnets. One might expect that Mössbauer spectroscopy on Ni is another boon for magnetic studies. Yet, the use of the 67.4 keV, 3/2–5/2, M1 transition in ^{61}Ni is hindered by three problems: sources are difficult to prepare, the temperature range is quite restricted and resolution is moderate at best. When analyzing very poorly resolved hyperfine spectra, the common approach of representing the spectral shape by a sum of Lorentzians may lead to some error in the parameter value. A precise analysis requires the numerical solution of the full transmission integral as discussed in [36].

Isomer shifts are small and the second order Doppler shift[4] is difficult to separate from the chemical shift. IS studies on some Ni compounds lead to a diagram of

---

[4]For explanations see book Chap. 14.

IS vs. electron configuration [37], analogous to the Walker-Wertheim-Jaccarino plot for iron. Yet the evaluation of the nature of valence orbitals remains in an elementary stage. Quadrupole spectra are usually not well resolved. The situation is somewhat more favorable if an additional magnetic splitting is present. Although the Zeeman pattern is complex, it can at least be resolved partially in favorable cases. The study of magnetic intermetallics, alloys and compounds constitute the main work with ^{61}Ni. An example is the measurement of variation of the hyperfine field with x in $Ni_{1-x}Pd_x$ alloy series [38].

8.2.3.2 ^{99}Ru

The generally used resonance is the 89.4 keV 3/2–5/2 M1 + 2.5%E2 transition which offers good resolution for hyperfine interactions. A substantial number of compounds have been investigated [39]. They contain ruthenium in seven charge states, Ru(II) to Ru(VIII). For each charge state, IS form a group. The groups are well separated for the higher charge states but come close to each other for Ru(III) and Ru(II). In that region, covalency effects are more pronounced while in the compounds containing higher charged ruthenium the electron configuration is close to the $4d^n$ ionic limit. IS have also been obtained for Ru as dilute impurity in essentially all $3d$-, $4d$-, and $5d$-metals, showing a decrease of ρ_0, the electron density at the nucleus, the heavier the host element. This can be correlated to an increase of the number of outer electrons. The decrease of ρ_0 means that outer $4d$- rather than $5s$-electrons are involved.[5]

The E2 admixture in the 89.4 keV γ-ray produces additional hyperfine lines since the transitions $\Delta m = \pm 2$, forbidden for pure dipolar radiation, are now allowed. The six line quadrupole pattern is usually resolved sufficiently to allow a good determination of the efg. As an example, the efg in ruthenium cyano complexes can be traced to an asymmetric expansion of t_{2g} electrons toward the ligand [39]. The hyperfine field of Ru in Fe is $\sim$50 T [40]. In compounds, fields between $\sim$35 T ($SrRuO_3$) and $\sim$60 T (Na_3RuO_4) were measured [41], covering about the same range found in the $3d$ series.

8.2.3.3 ^{193}Ir

The favorite resonance for research on iridium compounds and alloys is the 73 keV transition in ^{193}Ir, rather than the 129.4 keV transition in ^{191}Ir originally used by Rudolf Mössbauer. The advantage is a nearly two orders of magnitude narrower linewidth, a simpler spin sequence (1/2–3/2 vs. 5/2–3/2) and a larger f-factor.

The $4d$ element Ru and the $5d$ element Ir show rather analogous chemical behavior. Compounds are formed with iridium in four charge states, Ir(III) to Ir(VI),

---

[5]A plot is shown in book Chap. 10.

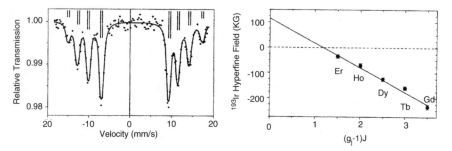

Fig. 8.6 *Left*: Magnetic hyperfine spectrum of IrF_6 measured with the 73 keV resonance in ^{193}Ir. The diagram gives positions and intensities of the eight Zeeman lines in the absorber. The *double bars* indicate the small quadrupole splitting in the ^{193}Os metal source. After [42]. *Right*: ^{193}Ir saturation hyperfine field in RIr_2 intermetallics as function of $(g_J - 1) J$ which is the projection of the spin of the R^{3+} ions onto their total angular momentum. After [44]

with the ionic configurations $5d^6$ to $5d^3$. As in Ru, the compounds containing differently charged Ir form groups of isomer shifts. The ranges of shifts for the lower charge states are wider due to stronger deviations from the ionic limit. In complexes with CN^- or NO_2^-, for example, the effective occupation number $5d^n$ is not only dependent on the Ir oxidation state, but also influenced by back bonding and σ-bond covalency effects [42]. Quadrupolar interactions are also pursued in terms of stereochemical properties. The inability to deduce the sign of the efg or a possible asymmetry parameter from pure quadrupole spectra limits somewhat their usefulness. The hyperfine field in Ir compounds can be quite large. In IrF_6, whose Zeeman spectrum is depicted as an example in Fig. 8.6, left, a value of $-185\,T$ was measured [42]. It arises from a contact field of $-165\,T$, a value at least 5 times larger than in iron and an orbital term of $-20\,T$. The field for dilute Ir in iron is $-45\,T$ [43], a value comparable to that of Ru in iron. Lower are the fields in Ir intermetallics with rare earths (R) of the form RIr_2 [44]. They range from about $-3\,T$ to $-25\,T$, indicating a partial compensation of the Ir hyperfine field by transferred fields from the R ions. The observation that the Ir hyperfine field is a linear function of $(g_J - 1)J$, i.e., the projection of the spin of the R ions onto their total angular momentum J (see Fig. 8.6, right) implies that the field from the $4f$ shell of the R ions is transferred to the Ir nucleus predominantly by spin polarization of the s-conduction electrons ($s - f$ interaction). Yet, the line B_{hf} *vs*. $(g_J - 1)J$ does not go through the origin which means that other transfer mechanism must also be present.

The magnetic hyperfine studies established a somewhat esoteric effect called "hyperfine anomaly" [42, 43]. The term refers to minute shifts in the measured values of the ratio $R_m = \mu_e/\mu_g$ caused by the different spatial distribution of the magnetic hyperfine field over the nuclear volume if either an externally applied field or the internal hyperfine field produces the Zeeman pattern. Details can be found in book Chap. 9 which also discusses tests of the time reversal invariance in electromagnetic interaction using magnetic spectra of [173]Ir [45] and [99]Ru [40] in polarized iron foils. No violation of time invariance could be detected.

8.2.3.4   [197]Au

The Mössbauer spectra of the 77.3 keV Mössbauer transition in [197]Au are quite similar to those of the 73 keV resonance in [193]Ir. The 100% abundance of [197]Au allows Mössbauer studies of materials with low gold content. The favorable Mössbauer properties make [197]Au the most used resonance in the d-transition series after [57]Fe. Gold forms Au^+ or Au^{3+} compounds. IS and quadrupole splitting have been measured for many compounds including organic materials. The essentially linear correlations between these parameters show that an increase in ρ_0 is accompanied by an increase of the quadrupole splitting [46]. Discussion is based on $6s6p$ and $5d6s6p^2$ hybridization. Under applied pressure monovalent and trivalent compounds both showed an increase of ρ_0, as expected. In contrast, the quadrupole splitting decreases strongly with pressure for Au^+ and increases weakly for Au^{3+}. The origin of that difference in behavior is rooted in the difference in geometry of their bond structures (linear vs. square planar). Intensive IS studies were also carried out for a large number of gold alloy systems (see [47]). A large hyperfine field of 160 T was observed in Au_2Mn [48]. Since gold is a diamagnetic atom, the field is a transferred field. One possible mechanism is the splitting of the $5d$-band into spin-up and spin-down parts which in turn polarizes the $5s$-electrons leading to a large Fermi contact field.

Gold precipitates comparatively easily into fine particles down to the nanometer scale. The famous gold ruby glass gets its color from small particles of metallic gold which is formed after the originally colorless gold containing glass is annealed. The chemical state of the gold before the glass gets its ruby color was a mystery. By [197]Au Mössbauer resonance it was shown that Au is originally monovalent forming linear bonds to two neighboring oxygen atoms from the basic glass ingredients like SiO_2 [49]. In some minerals like pyrite, the gold is too finely dispersed to be detected by standard methods (e.g. scanning electron micrography) and its exact chemical nature is poorly known. To help the extraction process, [197]Au Mössbauer spectroscopy was used to trace the state of the gold in these minerals. It was found that the ores contain chemically bound gold, not metallic gold [50]. Finally, since gold has been part of human culture throughout history, [197]Au Mössbauer spectroscopy can be applied to archeology. An example is the study of the gold alloys in Celtic coins [51].

8.2.4   *f*-Transition Elements

The 4f transition elements form the lanthanide (Ln) or rare earth (R) series. The numerous Mössbauer isotopes allow systematic studies of electronic properties and nuclear parameters. The latter is of importance since many of the rare earth nuclei fall into the region of "strongly deformed nuclei" for which detailed theoretical calculations exist as discussed in book Chap. 9. Concerning the electronic properties, the simplest approach is to consider all 14 rare earth elements to be chemically alike. Compounds are based on ionic bonding of the R^{3+} ions formed by donating the outer $6s$ electrons plus one $4f$-electron to the ligand. The remaining $4f$ orbitals are buried deep inside the atomic shell and not influenced by the bond. One speaks of f-electron localization, meaning that $4f$ valence or conduction band are, if formed, very narrow. Rare earth magnetism is termed local magnetism in contrast to the rather itinerant behavior of the $3d$ electrons. In a more subtle approach, noticeable differences between the various R elements appear. First, the light rare earths up to Eu are, with respect to their electronic structure, more complex than the heavy rare earths which come closer to the simple picture. Second, Eu atoms possess a half filled, Yb atoms a fully filled $4f$ shell. They form divalent and trivalent ionic states with a variety in bonding character and magnetic properties. The Mössbauer isotopes below Eu, plagued with technical difficulties, have mostly seen applications to nuclear physics. An exception, to a limited degree, is ^{149}Sm where some data on intermediate valence are reported. In contrast, Eu is a Garden of Eden. Its resonances are technically easy to handle and very powerful in unraveling the varied electronic structures of this element. It is the third most employed Mössbauer element, beating even antimony. The following heavy rare earths are mostly of interest with regard to magnetism, including especially dynamic processes (relaxation spectra).[6] Yb is electronically comparable to Eu, but the small IS of its resonances limits applications to magnetic studies mainly.

The $5f$ transition elements form the actinide (An) series. In the first half of the series the An differ in their electronic properties markedly from their rare earth counterparts in the sense that the $5f$-electrons are much less localized on the An atom and that in general the electronic properties of the actinides are stronger influenced by relativistic effects. They are able to hybridize more easily with $6d$ and $7s$ electrons. When originally only the elements up to uranium were known, they were considered $6d$ transition elements. The transuranic elements finally established the actinides as the $5f$ series. Yet true 5f character is found only in the second half of the series, americium being a borderline case. Except for some work on ^{243}Am, the heavy actinides are not accessible to Mössbauer studies because of their short life times. The workhorse of actinide Mössbauer research is ^{237}Np. Protactinium has found few applications. The three Mössbauer isotopes of uranium ($^{234,236,238}U$) are cumbersome to work with, due to severe source problems combined with poor resolution and the absence of IS. Hardly any Mössbauer data exists for this technically so important element. Plutonium where the $5f$-like nature of electron structure begins to develop would be an interesting candidate, yet technical problems are severe and resolution is poor. Even the Mössbauer work on ^{237}Np has largely come to a close, despite many open interesting problems. Safety

---

[6]Details and explanations can be found in book Chaps. 11 and 12.

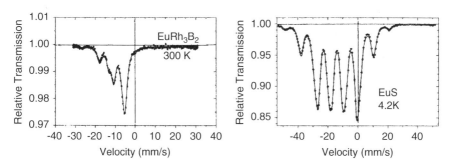

Fig. 8.7 Hyperfine spectra of ^{151}Eu in high resolution. *Left*: Quadrupole splitting. The interaction leads to eight lines, with three lines overlapping at the big peak and two lines at the peak around -17 mm s^{-1}. *Right*: Magnetic splitting. The interaction leads to 18 lines, but several of the lines overlap, leading in effect to an eight line Zeeman pattern. Visible is also a large IS because the absorbers are Eu^{2+}, the single line sources Eu^{3+} compounds. After [52,53]

regulations got so severe in the past decades that the use of this formidable resonance is now restricted to specially equipped laboratories like the European Transuranium Institute in Karlsruhe.

8.2.4.1 $^{151/153}$Eu

Of the two useful Mössbauer resonances, the 21.5 keV, 7/2–5/2, M1 transition in ^{151}Eu and the 103 keV, 3/2–5/2, M1 transition in ^{153}Eu, the former is generally used since it provides access to temperatures well above room temperature and has excellent resolving power for isomer shifts and hyperfine splittings, as demonstrated in Fig. 8.7.

The IS of divalent ($4f^7$) and trivalent ($4f^6$) europium compounds are located in two widely separated groups [54]. Between Eu^{2+} and Eu^{3+} ρ_0 increases since the loss of one f-electron to the ligand reduces the shielding of s-electron density at the Eu nucleus. The fairly narrow width of the two IS regions indicates weak covalency effects. Metallic compounds are usually slightly shifted to higher charge densities than the corresponding ionic compounds showing that the conduction band has a fair amount of s-character.

An interesting case is Eu$_3$S$_4$ which contains Eu^{2+} and Eu^{3+} on the same lattice site and is better described as (Eu^{2+}Eu$_2^{3+}$)S$_4$. Electrons hop back and forth between divalent and trivalent ions. As shown in Fig. 8.8, left, the temperature-dependent electron hopping rate severely influences the shape of the Mössbauer spectrum [55]. At low temperatures, hopping is so slow that on the Mössbauer time scale (given here mainly by the energy of the electric monopole interaction) the two ionic states appear fixed, resulting in a spectrum consisting of the well-separated Eu^{2+} and Eu^{3+} lines. At intermediate temperatures ($\sim$200 K), the timescale of the hoping process is comparable to the Mössbauer timescale and severe relaxation broadening

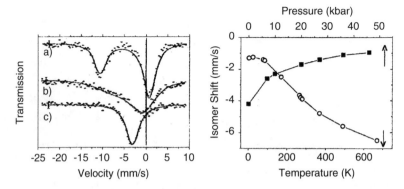

Fig. 8.8 *Left*: ^{151}Eu spectra of Eu$_3$S$_4$ at different temperatures showing electron hopping: (**a**) 83 K, $\tau = 10^{-7}$ s; (**b**) 228 K, $\tau = 1.7 \times 10^{-9}$ s; (**c**) 325 K, $\tau = 1.4 \times 10^{-11}$ s, with τ being the time constant used in the fits of the hopping process to the Mössbauer spectra. After [55]. *Right*: Isomer shift of EuCu$_2$Si$_2$ as function of temperature at ambient pressure (*open circles*, lower scale) and as a function of applied pressure at 300 K (*full squares*, upper scale). The lines are guides to the eye. More negative shifts indicate a move toward pure Eu^{2+}. After [57]

of the two resonance lines occurs which can no longer be separated. At room temperature and above, the hopping process is fast, initiating motional narrowing to a single resonance line positioned at the weighted average position of the two lines at low temperature. Using a model developed to describe Mössbauer spectra in the presence of spin fluctuations [56], the time constants of the electron hopping process could be determined.

A somewhat different situation exists in some Eu intermetallics. The charge fluctuation rate is at all temperatures fast, meaning that one is always in the "single line" motional narrowed limit. External parameters like temperature or pressure influence the ratio of the occupation numbers of the charge states involved. This ratio determines the effective valence v of the Eu ion which is directly related to the IS of the single resonance line. The measurements [57] depicted in Fig. 8.8, right show that temperature and pressure dependences of Eu valence are of opposite sign. Using the established IS scale [54], one finds that the valence decreases from $v \approx 2.8$ at 4.2 K to ≈ 2.35 at 673 K, while it increases at room temperature from 2.6 to 2.9 when 50 kbar pressure is applied. This behavior is discussed [58] in terms of an energy level and band diagram for metallic europium systems. Intermediate valence is found in other rare earths as well, especially in Yb, but also in Sm and Tm. The excellent features of the ^{151}Eu resonance makes this Mössbauer isotope the prime choice for detailed studies.

Investigations of magnetic properties is another important realm of ^{151}Eu Mössbauer spectroscopy, especially so, because the large neutron capture cross section of the Eu isotopes limits severely studies by neutron diffraction. Of course, Mössbauer spectra cannot give directly the spatial arrangement of magnetic moments (it works in real, not in reciprocal space), but render important information on local magnetic behavior and on temporal properties of the magnetic moments.

The two charge states of Eu differ fundamentally in their magnetic behavior. Eu^{2+} has a half-filled 4f shell and hence no orbital momentum. The atomic ground state is $^8S_{7/2}$, leading to pure spin magnetism. In Eu^{3+} the ground state is $^7F_0$ and, because of the zero total angular momentum J, non magnetic. A pertinent example is Eu_3O_4 containing divalent and trivalent Eu ions. The compound is an antiferromagnet. Below T_N, only the Eu^{2+} ions order ($B_{hf}(0) = 30.5\,T$) [59]. This is in contrast to Fe_3O_4, where both iron species order, resulting in a ferrimagnet. The europium iron garnet ($Eu_3Fe_5O_{12}$) with two Eu^{3+} sites shows magnetic ordering which involves not only the iron but also the europium. The large Eu hyperfine fields differ slightly for the two sites (63 T and 57 T) [60]. Magnetic Eu^{3+} ions result from an admixture of the energetically close excited atomic states F_1 and F_2 via exchange and crystalline electric field interactions.

Within the divalent compounds, the most important series are the monochalco-genides EuX (X = O, S, Se, Te), all ordering magnetically. Being simple cubic (NaCl structure) spin magnets, they are the best real system for Heisenberg magnets. With the help of Mössbauer data, particularly under high pressure and in high fields, a consistent model of their magnetism has been developed [61], based on the interplay between the nearest neighbor exchange J_1 and the next nearest neighbor superexchange J_2. The resulting dependence on lattice constant is shown in Fig. 8.9, left. Positive J_1 and J_2 are found for EuO which is a ferromagnet (FM) with $T_C = 69\,K$. J_1 in EuTe is still positive but weaker, J_2 slightly negative. The former wins and EuTe is still a FM but with reduced T_C (17 K). In EuSe the FM (J_1) and the antiferromagnetic (AFM) (J_2) exchange parameters are both weak, but nearly balance each other. The result is metamagnetic behavior, i.e., an AFM ($T_N = 4.6\,K$) whose spin arrangement is highly sensitive to applied fields. EuTe is a type I AFM with $T_N = 10\,K$ since J_1 and J_2 are negative. This model also explains the results of a high pressure ^{151}Eu Mössbauer study [62] on magnetic transition temperatures (Fig. 8.9, right). All exchange parameters increase with reduced lattice constant.

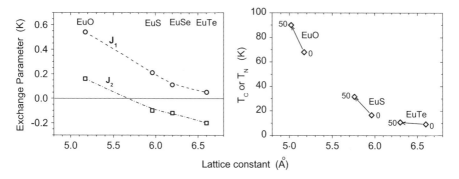

Fig. 8.9 *Left*: Dependence of nn exchange (J_1) and nnn superexchange (J_2) parameters on lattice constant. After [61]. *Right*: Dependence of the magnetic transition temperature in europium mono chalcogenides as function of lattice constant. The numbers indicate the applied pressure. Data were taken with the ^{151}Eu Mössbauer resonance. After [62]

The rise of the dominating J_1 exchange in EuO and EuS then causes the increase of T_C with pressure, the effect being a bit less pronounced in EuS because of the influence of J_2. The dominating J_2 in EuTe rises less steep with reduced volume and T_N shows only a small rise.

8.2.4.2 ^{155}Gd

Of the eight possible Mössbauer transitions among the stable Gd isotopes, only the 86 keV, E1, 3/2–5/2 transition in ^{155}Gd has found wider applications. Its E1 character requires a modification of the normally Lorentzian-shaped resonance lines by adding a dispersion term arising from an interference effect between photo- and conversion electrons. In ^{155}Gd, this dispersion term is small and often neglected without serious consequences. Dispersion, however, is important in low-energy E1 transitions like the 6.2 kev resonance in ^{181}Ta.[7]

Hyperfine splittings in ^{155}Gd are only moderately resolved and the IS scale is small in comparison to Eu. Gd salts cover an extremely narrow range of IS, confirming that they all contain ionic Gd^{3+} with little covalency. The range of IS for metallic compounds is larger than that of the salts, showing that a change from nearly pure $6s$ bands to bands including an admixture with $5d$- or $6p$-electron densities takes place [63]. Gd^{3+} is equivalent to Eu^{2+}, i.e., it has a half filled $4f$ shell and is a pure spin state ($S = 7/2$). Normally, the efg at the nucleus of a rare earth atom has its main source in the nonspherical open 4f shell charge distribution. This being absent in Gd makes the sole source of the efg at the Gd nucleus the charges on surrounding ions. Measured values of the efg in several Gd compounds could indeed be explained well by a simple point charge model. Magnetic studies are the dominating application, since here again large neutron absorption of Gd isotopes make diffraction studies practically impossible. The landscape of Gd magnetism is less colorful since only the trivalent ion exists. The main contribution to the hyperfine field at the Gd nucleus is the Fermi contact term. Systematic studies in combination with NMR data revealed for Gd^{3+} a contact field of 34 T, a value about three times smaller than found in the $5d$ series (e.g., Ir). Studies of the high T_c superconductor $GdBa_2Cu_3O_7$ ($T_c > 90$ K) confirmed magnetic ordering at 2.25 K and showed that the hyperfine field is parallel to the c axis, proving that exchange interaction is effective for the superconducting state.

8.2.4.3 ^{161}Dy

Generally used is the 25.7 keV, 5/2–5/2, E1 transition in ^{161}Dy. Single line sources (e.g. ^{161}GdF$_3$) give line widths $\sim$10 times the natural width, yet this still corresponds

---

[7] see book Chap. 9 where also a typical dispersion modified ^{181}Ta spectrum is shown.

to an energy spread of about one order of magnitude less than the large hyperfine splittings caused by the orbital contributions of the open $4f$ shell.

Isomer shifts for Dy compounds and alloys are small like in ^{155}Gd. Magnetic studies are once more the main applications. The ground state of Dy^{3+} is $^6H_{15/2}$. In magnetically ordered materials, one nearly always observes a hyperfine field around 580 T, which corresponds to the Dy^{3+} free ion value, and one also observes, even in cubic local symmetry, the so-called "induced quadrupole interaction" arising from the alignment of the non spherical $4f$-orbitals by the exchange field. Deviation from the free ion field occurs, if a $3d$ transition element is a ligand, because of the additional transferred field. A hyperfine field corresponding to the free ion value means that the electronic ground state is an essentially isolated $J_z = 15/2$ state, or, in other words, that the effect of the crystalline electric field is only a perturbation with respect to the strong exchange field. The high angular momentum hinders fluctuations between the $\pm 15/2$ levels separated by exchange interaction. If the rate of these fluctuations becomes comparable to the nuclear Zeeman energy, relaxation Mössbauer spectra are observed. The dominant effect is substantial asymmetric line broadenings [64]. This is commonly termed "ferromagnetic relaxation", but it applies to all sorts of long-range magnetic order. Mössbauer relaxation spectra can also develop in the paramagnetic regime. Fluctuations of the paramagnetic moments may slow down and motional averaging into a single resonance line (or a pure quadrupole pattern) will no longer be effective. The effect is called "paramagnetic hyperfine splitting". If the transition temperature from the paramagnetic into an ordered magnetic state is low, it can happen that paramagnetic relaxation has reached its slow fluctuation limit before. Then the spectra above and below the ordering temperature are indistinguishable and Mössbauer data are unable to pin the transition point. A pertinent case is the Mössbauer study of DyAg in its crystalline and amorphous form [65], where ferromagnetic and paramagnetic relaxation spectra are observed. A detailed discussion together with spectra can be found in book Chap. 12.

8.2.4.4 ^{166}Er

The workhorse is the 80.6 keV, 2–0, E2 transition in ^{166}Er fed by β-decay of 27 h, ^{166}Ho. The 80.6 keV state is a collective (rotational) nuclear excitation where the quantum numbers of the nucleons do not change. Consequently, no change of nuclear radius on excitation is expected, but higher order terms produce a tiny value of $\Delta <r^2>$. Nevertheless, the observed IS are too small to be used for elucidating chemical bond problems. E2 gamma radiation also gives rise to a dispersion term in the shape of the absorption line. Its negligence or improper inclusion may produce spurious IS.

Hyperfine splitting takes place only in the $I = 2$ excited state. This leads to a simple hyperfine spectra which can be resolved despite the fact that observable linewidths are $\sim 4 W_0$. The magnetic interaction pattern consists of five equally spaced lines. Studies of quadrupole splittings serve mainly to gain information on

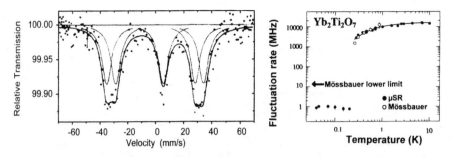

Fig. 8.10 *Left*: ^{166}Er hyperfine spectrum of $Y_{0.976}Er_{0.024}(C_2H_5SO_4) \cdot 9H_2O$ at 4.2 K. The *solid line* is the least squares fit to five Lorentzians of equal hight and width. After [66]. *Right*: Temperature variation of the Yb^{3+} fluctuation rate in $Yb_2Ti_2O_7$ obtained from ^{170}Yb Mössbauer spectroscopy and μSR measurements. The *solid line* follows a thermal excitation law described in [75]. The *arrow* gives the lowest fluctuation rate which can be resolved from ^{170}Yb Mössbauer spectra. After [75]

crystalline electric field interactions. As in Dy, magnetic materials give hyperfine fields very close to the free ion value, showing that the maximal J_z state is lowest. The Er^{3+} angular momentum is the same as that of Dy^{3+} (i.e. $J = 15/2$) and magnetic relaxation is also prominent in Er compounds. In fact, $ErFeO_3$ was the material where ferromagnetic relaxation was first observed [64] and interpreted.[8] Also encountered are paramagnetic hyperfine spectra. To be in their static limit similar in appearance to the effective field spectra in ordered magnets, a highly anisotropic hyperfine tensor is required. This condition is often fulfilled for the trivalent heavy rare earth ions (except Eu and Yb) due to the dominance of the 4f orbital term. Diluting erbium ethylsulfate highly with yttrium ethylsulfate produces at 4.2 K paramagnetic hyperfine spectra where the five Zeeman lines are distinctly unequally spaced (see Fig. 8.10, left). The origin is a breakdown of the effective field approximation, revealing a not fully anisotropic hyperfine tensor [66].

8.2.4.5 ^{169}Tm

The 8.4 keV, 3/2–1/2, M1 transition is also a rotational excitation and IS are too small to extract chemical information. Studies of quadrupole interactions rendered information on the crystalline electric field states. Magnetic materials, e.g., thulium metal, give the free ion field. The magnetic structure of thulium metal is complex. Below $T_N = 56$ K, a sinusoidally modulated structure with a seven lattice layer period is formed. Then, between 40 K and 25 K the sinusoidal structure squares up, ending below 25 K in a ferrimagnetic structure ($4\uparrow 3\downarrow$). In the ferrimagnetic regime, the magnitude of all Tm moments is the same, creating a sharp single

---

[8]Spectra are shown in book Chap. 11.

Zeeman pattern [67]. Above 25 K, the moment magnitudes are distributed according to the more and more sinusoidal modulation. The magnetic spectra now consist of broadened, only partially resolved lines. In particular, an additional single line grows in at the center, signaling that a portion of Tm nuclei sees no hyperfine field at all [68]. The shape of the spectra is quite similar[9] to ferromagnetic relaxation spectra. It is difficult, if not impossible, to distinguish in practice between hyperfine patterns created by a static distribution of magnetic moments, or by dynamic relaxation between different magnetic states. In fact, the spectra of Tm metal above 25 K have been more recently analyzed [69] within the the frame of relaxation theory, giving good results. In view of the neutron data, a static field distribution is certainly present, but additional relaxation phenomena most likely as well. To achieve a distinction between static and dynamic spectral broadening, application of an external field may be helpful, especially if single crystal specimen are available.

8.2.4.6 ^{170}Yb

The 84.3 keV transition in ^{170}Yb has similar properties as the 80.6 keV transition in ^{166}Er. A second possibility is the 3/2–1/2, M1+0.48E2, 66.7 keV resonance in ^{171}Yb (having higher isotopic abundance but a slightly poorer linewidth). It is, however, also a rotational excitation and no help in obtaining useful IS data. Ytterbium exists as divalent and trivalent ion, analogous to europium. The distinction between these charge states rests on different hyperfine parameters. In Yb^{2+} the $4f$ shell is filled ($4f^{14}$) leading to the nonmagnetic electronic ground state $^1S_0$. Also quadrupole splittings are small due to the absence of an efg from $4f$ electrons. In contrast, Yb^{3+} has an open $4f$ shell ($^2F_{7/2} : 4f^{13}$) and substantial field gradients as well as hyperfine fields can exist.

Ytterbium is also a candidate for intermediate valence behavior. An example is the study of the temperature and pressure dependence of the quadrupole interaction in YbCuAl [70]. At ambient pressure the quadrupole splitting decreases weakly when raising the temperature from 4.2 K to 80 K. Under 130 kbar, the three times larger decrease indicates a 4f contribution to the efg, meaning a valence shift toward Yb^{3+}.

Both Yb resonances have largely been used to study spin-lattice and spin-spin relaxation processes (in magnetism, the term "spin" is loosely used to mean the total angular momentum generating the atomic magnetic moment, rather than its spin part only). ^{171}Yb has the same nuclear spin sequence (3/2–1/2) as ^{57}Fe allowing the use of the extensive theoretical formalism developed for relaxation processes in iron compounds. The 2–0 spin sequence in ^{170}Yb leads to simple hyperfine couplings, making theoretical treatments easy. It became the workhorse for relaxation studies. In $YbCl_6 \cdot 6H_2O$, the variation of relaxation rates was studied over a wide temperature range [71]. At high temperatures the Raman

---

[9]The spectra are depicted in book Chap. 11.

spin-lattice relaxation with its typical $\tau \propto T^{-9}$ dependence dominates. Below
~20 K, temperature independent spin-spin relaxation ($\tau \approx 6 \times 10^{-9}$ s) wins. Unusual
results were found in $Cs_2NaYbCl_6$. Standard relaxation analysis returned a value
for the hyperfine parameter 4% greater than the free ion value, a nonphysical
result. The matter was explained as a break down of the commonly used "white
noise approximation" [72]. One of the consequences is, that the relaxation rate
becomes frequency dependent. Very strong sources of 130 d, ^{170}Tm can be made by
neutron activation of $TmAl_2$. In combination with high speed counting systems
[73], it was then possible to obtain very high quality spectra of $Cs_2NaYbCl_6$
which allowed to derive by direct inversion the frequency dependence of the
spin-spin relaxation rate [74] which possessed a distinct bimodal structure.[10] The
spin dynamics in the geometrically frustrated magnetic compound $Yb_2Ti_2O_7$ was
investigated [75] combining Mössbauer and μSR measurements. The two methods
cover different ranges of spin fluctuation rates [76] as is demonstrated in Fig. 8.10,
right. Characteristic for frustrated magnetic systems is a persistent slow fluctuation
rate at low temperatures, indicating quantum or tunneling relaxation processes. The
jump in rate at 0.24 K indicates a first order transition of the Yb^{3+} spin relaxation
rate into the quantum fluctuation regime [75].

A special relaxation process emerged when comparing the emission spectrum of
a ^{170}TmFe$_2$ source against a single line ^{170}Yb absorber to the ^{169}Tm absorption
spectrum of TmFe$_2$ [77]. The additional central resonance seen in the source
spectrum could be reproduced by a model calculation based on relaxation between
excited atomic levels populated non-thermaly by the preceding β-decay [78].

A number of metallic Yb compounds fall into the category of strongly correlated
electron systems. ^{170}Yb Mössbauer studies on YbSi, a heavy electron compound,
confirmed magnetic order below 1.6 K, albeit with a strong reduction in magnetic
moment. It was shown that a competition between RKKY exchange and Kondo
coupling exists, leading to Kondo-frustrated magnetism in the ordered regime [79].

8.2.4.7 ^{237}Np

In theory, the 5/2–5/2, E1, 60 keV transition in ^{237}Np has about the same resolution
as ^{57}Fe, yet in practice, the commonly used parent activity 458 y ^{241}Am in form of
Am metal gives about 30 times the minimum width. This is not a serious drawback,
since the hyperfine interaction energies are more than an order of magnitude larger
than in iron. Well-resolved isomer shifts and hyperfine split spectra are observed.
Actually, it would be difficult to trace such widely separated hyperfine lines with
resonances of natural width.

Neptunium comes in the valence states Np^{3+} to Np^{7+} with the ionic configurations $5f^4$ to $5f^0$. The basic results of IS studies [80] are depicted in Fig. 8.11, left.

---

[10]More details on non white noise relaxation and the case of $Cs_2NaYbCl_6$ can be found in book
Chap. 12 where the measured frequency variation of the relaxation rate is also shown.

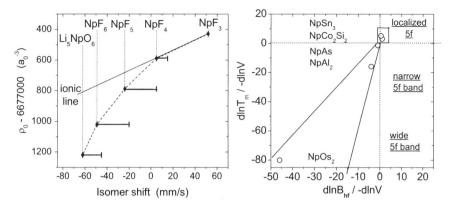

Fig. 8.11 *Left*: Isomer shifts (relative to Am metal) for several neptunium fluorides (except NpF$_7$ which does not exist) compared to theoretical calculations of ρ_0 for pure f^n states. The scales of ρ_0 and the isomer shift were matched by assuming that NpF$_3$ is fully ionic ($5f^4$). The horizontal lines indicate the spread of IS for a given valence due to covalency. After [81]. *Right*: Plot of the volume coefficient of magnetic ordering temperature against the volume coefficient of the hyperfine field for some Np intermetallics. After [82]

Compounds of different Np valence are located in groups on the IS scale. The groups Np^{3+} and Np^{4+} are narrow, those of the higher charge states become increasingly wider and overlap in part. Shown is further the change in ρ_0 for the pure $5f^n$ configurations obtained from Dirac-Fock calculations. Even for the simple fluorides NpF$_n$ which are regarded the most ionic compounds, the deviation from the ionic line is substantial for n > 4. In the higher valence states, only covalent bonds are possible, especially so for the heptavalent species. A $5f^0$ configuration is far from reality. Correlations between IS, quadrupole splitting, and actinide bond length have shown that valence electron structures in Np^{7+} species are basically similar to those of Np^{6+} neptunyls[11] minus the single f-electron in its non-bonding orbital [81].

Conducting materials are considered to have the Np^{3+} core. Through the influence of the conduction electrons their IS are displaced toward higher ρ_0, i.e., well into the Np^{4+} range. Values of hyperfine fields in magnetically ordered specimens are widely distributed. The control parameter is the actinide-actinide separation. The closer the actinides are to each other, the stronger the overlap of $5f$ orbitals. The 5f bands get wider and the magnetic behavior less localized. For example, the Laves phase intermetallics (e.g., NpAl$_2$), show a linear decrease of B$_{hf}$ with reduced d$_{Np}$, the Np–Np separation. Finally B$_{hf}$ vanishes altogether at d$_{Np} \approx 3.23$ Å, the so-called Hill-limit. Below the Hill limit, the $5f$ overlap generates bands too wide to support magnetic long-range order at all. Even in NpAl$_2$, the most localized Laves phase compound, B$_{hf}$ reaches only ∼2/3 of the Np^{3+} free ion value. The neptunium monochalcogenides have wider Np–Np separations. They also show an

---

[11]Neptunyls are a large series of materials where the ligand is (Np^{5+}O$_2$)$^+$ or (Np^{6+}O$_2$)$^{++}$.

increase of B_{hf} with the lattice constant a, attaining at $a = 5.8$ Å the free ion value. In keeping is the observation that the IS moves with rising B_{hf} toward the ionic Np^{3+} value. High pressure measurements on magnetic compounds are the ideal means to distinguish between localized and band-like behavior. Fig. 8.11, right summarizes the basic features [82]. In localized compounds like $NpCo_2Si_2$, the hyperfine field is essentially constant with reduced volume, while the transition temperature T_m rises somewhat due to the increase of RKKY exchange coupling. In contrast, when large decreases of B_{hf} and T_m are observed, one deals with increasingly itinerant band magnetism. An interesting case is $NpSn_3$ which originally was considered an itinerant magnet. Yet, high pressure data show unambiguously that $NpSn_3$ is a localized magnet (see Fig. 8.11, right). In a typical "modern" compound, superconducting $NpPd_5Al_2$, the absence of an ordered moment was detected and the charge state of the Np ion determined. The results were compared to theoretical electronic structure calculations based on Kondo behavior [83].

Relaxation phenomena are encountered in some magnetic neptunium compounds, especially in the Laves phases just below the ordering temperature [84]. Paramagnetic hyperfine splittings were observed, for example, in the hexavalent neptunyl compound $[(C_2H_5)_4N]_2(NpO_2)Cl_4$. At 4.2 K, the static limit is reached. The spectral shape reveals immediately a noneffective magnetic interaction. In contrast to EPR data on related neptunyls, Mössbauer resonance was able to determine the sign of all hyperfine parameters and detected further a small asymmetry parameter in the quadrupole interaction which points towards a slight deviation form the generally accepted rotational symmetry of the linear neptunyl (O−Np−O) ion [81].

8.2.5 High Resolution Mössbauer Resonances

Although [57]Fe offered at the time an unprecedented energy resolution, the quest for an even higher resolving power was nourished by the hope to get more precise results on relativistic problems or to find new effects in solid state physics. Available candidates are listed in Table 8.1 together with relevant parameters. Only ^{67}Zn has found applications in relativity as discussed in book Chap. 14. The Zn resonance, which belongs to the more commonly used Mössbauer isotope, has produced in

Table 8.1 Parameters of high resolution Mössbauer resonances. W_0 = minimum observable width

Isotope	E_γ (keV)	$t_{1/2}$	W_0 (mm s^{-1})	W_0 (eV)	$Q = E/W_0$
^{57}Fe	14.41	97 ns	0.2	1×10^{-8}	1.4×10^{10}
^{67}Zn	93.3	9 μs	3.2×10^{-4}	1×10^{-10}	9×10^{14}
^{181}Ta	6.24	7 μs	6.5×10^{-3}	1.2×10^{-10}	5×10^{13}
^{73}Ge	13.3	3 μs	7×10^{-3}	3×10^{-10}	5×10^{13}
^{109}Ag	88	40 s	8×10^{-11}	2.3×10^{-17}	4×10^{21}

addition important results in solid state physics and metallurgy. [181]Ta has also been of use for electronic structure investigations.

8.2.5.1 ^{67}Zn

High resolution is not the only attractive feature of the 93 keV ^{67}Zn resonance. The high transition energy in combination with the low atomic mass makes this Mössbauer transition most suited for elucidating lattice dynamical properties via the measurement of the recoil free fraction f and the second order Doppler shift (SOD). The f-factor and the SOD complement each other, since they weigh the phonon density of states $D(\omega)$ differently, and both differ in that respect from the specific heat C_V [85]. The drawback of high γ-ray energy is the upper temperature limit for data taking. Measurements usually extend from 1.5 K to $\sim$100 K. Zn metal has also been studied at dilution refrigerator temperatures. Single line sources giving only moderately broadened lines ($\sim$3 W_0) exist (e.g. ^{67}Ga in Cu). The Doppler velocity needed to scan a typical hyperfine spectrum is extremely small (100 μm s^{-1}). A special drive system based on the piezoelectric effect of quartz was developed. The piezo drive and the absorber holder, which can be extended to high pressure and heating devices, form one rigid unit. This is mounted with decoupling from outside vibrations inside the He cryostat [86]. These features allowed to investigate the temperature and volume dependences of hyperfine interactions and lattice dynamic properties in various Zn-containing systems, especially in Zn metal and alloys, the chalcogenides and oxide-spinels. The Mössbauer measurements were accompanied by ab initio theoretical calculations.

Studies of the IS are not the major part of Mössbauer work on ^{67}Zn. The reason is the same as in ^{61}Ni, namely that the observed Center Shift (CS) of a spectrum contains the true IS and the SOD in about equal parts. Their separation is in most cases not easy. Systematic studies of IS exist for the chalcogenides. They showed a linear correlation between decreasing IS and increasing electronegativity of the ligands. The most important contribution to this variation in the IS comes from a partial removal of 4s-electrons combined with a smaller contribution from 3s-electrons due to the repulsive influence of the ligand ion orbitals [87].

Hyperfine studies are limited to quadrupole interaction. Magnetic splitting by internal fields has not been observed. Only the Zeeman splitting in an external field of 21 mT has been reported and served to obtain the value of the excited state magnetic moment [88]. The 1/2–5/2 spin sequence of the 93 keV, E2 transition leads to rather simple quadrupole spectra which are easily resolved and allow the determination of the magnitude, the sign and the asymmetry parameter of the efg. Most thoroughly studied in this respect is, in combination with lattice dynamical measurements, hcp ZnO, using especially single crystalline specimen. The experiments included the volume dependences of the efg, the f-factor and the SOD. It was further possible to trace the incomplete phase transition of ZnO from hcp to fcc (NaCl) in the pressure range between 65 and 100 kbar [86]. In the oxide spinels, the changes in local symmetry when moving from the normal to the inverted configuration were traced. Mössbauer spectroscopy on ZnO and the normal spinel $ZnFe_2O_4$ has been extended to nanocrystalline samples together with neutron diffraction and μSR measurements [89].

The various crystallographic phases of the Cu–Zn alloy system (brass) are usually considered to be the model for the Hume–Rothery rules which connect structural changes with the average electron concentration. The Zn Mössbauer study of fcc α-brass gave the totally unexpected result that short range order is present, even at low Zn concentrations [85]. In addition to the expected single line spectrum, an additional quadrupole spectrum typical for the crystallographically ordered β'-phase was present. Increasing the Zn concentration within the range of the α-phase enhanced the relative strength of the quadrupole pattern but did not alter its parameters. The Mössbauer data demonstrate clearly that α-brass cannot be considered a uniform binary alloy with a random distribution of Cu and Zn atoms. It appears that in Cu–Zn alloys an inherent tendency exists to form a Cu_3Zn super-structure. Other commonly used methods to check for short-range crystallographic order in brass, like X-ray or neutron diffraction, will not work well, since for both methods, the scattering intensities from Cu and Zn atoms are about equal.

Lattice dynamical properties have extensively been studied in Zn metal via measurements of the temperature, angular and volume dependences of the f-factor and the SOD [85, 86]. Zn metal has the hcp structure with an unusually large c/a ratio (1.86) leading to highly anisotropic solid state properties and also to a large electric field gradient. Under ambient pressure, one finds for the efg at the ^{67}Zn nucleus $V_{zz} = 3.42 \times 10^{17}\,V\,cm^{-1}$ and $\eta = 0$ independent of temperature up to 47 K. Under pressure, the efg decreases linearly with reduced volume as shown in the lower panel of Fig. 8.12, left. This behavior reflects a move of the c/a ratio toward the ideal value $c/a = \sqrt{8/3} = 1.633$. Also plotted are the results of a relativistic Linearized Augmented Plane Wave (LAPW) calculation which produces the trend in variation of the efg quite well. The deviation in the absolute magnitude of the efg rests most likely in the 10% uncertainty of the value of the excited state quadrupole moment. As expected, the anisotropy in the f-factor is enormous. The ratio $f_\perp/f_\parallel$ with respect to the c-axis is $\sim$23 at 4.2 K but rises to $\sim$2100 at 47 K. The f-factor as well as the SOD data can be described by an extended Debye model based on two cut-off temperatures $\Theta_\perp = 242$ K and $\Theta_\parallel = 149$ K. However, the results are not well reproduced by the Axially Symmetric Model (MAS) of $D(\omega)$ derived from inelastic neutron scattering. The measurement of the volume dependence of the f-factor rendered the striking result depicted in the upper panel of Fig. 8.12, left. After a first sharp rise, the f-factor suddenly drops at $V/V_0 = 0.915$ by a factor of two. No crystallographic phase transition occurs at this point, since neither the CS nor V_{zz} (see lower panel of Fig. 8.12, left) show any irregularity. The cause lies in a drastic softening of low frequency acoustic and optical phonons. These phonons are hardened at low pressure by the so-called giant Kohn anomaly [91]. Calculations of the conduction band electronic structure [92] indicate the destruction of the giant Kohn anomaly at reduced volume by an electronic topological transition. Finally, we mention the measurement on Zn metal below its superconducting transition

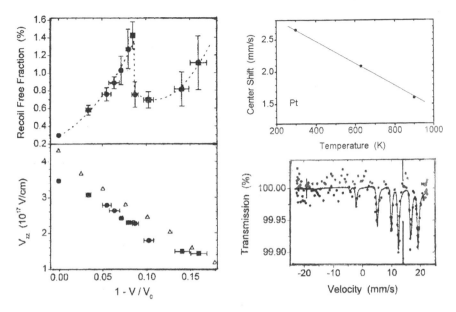

Fig. 8.12 *Left*: ^{67}Zn – Volume dependences of the recoil free fraction (*top*) and of the main component of the electric field gradient (*bottom*) in Zn metal. *Filled symbols* are measured data. The *dashed line* is a guide to the eye. Open triangles are theoretical calculations (LAPW). After [90]. *Right*: ^{181}Ta – *Top*: Temperature dependence of the center shift of ^{181}W in Pt as a source against a Ta metal absorber. *Bottom*: Quadrupole spectrum of ^{181}W in single crystalline Rhenium metal source against a Ta metal absorber. The *vertical line* indicates the IS. After [94]

temperature of 835 mK. No difference in CS or V_{zz} was seen between the normal and superconducting states.

8.2.5.2 ^{181}Ta

121 d ^{181}W is the commonly used parent activity for the 6.2 keV, 7/2–9/2, E1 resonance in ^{181}Ta. Even if natural line width W_0 were obtainable for ^{181}Ta, it remains inferior in resolution to the 93 keV transition in ^{67}Zn. In practice line, broadenings by a factor 10 (best result) to over 100 are encountered. This loss of resolution is not serious for condensed matter studies since the nuclear hyperfine parameters involved are large. IS extend to $\sim$100 mm s^{-1}. Hyperfine splittings like the quadrupole spectrum shown in Fig. 8.12, right can also be sufficiently resolved. The real problem of the broadened lines is the corresponding severe reduction in resonance absorption. The line shape of the 6.2 keV resonance is characterized by a strong dispersion term [93] which needs to be taken into account.

Most studies using the ^{181}Ta resonance centered on IS measurements [94], prominently on Ta as dilute impurity in various d-elements in form of source experiments. A similar relation between IS and the number of outer electrons as already discussed in the paragraph on ^{99}Ru was established. It appears that this behavior is a general feature in all d-transition elements. Hydrogen dissolved in Ta metal is another subject of interest [94]. A strong linear dependence of IS with hydrogen content (up to 15 at.%) was seen. One mechanism is the lattice expansion induced by hydrogen absorption. Comparison to high pressure measurements (up to 5 kbar in an oil pressure cell) showed that lattice expansion accounts only for 2/3 of the observed effect. The rest reflects changes in the electronic structure of the Ta host, possibly by filling up empty states in the valence band by hydrogen electrons. ^{181}Ta spectra also exhibit a strong temperature dependence of CS [95]. A temperature-dependent shift is of course present in all Mössbauer resonances, but reflects essentially the SOD. In ^{181}Ta, the SOD is small due to the large nuclear mass and the low transition energy, therefore the true temperature dependence of the IS dominates. Fig. 8.12, right shows the case for a source experiment with dilute Ta in Pt. Here the SOD accounts only for about 10% of the observed variation, the rest being the temperature dependence of IS (e.g. ρ_0). The trivial effect is the change in s-electron density due to thermal lattice expansion, which can be obtained from high pressure experiments. The remaining variation (amounting to $\sim$50% of the total shift) must be due to a change in electron structure. The most probable explanation is that the valence bands become more free-electron like due to electron–phonon coupling [96].

Quadrupole spectra like the one depicted in Fig. 8.12, right have been observed in some Ta compounds (e.g., LiTaO$_3$) as well, but were not specifically interpreted. The magnetic hyperfine spectrum of Ta in Ni gave B$_{hf}$ $\approx$9 T, a value comparable to the results in other d-elements.

8.2.5.3 ^{73}Ge

The 13 keV, 9/2–5/2, E2 transition in ^{73}Ge also has a lower resolving power than ^{67}Zn. Yet, one should think this Mössbauer isotope to be a bonanza allowing to study the technologically so important germanium. This never materialized. Nearly all work with ^{73}Ge centered on taming this stubborn resonance [97]. A major problem is the huge internal conversion coefficient ($\alpha = 1100$) making even a strong sources of 110 d ^{73}As a weak emitter of the resonant γ-radiation. Furthermore, it reduces severely the resonance absorption cross section, while the electronic absorption cross section for the 13 keV radiation, which is close in energy to the K-edge of Ge, is high. Absorbers must be very thin (e.g., epitaxial deposition) to get any resonant radiation transmitted, which reduces the Mössbauer absorption strength. Also, even with high resolution γ-ray detectors, the signal to noise ratio for the 13 keV γ-line is poor. An electromechanical velocity drive can be used if decoupled carefully from vibrations. Without special care as to the quality of source and absorber materials severe line broadenings occur, but a resonance line close to natural width could be realized. Isomer shifts have been observed, but only for different source matrices (e.g., Ge vs. Cu) [98].

8.2.5.4 $^{107/109}$Ag

Normally used is ^{109}Ag which has the longer lived parent (461 d ^{109}Cd). The extremely narrow width rules out a normal Doppler shift experiment. One has to rely on measurements of resonant self-absorption. Whether an effect has been found is controversial. The most recent works using ^{109}Cd-doped silver report a decrease of self-absorption of the 80 keV radiation when changing from horizontal to vertical beam geometry [99], similarly for heating from 4.2 K to 77 K and finally as a function of inclination angle [100]. The observed effects are barely outside the one sigma error. In the inclination experiment, a sensitivity to the gravitational red shift eight orders of magnitude higher than for ^{57}Fe is claimed.

References

1. L.L. Lee Jr., L. Meyer-Schutzmeister, J.P. Schiffer, D. Vincent, Phys. Rev. Lett. **3**, 223 (1959)
2. F.E. Wagner, H. Spieler, D. Kucheida, P. Kienle, R. Wäppling, Z. Physik **254**, 112 (1972)
3. Mössbauer Effect Data Center, Dalian Institute of Chemical Physics, Dalian, China
4. G.M. Kalvius, F.E. Wagner, W. Potzel, J. Phys. (Paris) Colloq. **37**, C6-657 (1976)
5. G.K. Shenoy, F.E. Wagner (eds.), *Mössbauer Isomer Shifts* (North Holland, Amsterdam, 1978)
6. G.J. Lang (ed.), *Mössbauer Spectroscopy Applied to Inorganic Chemistry* vols. 1 and 2 (Plenum Press, New York, 1984/1987)
7. R. Barloutaud, J.L. Picou, C. Tsara, Compt. Rend. **250**, 2705 (1960)
8. A.J.F. Boyle, D.S.P. Bunburry, C. Edwards, H.E. Hall, Proc. Phys. Soc. London **76** 165 (1960)
9. N.N. Delyagin, V.S. Shpinel, V.A. Bryukhanov, B. Zvenglinski, JETP **12**, 159 (1961)
10. A.J.F. Boyle, private communication (2009)
11. A.J.F. Boyle, H.E. Hall, *Reports on Progress in Physics*, vol. 25, p. 441ff (The Institute of Physics, London, UK, 1962)
12. K.S. Singwi, A. Sjölander, Phys. Rev. **120**, 1093 (1960)
13. H.H. Wickman, G.K. Wertheim, in *Chemical Applications of Mössbauer Spectroscopy*, Chap. 11, ed. by V.I. Goldanskii, R.H. Herber (Academic, New York, 1968)
14. V.A. Bryukhanov, N.N. Delyagin, V.S. Spinel, JETP **20**, 55 (1965)
15. R.V. Parish, in *Mössbauer Spectroscopy Applied to Inorganic Chemistry*, vols. 1 and 2, Chap. 16, ed. by G.J. Lang (Plenum Press, New York, 1984)
16. J.G. Stevens, Mössbauer Effect Reference and Data Journal **33**(5), 86 (2010)
17. H.S. Möller, R.L. Mössbauer, Phys. Lett. **24A**, 416 (1967)
18. V.N. Samilov, V.V. Skliarevsky, E.P. Stepanov, JETP **11**, 261 (1960)
19. V.I. Goldanskii, V.A. Trukhtanov, M.N. Devisheva, V.F. Belov, JETP Lett. **1**, 19 (1965)
20. S.L. Ruby, G.K. Shenoy, in *Mössbauer Isomer Shifts*, Chap. 9b, ed. by F.E. Wagner (North Holland, Amsterdam, 1978)
21. J.G. Stevens, Mössbauer Effect Reference and Data Journal **33**(4), 72 (2010)
22. F.J. Berry, in *Mössbauer Spectroscopy Applied to Inorganic Chemistry*, vols. 1 and 2, Chap. 16, ed. by G.J. Lang (Plenum Press, New York, 1987)
23. R.B. Frankel, J.J. Huntzicker, D.A. Shirley, N.J. Stone, Phys. Lett. **26A** 452 (1968)
24. M. Pasternak, A.L. Spijvet, Phys. Rev. **181**, 574 (1969)
25. P. Jung, W. Triftshäuser, Phys. Rev. **175**, 512 (1968)
26. M. Pasternak, A. Simopoulos, Y. Hazony, Phys. Rev. **140**, A1892 (1965)
27. R.V. Parish, ref in *Mössbauer Spectroscopy Applied to Inorganic Chemistry*, vols. 1 and 2, Chap. 9, ed. by G.J. Lang (Plenum Press, New York, 1987)

28. G.J. Perlow, M.R. Perlow, J. Chem. Phys. **45**, 2193 (1966)
29. S. Bukhspan, C. Goldstein, T. Sonnio, J. Chem. Phys. **49**, 5477 (1968)
30. H. De Waard, in *Mössbauer Spectroscopy and its Applications* p. 123ff (International Atomic Energy Agency, Vienna, 1972)
31. P.K. Tseng, S.L. Ruby, Phys. Rev. **172**, 249 (1968)
32. L.E. Campbell, in *Mössbauer Isomer Shifts*, Chap. 12, ed. by F.E. Wagner (North Holland, Amsterdam, 1978)]
33. L.E. Campbell, G.L. Monet, G.J. Perlow, Phys. Rev. B **15**, 3318 (1977)
34. J.P. Biberian, M. Bienfait, J.B. Theetin, Acta Cryst. **A29**, 221 (1973)
35. P. Gütlich, R. Link, A. Trautwein, *Mössbauer Spectroscopy and Transition Metal Chemistry* (Springer, Berlin, 1978)
36. T.E. Cranshaw, J. Phys. E **7**, 122 and 497 (1974)
37. J.C. Travis, J.J. Spijkerman, in *Mössbauer Effect Methodology*, vol. 4, ed. by I.J. Gruverman (Plenum Press New York, 1970), p. 237
38. I. Tansil, F.E. Obenshain, G. Czyzek, Phys. Rev. B **6**, 4657 (1972)
39. F.E. Wagner, U. Wagner, in *Mössbauer Isomer Shifts*, Chap. 8a (North Holland, Amsterdam, 1978)
40. O.C. Kistner, Phys. Rev. Lett. **19**, 872 (1967)
41. T.C. Gibb, R. Greatrex, N.N. Greenwood, P. Kaspi, Chem. Comm. 319 (1971)
42. F. Wagner, U. Zahn, Z. Phys. **233**, 1 (1970)
43. G.J. Perlow, W. Henning, D. Olson, G.L. Goodman, Phys. Rev. Lett. **23**, 680 (1969)
44. A. Heuberger, F. Pobell, P. Kienle, Z. Phys. **205**, 503 (1967)
45. E. Zech, F. Wagner, H.J. Körner, P. Kienle, in *Hyperfine Structure and Nuclear Radiations*, ed. by E. Matthias, D.A. Shirley (North Holland, Amsterdam, 1968) p. 314
46. H.D. Bartunik, G. Kaindl in *Mössbauer Isomer Shifts*, Chap. 8b (North Holland, Amsterdam, 1978)
47. R.L. Cohen in *Mössbauer Isomer Shifts*, Chap. 8c (North Holland, Amsterdam, 1978)
48. D.O. Patterson, J.O. Thomson, P.G. Hurray, L.D. Roberts, Phys. Rev. B **20**, 2440 (1970)
49. F.E. Wagner, S. Haselbeck, L. Stievano, S. Calogero, Q.A. Pankhurst, K-P. Martinek, Nature **407**, 691 (2000)
50. F.E. Wagner, Ph. Marion, J-R. Regnard, Hyperfine Interactions **41**, 851 (1988)
51. A. Kyek, F.E. Wagner, G. Lehrberger, Q.A. Pankhurst, B. Ziegaus, Hyperfine Interactions **126**, 235 (2000)
52. S.K. Malik, G.K. Shenoy, S.M. Heald, J.M. Tranquada, Phys. Rev. Let. **55**, 316 (1985)
53. C. Crecelius, S. Hüfner, Phys. Lett. **30A**, 124 (1969)
54. E.R. Bauminger, G.M. Kalvius and I. Nowik, *Mössbauer Isomer Shifts*, Chap. 10 (North Holland, Amsterdam, 1978)
55. O. Berkooz, M. Malamud, S. Shtrikman, Solid State Commun. **6**, 185 (1968)
56. H.H. Wickmann, M.P. Klein, D.A. Shirley, Phys. Rev. **152**, 345 (1966)
57. J. Röhler, D. Wohlleben, G. Kaindl, H. Balster, Phys. Rev. Lett. **49**, 65 (1982)
58. I. Nowik, Hyperfine Interactions **13**, 89 (1983)
59. H.H. Wickmann, E. Catalano, J. Appl. Phys. **39**, 1248 (1968)
60. M. Stachel, S. Hüfner, C. Crecelius, D. Quitmann, Phys. Rev. **186**, 355 (1969)
61. W. Zinn, J. Magn. Magn. Mat. **3**, 23 (1976)
62. U.F. Klein, J. Moser, G. Wortmann, G.M. Kalvius, Physica B **86–88**, 118 (1977)
63. G. Czjzek, in *Mössbauer Spectroscopy Applied to Magnetism and Material Science*, ed. by G.J. Long, F. Granjean, vol. 1, Chap. 9 (Plenum Press, New York, 1993)
64. I. Nowik, H.H. Wickman, Phys. Rev. Lett. **17**, 949 (1966)
65. J. Chappert, L. Asch, M. Bogé, G.M. Kalvius, B. Boucher, J. Magn. Magn. Mat. **28**, 124 (1982)
66. E.R. Seidel, G. Kaindl, M.J. Clauser, R.L. Mössbauer, Phys. Lett. **25A**, 328 (1967)
67. M. Kalvius, P. Kienle, H. Eicher, W. Wiedemann, C. Schüler, Z. Physik **172**, 23 (1963)
68. R.L. Cohen, Phys. Rev. **169**, 432 (1968)
69. B.B. Triplett, N.S. Dixon, L.S. Fritz, Y. Mahmud, Hyperfine Interactions **72**, 97 (1992)

70. M. Schöppner, J. Moser, A. Kratzer, U. Potzel, J.M. Mignod, G.M. Kalvius, Z. Phys. B **63**, 25 (1986)
71. B.D. Dunlap, G.K. Shenoy, G.M. Kalvius, Phys. Rev. B **10**, 26 (1974)
72. S. Dattagupta, G.K. Shenoy, B.D. Dunlap, L. Asch, Phys. Rev. B **16**, 3893 (1976)
73. G.M. Kalvius, W. Potzel, W. Koch, A. Forster, L. Asch, AIP Conf. Ser. **38**, 93 (1977)
74. A.M. Afanas'ev, E.V. Onishenko, L. Asch, G.M. Kalvius, Phys. Rev. Lett. **40**, 816 (1978)
75. J.A. Hodges, P. Bonville, A. Forget, A. Yaouanc, P. Dalmas de Réotier, G. André, M. Rams, K. Królas, C. Ritter, P.C.M. Gubbens, C.T. Kaiser, P.J.C. King, C. Baines, Phys. Rev. Lett. **88**, 077204 (2002)
76. G.M. Kalvius, D.R. Noakes, O. Hartmann, in *Handbook on the Physics and Chemistry of Rare Earth*, vol. 32, ed. by K.A. Gschneidner Jr., et al. (Elsevier Science, Amsterdam, 2001), p. 55
77. L.L. Hirst, J. Stöhr, E. Zech, G.K. Shenoy, G.M. Kalvius, in *Magnetic Resonance and Related Phenomena (Proc. 18th Cong. Ampere)*, vol. 1, p. 67 (University of Nottingham, Nottingham, 1974)
78. L.L. Hirst, J. Stöhr, G.K. Shenoy, G.M. Kalvius, Phys. Rev. Lett. **33**, 198 (1974)
79. P. Bonville, F. Gonzalez-Jimmenez, P. Imbert, G. Jéhano, J. Sierro, J. Phys. Cond. Matter **1**, 8567 (1989)
80. B.D. Dunlap in *Mössbauer Spectroscopy Applied to Inorganic Chemistry* vols. 1 and 2, Chap. 11 (Plenum Press, New York, 1987)
81. G.M. Kalvius, B.D. Dunlap, L. Asch, F. Weigel, J. Solid State Chem. **78**, 545 (2005)
82. W. Potzel, G.M. Kalvius, J. Gal, in *Handbook of the Physics and Chemistry of Rare Earths*, vol. 17, ed. by K.A. Gschneidner Jr., et al., (Elsevier, Amsterdam, 1993), p. 539
83. K. Gofryk, J-C. Griveau, E. Collineau, J.P. Sanchez, J. Rebizant, R. Caciuffo, Phys. Rev. B **79**, 134525 (2009)
84. J. Gal, Z. Hadari, E.R. Bauminger, I. Nowik, S. Ofer, M. Perkal, Phys. Lett. **31A**, 511 (1970)
85. Th. Obenhuber, W. Adlassnig, J. Zänkert, U. Närger, W. Potzel, G.M. Kalvius, Hyperfine Interactions **33**, 69 (1987)
86. W. Potzel, in *Mössbauer Spectroscopy Applied to Magnetism and Material Science*, vol. 1, Chap. 8, ed. by G.J. Long, F. Granjean (Plenum Press, New York, 1993)
87. D.W. Mitchel, T.P. Das, W. Potzel, G.M. Kalvius, H. Karzel, W. Schiessl, M. Steiner, M. Köfferlein, Phys. Rev. B **48**, 499 (1993)
88. G.J. Perlow, L.E. Campbell, L.E. Conroy, W. Potzel, Phys. Rev. B **7**, 4044 (1973)
89. W. Potzel, W. Schäfer, G.M. Kalvius, Hyperfine Interactions **130**, 241 (2000)
90. W. Potzel, M. Steiner, H. Karzel, M. Köfferlein, G.M. Kalvius, P. Blaha, Phys. Rev. Lett. **74**, 1139 (1995)
91. W. Kohn, Phys. Rev. Lett. **2**, 393 (1959)
92. Yu. Kagan, V.V. Pushkarev, A. Holas, JETP **57**, 870 (1983)
93. C. Sauer, E. Matthias, R.L. Mössbauer, Phys. Rev. Lett. **21**, 961 (1968)
94. G. Kaindl, D. Salomon, G. Wortmann, *Mössbauer Isomer Shifts*, Chap. 8d (North Holland, Amsterdam, 1978)
95. G. Kaindl, D. Salomon, Phys. Rev. Lett. **30**, 579 (1973)
96. R.V. Kasakovski, L.M. Falicov, Phys. Rev. Lett. **22**, 1001 (1969)
97. L. Pfeiffer, R.S. Raghavan, C.P. Lichtenwallner, A.G. Cullis, Phys. Rev. B **12**, 4793 (1975)
98. L. Pfeiffer, T. Kovacs, Phys. Rev. B **23**, 5725 (1981)
99. V.G. Alpatov, Yu.D. Bayukov, A.V. Davidov, Yu.N. Isaev, G.R. Kartashov, M.M. Korotkov, D.V. L'vov, Laser Phys. **15**, 1680 (2005)
100. Yu.D. Bayukov, A.V. Davidov, Yu.N. Isaev, G.R. Kartashov, M.M. Korotkov, V.V. Migachev, JETP Lett. **90**, 499 (2009)

Part II
Highlights of Applications of the Mössbauer Effect

Unique applications of the Mössbauer effect in many fields spread very quickly and led to highlights in frontier sciences. In nuclear structure physics, the observation of hyperfine splitting of γ-lines enabled to measure electromagnetic moments of excited nuclear states, which when known also allowed determining internal magnetic fields in magnetic materials and electric field gradients for the study of the electronic structure of chemical compounds and biological materials. Furthermore, the study of effects of time changing fields on Mössbauer lines led to the observation of dynamic processes in condensed matter. Unique results reveal the studies of the isomer or chemical shifts of γ-lines in various chemical compounds because they disclosed on one hand tiny changes of the nuclear charge distribution in a γ-transition and on the other hand on the change of the electron density in different chemical compounds. Various relativistic effects on the γ-lines were observed such as the "weight of photons" in a gravitational field. Mössbauer spectroscopy of extraterrestal samples, such as from the Mars, are very important probes of the structure of the planetary system of the sun. Coherence phenomena play an important role in quantum absorption and emission processes such as the Mössbauer effect in crystals. The observation of the Mössbauer effect using synchrotron radiation has become a very effective new tool for many interesting applications.

Chapter 9
Nuclear Physics Applications of the Mössbauer Effect

Walter F. Henning

9.1 Introduction: Some Historical Remarks

If somewhere in the mid 1950s someone had made the comment to nuclear physicists studying nuclear decays that in 1958 a graduate student would discover a method that would ultimately allow measurements of certain γ-ray transitions in nuclei with resolutions as high as perhaps 10^{-10} eV, the reaction would have been most likely a ridicule or worse. So unbelievable would such a prediction have seemed, but it did happen and the Mössbauer effect [1] provided yet another testimony to the wonders of physics!

As this book illustrates, the Mössbauer effect by now has had an enormous impact on a broad range of areas in research and application. For nuclear physics, after the first flurry of exciting nuclear structure studies for perhaps a decade or two, the research involving the Mössbauer effect came essentially to an end. With the number of nuclear transitions that exhibit the Mössbauer effect being finite, this was perhaps no surprise. This chapter is, therefore, more of a reminiscence, a look back at exciting times, and remembrance of the seminal period of nuclear physics studies with the Mössbauer effect. Where and when, for example, a certain graduate student could have as many as 20 publications during his three years of Masters and PhD work.

From a nuclear physics point of view, it all started a number of years before the mid 1950s. The original motivation for studies in nuclear resonance fluorescence, i.e., the process whereby electromagnetic radiation is emitted from one atomic nucleus and resonantly scattered or absorbed by another, was to develop a method for the determination of excited-state properties, in particular of lifetimes. The idea obviously leaned on the successful method of atomic fluorescence.

W.F. Henning (✉)
Physics Department, Argonne National Laboratory, Argonne, IL 60439, USA
e-mail: wfhenning@anl.gov

M. Kalvius and P. Kienle (eds.), *The Rudolf Mössbauer Story*,
DOI 10.1007/978-3-642-17952-5_9, © Springer-Verlag Berlin Heidelberg 2012

Conceptually simple, the experiments turned out to be rather difficult and led to many formidable challenges, from physics to technology. In contrast to the eV energies of atomic transitions, nuclear transitions have typical energies of tens of keV, up to several MeV. Consequently, the energy shift of γ-rays due to the nuclear recoil is generally much larger than the resonance width, thus strongly reducing resonant scattering and absorption by another nucleus. To overcome this difficulty it was thought necessary to compensate for the energy loss in the emission and absorption processes [2].

Several methods were developed: centrifuges [3], thermal broadening at high temperatures [4], and nuclear recoils from preceding β or γ decays in gaseous γ-ray sources [5]. From a technical point of view, the limited resolution of detectors for high-energy γ-rays, for example, often led to overlapping lines and complicated analysis. In addition, Compton scattering and Bremsstrahlung, pair production, positron annihilation, and simple elastic scattering from many parts of the experimental set-up created large backgrounds.

But starting with the early experiments in the 1940s, where initially often no resonant effect was seen and the best that could be learned was perhaps a lower limit on the lifetime and thus some speculation about the multipolarity of the nuclear transition [2], the field eventually developed. In the 1950s, quite a number of successful experiments were performed providing lifetimes, spins, and parities of nuclear levels, and multipolarities and mixing ratios of the decay radiation. These in turn provided important input to nuclear theory and model calculations. In some selected cases, large resonance effects could be observed with little background as illustrated by Fig. 9.1 [6]. The latter in turn allowed systematic studies to

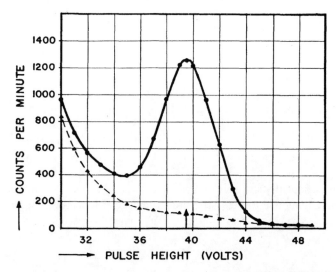

Fig. 9.1 Pulse height spectra for γ-rays emitted following the decay of ^{75}Se into ^{75}As and resonantly scattered from a solid arsenic absorber. The *arrow* indicates the position of the 265 keV line in ^{75}As. *Full dots* are for a gaseous H_2Se source, and *triangles* for a solid selenium source of equal strength [6]

learn more about the process itself, with innovative experimental configurations such as line shaping through resonant absorption followed by scattering (perhaps with some similarity to methods developed recently for synchrotron-radiation based Mössbauer studies) and, e.g., the attempt [6] to test Lamb's prediction [7] of effective versus real temperatures by comparing measurements of nuclear fluorescence in solids at room and liquid nitrogen temperatures. But the highlight of such experiments was the determination of the left-handedness of neutrinos by a combined analysis of the circular polarization and resonant scattering of the 960 keV γ-rays of ^{152}Sm following the orbital electron capture decay of ^{152m}Eu [8].

It is into this environment that the bombshell exploded: Mössbauer's work and publication in 1958 [1]. The discovery of recoilless γ-ray emission at low temperatures in the ^{191}Ir experiment, i.e., resonance absorption and scattering at – in principle – natural line width and the large resulting cross-sections that this implied, opened the flood gates: for experiments from fundamental studies to applications; from materials and their properties to new special effects and methods; from singular studies to systematic explorations and surveys; and led to what is impressively described in the present anniversary volume, and in many other previous books about the Mössbauer effect, often dedicated to a specific research area and/or application.

9.2   The Nuclear Landscape and the Mössbauer Effect

The vast range of applications for the Mössbauer effect is discussed in detail in book Chap. 7. It is commonly represented by marking the chemical elements where Mössbauer studies are possible in the periodic system. In Fig. 9.2, a version different from that in Chap. 7 is shown [9]. One finds 45 elements exhibiting low-energy nuclear transitions suitable for Mössbauer effect studies, with the lightest being potassium and americium the heaviest. For the 45 chemical elements, there are 86 (stable or very long-lived) isotopes, with altogether 106 low-energy ground-state γ-ray transitions since several isotopes exhibit more than one such γ-ray. Figure 9.2 gives the number of Mössbauer isotopes and the total number of Mössbauer transitions for each element.

A striking feature is the well-known fact that a large number of these Mössbauer transitions (actually 45, or more than 40% of all existing ones) are in rare-earth nuclei. In fact, for the rare earths each chemical element, with the exception of cerium, exhibits suitable γ-rays.

From a nuclear physics point of view, this behavior is fairly easily understood. Figure 9.3 shows the chart of nuclides, color-coded according to the lowest-energy ground-state transition for each isotope [10]. (The chart shows all known nuclei including the radioactive, short-lived ones that are not available for Mössbauer studies, of course). The overall trends are clearly visible. First, in the simplest picture (Fermi gas model), the level density at a given excitation energy scales with mass and the energy of the first excited state will, on average, decrease with mass.

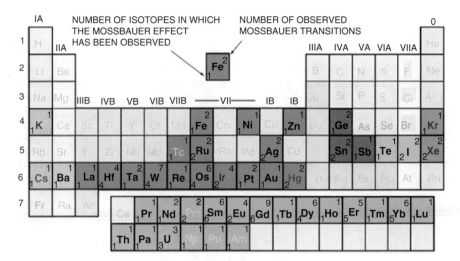

Fig. 9.2 Periodic system of the chemical elements, highlighting those that exhibit the Mössbauer effect [9]

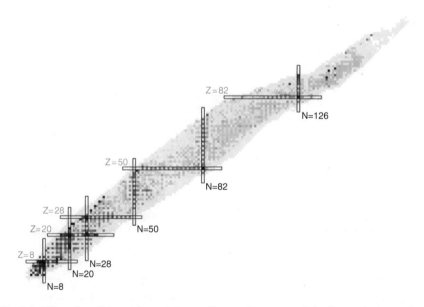

Fig. 9.3 Chart of nuclides, color-coded according to the energy of the first-excited state in a nucleus [10]

Second, the regions of deformed nuclei, located between the closed shells indicated by the horizontal and vertical bars, exhibit low-lying collective excitations. So excitation energies of less than 150 keV (green) dominate for the heavier nuclei and the deformed regions. The lightest nuclei exhibit first excited states often well above 1 MeV. In stable nuclei below potassium-40, only ^{19}F exhibits states below 200 keV.

(This raises an interesting challenge, given the unique role fluorine often plays in chemistry and in many materials. Unfortunately the excitation energies, 110 keV and 197 keV, are very high and thus the recoilless fractions of the γ transitions seem prohibitively low for such a low-mass nucleus; the intensity for the more favorable 110 keV transition is quite weak; and the parent nuclei feeding ^{19}F (either via β-decay of ^{19}O or electron-capture decay of ^{19}Ne) are short-lived, with half-lives of 27 s and 17 s, respectively. So any prospect of actual Mössbauer spectroscopy is essentially zero. However, the recently available intensities of radioactive ion beams produced in flight might warrant a new look.)

9.3 Mössbauer Studies of Nuclear Properties

The discovery of the Mössbauer effect allowed nuclear spectroscopy of unprecedented precision for electromagnetic transitions between the ground-state and low-lying excited states in nuclei. The recoilless emission and absorption of γ-rays provides, in principle, for energy precisions equal to fractions of the natural line widths of such transitions. This means, for low-energy γ-ray transitions of typically 100 keV and, say, 1 ns half-life an absolute energy resolution of about 10^{-9} eV, or a relative energy resolution of perhaps 10^{-14}.

Because of this high resolution of the Mössbauer effect, many of the ensuing applications were concerned with the study of resolved hyperfine interactions and nuclear moments, i.e., the detection of the subtle changes in the nuclear environment due to the atomic (or conduction) electrons or externally applied fields. Typically, there are three types of hyperfine interactions that are observed, isomer (or chemical) shift, quadrupole splitting, and magnetic (Zeeman) splitting. Figure 9.4 shows the landmark measurement [11] for the most important Mössbauer isotope, ^{57}Fe, which for the first time clearly demonstrated all three effects in one spectrum and discovered the isomer shift. ^{57}Fe had been proposed already in 1959 [12, 13] as an ideal candidate for Mössbauer studies due to its unique nuclear properties. And first measurements had demonstrated Zeeman splitting in internal and external fields. But the fully developed spectrum for iron oxide in Fig. 9.4, measured with a single-line stainless steel source, beautifully illustrated the complete hyperfine structure for ^{57}Fe.

Subsequently, there was a flurry of studies of nuclear properties carried out with the use of the Mössbauer effect over the first two decades after its discovery. Because of the inverse exponential dependence on the square of the momentum of the γ-rays emitted, the recoil free fraction decreases strongly with γ-ray energy and becomes experimentally essentially impossible to observe at energies above 200 keV or so.

The isomer shift reflects the chemical bonding of the atoms and is related to the electron density at the nucleus and to changes in the size of the nucleus. The Mössbauer effect is the only method to reliably measure this quantity between nuclear states. Isomer shift measurements have also been performed in muonic atoms but, as a rule, are plagued by the complexity of additional contributions

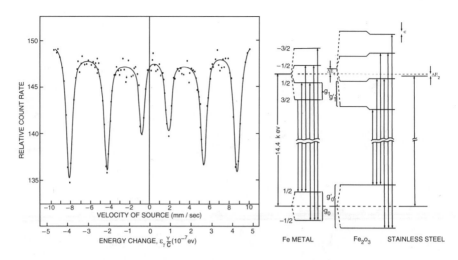

Fig. 9.4 The benchmark Mössbauer experiment for ^{57}Fe by Kistner and Sunyar [11], which fully displayed for the first time all three hyperfine interactions: isomer shift, magnetic, and (less pronounced) quadrupole splitting

to the energy shifts which makes the extraction of the nuclear size changes more ambiguous. The results of isomer shift studies using the Mössbauer effect have had significant impact on nuclear models concerning both, collective and single-particle degrees of freedom.

The quadrupole splitting reflects the interaction between the nuclear quadrupole moment of the nucleus and the surrounding electric field gradient. Together with other measurements of nuclear collectivity, the direct measurement of the spectroscopic quadrupole moment provides independent information on nuclear shapes and deformations.

The magnetic hyperfine splitting is a result of the interaction between the nuclear magnetic moment and any surrounding magnetic field. The magnetic moment measurement provides critical information on the nature of the nuclear state, from the nuclear spin via the multiplicity of the hyperfine states to the composition and model quantum numbers of the respective nuclear wave function and the multipolarity of the transition.

There were other significant nuclear measurements using the Mössbauer effect: (1) the demonstration of the nuclear hyperfine anomaly between excited and ground state in a nucleus, i.e., minute shifts in nuclear g-factor measurements depending on whether the magnetic interaction arises from external (or electron orbital) magnetic fields or from electron contact fields; (2) the discovery of the quantum interference effect between nuclear electromagnetic decay via electron conversion and the photoelectric effect, both involving the same atomic electrons and thus quantum mechanically indistinguishable; (3) the measurements to test limits of time-reversal violation in the electromagnetic decay of nuclei. The mixed electromagnetic nature of some nuclear transitions, for example $M1$ plus $E2$, allows configuring aligned (polarized) ensembles of emitting and absorbing Mössbauer nuclei such that a violation in time reversal becomes visible in the intensity asymmetry between hyperfine lines in the Mössbauer spectrum.

In the following, we address in some more detail the areas of nuclear studies just outlined. This cannot be a comprehensive review of nuclear physics experiments with the Mössbauer effect given the limited space available. The examples are chosen to bring out some of the key characteristics but, in most cases, are somewhat arbitrary. They are not necessarily meant to represent the original or most important work in that area but more often reflect the author's familiarity with the specific work done. The author apologizes to all his colleagues for work not mentioned. But since much of it was generally done during the early phase of the Mössbauer effect the extensive literature and many earlier books cover them well.

9.3.1 Measurement of Nuclear Lifetimes

As was discussed in the Introduction, a major motivation for the very early studies in nuclear resonance fluorescence was the hope to determine nuclear properties, in particular the lifetimes of nuclear states. Given that the recoilless emission and absorption of the Mössbauer effect can, in principal, measure the natural linewidth it would appear that lifetime measurements might initially constitute a major application. In particular also since the Mössbauer effect allows the measurement of resolved hyperfine spectra so that individual transition lines between magnetic substate should be less susceptible to various line-broadening effects, including those from absorber thickness. But as it turns out, the lifetimes of the low-energy transitions in stable nuclei to which the Mössbauer effect generally can be applied are mostly in the nanosecond region and thus accessible with precision delayed-coincidence measurements.

In some selected cases, though half-life measurements turned out to be quite useful; in particular for relatively short-lived states not easily accessible with electronic coincidences but, because of the broad (natural) linewidths, less susceptible to unknown field broadening effects. In many cases, tests were performed to check the consistency or improve on half-lives obtained electronically; in other cases, resolved hyperfine lines [14] or shortlived states of a few hundred picoseconds half-life allowed new quantitative half-life measurements. An example for the latter is shown in Fig. 9.5 for the case of the 140 keV transition in ^{99}Tc [15].

9.3.2 Isomer Shifts in Nuclei

The Coulomb interaction between the charge distribution of a nuclear state and the surrounding electron cloud contributes to the energy of the total system. Between two nuclear states with different charge distributions, this leads to energy shifts in atomic as well as nuclear observables. One such effect, the isotope shift, has

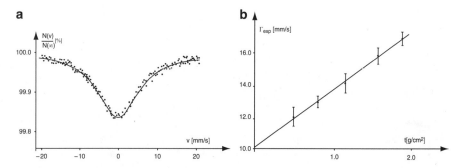

Fig. 9.5 A single-line Mössbauer spectrum (*left panel*) was observed for the 140 keV transition in ^{99}Tc with a molybdenum metal source and a technetium metal absorber [15]. From the dependence of the line width on absorber thickness (*right panel*), a mean lifetime of $\tau = 277 \pm 14$ ps was determined

been extensively studied in atomic spectra between different isotopes of the same chemical element in their ground states. A similar effect will occur between isomeric states in the same nucleus. But the effect will be considerably smaller since the major causes for the isotope shifts, i.e., the direct mass and volume effects from the different number of nucleons in different isotopes, are not present between isomers.

In principle, if the transition energy between the two nuclear states were exactly known for a bare nucleus, the isomer shift introduced by electrons could be directly determined from the transition energy in any chemical environment. But, of course, there is little chance of measuring γ-ray transitions in bare nuclei and, in any case, no conventional method of γ-ray spectroscopy can provide the necessary energy resolution.

By measuring the energy shift between source and absorber made of different chemical compounds, the electron density difference enters between, in principle, two known electron configurations. This isomer shift S, for constant electron density over the nuclear volume, is then given by the following expression:

$$S = (2\pi/3) \left(Ze^2 \Delta\rho_e(0) \times \Delta <r^2> \right). \qquad (9.1)$$

Here, $\Delta <r^2>$ is the difference in mean square nuclear charge radius between the nuclear states, $\Delta\rho_e(0)$ is the difference in electron density at the nucleus. As shown, e.g., in [16], higher order terms in the nuclear charge radius enter if the electron density varies over the nuclear radius. But such terms are small and, in the numerical calculations performed by the authors, contributed less than a few percent to the isomer shift even for the heaviest nuclei.

As already mentioned in Sect. 9.2, an isomer shift was first observed in the landmark ^{57}Fe measurement of Kistner and Sunyar [11]. By now, energy shifts of 10^{-8} eV or less can be reliably measured. Consequently, isomer shifts have become a well-established tool in solid state chemistry (see Chap. 10) yielding valuable information on chemical bonding and on different oxidation and spin states. An example for this is shown in Fig. 9.6 [17].

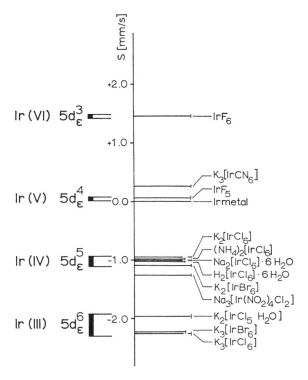

Fig. 9.6 Systematic of isomer shifts (in mm/s) for iridium compounds relative to iridium metal, reflecting chemical bonding and valence states [17]

Of course, as seen from (9.1), determining information on electron densities requires knowledge of the *nuclear calibration* factor, and vice versa. As we will see, the quantitative theoretical calculation of the nuclear factor, i.e., of the change in mean square nuclear charge radius is extremely difficult and may depend sensitively on very specific nuclear structure aspects. As a result, the accepted approach to calibrate the nuclear factor is rather empirical, and theoretical at the same time: a linear calibration of the isomer shift on the electron density is assumed and the *calibration constant* is then extracted from a linear regression between measured isomer shifts and sophisticated theoretical calculations of the electron densities, such as those obtained from specific density functional calculations. This is the subject of Chap. 10 and will not be further discussed here.

From a nuclear physics point of view, the measurement of isomer shifts with the Mössbauer effect opened the possibility to determine small changes of nuclear radial matrix elements (one finds values of typically a few to a few tens of 10^{-3}fm^2). This was a true challenge and a severe test of relevant nuclear structure theories. Moreover, very distinctive features of nuclear phenomena became apparent from these extremely sensitive measurements. Data for rotational transitions of deformed nuclei suggested the marked dominance of pair breaking via Coriolis effects and led subsequently to a microscopic theory of nuclear rotation. Similarly, the importance of core polarization effects in single-particle transitions was observed for proton, and in particular, for neutron transitions and led to their theoretical explanation.

Before going any further and discussing some examples, it is useful to return to the method mentioned above that in principle can provide similar information: the excitation of nuclear levels in muonic atoms. This was predicted in early theoretical studies [18, 19] which considered the mixing of nuclear and muonic states by a strong quadrupole interaction between the bound muon and a deformed nucleus. The γ-rays de-exciting the nuclear levels are emitted in the presence of the muon in its $1\,s_{1/2}$ ground state. Because of the muon's presence, the γ-rays were predicted to be substantially shifted (and also magnetically split), with typical values for such shifts of several hundred eV so that high-resolution spectrometers should be able to observe these shifts directly. Extensive studies in muonic atoms have confirmed this expectation [20].

Compared to Mössbauer effect studies it was, in fact, initially expected that the interpretation of muonic isomer shifts had the big advantage that the muon eigenfunction can be calculated with high accuracy. This would avoid the difficulties encountered with electron density differences in Mössbauer isomer shifts. In fact, the muonic shifts might provide the accurate calibration for Mössbauer data.

But a detailed analysis revealed several problems: (1) the magnetic hyperfine splitting of the nuclear levels introduced a major uncertainty in that the population of sub-levels turned out not to be statistical but depended on the excitation mechanism with, in addition, fast inter-doublet transitions; (2) muonic shifts do not measure the change of the mean square radius; model-independent radial moments can be determined, but a comparison with Mössbauer data is always model dependent; (3) the polarization of the nucleus is an important effect of several keV; here it enters only through the difference in excited and ground state but differential calculations have considerable uncertainties; (4) weak muonic lines are plagued by backgrounds from muonic cascades, background from the surroundings, delayed lines from muon capture, feeding from higher nuclear levels, X-ray backgrounds, etc. All these effects add up to uncertainties of the same order as the extracted muonic shifts. On the other hand, substantial progress has been made in the electron density calibrations of Mössbauer isomer shifts (next chapter) such that the latter have been established as the primary method.

It should also be mentioned that due to advances in optical spectroscopy with lasers, it has been possible to observe isomer shifts in atomic transitions between long-lived isomers and the ground state in several nuclei including radioactive ones [21, 22].

9.3.3 Isomer Shifts and Nuclear Models

For the isomer shift between single-particle states in a nucleus, the simplest microscopic model is the independent particle model. The definition of the change of the mean square charge radius is

$$\Delta <r^2> = 1/Z \left(< \Psi_1 \left|r^2\right| \Psi_1 > - < \Psi_0 \left|r^2\right| \Psi_0 >\right). \tag{9.2}$$

Here $|\Psi_1 >$ and $|\Psi_0 >$ are the many-body nuclear wave-functions of excited and ground state of a nucleus, Z the number of protons, and r^2 the operator of the square of the charge radius:

$$r^2 = \Sigma_{\lambda 1,\lambda 2} < \lambda_1 \left|r^2\right| \lambda_2 > a_{\lambda 1}^+ a_{\lambda 2}. \tag{9.3}$$

The sum runs only over the proton states. The $a_\lambda^+ a_\lambda$ denote fermion creation and annihilation operators, $|\lambda >$, are single-particle wave functions of a shell model and $< \lambda_1|r^2|\lambda_2 >$ are the single-particle matrix elements. In the independent particle model, the many-body wave function of an odd-mass nucleus with $(A + 1)$ nucleons can be described as the product of a shell-model creation operator a_λ^+ operating on the uncorrelated ground state $|0, A >_{\text{unc}}$ of the neighboring even–even nucleus with A nucleons. In the harmonic oscillator approach, the expectation value of $\Delta <r^2>$ is then proportional to the difference ΔN in the main quantum numbers of ground- and excited state for protons and zero for neutrons.

This obviously is not consistent with the experimental data. If one uses the more realistic Wood–Saxon potential, then the major N shells are slightly mixed and, as a result, $\Delta <r^2>$ is non-zero within a major shell. But it is substantially smaller than for a *trans*-shell transition. At the time of these studies, it was well-known that pairing correlations play a significant role for nuclei away from closed shells. The BCS model [23, 24] takes into account the pairing interaction by transforming the shell-model creation and annihilation operators into BCS quasi-particle operators. However this approach still does not take into account the interaction between the core and the valence nucleon. Also the blocking effect of the odd nucleon is not included which, however, can be approximated by solving the BCS gap equation without the level of the unpaired nucleon. Overall though, the pairing correlations change the expectation values of $\Delta <r^2>$ quite substantially.

This became obvious in subsequent studies where the crucial interaction between valence nucleon and core was also taken into account giving rise to changes of the charge distribution of the core. For odd-neutron nuclei, this is the only contribution to the isomer shift, but also for protons this leads to substantial changes of the single-particle value. This was studied by two different theoretical approaches: the extended pairing plus quadrupole model developed by Uher and Sorensen [25], and the theory of finite Fermi system [26] applied to isomer shift calculations by Belyakov [27] and Speth [28]. Major progress was achieved, approaching the experimentally observed values although there were some caveats with regard to the work of Uher and Sorensen and that of Belyakov. For the first, a number of parameters had been adjusted to fit to a set of earlier isomer shift data and the predictive power to newer cases was unconvincing. In Belyakov's calculations, several approximations were made, including the omission of the particle–particle

Table 9.1 Experimental and theoretical ratios of $\Delta <r^2>$ between transitions in isotopes of the same chemical element. The theoretical calculations are obtained within the Midgal model of finite Fermi systems extended to single-particle states in deformed odd-A nuclei [16]

Isotopes (E_γ)	Experiment	Theory
^{153}Eu(103 keV) ^{151}Eu(21 keV)	-5.56 ± 0.03	-5.25
^{153}Eu(103 keV) ^{153}Eu(97 P keV)	$+1.09 \pm 0.01$	$+1.14$
^{155}Gd(105 keV) ^{155}Gd(87 keV)	$+1.38 \pm 0.02$	$+1.68$
^{157}Gd(64 keV) ^{155}Gd(87 keV)	-2.45 ± 0.16	-1.73
^{161}Dy(75 keV) ^{161}Dy(26 keV)	$+1.07 \pm 0.10$	$+0.89$

interaction which thus ignores the blocking effect. Secondly, the gap equation was not solved but experimental state-independent energy gaps used. The results were of the right order of magnitude, but sometimes of opposite sign as the experimental values. Overall agreement with experiment was in the end not significantly better than for the simple single-particle model.

In calculations of Meyer and Speth [29] essentially all of these approximations were avoided. In particular, for single-particle states in deformed nuclei, it was shown that the theory of finite Fermi systems is applicable if one uses Nilsson single-particle wave functions. The complete theory including BCS plus particle conservation as well as particle–hole interactions led to much improved results. In particular, for ratios of $\Delta <r^2>$ between transitions in isotopes of the same elements, surprisingly, good agreement was found with experiment (Table 9.1) [16].

A major nuclear physics interest in isomer shifts has been for rotational transitions in deformed nuclei. Much of the early interest in the resulting values of $\Delta <r^2>$ arose from the ability to interpret the data in terms of the phenomenological model of Bohr and Mottelson [30]. The model describes the low-energy excitations of a deformed, axially symmetric even–even nucleus as collective rotations. Changes of the mean square charge radius are expected only when the intrinsic shape of the nucleus changes due to the rotation. Naively one might expect some stretching of the nucleus. In the original model such effects (which also increase the moment of inertia) are treated as interactions between rotational and vibrational modes [31], similar to those observed in molecules.

The ease with which the model can be adapted to a calculation of $\Delta <r^2>$ comes from the assumption that the change in energy (moment of inertia) can be attributed completely to shape changes. However, the results are not at all in agreement with this picture (Fig. 9.7). This led to microscopic calculations of $\Delta <r^2>$ based on the self-consistent cranking model extended to second-order by Marshalek [32] and Meyer and Speth [33]. The two approaches differed in the assumed residual interactions. Marshalek from the beginning assumes a separable pairing-plus-quadrupole force, which simplifies the formulas considerably. Meyer and Speth derive their formulas without such approximation.

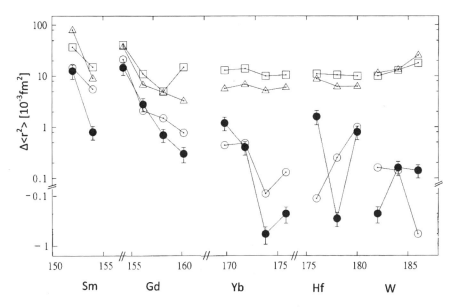

Fig. 9.7 Changes in mean square nuclear charge radius $\Delta<r^2>$ between the ground and first excited 2^+ rotational states in strongly deformed nuclei. The experimental data (*solid dot*; [34, 36–38]) are best described by a fully microscopic cranking model calculation (*open circles*; [33]), whereas a restricted cranking model (*open triangles*; [32]) and calculations (*open squares*; [34]) in a phenomenological rotational-vibrational model [31, 35] widely fail, largely due to an over-emphasis of rotational stretching

The results are quite different (Fig. 9.7). In Marshalek's work, stretching is the most important contribution. Coriolis anti-pairing effects are very small. Theory and experiment in most cases agree within the right order of magnitude, but the theoretical values are generally too large by factors of 3 to 5. At the upper end of the rare-earth nuclei, the predictions disagree considerably with experiment (also never predict negative values as experimentally observed). This failure is due to the fact that with the quadrupole force used, the stretching was overestimated. Moreover, with the choice of the pairing force, the Coriolis anti-pairing effect is severely underestimated.

The latter, however, plays an important role for all deformed nuclei. The role and nature of anti-pairing are demonstrated in Fig. 9.8 [16, 33]. The calculated values of $\Delta<r^2>$ for rotational $2^+ \to 0^+$ transitions are mainly determined by the change of occupation probabilities of single-particle levels near the Fermi surface. The anti-pairing effect characteristically depopulates the levels just above the Fermi energy (for rare earth and heavier transitional nuclei, these are $N = 5$ states with larger radii) in favor of the levels just below ($N = 4$ states with comparatively smaller radii). This leads to a negative contribution to the change in the nuclear charge radius.

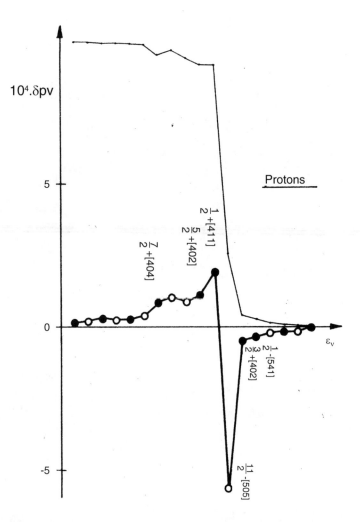

Fig. 9.8 Calculated change $\delta\rho_\upsilon$ of the occupation probabilities of Nilsson levels as a function of single-particle energy ε_υ, near the Fermi surface for protons in ^{188}Os (*open circles* $N = 5$ and *full circles* $N = 4$ levels), illustrating Coriolis anti-pairing and particle–particle interaction effects in rotational excitations [16, 33]

The results of Meyer and Speth [33] showed much better overall agreement with the data and led to a deeper understanding of isomer shifts in rotational nuclei. The most significant feature of Fig. 9.7 is the rapid drop in $\Delta<r^2>$ at $N = 90$. While not discussed in those terms at the time, this is exactly the behavior discussed further below in the context of quantum phase transitions in recent microscopic mean-field calculations of heavy nuclei.

The isomer shift studies showed that absolute theoretical predictions of $\Delta<r^2>$ are extremely difficult. Specifically: (1) stretching is important for soft $N = 90$

nuclei, but is nearly absent in well-deformed nuclei; (2) $\Delta<r^2>$ in rotational excitations in well-deformed nuclei displays a fine structure that is generated by the Coriolis anti-pairing effect; even negative values of $\Delta<r^2>$ arise in some cases; (3) the fine structure is a genuine shell effect; the changes in nuclear radii are determined by the redistribution of only a few particles near the Fermi energy; (4) the results support the choice of the energy-dependent δ-force instead of the pairing-plus-quadrupole force.

Much of the work on the nuclear physics aspects in isomer shifts happened in the first two or three decades of the Mössbauer effect. Now the emphasis is unambiguously on its use as a tool of chemical analysis capable of providing information about the chemical environment of the resonating nucleus on an atomic scale. The most well-known application is perhaps the determination of the ^{57}Fe isomer shift in crystalline and in disordered solid samples which finds an increasing number of applications in biochemistry and biophysics as well as in materials science, nanoscience, and catalysis. However, there are some recent developments in isomer shift studies with regard to nuclear physics that are worth mentioning.

The first concerns theoretical developments towards calculations of $\Delta<r^2>$. Increased computational capabilities as well as model developments and the search for new physics phenomena have led to microscopic nuclear structure studies relevant to the present discussion. A recent paper [39] proposes, and presents calculations of isomer shifts as observables in the quest to identify changes in equilibrium shapes of nuclei corresponding to first- and second-order quantum phase transitions (QPTs), i.e., spherical to axially deformed shapes. Iachello and Zamfir [40] have shown in a study of QPT in mesoscopic systems that the main features of phase transitions, defined for an infinite number of particles, $N \rightarrow \infty$, persist even for a moderate $N \approx 10$. This was followed by several studies of shape phase-transitional patterns in nuclei as a function of particle number. According to [40], there are two approaches, the method of Landau based on potentials or the direct computation of order parameters. For nuclei, however, a quantitative analysis of QPT must go beyond a simple study of, e.g., potential energy surfaces. This is because potentials or more specifically deformation parameters are not observables and can only be related to observables via specific model assumptions.

In the work considered here [39], both approaches are combined in a consistent microscopic framework, i.e., a five-dimensional Hamiltonian for quadrupole vibrational and rotational degrees of freedom with parameters determined by constrained microscopic self-consistent mean field calculations. The diagonalization of the Hamiltonian yields the excitation energies and the collective wave functions that are used to calculate observables. They include predictions of isomer shifts between the first 2^+ state and 0^+ ground state as well as the isomer shifts between the second and first 0^+ states, both as a function of the neutron number for even–even neodymium isotopes (Fig. 9.9). The neodymium isotope chain was chosen for these calculations because changes of structure data at neutron number 90 had been previously observed. The predicted isomer shifts display behavior characteristic for a first-order phase transition. The authors draw attention to the particular feature that, although

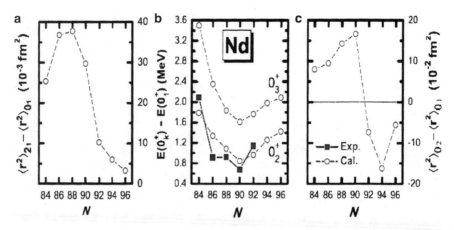

Fig. 9.9 Microscopic nuclear model prediction of $\Delta<r^2>$ between 2^+ and 0^+ states of the ground-state band (**a**), and between the second and first excited 0^+ states (**c**) in neodymium isotopes as a function of neutron number [39]. Both indicate the rapid phase transition at $N = 90$ while the energies of the experimental and theoretical excited 0^+ states (**b**) vary more smoothly

the overall results of their calculations correspond to a realistic prediction of ground states and collective excitation spectra of neodymium isotopes, both isomer shifts exhibit very sharp discontinuities at $N = 90$. A direct comparison to isomer shift data though will be essentially impossible for neodymium isotopes since at most one of the even–even isotopes might allow Mössbauer studies. However, as already illustrated above (Fig. 9.7), the phase transition-like behavior is observed in neighboring rare earth nuclei near $N = 90$.

Finally, there is an interesting recent development of isomer shifts for very long-lived excited nuclear states. Optical hyperfine studies on several multi-quasi particle isomers [22], such as the $I^\pi = 16^+$ state in ^{178}Hf which was the first where such studies were carried out [21], have consistently revealed a decrease in mean square charge radius between the nuclear isomer and the ground state. The method used was collinear laser spectroscopy of nuclear reaction products produced in transfer and fusion reactions and collected in an ion source configuration; then extracted, mass separated and cooled, and bunched in a linear Paul trap, and then subjected to collinear atomic spectroscopy. The negative values of $\Delta<r^2>$ observed for all of the six isomeric states investigated up to now, together with theoretical calculations hint at both, increased rigidity and loss of pairing for these multi-quasi particle configurations.

9.3.4 Hyperfine Splitting and Nuclear Moments

The recoilless emission and absorption of γ-rays allows observing low-energy nuclear groundstate transitions with often fully resolved hyperfine structure. The best known example is the $14\,\mathrm{keV}$ $3/2^+ \rightarrow 1/2^+$ transition in $^{57}\mathrm{Fe}$ (Fig 9.2). But many measurements of hyperfine spectra of highest precision exist; some are illustrated in the remainder of this chapter.

The hyperfine splitting between the substates m of a nuclear state with total angular momentum I ("nuclear spin") is in general a complex expression. For the specific but often employed field geometries of a vector magnetic field and an axially symmetric electric field gradient, both having the same z-axis, it is given by:

$$E(m, I) = mg\mu_\mathrm{n}H + eQV_{zz}[3m^2 - I(I + 1)]. \qquad (9.4)$$

Here H and V_{zz} are the magnetic field value and the electric field gradient that the nucleus sees, μ_n is the nuclear magneton, g the gyromagnetic ratio, and Q the electric quadrupole moment. In a Mössbauer spectrum we measure the difference in hyperfine splitting between the m substates of the nuclear excited and ground state (we omit here the isomer shift discussed in detail in Sect. 9.3.2). This allows the determination of the nuclear moments and/or the values of the magnetic and electric fields. The latter is today, of course, the primary area of Mössbauer effect studies as discussed in the other chapters of this book. During perhaps the first decade or two of the Mössbauer effect, measurements of the nuclear moments provided important information for nuclear structure studies and nuclear theory. This is illustrated in this section with two brief examples. (In several of the other sections of this chapter, hyperfine structures are presented in the context of studies with different goals).

The first example summarizes measurements of magnetic properties with the Mössbauer effect for two states in the same nucleus, allowing a precision study of the nuclear structure and the underlying unified rotational model. The second example provides data for another interesting nuclear situation: while the nucleus under study is highly deformed, with a large intrinsic quadrupole moment, the actual quadrupole interaction is close to zero for the excited state because of an unusual Coriolis mixing of the single-particle wave function. Figure 9.10 shows Mössbauer spectra [14] for the $66.7\,\mathrm{keV}$ γ-ray transition in $^{171}\mathrm{Yb}$, one of two groundstate transitions within the same rotational band, i.e. the $K = 1/2$ single-neutron band building on the groundstate [521] Nilsson level. Without going into the details we list the nuclear structure information that was obtained from a rather comprehensive study of the two Mössbauer transitions from the $66.7\,\mathrm{keV}$ $3/2^-$ state [14] and the $75.9\,\mathrm{keV}$ $5/2^-$ state [41] to the $1/2^-$ ground state, respectively.

First, because of the fully resolved hyperfine structure and the absence of quadrupole splitting in the absorber as well as in the single-line erbium (for the $66.7\,\mathrm{keV}$ transition), respectively lutetium sources (for the 75.9 keV transition), good halflife values were obtained after careful correction for absorber thickness broadening. From the intensities of the various sub-state transitions the $E2/M1$ mixing ratio ($\delta^2 = 0.63 \pm 0.06$) was rather precisely measured for the $66.7\,\mathrm{keV}$ γ-rays; this allowed also an experimental check on the partial conversion coefficients for $M1$ and $E2$ transitions and absolute multipole transition probabilities

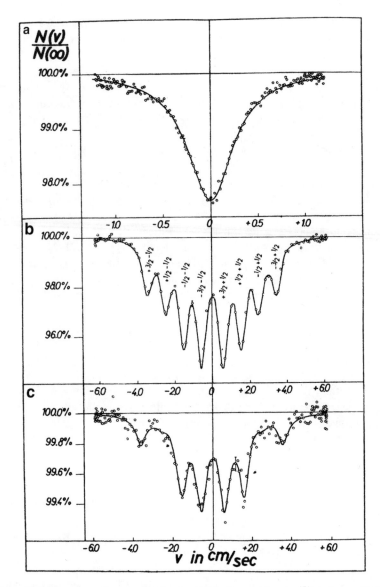

Fig. 9.10 Mössbauer hyperfine spectra for the 66.7 keV transition in [171]Yb for an Er(Tm)Al source and various absorbers: (**a**) Yb metal; (**b**) YbCl$_3$ · 6H$_2$O, polycrystalline; (**c**) YbCl$_3$ · 6H$_2$O, single crystal [14]

relative to single-particle Weisskopf units ($M1$ hindrance factor 170 and $E2$ enhancement factor 140).

The determination of the magnetic moments for the two excited states and reduced multipole transition probabilities, plus the known groundstate (and band head) magnetic moment allowed a precise test of the unified model for single-particle rotational excitation, i.e. the validity of a separation of the wave function into an intrinsic and a rotational part. The data are consistent with the model and allow a precise determination of the model parameters. This is further supported by other Mössbauer measurements in ^{171}Yb, in particular measurements of quadrupole splittings confirming a constant intrinsic quadrupole moment within the $K = 1/2$ band [42].

Finally, the sub-state transitions for the 66.7 keV state measured for the highly polarized single-crystal YbCl$_3$ · 6H$_2$O absorber (Fig 9.10c) provided a value of $\cos \eta = 1.00 \pm .03$ for the phase angle η between the $M1$ and $E2$ amplitudes. As will be seen in Sect. 9.3.8, η can provide a limit on the time reversal violating amplitude for a nuclear transition but, of course, only a $\sin \eta$ term can provide a significant sensitivity; the cos term is too insensitive.

For a second example of a hyperfine study we consider the gadolinium nucleus ^{155}Gd. This is an interesting case because of an unusually small spectroscopic quadrupole moment of the 86.5 keV state despite the fact that the nucleus is highly deformed, and because of the $E1$ nature of the γ-ray transition. Also it is perhaps the most convenient γ-ray transition in the gadolinium isotopes for Mössbauer studies so it was important to determine its nuclear moments.

From the energy levels one deduces that ^{155}Gd is strongly deformed and all low-lying levels can be wellgrouped into rotational bands. On the other hand, the detailed analysis of the nuclear level scheme and direct reaction studies with light ions had indicated that the 86.5 keV level exhibits extremely large band mixing through the Coriolis force [43]. Figure 9.11 shows Mössbauer spectra from [44] from which rather precise values were obtained for the nuclear moments. They were indeed consistent with the strong mixing scenario. But a particularly interesting result was the fact that, although strongly deformed, the spectroscopic quadrupole moment of the 86.5 keV $5/2^+$ state was only about one tenth of that of the $3/2^-$ ground state. This simplifies the determination of electric field gradients in solid state studies since the absence of the quadrupole splitting of the 86.5 keV leads to a simpler absorption pattern which is helpful when, in particular, several different chemical compounds are present in a sample.

In this context the following observation is worth mentioning. The 68.5 keV state decays via an $E1$ transition where an interference term is always present (see discussion in Sect. 9.3.7). This makes the spectral lines asymmetric. And indeed inspection of the spectra in Fig 9.11 clearly reveals such an asymmetry. In fact, as reported in Sect. 9.3.7, the amplitude of the interference term is about 5% for ^{155}Gd, very much of the magnitude of, for example, the asymmetry seen in the peak heights (also about 5%) of the two-line quadrupole spectrum of the SmH$_2$ source. While this requires a detailed and quantitative analysis and new fitting of the spectra, the similarity of these two values suggests that the quadrupole moment of the 86.5 keV state might actually be very close to zero. A very unusual, and perhaps amusing finding, but that might have some use in certain studies.

It seems perhaps obvious from these last remarks that the dispersion should be taken into account in precision hyperfine structure and nuclear moment measurements for $E1$ transitions.

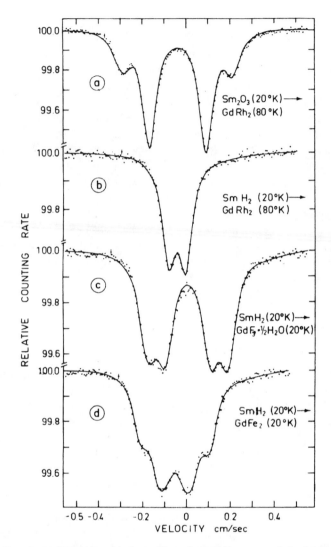

Fig. 9.11 Experimental and theoretical recoilless absorption spectra of the 86.5 keV γ-rays of
^{155}Gd with the indicated source-absorber combinations [44]

9.3.5 Hyperfine Anomaly

It was well-known from atomic hyperfine spectroscopy of nuclear ground states
(and of very long-lived isomeric nuclear states that are accessible by atomic
spectroscopy) that high-precision values for ratios of nuclear magnetic moments
may differ at the percent level between different chemical systems. This is in
particular also the case when compared to g-factor ratios measured in external

magnetic fields. It was pointed out by Bitter [45] that anomalies of this magnitude might be expected if the nuclear magnetic moments are considered as a distribution of magnetism over the nuclear volume rather than as point dipoles.

This effect, which became known as magnetic hyperfine anomaly, was studied in detail by Bohr and Weisskopf [46] who felt that such effects might offer interesting information regarding the structure of nuclear moments. By definition, nuclear magnetic moments or g-factors are derived from the magnetic hyperfine splitting or Larmor precession frequency observed in a uniform magnetic field such as an externally applied, or a hyperfine field from the orbital motion of electrons. Hyperfine fields in solids, however, usually arise at least in part from the Fermi contact interaction, i.e. from a non-zero spin density of s-electrons inside the nucleus. The hyperfine anomaly arises because this spin density increases towards the center of the nucleus, so that contact fields probe the distribution of nuclear magnetization inside the nucleus in a different way than a uniform field. Experimentally, since a hyperfine field is observable only in a product with a nuclear moment, the anomaly Δ is observed by comparing the ratio of two moments or nuclear g factors (g_1 and g_2) measured in two types of fields:

$$\Delta(1 - 2) = (g_1/g_2)_\text{u} / (g_1/g_2)_\text{c} - 1, \qquad (9.5)$$

where the subscripts refer to, for example, a uniform (u) external or orbital, and a solid compound's (c) mixed hyperfine magnetic field.

The high resolution of the Mössbauer effect offered an opportunity to search for the hyperfine anomaly between excited and ground states in the same nucleus. The first attempt was made by Grodzins and Blum on ^{57}Fe [47] but the result exhibited only a 0.5% deviation from unity, equal to the 1σ error bar. The first unambiguous observation with the Mössbauer effect was made by Perlow et al. [48] for the 73 keV resonance in ^{193}Ir where a 7% anomaly was found (Fig 9.12). Additional Mössbauer measurements with the 73 keV and 82 keV transitions in ^{193}Ir and ^{191}Ir, respectively, provided a range of data from which hyperfine anomalies were extracted, as summarized by Wagner and Zahn [17]. The results provided information on both the magnetic structure of the nuclear states involved as well as the nature of the hyperfine fields.

As an example we discuss the original experiment [48]. It was expected at the time that a relatively large anomaly might be present in the iridium nuclei ^{191}Ir and ^{193}Ir. In both cases the ground states with spin $3/2^+$ are deformed [402] Nilsson orbitals originating predominantly from the $d_{3/2}$ state of the spherical shell model. In such states the spin and orbital magnetic moments are antiparallel and nearly cancel each other. This behavior is qualitatively unchanged in a deformed nucleus as seen, for example, from the small g-factor of 0.106 for the ground state of ^{191}Ir. Insofar as the orbital and spin contributions to the nuclear moment tend to cancel, a minor difference in the separate interactions shows up as a relatively large effect. In the experiment, the g-factor ratio between the 73 keV excited state and the ground state were measured in iridium metal in an external magnetic field of 73 kG and in IrF$_6$. In addition, the g-factor ratio obtained in an alloy of 2.7 atomic percent iridium in iron by Wagner et al. [49] was used in the analysis. Figure 9.12 shows the MBE spectra

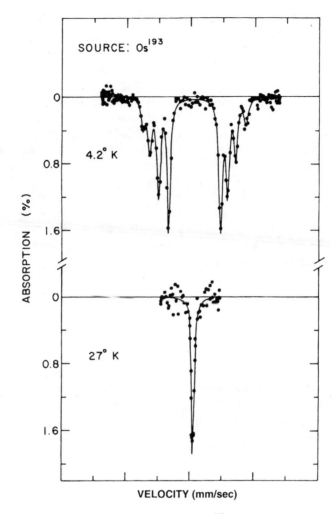

Fig. 9.12 Mössbauer spectra for the 73 keV transition in ^{193}Ir with an osmium metal source and IrF$_6$ absorber, taken below (*top*) and above (*bottom*) the anti-ferromagnetic transition temperature [48]. The g-factor ratio deduced from the magnetic splitting in IrF$_6$ reveals a 5% hyperfine anomaly when compared to the ratio obtained in a uniform external magnetic field

observed for the IrF$_6$ absorber, an antiferromagnet with transition temperature at about 8°K, well below and above its transition temperature. The narrow width of the single line at the higher temperature shows that, e.g., quadrupole interactions are extremely small and can be neglected. The observed anomalies for IrF$_6$ and the Ir–Fe alloy, in comparison to the external field measurements, are 5% and 7% respectively. The measured values provide constraints on the nuclear magnetization and sensitive nuclear structure aspects as well as on the solid state aspects of the hyperfine fields. For the latter, for example, the results allow the partition of the hyperfine field in the Ir–Fe alloy into a contribution due to a Fermi contact term and one due to non-contact fields.

9.3.6 Tests of Time Reversal Invariance

The decay of the long-lived component of neutral K-mesons [50] into two charged pions can only be explained by CP violation. Bernstein et al. [51] explained it as a maximal violation of CP- and T-invariance of the electro-weak interaction of hadrons. In such a case, one also expects time reversal violating terms of the order 10^{-3} in nuclear decays. Lloyd [52] first pointed out that in a nuclear transition in which a γ-ray of mixed multipolarity (e.g., $M1$–$E2$) is emitted or absorbed, the reduced matrix elements of the current multi-pole moments which give the amplitudes of the respective radiations are in phase (or 180° out of phase) if the strong (and electromagnetic) interactions are T invariant. On the other hand, if there is a small T non-invariant admixture in the interactions, then the reduced matrix elements are no longer relatively real. For a mixed $M1$–$E2$ transition, the mixing ratio Δ is defined as:

$$\Delta = < ||E2|| > / < ||M1|| > = |\delta| \, e^{i\eta}. \tag{9.6}$$

Under time reversal invariance, Δ is real and thus $\eta = 0$ or π and $\sin \eta = 0$. To detect time reversal invariance, one has to find an observable which is invariant against parity reversal but not against a time reversal transformation. Jacobson and Henley [53] and Stichel [54] have determined such relations. One of these is given by the following expression:

$$(k \cdot < j > \times \varepsilon)(k \cdot < j >)(\varepsilon \cdot < j >), \tag{9.7}$$

with terms $< j >$ which can be any arbitrary combination of the ground state or excited state nuclear spins, and linear polarization ε of radiation emitted in the direction of k. Figure 9.13 (left side) illustrates such a configuration [55]. Several Mössbauer studies were carried out in search of time reversal invariance: one with ^{99}Ru [56] and two for ^{193}Ir [55, 57]. The right side of Fig. 9.13 shows results from [55]: the sum of two spectra with polarization in source and absorber reversed (top), and the difference spectrum (bottom). A finite value of $\sin \eta$ would result in yields different from zero in the bottom spectrum. The fit results of the relevant hyperfine structure to the data, averaged over several measurements, provide a value $\sin \eta = (1.3 \pm 2.5) \times 10^{-3}$.

At such a level of precision, however, it was shown by Hannon and Tremmel [58] that other effects than the nuclear matrix elements can contribute a phase factor ζ which cannot be distinguished from η. The transition radiation can induce currents in inner-shell electrons, which are of different magnitude for $M1$ and $E2$ amplitudes. For the example of the time reversal measurement described above, this

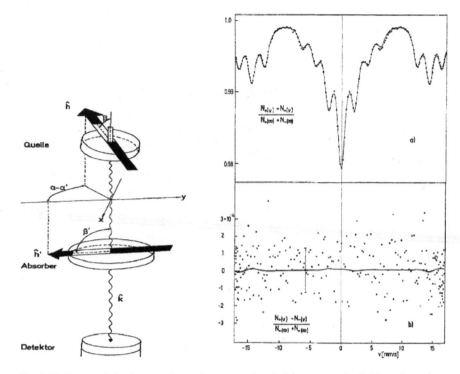

Fig. 9.13 On the *left*, the experimental setup and polarizing magnetic field configurations are illustrated for the Mössbauer time reversal test of [55]. The massive *arrows* indicate the magnetization axes for source and absorber with angles β and β' relative to the photon momentum vector k, and their relative azimuthal orientation $(\alpha - \alpha')$. On the right the experimental results are shown as a sum spectrum (**a**) and a difference spectrum (**b**) for two opposite magnetic field settings

additional phase factor ζ is calculated to be 0.9×10^{-3}. Clearly, such a contribution limits the accuracy, which can be obtained for a time reversal violating phase η. In addition to this direct contribution to the phase, Hannon and Trammel also predict dispersion terms from the radiation induced currents in inner-shell electrons. With regard to the time reversal measurement, these do not play a role since they disappear in the difference spectrum. However, as discussed in the next section, such dispersion terms can be quite large in regular Mössbauer spectra, in particular for $E1$ nuclear transitions.

9.3.7 Dispersion Phenomena in Mössbauer Spectra

In several earlier Mössbauer studies, unexpected and quite noticeable line shape asymmetries had been observed, such as the one seen in Fig. 9.14 for the 6.4 keV $E1$ γ-transition in ^{181}Ta [59]. There were several attempts for an explanation of

Fig. 9.14 Mössbauer spectrum for the 6.2 keV $E1$ γ-ray transition in ^{181}Ta for a tungsten metal source and tantalum metal absorber. A dispersion term was added to explain the asymmetric line shape [59]

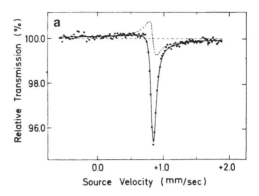

the line asymmetries, including the interference between Rayleigh and Mössbauer forward scattering for the spectrum in Fig. 9.14. The actual cause for the dispersion phenomenon turned out to be the interference between photoelectric absorption and internal conversion following nuclear excitation [58, 60, 61]. This is perhaps most easily understood in that the final states for both processes are identical and thus indistinguishable and the amplitudes add coherently.

The interference effect should be most evident for $E1$ transitions because of the dominance of $E1$ in photoelectric excitation. This is indeed the case as seen for ^{181}Ta (Fig. 9.14; see also the discussion in Sect. 9.3.4). Borobchencko et al. [62] reported an asymmetry for the 26 keV $E1$ transition of ^{161}Dy, and attributed it to the interference effect. Yet the earlier data [59, 62] were obtained for absorption lines more than ten times the natural line width due to unresolved quadrupole splitting.

A systematic study was performed with most $E1$ Mössbauer transitions known at that time [63], except for ^{181}Ta and nuclei in the actinide region. The main goal was to establish the existence and determine the magnitude of the interference amplitude for unbroadened transitions [^{155}Gd (105 keV) and ^{153}Eu (97 keV)] and transitions with line widths from two to three times natural widths [^{155}Gd (87 keV) and ^{161}Dy (75 keV)]. The 26 keV transition of ^{161}Dy was reexamined.

An overall number of about 50 absorption spectra were measured and dispersion terms were observed for all transitions. Some of the results are shown in Fig. 9.15 and in Table 9.2. The detailed analysis is described in [63]. Overall, the results indicate sizable asymmetries, up to 7%, which need to be considered in the analysis of any such spectra if, e.g., precision values for hyperfine parameters and nuclear moments are the goal.

9.4 Conclusion and Perspectives

As discussed at the outset of this chapter, most of the research in nuclear physics using the Mössbauer effect was essentially carried out in the first two decades after its discovery. For the nuclear transitions amenable to the Mössbauer effect,

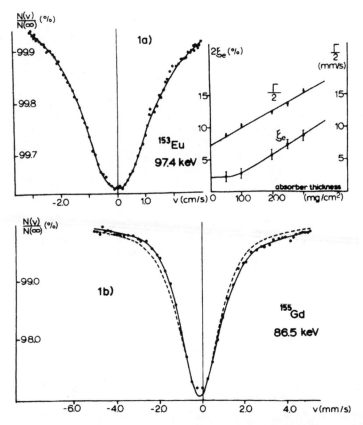

Fig. 9.15 Mössbauer studies to investigate the $E1$ dispersion term 2ξ [58]. Shown are results for the 97.4 keV transition in ^{153}Eu and the 86.5 keV transition in ^{155}Gd [63]. The *inset* on the *upper right* shows absorber thickness dependences for the line width Γ and the dispersion term

Table 9.2 Experimental and theoretical dispersion amplitudes 2ξ for $E1$ Mössbauer transitions [62]

Isotope	E_γ (keV)	$2\xi_{exp}$	$2\xi_{th}$
^{153}Eu	97.4	0.022 ± 0.005	0.029
^{155}Gd	96.5	0.05 ± 0.01	0.035
^{155}Gd	105.3	0.035 ± 0.010	0.026
^{161}Dy	25.6	0.07 ± 0.01	0.067
^{161}Dy	74.5	0.06 ± 0.01	0.050

most had been studied in their nuclear properties by then. The results provided important precision data for model calculations and nuclear structure theory. After the early 1980s, only perhaps five new transitions were added to the more than 100 previously established. Some nuclear problems may still warrant further studies, such as an improved test of time reversal invariance in nuclear transitions, perhaps even at a light source with highly polarized γ-rays and an emitter plus absorber plus scatterer setup as developed recently [64]. The general thrust of the Mössbauer effect is, however, without doubt towards application for a wide range of areas, as impressively demonstrated in this book.

However, nuclear techniques are likely to continue some role in helping to expand on some of the applications. This is, for example, the case with Mössbauer sources that are generated online in a nuclear reaction. This can be for different reasons, such as the lifetime of the mother nucleus populating the Mössbauer state being too short for conventional Mössbauer spectroscopy; or the exotic chemical and physical changes induced by the nuclear reaction that are not achievable by ordinary thermal, electromagnetic, or mechanical excitations; or the properties of single, fully isolated atoms implanted in special materials; or the relaxation of highly excited processes far from equilibrium on times scales determined by the half-life of the mother nucleus, thus much longer than the half-life of the Mössbauer state; and possibly other aspects.

In-beam nuclear reactions have already been used in the early phase of the Mössbauer studies. To produce sources of very short-lived isotopes, thermal neutron capture was invoked online [65–67]. While this approach was little pursued in the interim years, new in-beam Mössbauer studies with neutrons have been performed recently [68, 69]. Thermal neutron capture is believed to be unique and different from other nuclear processes since neutron propagation, as well as the neutron capture reaction, are free from ionization induced by the passage of charged particles. On the other hand, the capture reaction liberates typically 8 MeV energy in γ-rays which leads to several hundred electron volt energy deposited by the recoiling nucleus which experiences unusual chemical and physical state changes. One can observe relaxation products of a highly excited atom far from equilibrium with the medium.

A second approach is Coulomb excitation of the Mössbauer state in a heavy ion collision [70–73]. This generally leads to the recoiling Mössbauer nucleus implanted as a single isolated atom. Recent measurements with ^{57}Fe were aimed at the study of the magnetic properties of such atoms in graphite [74]. Indeed magnetism at the atomic scale was observed in the nominally diamagnetic highly oriented pyrolytic graphite. Without going into the details of this specific study, we just mention that there are strong efforts to develop materials in which the magnetic ion constitutes only a minor fraction of the material. Driving force for this research is both, interest in the fundamental physics of magnetism as well as possible applications, e.g., for spintronics or the development of lighter magnetic materials.

Finally, the availability recently of energetic short-lived radioactive beams has opened new opportunities for the Mössbauer effect. The high energy of beams from in-flight fragmentation reactions provide for wide spatial distribution and extremely dilute atoms in materials. For ^{57}Fe studies, 85.4 s ^{57}Mn becomes available as Mössbauer source nucleus. Manganese, for example, takes various oxidation states from 0 to 7^+ in ordinary solids, and the higher valence states of Mn^{6+} and Mn^{7+} are stable under normal conditions. Analytical and nuclear chemists have been much interested in such *hot atom* chemistry. Interesting Mössbauer studies

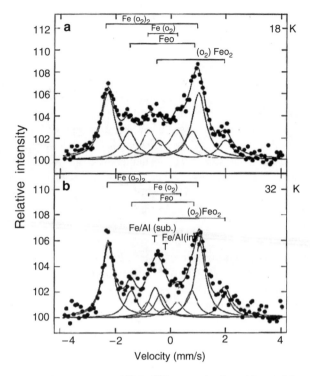

Fig. 9.16 In-beam Mössbauer spectra of $^{57}Mn/^{57}Fe$ atoms implanted into solid oxygen at 18 and 32 K [74]. The various sub-spectra indicated in the analysis of spectrum (**a**) are all assumed to be quadrupole doublets

have been performed, for example, on ^{57}Mn implanted into solid oxygen [75] and in n-type silicon [76]. Figure 9.16 shows, as an example, Mössbauer spectra of $^{57}Mn/^{57}Fe$ implanted into solid oxygen at 18 K and 32 K.

With the description of some newer studies involving nuclear physics aspect, we have nevertheless quickly moved from the nuclear physics to the application aspects of the Mössbauer effect. This is the central area where the biggest benefits of the Mössbauer effect lie. Returning to the beginning of this chapter which describes the original attempts in nuclear resonance fluorescence, it may be worth to mention in conclusion that significant progress has also been achieved in the conventional γ-ray scattering studies of nuclei. With the use of narrow band synchrotron-based γ radiation from laser back scattering or high-resolution tagged Bremsstrahlung, inelastic γ-ray scattering has provided extensive and important structure information for nuclear physics.

References

1. R.L. Mössbauer, Z. Phys. **151**, 124 (1958)
2. See, e.g., E. Pollard, D.E. Alburger, Phys. Rev. **74**, 926 (1948)
3. P.B. Moon, Proc. Phys. Soc. **A64**, 76 (1951)
4. K.G. Malmfors, Arkiv Fysik **6**, 49 (1952)
5. F.R. Metzger, Phys. Rev. **101** 286 (1956)
6. See, e.g., F.R. Metzger, Phys. Rev. **110**, 123 (1958)
7. W.E. Lamb, Phys. Rev. **55**, 190 (1939)
8. M. Goldhaber, L. Grodzins, A.W. Sunyar, Phys. Rev. **109**, 1015 (1958)
9. Periodic Table/*Properties of Isotopes Relevant to Mössbauer Spectroscopy*. Mössbauer Effect Data Center (MEDC)
10. A. Sonzogni, *National Nuclear Data Center* (Brookhaven National Laboratory, USA)
11. O.C. Kistner, A.W. Sunyar, Phys. Rev. Lett. **4**, 412 (1960)
12. R.V. Pound, G.A. Rebka, Phys. Rev. Lett. **3**, 439 (1959)
13. J.P. Schiffer, W. Marshall, Phys. Rev. Lett. **3**, 556 (1959)
14. W. Henning, P. Kienle, E. Steichele, F. Wagner, Phys. Lett. **22**, 446 (1966)
15. P. Steiner, E. Gerdau, W. Hautsch, D. Steenlen, Z. Phys. **221**, 281 (1969)
16. J. Speth, W. Henning, P. Kienle, J. Meyer, in *Isomer Shifts and Nuclear Models. Mössbauer Isomer Shifts*, ed. by G.K. Shenoy, F.E. Wagner (North-Holland, 1978), p. 795
17. F. Wagner, U. Zahn, Z. Phys. **233**, 1 (1970)
18. L. Wilets, Fys. Medd. Dan. Vid. Selsk **29**, no. 3 (1954)
19. B.A. Jacobsohn, Phys. Rev. **96**, 1637 (1954)
20. H. Backe, E. Kankeleit, H.K. Walter, in *Muonic Isomer Shifts. Mössbauer Isomer Shifts*, ed. by G.K. Shenoy, F.E. Wagner (North-Holland, 1978), p. 833
21. N. Boos et al., Phys. Rev. Lett. **72**, 2689 (1994)
22. Further examples and references in: M.L. Bisell et al., Phys. Lett. B **645**, 330 (2007)
23. J. Bardeen, L.N. Cooper, J.R. Schrieffer, Phys. Rev. **108**, 1175 (1957)
24. N.N. Bogoliubov, Sov. Phys. JETP **22**, 578 (1958)
25. R.A. Uher, R.A. Sorensen, Nucl. Phys. **86**, 1 (1966)
26. A.B. Migdal, *Theory of Finite Fermi Systems and Applications to Atomic Nuclei* (Interscience, New York, 1967)
27. V.A. Belyakov, Sov. Phys. JETP **22**, 578 (1966)
28. J. Speth, Nucl. Phys. A **135**, 445 (1969)
29. J. Meyer, J. Speth, Phys. Lett. B **39**, 330 (1972)
30. A.N. Bohr B.R. Mottelson, Kgl. Dan. Mat. Fys. Medd. **27**, 7 (1953)
31. A. Faessler, W. Greiner, Z. Phys. **168**, 425 (1962)
32. E.R. Marshalek, Phys. Rev. Lett. **20**, 214 (1968)
33. J. Meyer, J. Speth, Nucl. Phys. A **203**, 17 (1973)
34. W. Henning, Z. Phys. **217**, 438 (1968)
35. E.R. Marshalek, J.B. Milazzo, Phys. Rev. Lett. **16**, 190 (1966)
36. G.K. Shenoy, G.M. Kalvius, in *Hyperfine Interactions in Excited Nuclei*, ed. by G. Goldring, R. Kalish (Gordon and Breach, New York, 1971), p. 1201
37. G.M. Kalvius, G.K. Shenoy, *Atomic Data and Nuclear Data Tables* (1974)
38. P.B. Russel, G.L. Latshaw, S.S. Hanna, G. Kaindl, Nucl. Phys. A **210**, 133 (1973)
39. Z.P. Li, T. Niksic, D. Vretenar, J. Meng, Phys. Rev. C **80**, 061301(R) (2009)
40. F. Iachello, N.V. Zamfir, Phys. Rev. Lett. **92**, 212501 (2004)
41. W. Henning, P. Kienle, H.J. Körner, Z. Phys. **199**, 207 (1967)
42. M. Kalvius, J.K. Tison, Phys. Rev. **152**, 829 (1966)
43. M.E. Bunker, C.W. Reich, Phys. Lett. B **25**, 396 (1967)
44. E.R. Bauminger et al., Phys. Lett. B **30**, 531 (1969)
45. F. Bitter, Phys. Rev. **76**, 150 (1949)
46. A. Bohr, V.F. Weisskopf, Phys. Rev. **77**, 94 (1950)

47. L. Grodzins, N. Blum, Rev. Mod. Phys. **30**, 528 (1964)
48. G.J. Perlow, W. Henning, D. Olson, G.L. Goodman, Phys. Rev. Lett. **23**, 680 (1969)
49. F. Wagner, G. Kaindl, P. Kienle, H.J. Körner, Z. Phys. **207**, 500 (1967)
50. J.H. Christenson, J.W. Cronin, V.L. Fitch, R. Turlay, Phys. Rev. Lett. **13**, 138 (1964)
51. J. Bernstein, G. Feinberg, T.D. Lee, Phys. Rev. **139**, 1650 (1965)
52. S.P. Lloyd, Phys. Rev. **81**, 161 (1951)
53. B.A. Jacobson, E.M. Henley, Phys. Rev. **113**, 234 (1959)
54. P. Stichel, Z Phys. **150**, 264 (1958)
55. E. Zech, Z. Phys. A **239**, 197 (1970)
56. O.C. Kistner, Phys. Rev. Lett. **19**, 872 (1967)
57. M. Atac, B. Chrisman, P. Debrunner, H. Frauenfelder, Phys. Rev. Lett. **20**, 691 (1968)
58. J.P. Hannon, G.T. Trammel, Phys. Rev. Lett. **21**, 726 (1968)
59. C. Sauer, E. Matthias, R.L. Mössbauer, Phys. Rev. Lett. **21**, 961 (1968)
60. Y. Kagan, A.M. Afans'ev, V.K. Voitovetski, Zh. Eksp. i Teor. Fiz. Pis'ma Redak **8**, 342 (1968)
61. H. Lipkin, L. Tassie, *Report at the First Mössbauer Conference* (Allerton House, 1960)
62. D.V. Borobchencko, I.I. Lukashevich, V.V. Sklyarevskii, N.I. Filippov, Zh. Eksp. I Teor. Fiz. Pis'ma Redakt. **9**, 237 (1969)
63. W. Henning, G. Bähre, P. Kienle, Phys. Lett. B **31**, 203 (1970)
64. M. Seto et al., Phys. Rev. Lett. **102**, 217602 (2009)
65. D.W. Hafemeister, E. Brooks Shera, Phys. Rev. Lett. **14**, 593 (1965)
66. J. Fink, P. Kienle, Phys. Lett. **17**, 326 (1965)
67. W. Henning, D. Heunemann, W. Weber, P. Kienle, H.J. Körner, Z. Phys. **207**, 505 (1967)
68. Y. Kobayashi et al., J. Rad. Nucl. Chem. **272**, 623 (2007)
69. Y. Kobayashi et al., Hyperfine Int. **187**, 49 (2008)
70. G.D. Sprouse, G.M. Kalvius, F.E. Obenshain, in *Mössbauer Effect Methodology*, ed. by Gruverman (Plenum, New York, 1968)
71. M. Menningen et al., Europhys. Lett. **3**, 927 (1987)
72. S. Laubach et al., Z. Phys. B **75**, 173 (1989)
73. P. Schwalbach et al., Phys. Rev. Lett. **64**, 1274 (1990)
74. R. Sielemann et al., Phys. Rev. Lett. **101**, 137206 (2008)
75. Y. Kobayashi et al., Hyperfine Int. **166**, 357 (2005)
76. Y. Yoshida et al., Physica B **401**, 101 (2007)

Chapter 10
Isomer Shifts in Solid State Chemistry

F.E. Wagner and L. Stievano

10.1 Introduction

The isomer shift of the Mössbauer resonance is a rather unique quantity that cannot be obtained by any of the other techniques used for measuring hyperfine interactions in solids, such as NMR or perturbed angular correlations (TDPAC). It *shifts* the resonance pattern as a whole without affecting the magnetic dipole and electric quadrupole hyperfine *splittings*. Methods that measure only these hyperfine splittings are insensitive to the isomer shift. The magnitude of the observed shift is proportional to the product of a nuclear parameter, the change $\Delta \langle r^2 \rangle$ of the nuclear radius that goes along with the Mössbauer transition, and to an electronic property of the material, the electron density $\rho(0)$ at the Mössbauer nucleus or, more precisely, to the difference $\Delta \rho(0)$ of the electron densities at the Mössbauer nuclei in the materials of which the source and the absorber are made. The electron density at the nucleus is due to s-electrons and, to a lesser extent and mainly in heavy nuclei, to relativistic $p_{1/2}$-electrons. All the other electrons have a vanishing density inside the nucleus and do not contribute. Thus, to a very good accuracy, the Mössbauer isomer shift enables one to obtain information on the s-electron density at the Mössbauer nuclei in solids.

The isomer shift thus yields important insights into the chemical bonding in solids. For instance, it often allows an easy distinction between different oxidation states of an element, e.g., between divalent and trivalent iron or divalent and tetravalent tin. The s-electron densities are also sensitive to covalency in chemical compounds and to effects of band structure in metals. For this reason, the isomer

F.E. Wagner (✉)
Physics Department, Technical University Munich, 85747 Garching, Germany
e-mail: friedrich.wagner@ph.tum.de

L. Stievano
Université Montpellier 2, Case 1502, 34095 Montpellier cedex 5, France
e-mail: lorenzo.stievano@univ-montp2.fr

M. Kalvius and P. Kienle (eds.), *The Rudolf Mössbauer Story*,
DOI 10.1007/978-3-642-17952-5_10, © Springer-Verlag Berlin Heidelberg 2012

shift is of outstanding importance in chemical applications of the Mössbauer effect. In 1978, a special volume on Mössbauer isomer shifts was edited by Shenoy and Wagner [1], which summarized the knowledge in this field existing 20 years after the discovery of the Mössbauer effect. In the meantime, considerable progress has been made in the calibration of the isomer shift, i.e., in the determination of values for $\Delta \langle r^2 \rangle$, and the applications have widened, though without any really basic new developments.

It should not go unmentioned that, in addition to the isomer shift, there is another effect that shifts the Mössbauer resonance, namely the so-called second-order Doppler shift (SOD). This is a relativistic effect resulting from the time dilatation going along with the vibrational motion of atoms bound in a solid. The SOD is treated in chapter 14 of this volume [2]. It adds to the isomer shift, which means that one always measures the sum of both shifts and then needs to unravel the two independent contributions in some manner. In doing this, it is often helpful that the isomer shift is practically independent of temperature, while the SOD depends on temperature because the mean square velocity of the atoms in a solid does. However, the SOD does not vanish completely even at very low temperaures because of zero-point motion. Fortunately in many cases, particularly in heavy elements, the SOD is often much smaller than the isomer shift. For iron Mössbauer spectroscopy, it plays an important role, however, and must be taken into account particularly when one compares shifts measured at different temperatures.

In the following, the development and concepts of Mössbauer isomer shifts will be discussed. It is impossible to give anything close to a description of all aspects of the isomer shift, many of which have already been summarized in detail three decades ago [1]. Only for the $3d$ transitions elements and the main group elements, among which there are several much used Mössbauer resonances, will actual examples and applications be discussed in this review.

10.2   History and Basic Concepts

The isomer shift in Mössbauer spectroscopy is closely related to the isotope shift in optical hyperfine spectra of free atoms that has been known since the 1920s [3]. While the optical isotope shift in light elements is mainly due to the kinematic effect of finite nuclear mass, the isotope shift in the heavier elements results mainly from the different nuclear volume of the different isotopes of an element, as Pauli and Peierls first suggested in 1931 [4]. As a consequence, the binding energy of the electrons is slightly different for the different isotopes of an element. This difference in binding energy arises because inside the nucleus the electrostatic potential seen by the electrons is no longer a Coulomb potential proportional to $1/r$. Instead, if the nucleus is assumed to be a uniformly charged sphere, the potential inside the nucleus varies proportional to r^2, with the potential inside and outside the nucleus joining smoothly at the nuclear surface. It is practically only the s-electrons that sense this difference, because only they have a substantial probability of being inside the nucleus. In optics therefore mainly spectral lines arising from transitions between electronic states of which at least one involves an s-electron configuration show large isotope shifts [3].

In Mössbauer spectroscopy, only one isotope is involved, but the excited state and the groundstate of the nucleus usually have different sizes. The binding energy of the s-elctrons is therefore different in the excited state and the groundstate. If the nucleus is larger in the excited state than in the groundstate, the binding energy of the s-electrons is smaller in the former than in the latter, because the electrostatic potential inside the nucleus is less deep when the nucleus is larger. Consequently, the total electronic energy of the atom is higher in the excited state than in the groundstate. When the nucleus decays, the emitted Mössbauer γ-ray photon carries away this excess electronic energy, which increases its energy.

There is, however, an important experimental difference between measurements in optics and Mössbauer experiments. In the former, one measures the wavelength, and hence the quantum energy, of the light absolutely by means of a grating spectrograph or an interferometer. In Mössbauer spectroscopy, the energy of the emission line in the source is not measured absolutely, but merely compared with that of the resonance energy in the absorber; only the difference between the transition energy in the source and the absorber has to be compensated by the Doppler velocity of the Mössbauer spectrometer. As long as the energy shifts caused by the changing nuclear size are the same in the source and the absorber, one will not observe them in a Mössbauer experiment. Only the difference of the shifts in the source and the absorber is accessible experimentally. Such differences arise if, and only if, the electron densities at the nuclei in the source and in the absorber are different, as they will be when two materials with different chemical bonding are used. Excited states of nuclei, particularly long-lived ones, are often called isomeric states. It therefore makes sense to call the ensuing shift of the Mössbauer line an isomer shift in analogy to the isotope shift in optical spectroscopy. Alternatively, the expression chemical shift is sometimes used instead. The term isomer shift, however, appears more adequate, not only because the impact of this phenomenon is not restricted to chemical applications but also because the term chemical shift is also used, for physically different phenomena, in NMR and XPS (ESCA), which may give cause for confusion.

The isomer shift in Mössbauer spectra was first observed for the 14.4 keV γ-rays of ^{57}Fe: Kistner and Sunyar [5] in the spring of 1960 reported an energy shift of the 14.4 keV γ-rays of $\Delta E = 2.3 \cdot 10^{-8}$ eV between the source of ^{57}Co in stainless steel and ferric iron in hematite, Fe_2O_3. This is a relative energy difference of only $1.6 \cdot 10^{-12}$, which is very small but easily measurable owing to the high energy resolution of Mössbauer spectroscopy. Kistner and Sunyar ruled out the second-order Doppler effect as the main reason for this shift, which they correctly attributed to the effect of the different nuclear sizes in the ground and excited state of the Mössbauer transition.

Incidentally, Mössbauer spectroscopy just barely lost out to optical spectroscopy for the first observation of an isomer shift: In 1959, Melissinos and Davis [6] reported on an observation of optical "isotope" shifts of several spectral lines of mercury between the long-lived isomeric state ($T_{1/2} = 23$ h) and the groundstate of the radioactive ^{197}Hg isotope. Since these shifts were not between different isotopes of mercury, but between the groundstate and an isomeric state of the same isotope, Melissinos and Davis called this shift "isomeric isotope shift".

At this point, a numerical example may be of interest: The energy shift of $\Delta E = 2.3 \cdot 10^{-8}$ eV mentioned above, which in velocity units corresponds to a shift of $S = 0.46$ mm s^{-1}, is several orders of magnitude smaller than the total energy shift of the γ-rays caused by the change of the nuclear volume, which is about $\Delta E = 2 \cdot 10^{-4}$ eV, i.e., the isomer shift observed in Mössbauer spectra is only a small fraction of the total energy change caused by the different electron binding energies in the excited and the ground state of the nucleus. This is the energy by which the energy of the Mössbauer photon is bigger than the transition energy in a "naked" nucleus, one without any electrons at all, would be.[1] The reason for this difference is that the total s-electron density at the nucleus is about four orders of magnitude larger than the density of a valence s-electron, the density of which determines order of magnitude of chemical changes of the electron density at the nucleus. For iron, for instance, the total s-electron density at the nucleus is about 15,000 a.u.,[2] whereas the density of a single $4s$-electron is only 3.1 a.u. according to Dirac–Fock calculations for free iron atoms with a $3d^7 4s^1$ electron configuration [7]. Thus, the chemical electron density changes are about $2 \cdot 10^{-4}$ of the total electron density at the nucleus. Using $\Delta \langle r^2 \rangle = -20 \cdot 10^{-3}$ fm^2 for the change of the mean square nuclear charge radius (cf. Table 10.1), the total electron density at the nucleus causes an energy shift of the 14.4 keV gamma rays of $1.8 \cdot 10^{-4}$ eV, while the observable isomer shift corresponding to an electron density difference of one valence $4s$-electron would be

Table 10.1 Some recent values of $\Delta \langle r^2 \rangle$, $\Delta R/R$, and α derived from calculated electron densities. Errors are generally omitted since the error limits given in the references are often based on linear regression analysis and reflect only the minor part of the true uncertainties

Isotope	E_γ (keV)	$10^3 \cdot \Delta \langle r^2 \rangle$ (fm^2)	$10^3 \cdot \Delta R/R$	α (mm · s^{-1} · a_0^3)	Reference
^{57}Fe	14.4	−20	−0.78	−0.22	[34]
^{57}Fe	14.4	−26	−1.01	−0.29	[35]
^{119}Sn	23.9	7.2	0.165	0.092	[36]
^{119}Sn	23.9	6.3	0.154	0.081	[37]
^{119}Sn	23.9	7.1	0.163	0.091	[37]
^{121}Sb	37.1	−52	−1.77	−0.44	[38]
^{127}I	57.6	−10.1	0.33	−0.057	[39]
^{197}Au	77.3	4.6	0.113	0.029	[40]
^{197}Au	77.3	10.6	0.26	0.067	[41]

---

[1]The energy, indeed, becomes bigger for ^{57}Fe because the nucleus in this case is smaller in the excited state than in the groundstate.

[2]Electron densities are often given in atomic units (a.u.). An atomic unit for the electron density is the number of electrons found within a volume of a_0^3, where $a_0 = 0.0529$ nm is Bohr's radius.

about $3.3 \cdot 10^{-8}$ eV, or 0.66 mm s^{-1}, which is the order of magnitude of the actually observed shifts.

The first quantitative interpretation of the isomer shifts in compounds and alloys of ^{57}Fe was attempted by Walker, Wertheim, and Jaccarino in 1961 [8]. These authors attributed the isomer shift observed between compounds of divalent and trivalent iron, with formal $3d^6$ and $3d^5$ electron configurations, to the shielding effect, which the varying number of $3d$-electrons exert on the s-electrons of the core, i.e., mainly the $3s$-electrons. They assumed that the most ionic ferrous and ferric compounds, e.g., $FeSO_4 \cdot 7H_2O$ and $Fe_2(SO_4)_3 \cdot 6H_2O$ could be associated with the $3d^6$ and $3d^5$ electron configurations of free Fe^{2+} and Fe^{2+} ions. To obtain a quantitative value for the electron density difference at the iron nuclei caused by this shielding effect, they made use of self-consistent Hartree–Fock calculations for free iron ions with $3d^6$ and $3d^5$ electron configurations, which had been performed shortly before by Watson at MIT [9]. This electron density difference is mainly caused by the shielding effect, which an additional $3d$-electron exerts on the $3s$-electrons. The electron density at the nucleus is therefore smaller for the $3d^6$ than for the $3d^5$ electron configuration. Combining the measured shifts with the calculated electron density differences, they obtained the change of the nuclear charge radius accompanying the 14.4 keV transition. In terms of the change of the mean square nuclear charge radius, they obtained $\Delta \langle r^2 \rangle = -38 \cdot 10^{-3}$ fm^2.

The accuracy of the magnitude of this value is certainly doubtful since the assumptions made in its derivation are rather crude, but the negative sign and the order of magnitude are correct. Walker et al. were already aware of the importance of covalency effects in the less ionic compounds of iron like, for instance, the hexacyanide complexes, in which direct $4s$ contributions to $\rho(0))$ play a major role, and a similar situation arises in metallic systems owing to the s conduction electrons. The bearing of covalency effects on the determination of $\Delta \langle r^2 \rangle$ was soon pointed out. Danon [10], for instance, argued that even in the most ionic compounds of iron Fe^{3+}, because of its higher ionic charge, would inevitably be more covalent than Fe^{2+}. By neglecting this, Walker et al. certainly have underestimated the electron density difference and hence have obtained too large a value for $\Delta \langle r^2 \rangle$. The value of $\Delta \langle r^2 \rangle = -15 \cdot 10^{-3}$ Danon obtained by his arguments is, indeed, much smaller than that of Walker et al.

The efforts to improve the reliability of the $\Delta \langle r^2 \rangle$ value for ^{57}Fe and other Mössbauer resonances will be described in more detail further below, together with the present state of the art. But even though the quantitative understanding was still poor in the early days, the importance of the isomer shifts soon became obvious: One could obtain information on chemical bonding, the band structure of metals and other solid-state effects even without a perfect quantitative understanding. Even today, most applications of isomer shifts do not rely on an accurate quantitative understanding of the underlying electron density differences.

10.3   Theory of the Isomer Shift

As has been mentioned, the isomer shift S depends on the product of a nuclear quantity describing the change of the nuclear size going along with the Mössbauer transition and an electronic quantity, namely the difference of the electron charge density inside the nucleus in the source and the absorber. The theoretical calculation of the isomer shift is basically a case of simple electrostatics that can be performed without resorting to quantum mechanics, which play a decisive role in the understanding of the hyperfine splittings of the Mössbauer resonance. The theory of the isomer shift has, for instance, been given by Kalvius and Shenoy [11] and by Dunlap and Kalvius [12]. Here, only some of the resulting relevant formulae will be given.

Assuming that the electron charge density is constant inside the nucleus, one obtains an energy shift (in SI units)

$$\Delta E_{\mathrm{IS}} = \frac{1}{6\epsilon_0} Z e^2 \cdot \Delta\rho(0) \cdot \Delta\langle r^2 \rangle. \tag{10.1}$$

The isomer shift in velocity units as measured in a Mössbauer experiment is then

$$S = \frac{c}{E_\gamma} \cdot \Delta E_{\mathrm{IS}}, \tag{10.2}$$

where E_γ is the energy of the Mössbauer transition. The mean square nuclear charge radius is defined as

$$\langle r^2 \rangle = \frac{\int \rho_n(r) r^2 \mathrm{d}^3 r}{\int \rho_n(r) \mathrm{d}^3 r} = \frac{1}{Ze} \int \rho_n(r) r^2 \mathrm{d}^3 r, \tag{10.3}$$

where the integration is over all space where the nuclear charge differs from zero.

$$\Delta\langle r^2 \rangle = \langle r^2 \rangle_{\mathrm{ex}} - \langle r^2 \rangle_{\mathrm{g}} \tag{10.4}$$

is the difference of the mean square nuclear charge radii in the excited and the ground state of the Mössbauer nucleus, and

$$\Delta\rho(0) = \rho_{\mathrm{s}}(0) - \rho_{\mathrm{a}}(0) \tag{10.5}$$

is the difference between the electron densites at the nuclear site in the source and the absorber.

$\Delta\langle r^2 \rangle$ as defined above is the appropriate nuclear parameter entering into the equation for the isomer shift (10.1), since it is independent of specific assumptions on, or models of the nuclear charge distribution. However, sometimes different parameters are used to correlate the isomer shift and the electron density differences at the nuclear site.

Using the liquid drop model of atomic nuclei, where one considers the nucleus as a uniformly charged sphere with a radius

$$R = 1.2 \cdot A^{1/3} \cdot 10^{-15} \, \text{m},\tag{10.6}$$

with A being the nuclear mass number, one obtains

$$\langle r^2 \rangle = \frac{3}{5} \cdot R^2 \tag{10.7}$$

and

$$\Delta \langle r^2 \rangle = \frac{3}{5} \cdot (R_{ex}^2 - R_g^2) = \frac{3}{5} \cdot \Delta(R^2).\tag{10.8}$$

Based on this model, the relative change of the nuclear radius, $\Delta R/R$, is sometimes used in Mössbauer spectroscopy to describe the relative change of the nuclear charge radius. The relation between this quantity and $\Delta \langle r^2 \rangle$ is

$$\Delta \langle r^2 \rangle = \frac{6}{5} R^2 \cdot \left(\frac{\Delta R}{R} \right).\tag{10.9}$$

One should note that this expression is model dependent; it is not only based on the assumption of a hard-sphere nucleus, but also on a specific value for the nuclear radius (10.6). However, the use of $\Delta R/R$ has the charm of giving an idea of how small the relative changes of the nuclear size going along with isomeric transitions really are (see Table 10.1).

Since most applications of isomer shifts aim at obtaining information on $\Delta \rho(0)$[3], one often gives the relation between the isomer shift S and the electron density difference at the Mössbauer nuclei simply as

$$S = \alpha \cdot \Delta \rho(0) \quad \text{with} \quad \alpha = \frac{cZe^2}{6\epsilon_0 E_\gamma a_0^3} \cdot \Delta \langle r^2 \rangle,\tag{10.10}$$

where S is usually given in $\text{mm} \, \text{s}^{-1}$, $\Delta \rho(0)$ in atomic units (a.u.) and hence the calibration constant α is in $\text{mm} \cdot \text{s}^{-1} \cdot a_0^3$.

In deriving (10.1), the electron density difference $\Delta \rho(0)$ was assumed to be constant throughout the nuclear volume, which is an approximation. In light elements such as iron, the decrease of the electron density from the center to the surface of the nucleus is small, but for heavy elements, it is quite substantial [20] and needs to be taken into account.

When the electron density difference $\Delta \rho(0)$ is obtained by calculations of the electronic structure in atoms or molecules, relativistic calculations must be

---

[3]The rarer case of evaluating $\Delta \langle r^2 \rangle$ in terms of nuclear models is described elsewhere in this volume [13].

used, since relativistic effects increase the electron density at and near the nucleus compared to values calculated non-relativistically. The calculations must also take into account the finite size of the nucleus; assuming a point nucleus, one obtains densitites that are about 20% too high for heavy nuclei like Ir, though the difference is minor for iron [7].

For iron, the difference between relativistic and non-relativistic electron densities is still moderate, but for heavy elements such as gold or the actinides it is nearly an order of magnitude. In the early days, estimates of the true electron densities were often made by multiplying non-relativistic densities with a correction factor $S'(Z)$, such that the true relativistic electron density at the nucleus becomes $\rho(0) = |\Psi(0)^2| \cdot S'(Z)$, where $|\Psi(0)^2|$ stands for the non-relativistic density. Values of S' have been tabulated by Shirley [14]. For iron, these corrections are only moderate ($S' = 1.29$), but they are large for heavy elements (e.g., $S' = 6.84$ for Au and $S' = 13.6$ for Np). Such an approach should, however, nowadays be avoided, particularly for the heavier elements.

10.4 The Measurement of Isomer Shifts

As has been said, in a single Mössbauer experiment, one measures the isomer shift between the source and the absorber. When one is interested in the isomer shift between different absorber materials a and b, one gets this as the difference between two measurements,

$$\Delta S_{ab} = S_a - S_b, \tag{10.11}$$

though two individual measurements may not be necessary when the absorber contains both a and b, e.g., iron as Fe^{2+} and Fe^{3+}. The source properties, including the source temperature which affects the second-order Doppler shift, then cancel. Usually, one refers the isomer shift of a material of interest to some appropriately chosen standard material, whose shift will then take the role of S_b in (10.11). Preferably, the reference material should be one that can easily be measured and does not exhibit hyperfine splittings. For ^{57}Fe, however, the commonly used reference standard is metallic bcc iron as the reference standard, even though it exhibits a magnetic hyperfine splitting, which is even an advantage because it can also be used for the velocity calibration of the Mössbauer spectrometer. A list of other reference standards, not only for iron but also for other Mössbauer isotopes, has been compiled by Stevens and Gettys [15]. Often isomer shifts are simply given as measured, i.e., with respect to the source.[4] The isomer shift one finds at ambient temperature for α-Fe with a ^{57}Co:Rh source is -0.11 mm s^{-1}. When the source is at 4.2 K while the metallic iron absorber is at ambient temperature, one finds a shift

---

[4]The source matrix should be cubic and non-magnetic to yield a single line, and it should not show radiochemical effects of the nuclear decay, which is best guaranteed by a metallic matrix. For ^{57}Fe Mössbauer spectroscopy, it has become customary to use metallic rhodium as the source matrix.

of $-0.23\,\mathrm{mm\,s^{-1}}$. The difference of $0.12\,\mathrm{mm\,s^{-1}}$ is the second-order Doppler shift (SOD) in the source. Even though the exact value of the SOD depends on lattice-dynamical properties of the materials under study, the SOD observed for a ^{57}Co:Rh source is representative for other materials and gives an impression of the extent to which the SOD may affect isomer shift measurements, except for materials having a particularly low Debye temperature like biological compounds.

The lesson to be learned is that (1) isomer shifts must always be given with respect to a specific reference standard, and that (2) the temperatures of the absorber and of the standard are important and should be given.

Fortunately, in many cases the isomer shifts are big enough to be measured with relative ease. This will, very roughly, be the case when the shifts are at least 10% of the experimental FWHM linewidth. Shifts smaller than that may be due to very large linewidths arising from a short mean lifetime of the excited state, or to very small changes of the nuclear charge radius. Notoriously, small $\Delta\langle r^2 \rangle$ values are found for Mössbauer transitions that involve rotational excitations of the Mössbauer nuclei. Such excitations occur in the rare earths and the lighter $5d$ transition elements and in the actinides [16]. Much effort has been put into the measurement of such small isomer shifts, often without yielding really convincing results (e.g., [17]). If one uses the criterion of (1) the ease of observation of the resonance as such and (2) the size of the expected isomer shifts, good cases are, except ^{57}Fe, the resonances in ^{99}Ru (89.4 keV), ^{119}Sn (23.9 keV), ^{121}Sb (37.1 keV), ^{127}I (57.6 keV), ^{129}I (27.7 keV), ^{151}Eu (21.6 keV), ^{193}Ir (73.0 keV), ^{197}Au (77.3 keV), and ^{237}Np (59.5 kev).

10.5   The Calibration Problem

The Mössbauer isomer shift yields only the product of the nuclear parameter $\Delta\langle r^2 \rangle$ and of the electronic property $\Delta\rho(0)$. Since Mössbauer spectroscopy is nowadays mainly driven by the desire to obtain information on materials and to address problems in solid-state chemistry or physics, $\Delta\rho(0)$ is the quantity of interest in practically all applications. The situation is similar to magnetic hyperfine interactions, where one is interested in the magnitude of magnetic hyperfine fields at the nucleus, while the nuclear magnetic moments are of little interest but need to be known if the values of the hyperfine fields are to be determined. An accurate determination of nuclear magnetic moments is, at least in principle, easy because one can apply external fields of known magnitude and measure the magnetic hyperfine interaction in such fields.

With isomer shifts, the determination of $\Delta\langle r^2 \rangle$ requires that the electron density difference $\Delta\rho(0)$ between any two solids containing the respective Mössbauer nuclei can be determined independently. To achieve this is what one calls the calibration of the isomer shift. The difficulty is that there is practically no way to measure electron densities at the nuclei in solids except Mössbauer spectroscopy. The calibration of the Mössbauer isomer shift is therefore heavily relying on theoretical calculations of $\rho(0)$ values for selected solids and on comparing these with the experimental values of the isomer shifts. In fact, the problem of determining $\Delta\langle r^2\rangle$ is largely equivalent to the determination of nuclear quadrupole moments, which also relies largely on the calculation of the electric field gradient tensor in selected solids. Whereas for the determination of the nuclear quadrupole moment an accurate calculation for a single material is, at least in principle, sufficient, for the isomer shift one must independently calculate the electron densities $\rho(0)$ in at least two suitable solids to obtain $\Delta\rho(0)$. Since the relative electron density differences are only of the order of 10^{-4} of the total densities, the calculations need to of high precision in order to yield meaningful results.

As has already been mentioned, the first effort of a quantitative interpretation of isomer shifts of ^{57}Fe pointed the way to go to solve the problem when Walker et al. [8] used self-consistent, but still nonrelativistic, Hartree–Fock calculations for various free ion electron configurations of iron [9] and compared these with the isomer shifts for compounds that were supposed to have, at least approximately, these electron configurations in solids. It was difficult to assess, however, how exactly even accurate calculations for free atoms represent the actual electron density differences in solids.

During the first decades of Mössbauer spectroscopy, many efforts were made to derive $\Delta\rho(0)$ values by improving the approach of Walker et al., first by resorting to relativistic Dirac–Fock calculations for appropriate electron configurations of free atoms that also took the finite nuclear size into account [7,18,19]. Such calculations were the basis of many determinations of $\Delta\langle r^2\rangle$. For instance, Kalvius and Shenoy calculated $\Delta\langle r^2\rangle$ values for all known Mössbauer transitions using Dirac–Fock–Slater calculations for free atoms [11,20]. The next step of improvement was to estimate appropriate electron configurations for the Mössbauer atoms solids by molecular orbital arguments and calculations. Another was to perform self-consistent field calculations for simple clusters including only the Mössbauer atom and its nearest neighbors. Band structure calculations for the outer electrons were used for metals, and estimates of overlap integrals were made. Some of these early theoretical efforts have been reviewed by Freeman and Ellis [7].

The success of these methods was at best moderate. By the middle of the 1970s, more than 20 different approaches to the calibration of the ^{57}Fe isomer shift had been published, yielding $\Delta\langle r^2\rangle$ values between -38 and $-8.2\cdot10^{-3}$ fm^2. Reviews of these results have been given by Duff [21] and by Ingalls et al. [22]. The values thus differed by nearly a factor of five, which definitely was not a satisfactory situation.

Efforts were also made to solve the calibration problem by an independent experimental determination of the electron density at the iron nuclei in suitable pairs of compounds. The first such approach was based on internal conversion. The 14.4 keV transition in ^{57}Fe is, indeed, highly converted, with a total conversion coefficient of $\alpha_{tot} = 8.2$ [23,24]. Since internal conversion opens an additional decay channel for the excited nuclear state, the lifetime of the 14.4 keV state depends on the conversion coefficient, which in turn depends on the density of electrons near the nucleus, though not strictly inside the nucleus. Since the electron distribution near the nucleus is affected by chemical bonding, so is the lifetime of the 14.4 keV state. Rüegsegger and Kündig [25] exploited this to obtain an isomer shift calibration for ^{57}Fe by measuring the lifetime of the 14.4 keV excited state of ^{57}Fe in various iron compounds and, indeed, found relative changes of the order of about 10^{-4}, which they correlated with the respective isomer shifts. This yielded one of the lowest values ever derived, namely $\Delta\langle r^2\rangle = -8.9 \cdot 10^{-3}$ fm^2, for which the authors quote an error of about 20%. If one accepts the value of $-20 \cdot 10^{-3}$ fm^2 (Table 10.1) as realistic, the value of Rüegsegger and Kündig is by far too small.

Perhaps the most elegant, though unfortunately not generally applicable, experimental approach is to derive the electron densities at the nuclear site in different compounds of an element from the decay constants for the electron capture decay, a special type of β-decay in which a nucleus captures an electron from its own electron shell, emits a neutrino and converts to a nucleus of lower Z. Owing to the short range of the weak interaction, only electrons inside the nucleus, i.e., s-electrons can be captured. The decay rate therefore is proportional to the electron density at the nucleus and not, like in the case of internal conversion, merely to the electron distribution in the vicinity of the nucleus. The electron capture probability depends on chemical bonding in the same way as the electron density at the nucleus does, and hence in the same way as the isomer shift. If one measures the decay constants, which are the inverse of the lifetimes of the decaying nuclei, in different compounds of the parent element, one obtains differences of the electron densities, which one can compare with Mössbauer isomer shifts, and thus obtain $\Delta\langle r^2\rangle$ without systematic errors, except for minor corrections [26]. The relative changes of the lifetime are, however, again only of the order of 10^{-4} and hence difficult to measure. Moreover, the method can only be applied if the Mössbauer element has an isotope that decays by electron capture with an appropriate lieftime, which should best be of the order of days.

This method was first used by Hoffmann-Reinecke et al. [27] to determine $\Delta\langle r^2\rangle$ for the 89 keV Mössbauer transition in ^{99}Ru. The radioactive isotope they used was ^{97}Ru ($T_{1/2} = 2.9$ d), and a value of $\Delta\langle r^2\rangle = 12.8 \cdot 10^{-3}$ fm^2 was obtained, for which the authors quote an experimental error of 8%. This value is about a factor of two smaller than the value of $\Delta\langle r^2\rangle = 20 \cdot 10^{-3}$ fm^2 estimated by Shenoy and Kalvius [11,20]. Unfortunately, there are no other determinations of $\Delta\langle r^2\rangle$ for ^{99}Ru with which this value could be compared.

For ^{57}Fe, the electron capture decay of ^{52}Fe ($T_{1/2} = 8.3$ h) has been used by Meykens et al. [28] for an isomer shift calibration. They obtained $\Delta\langle r^2\rangle = -(33 \pm 3) \cdot 10^{-3}$ fm^2, a value that is rather on the high side of the distribution of values quoted at that time in the literature [21, 22] and also quite large compared to the values based on more recent calculations (Table 10.1).

A variety of other approaches to solve the calibration problem were suggested during the first two decades of Mössbauer spectroscopy [1], but eventually it transpired that the best and only way to solve the calibration problem was through reliable and sufficiently accurate calculations of $\rho(0)$ in appropriate solids. What is needed are relativistic self-consistent ab initio calculations of electron densities at the nuclei of the Mössbauer isotope of interest in selected, structurally simple compounds for which isomer shifts can also be measured. A variety of computational methods have been used in the past two decades, many of them using the augmented plane wave method and density functional theory.

The various theoretical approaches and their results have recently been reviewed by Filatov [29]. A detailed discussion of different approaches can also be found in the recent textbook by Gütlich, Bill, and Trautwein [30]. A much used program for the self-consistent calculation solid-state properties has been developed by a group at the University of Vienna [31–33]. This program, called Wien2k in its most recent version, uses density functional theory (DFT) and the full potential linearized augmented plane wave (LAPW) method to calculate a variety of properties of solids. It is available from the authors[5] and can be used by anybody seriously interested in calculations of electron densities at Mössbauer nuclei and of other hyperfine parameters like electric field gradients. To derive $\Delta\langle r^2\rangle$ for a Mössbauer isotope, one will not only calculate the electron densities at the Mössbauer nuclei in a single pair of compounds, but in as large a number of systems as possible. If one plots the experimental isomer shifts versus the calculated electron densities at the Mössbauer nuclei, the data points are expected to lie on a straight line, the slope of which yields $\Delta\langle r^2\rangle$. Deviations of the individual points from the line of best fit indicate uncertainties in the calculations and the standard deviation may be taken as an estimate of the uncertainty of the $\Delta\langle r^2\rangle$ value. An example is shown in Fig. 10.1. In Table 10.1, some recent results based on calculations with Wien2k and other computer codes are presented.

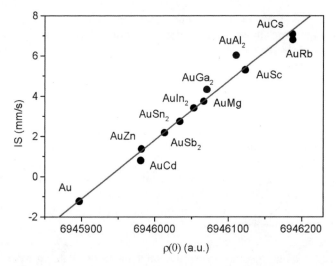

Fig. 10.1 Plot of calculated electron densities and measured isomer shifts for various compounds of gold. The resulting value of $\Delta\langle r^2\rangle$ for the 77 keV transition in ^{197}Au is given in Table 10.1. From [40]

---

[5]Details can be found on the Wien2k web site (www.wien2k.at).

The usual manner of deriving $\Delta\langle r^2\rangle$ values is thus to calculate the electron densities at the Mössbauer nuclei and compare these with the respective experimental isomer shifts. Filatov [42] recently suggested another interesting approach that is based on the idea that the isomer shift can be considered as a result of the dependence of the s-electron binding energy on nuclear size. Thus, Filatov calculated the total electronic energy of different solids as a function of the nuclear size and compared these results with experimental isomer shift energies. Applied to ^{57}Fe, this approach yielded reasonable values for $\Delta\langle r^2\rangle$ [42].

However if one does the calculations, one cannot be certain that the theoretical approach or the computer program one uses is not subject to systematic deficiencies that yield an erroneous $\Delta\langle r^2\rangle$ value even when a good fit with a small standard deviation is obtained in a plot like that of Fig. 10.1. The only way of getting some degree of certainty is probably to use different theoretical approaches, which should all yield the same $\Delta\langle r^2\rangle$ value, but do not necessarily do so. The best-studied case is that of the 14.4 keV transition in ^{57}Fe. The considerable variance of the $\Delta\langle r^2\rangle$ values obtained by different methods, both theoretical and experimental, for this case in the past 5 decades is shown in Fig. 10.2. The more recent values based on ab initio calculations do, however, converge toward a value of about $-1.1 \cdot 10^{-3}$ for $\Delta R/R$ or $-28\cdot10^{-3}$ fm^2 for $\Delta\langle r^2\rangle$, which is probably close to the true value, though with an uncertainty that is difficult to asses but certainly not less than 10%. However reliable calculations of electron densities at the nuclei may be, most methods give numerical values but convey little intuitive understanding of the causes of different electron densities in different materials. These causes are thus still an interesting and controversial matter [43].

In the following, a few practical examples of the use of isomer shifts will be reported, the selection of examples being quite arbitrary. These examples will deal with the two groups of elements for which the most Mössbauer work has been done,

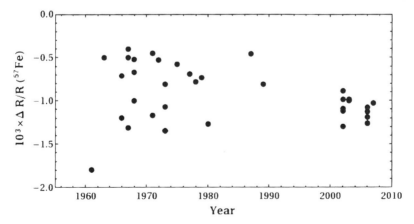

Fig. 10.2 Values of $\Delta R/R$ for the 14.4 keV transition in ^{57}Fe obtained by various methods since 1961. From [29]

namely the d transition elements and the main group elements. Two other groups of elements, for which a substantial body of Mössbauerstudies exists [44] namely the rare earths and the actinides, will be omitted for the sake of brevity.

10.6   Isomer Shifts in d Transition Elements

The 3d, 4d, and 5d transition elements have a number of Mössbauer resonances for which isomer shifts can be and have been measured. General features of these transitions are discussed elsewhere in this volume [44]. The isomer shifts observed in these resonances have been reviewed some time ago [22, 45–47, 49]. More recent results added some additional aspects but did not seriously change the general picture of isomer shifts in the d transition elements, which show some rather general behavior. An account of Mössbauer spectroscopy in transition elements in general has recently been given by Gütlich et al. [30].

In the $3d$ series the only resonance except ^{57}Fe is ^{61}Ni (67 keV), which is of minor importance since it is difficult to observe and yields only small isomer shifts that cannot be separated easily from the second-order Doppler shift [30]. For the $4d$ transition elements, the only useful Mössbauer resonance is ^{99}Ru (89 keV). For this resonance, isomer shifts have been measured in many compounds and alloy systems. The same is true for the two important transitions in the $5d$ isotopes, ^{193}Ir (73 keV) and ^{197}Au (77 keV). For the resonances in ^{189}Os (36 and 69 keV) and ^{195}Pt (99 keV), there are also some isomer shift data. They fit well into the general picture [45]. The 6.2 keV resonance in ^{181}Ta yields large isomer shifts compared to the narrow linewidth, but due to experimental difficulties mainly metallic systems have been studied [49].

The general common feature of isomer shifts in d transition elements is the systematic dependence of the shifts in reasonably ionic compounds on the number of valence d-electrons: The electron density at the nucleus decreases systematically as the number of valence d electrons increases owing to the shielding effect of the d-electrons on the s-electrons of the outer core and on the valence s-electrons, as has already been described above. For ^{57}Fe, this behavior can be followed from Fe^{2+} ($3d^6$) to Fe^{6+} ($3d^2$), as is shown in Fig. 10.3, which includes data for the oxidation state 0, i.e., for neutral iron atoms with a $3d^64s^2$ electron configuration. These data stem from Mössbauer measurements on iron atoms matrix isolated in solid noble gases [50]. The electron density at the nuclei of neutral iron atoms is higher (and hence the isomer shifts are smaller) than for Fe^{2+} with a $3d^6$ electron configuration because of the direct contribution of the $4s$-electrons.

Those $4d$ and $5d$ elements for which an adequate number of compounds could be studied show a similar behavior of the isomer shifts in compounds with the same formal electron configurations. This is shown in Fig. 10.4. The ionic electron configurations that can be observed in the various elements range from d^0 for Ru^{8+} or Os^{8+} to d^{10} for Au$^+$. Compounds of different d elements which are isoelectronic,

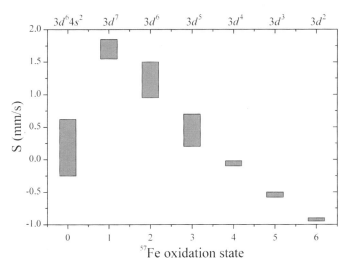

Fig. 10.3 Ranges of isomer shifts with respect to metallic α-iron observed at ambient temperature for ionic compounds of iron in different oxidation states. Note that increasing isomer shifts correspond to decreasing electron densities at the nucleus. A more detailed plot can be found in [30]

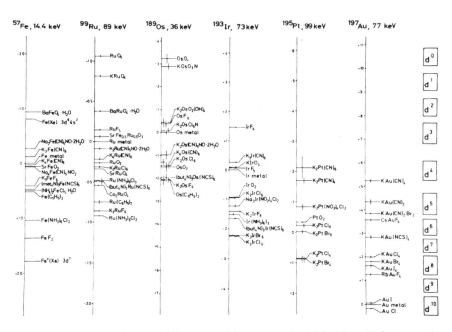

Fig. 10.4 Comparison of isomer shifts observed in compounds of $3d$, $4d$, and $5d$ elements in different oxidation states and different degrees of covalency. The isomer shift scales have been adjusted in such a way that compounds with the same number of valence d electrons lie on *horizontal lines* and that the electron densities increase from *bottom* to *top*. The number of valence d electrons is given on the *right*. From [45]

i.e., which have the same formal electron configurations and the same kind of ligands, indeed, show a very similar behavior of their isomer shifts [45].

The isomer shift regions for the individual valence states given in Fig. 10.3 are for reasonably ionic systems. In strongly covalent compounds, the electron densities are generally larger than in ionic ones, mainly because of the contributions from valence s electrons that are absent in fully ionic cases. For instance, divalent iron in the highly ionic compound $FeSO_4 \cdot 7H_2O$, in which the iron is octahedrally coordinated to six H_2O ligands, is $1.27 \, \text{mm s}^{-1}$, whereas $K_4Fe(CN)_6$, in which the bonding to the CN ligands is highly covalent, exhibits an isomer shift of $-0.12 \, \text{mm s}^{-1}$, the difference being bigger than the difference between ionic Fe^{2+} and Fe^{3+} (Fig. 10.3). The isomer shift thus yields information both on the valence state and on the covalency of compounds of iron. The isomer shifts of the other d transition elements behave in a similar manner (Fig. 10.4), reflecting the similarities in the bonding conditions in $3d$, $4d$, and $5d$ elements.

Such similarities are not restricted to chemical compounds. In alloy systems, one also finds close similarities. Figure 10.5 shows isomer shift data for ^{99}Ru, ^{193}Ir and ^{195}Pt alloyed into $3d$, $4d$, and $5d$ metals [45]. Similar data exist for ^{57}Fe and ^{197}Au [45, 48], and to a lesser extent for ^{181}Ta [49]. Taking the sign and magnitude of $\Delta \langle r^2 \rangle$ into account, one finds that the electron densities at all these Mössbauer nuclei depend in much the same way on the host lattice, whose electron structure and lattice volume appear to be the main properties affecting the isomer shifts.

Coming back to chemical compounds, an example of the importance of covalency for the isomer shift in iron compounds can be found in metalorganic complexes that exhibit high-spin (HS) $\rightleftharpoons$ low-spin (LS) transitions, as is the case, for instance, in some octahedrally coordinated Fe(II) compounds. In such compounds, the iron may undergo a transition between the $^5T_2$ ($t_2^4 e^2$) HS state and the $^1A_1$ (t_2^6) LS state. In the latter, all six $3d$ electrons occupy the lower t_{2g} states with paired spins, yielding zero total spin and a vanishing magnetic moment. In the HS state, five $3d$ electrons occupy both the lower t_{2g} and the upper e_g states with parallel spins according to Hund's first rule, while the sixth electron enters a t_{2g} state with antiparallel spin, yielding a total spin $S = 2$ and a magnetic moment of $\mu = 4\mu_B$. Whether the HS or the LS state is the electronic groundstate depends grossly on the relative magnitude of the ligand field splitting $\Delta = 10Dq$ between e_g and t_{2g}, which is about $12000 \, \text{cm}^{-1}$, and the exchange energy gained when the electron spins align parallel to one another. If the former is larger than the latter, the LS spin state is energetically preferred, in the opposite case the HS state is the preferred one. HS $\rightleftharpoons$ LS transitions may occur when the two energies are comparable. Temperature or pressure changes or other effects [30, 51] may then induce the transition. The HS state is usually preferred at higher temperatures and the LS state at lower ones [51, 52]. The transitions occurs without a serious change of the geometry of the complex, except for a decrease of the metal-ligand bond length, which may be up to 10% shorter in the LS state than in the HS state [52]. The LS state is thus expected to be more covalent than the HS state, despite the fact that the ligands are the same. This is impressively reflected in the change of the isomer shift. In the iron complex $(Fe(2\text{-pic})_3)Cl_2 \cdot EtOH$, for instance [53], one observes the pure HS state at

Fig. 10.5 Comparison of isomer shifts observed in dilute alloys of $3d$, $4d$, and $5d$ elements. The data are for various Mössbauer isotopes that were alloyed in low concentrations into $3d$, $4d$, and $5d$ metal host lattices. From [45]

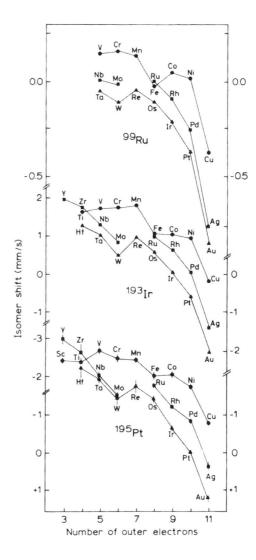

ambient temperature and the pure LS state at 82 K. In this complex, the iron has three bidentate 2-picolylamine ligands with nitrogen atoms bonding to the iron. The HS state has an isomer shift of $0.94 \, \text{mm s}^{-1}$ with respect to metallic iron, which is somewhat smaller than that for $FeSO_4 \cdot 7H_2O$ ($1.27 \, \text{mm s}^{-1}$), because the bonds are more covalent. In the LS state at 82 K, however, the bonding is much more covalent, resulting in a room temperature isomer shift of only $0.45 \, \text{mm s}^{-1}$ obtained by correction of the 82 K value for the second-order Doppler shift. This difference is less than that between $FeSO_4 \cdot 7H_2O$ and $K_4Fe(CN)_6$ given above, but it is a fine demonstration of the effect of bond lengths and the ensuing change of covalency on the electron density at the nucleus.

Fig. 10.6 Plot of electric quadrupole splittings Δ versus isomer shifts S for ^{197}Au. The regions for Au(I) and Au(III) compounds are sufficiently distinct to allow an attribution of the gold valence state. The *dot* represents metallic gold, which is usually chosen as the reference for the isomer shifts

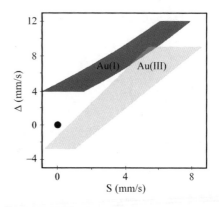

The 77 kev Mössbauer resonance in ^{197}Au has seen frequent use, mainly for studying chemical compounds. Gold is most often found in the oxidation states Au(I) and Au(III). Au(I) compounds are usually linearly coordinated to two ligands, while Au(III) forms four-coordinated planar complexes. The bonds can be reasonably ionic, like in AuCl, or highly covalent, like in KAu(CN)$_2$. The isomer shifts therefore span a wide range with low values of electron densities and isomer shifts for the more ionic compounds and with higher densities and shifts for the more covalent ones [30, 46]. The isomer shift ranges for Au(I) and Au(III) overlap, allowing no reliable distinction between the two valence states. When one also takes the electric quadrupole splitting into account, however, it is fairly easy to make that distinction. Figure 10.6 shows the regions in which Au(I) and Au(III) compounds are found on a plot of quadrupole splittings versus isomer shifts. In Fig. 10.6, the magnitude of the quadrupole splittings is plotted, irrespective of the sign. In fact, the sign of the electric field gradient is negative in Au(I) and positive in Au(III) compounds. One could therefore easily make a distinction between Au(I) and Au(III) on the basis of the quadrupole splitting alone, if its sign could be measured. This is, however, not possible for the quadrupole doublets observed in ^{197}Au, because the splitting of the $I = 3/2^+$ groundstate gives rise to a symmetric quadrupole doublet that is independent of the sign of the electric field gradient. The sign can only be obtained when single crystals are available [54], but this is rarely the case.

10.7   Isomer Shifts in Main Group Elements

Of the main group elements, only tin has been used widely. The 23.9 keV Mössbauer resonance in ^{119}Sn is, indeed, the second most used of all Mössbauer resonances with about 18% of all publications on Mössbauer spectroscopy, after ^{57}Fe with 67%. The 37.2 keV resonance in ^{121}Sb accounts for about 2%, the 35.5 keV resonance in ^{125}Te for 1.4%, and the resonances in ^{127,129}I for 1.2% [44]. We shall therefore only deal with tin; a review of the situation in the other isotopes mentioned above has, for instance, been given in [55]. The principles that determine the isomer shifts in tin can, however, be applied to some extent to the other main group elements as well. They have, for instance, been discussed in [56].

Tin occurs in the oxidation states $+2$ and $+4$, in compounds that may be either reasonably ionic or highly covalent. The $4d$ shell is filled with ten electrons in both Sn(II) and Sn(IV), hence shielding effects due to a varying number of d-electrons play no role. While ionic Sn(IV) has no additional electrons, ionic Sn(II) is expected to have two $5s$-electrons in addition. It therefore has a higher electron density at the nucleus and hence a larger isomer shift than Sn(IV), since $\Delta \langle r^2 \rangle$ is positive in this case (Table 10.1). For Sn(IV) $5s$ and $5p$-electrons are present in covalent compounds, which therefore have increasing isomer shifts as the covalency increases. The usual reference material for the isomer shifts is tetravalent tin in $CaSnO_3$, which has virtually the same isomer shift as SnO_2 or $BaSnO_3$. Since these compounds are reasonably ionic, most inorganic Sn(IV) compounds are more covalent and have positive isomer shifts up to about 1 mm s^{-1} (Fig. 10.7). SnF_4 and $(SnF_6)^{2-}$ are more ionic than the reference and have substantially smaller isomer shifts (about -0.4 mm s^{-1}). Metal organic compounds, owing to their high degree of covalency, have positive isomer shifts between about 1 and 2 mm s^{-1}. Covalency and the ligand geometry also cause electric quadrupole splittings which, together with the isomer shift, result in characteristic Mössbauer patterns for different compounds.

For Sn(II), the situation is different. Ionic Sn(II) is expected to have two $5s$-electrons outside the filled $4d$ shell, which would correspond to a positive isomer shift of about 5 mm s^{-1}. Highly ionic compounds such as $Sn(SbF_6)_2$ ($S = 4.44$ mm/s) come close to this value. Increasing covalency gives rise to $s-p$ hybridisation

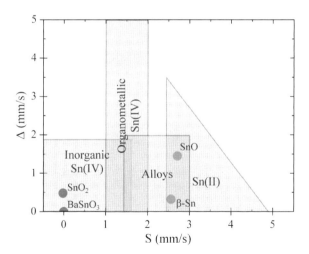

Fig. 10.7 Ranges of isomer shifts S and electric quadrupole splittings Δ in compounds and alloys of tin. Since $\Delta \langle r^2 \rangle$ is positive in this case, larger isomer shifts correspond to larger $\rho(0)$ values

and hence tends to reduce the s-electron density at the tin nuclei. At the same time, increasing electric field gradients arise, which leads to an anticorrelation of isomer shift and quadrupole splitting. However, the quadrupole splitting Δ also depends sensitively on the symmetry of the ligands around the tin. One therefore finds a wide distribution of quadrupole splittings for the same isomer shift. The smallest shifts occur in SnO ($S = 1.64\,\mathrm{mm\,s^{-1}}$; $\Delta = 1.36\,\mathrm{mm\,s^{-1}}$) and Sn(OH)$_2$ ($S = 2.30\,\mathrm{mm\,s^{-1}}$; $\Delta = 3.05\,\mathrm{mm\,s^{-1}}$). It is therefore necessary to consider both the isomer shift and the quadrupole splitting to characterize tin compounds. The situation is depicted in Fig. 10.7, which also gives the range of shifts one finds in metallic systems.

An interesting and timely application of ^{119}Sn isomer shifts can be found in the field of lithium-ion batteries, where tin has been proposed as a viable substitute for graphite as the negative electrode material. In fact, while graphite offers a relatively poor theoretical capacity of 372 mAh/g corresponding to the formation of LiC$_6$, Li insertion into metallic tin provides theoretical capacities as high as 991 mAh/g. This capacity corresponds to the formation of Li$_{22}$Sn$_5$ [57], the end member of a series of increasingly Li-rich intermetallic phases in the Li-Sn binary phase diagram, in which seven intermetallic phases exist, namely Li$_2$Sn$_5$, LiSn, Li$_7$Sn$_3$, Li$_5$Sn$_2$, Li$_{13}$Sn$_5$, Li$_7$Sn$_2$, and Li$_{22}$Sn$_5$ [58]. Tin-based electrodes, however, suffer a rapid loss of specific capacity due to the large volume changes of up to about 300% occurring during lithium insertion and extraction [59]. Strategies to overcome this difficulty are based on reducing the metal particle size, for instance by supporting the tin metal particles in an electrochemically inert buffer matrix [60–62]. Still, the result of the Li insertion is always the formation of Li-Sn alloy particles with stoichiometries corresponding to those of the known intermetallic compounds.

^{119}Sn Mössbauer spectroscopy has been used extensively in the study of tin-based electrode materials, with Mössbauer studies of the Li-Sn alloy system [63,64] serving as a basis for the interpretaion of the data. From the point of view of the crystal structure, six of the seven known Li-Sn intermetallics, namely Li$_2$Sn$_5$, LiSn, Li$_7$Sn$_3$, Li$_{13}$Sn$_5$, Li$_7$Sn$_2$, and Li$_{22}$Sn$_5$ contain several different crystallographic tin sites, which results in complicated Mössbauer spectra. Only Li$_5$Sn$_2$ shows a simple quadrupole doublet. There is, however, a characteristic decrease of the average isomer shift, i.e., the average electron density at the tin nuclei with increasing lithium content (Fig. 10.8). This decrease of the isomer shift with Li content, which is moderate between β-Sn and LiSn and then becomes steeper, has been used to study the lithium uptake and release during the electrochemical charge/discharge cycles in tin-containing negative electrodes for Li-ion batteries [65, 66] even when details on the phases formed cannot be derived. With the help of the Mössbauer hyperfine parameters, it has become possible not only to analyze but also to predict Li reaction mechanisms in tin-based negative electrodes for Li-ion batteries [67]. The Li-insertion mechanism could even be followed in situ in an electrochemical cell allowing the measurement of the Mössbauer spectra during the reaction with Li metal. It would go too far to describe the results in detail, but this is a fine example of the use of Mössbauer spectroscopy, and of isomer shifts in particular, in efforts to solve one of the pressing problems of our time.

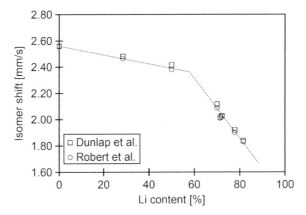

Fig. 10.8 Dependence of the [119]Sn isomer shifts in intermetallic compounds of tin and lithium on the Li content. The data are from [63,64]

10.8   Outlook

The isomer shift is just one of the aspects of Mössbauer spectroscopy, but it is certainly an important quantity that provides considerable insight into chemical bonding. It also helps empirically, as one of the hyperfine parameters that characterize a Mössbauer spectrum, to identify the chemical and physical state of the Mössbauer element. For materials with unknown components, one often simply compares the Mössbauer spectra – which may be complicated and may contain more than one Mössbauer pattern – with reference spectra of known and well-characterized reference materials. Such fingerprint applications of Mössbauer spectroscopy are often used in practice. The isomer shift is then often the important feature in a hyperfine pattern that helps to make it distinctive.

In such fingerprint applications of Mössbauer spectroscopy, a problem that often arises is that there are no suitable reference materials for the systems one wants to investigate and perhaps identify. Here, ab initio calculations of hyperfine paramters – not only the isomer shift, but also the electric quadrupole interaction and magnetic properties whenever applicable – of the systems one has in mind may help to identify hitherto unknown phases. First efforts to calculate the Mössbauer parameters for as yet unobserved, perhaps even hypothetical, materials which one hopes to identify in Mössbauer spectra have already been made [68] and will certainly become more important in the future.

References

1. G.K. Shenoy, F.E. Wagner, (eds.), *Mössbauer Isomer Shifts* (North Holland, Amsterdam, 1978)
2. W. Potzel, Relativistic Phenomena Investigated by the MössbauerEffect, The Rudolf MössbauerStory

3. H. Kopfermann, *Nuclear Moments* (Academic Press, New York, 1958)
4. W. Pauli, R.E. Peierls, Phys. Z. **32**, 670 (1931)
5. O.C. Kistner, A.W. Sunyar, Phys. Rev. Lett. **4**, 412 (1960)
6. A.C. Melissinos, S.P. Davis, Phys. Rev. **115**, 130 (1959)
7. A.J. Freeman, D.E. Ellis, in *Mössbauer Isomer Shifts*, ed. by G.K. Shenoy, F.E. Wagner (North Holland, Amsterdam, 1978), p. 111
8. L.R.Walker, G.K. Wertheim, V. Jaccarino, Phys. Rev. Lett. **6**, 98 (1961)
9. R.E. Watson, Technical Report No. 12, Massachusetts Institute of Technology, 1959
10. J. Danon, in *Applications of the Mössbauer Effect in Chemistry and Solid State Physics*, Technical Report No. 50 (International Atomic Energy Agency, Vienna, 1966), p. 89
11. G.M. Kalvius, G.K. Shenoy, Atomic Data and Nuclear Data Tables **14**, 639 (1974)
12. B.D. Dunlap, G.M. Kalvius, in *Mössbauer Isomer Shifts*, ed. by G.K. Shenoy, F.E. Wagner (North Holland, Amsterdam, 1978), p. 15
13. W.F. Henning, Nuclear Physics Application of the MössbauerEffect, The Rudolf Mössbauer-Story
14. D.A. Shirley, Rev. Mod. Phys. **36**, 339 (1964)
15. J.G. Stevens, W.L. Gettys, in *Mössbauer Isomer Shifts*, ed. by G.K. Shenoy, F.E. Wagner (North Holland, Amsterdam, 1978), p. 901
16. J. Speth, W. Henning, P. Kienle, J. Meyer, in *Mössbauer Isomer Shifts*, ed. by G.K. Shenoy, F.E. Wagner (North Holland, Amsterdam, 1978), p. 795
17. F.E. Wagner, M. Karger, M. Seiderer, G. Wortmann, in AIP Conference Proceedings, vol. 38, ed. by G.J. Perlow (American Institute of Physics, New York, 1977), p.93
18. J.B. Mann, Los Alamos National Laboratory Scientific Report LA 3691 (1968)
19. J.B. Mann, J. Chem. Phys. **51**, 841 (1969)
20. G.K. Shenoy, B.D. Dunlap, in *Mössbauer Isomer Shifts*, ed. by G.K. Shenoy, F.E. Wagner (North Holland, Amsterdam, 1978), p. 869.
21. K.J. Duff, Phys. Rev. B **9**, 66 (1974)
22. R. Ingalls, F. Van der Woude, G.A. Sawatzky, in *Mössbauer Isomer Shifts*, ed. by G.K. Shenoy, F.E. Wagner (North Holland, Amsterdam, 1978), p. 361
23. D.P. Johnson, Phys. Rev. **B1**, 3551 (1970)
24. H.U. Freund, J.C. McGeorge, Z. Physik **238**, 6 (1970)
25. R. Rüegsegger, W. Kündig, Phys. Lett. **39B**, 620 (1972)
26. E. Vatai, Nucl. Phys. **A156**, 541 (1970)
27. H. Hofmann-Reinecke, U. Zahn, H. Daniel, Phys. Lett. **47B**, 494 (1973)
28. A. Meykens, R. Coussement, J. Ladrière, M. Cogneau, M. Bogé, P. Auric, R. Bouchez, A. Bernabel, J. Godard, Phys. Rev. B **21**, 3816 (1980)
29. M. Filatov, Coord. Chem. Rev. **253**, 594 (2009)
30. P. Gütlich, E. Bill, A.X. Trautwein, *Mössbauer Spectroscopy and Transition Metal Chemistry: Fundamentals and Applications* (Springer, Berlin, 2011)
31. K. Schwarz, P. Blaha, G.K.H. Madsen, Comp. Phys. Commun. **147**, 71 (2002)
32. K. Schwarz, J. Solid State Chem. **176**, 319 (2003)
33. P. Blaha, J. Phys. Conf. Series **217**, 012009 (2010)
34. O. Eriksson, A. Svane, J. Phys. Cond. Matter **1**, 1589 (1989)
35. U.D. Wdowik, K. Ruebenbauer, Phys. Rev B **76**, 155118 (2007)
36. A. Svane, N.E. Christensen, C.O. Rodriguez, M. Methfessel, Phys. Rev. B **55**, 12572 (1997)
37. R. Kurian, M. Filatov, J. Chem. Phys. **130**, 124121 (2009)
38. A. Svane, Phys. Rev. **68**, 064422 (2003)
39. U.D. Wdowik, D. Legut, K. Ruebenbauer, J. Chem. Phys. A **114**, 7146 (2010)
40. P. Palade, F.E. Wagner, G. Filoti, Hyperfine Interactions (C) **5**, 195 (2002)
41. U.D. Wdowik, K. Ruebenbauer, J. Chem. Phys. **129**, 104504 (2008)
42. M. Filatov, J. Chem. Phys. **127**, 084101 (2007)
43. F. Neese, Inorg. Chim. Acta **337**, 181 (2002)
44. G.M. Kalvius, The World beyond Iron, The Rudolf MössbauerStory

45. F.E. Wagner, U. Wagner, in *Mössbauer Isomer Shifts*, ed. by G.K. Shenoy, F.E. Wagner (North Holland, Amsterdam, 1978), p. 431
46. H.D. Bartunik, G. Kaindl, in *Mössbauer Isomer Shifts*, ed. by G.K. Shenoy, F.E. Wagner (North Holland, Amsterdam, 1978), p. 515.
47. R.L. Cohen, in *Mössbauer Isomer Shifts*, ed. by G.K. Shenoy, F.E. Wagner (North Holland, Amsterdam, 1978), p. 541.
48. F.E. Wagner, G. Wortmann, G.M. Kalvius, Phys. Lett. **42A**, 483 (1973)
49. G. Kaindl, D. Salomon, G. Wortmann, in *Mössbauer Isomer Shifts*, ed. by G.K. Shenoy, F.E. Wagner (North Holland, Amsterdam, 1978), p. 561
50. T.K. NcNab, H. Micklitz, P.H. Barrett, in *Mössbauer Isomer Shifts*, ed. by G.K. Shenoy, F.E. Wagner (North Holland, Amsterdam, 1978), p. 223
51. P. Gütlich, J. Garcia, J. Phys. Conf. Ser. **217**, 012001 (2010)
52. E. König, G. Ritter, S.K. Kulshreshtha, Chem. Rev. **85**, 219 (1985)
53. M. Sorai, J. Ensling, K.M. Hasselbach, P. Gütlich, Chem. Phys. **20**, 197 (1977)
54. H. Prosser, F.E. Wagner, G. Wortmann, G.M. Kalvius, Hyperfine Interactions **1**, 25 (1975)
55. S.L. Ruby, G.K. Shenoy, in *Mössbauer Isomer Shifts*, ed. by G.K. Shenoy, F.E. Wagner (North Holland, Amsterdam, 1978), p. 617
56. P.A. Flinn, in *Mössbauer Isomer Shifts*, ed. by G.K. Shenoy, F.E. Wagner (North Holland, Amsterdam, 1978), p. 593
57. J. Wang, I.D. Raistrick, R.A. Huggins, J. Electrochem. Soc. **133**, 457 (1986)
58. C.J. Wen, R.A. Huggins, J. Electrochem. Soc. **128**, 1181 (1981)
59. S. Machill, T. Shodai, Y. Sakurai, J.-I. Yamaki, J. Power Sources **73**, 216 (1998)
60. Z. Chen, Y. Cao, J. Qian, X. Ai, H. Yang, J. Mater. Chem. **20**, 7266 (2010)
61. A. Aboulaich, M. Womes, J. Olivier-Fourcade, P. Willmann, J.-C. Jumas, Solid State Sci. **12**, 65 (2010)
62. X.-L. Wang, W.-Q. Han, J. Chen, J. Graetz, ACS Appl. Mater. Interfaces **2**, 1548 (2010)
63. R.A. Dunlap, D.A. Small, D.D. MacNeil, M.N. Obravac, J.R. Dahn, J. Alloys Comp. **289**, 135 (1999)
64. F. Robert, P.-E. Lippens, J. Olivier-Fourcade, J.-C. Jumas, F. Gillot, M. Morcrette, J.-M. Tarascon, J. Solid State Chem. **180**, 339 (2007)
65. A. Aboulaich, F. Robert, P.-E. Lippens, L. Aldon, J. Olivier-Fourcade, P. Willmann, J.-C. Jumas, Hyperfine Interactions **167**, 733 (2006)
66. J.-C. Jumas, M. Womes, L. Aldon, P.-E. Lippens, J. Olivier-Fourcade, Mössbauer Effect Data Reference J. **33**, 46 (2010)
67. S. Naille, J.-C. Jumas, P.-E. Lippens, J. Olivier-Fourcade. J. Power Sources **189**, 814 (2009)
68. P.E. Lippens, J.-C. Jumas, J. Olivier-Fourcade, Hyperfine Interactions **156/157**, 327 (2004)

Chapter 11
The Internal Magnetic Fields Acting on Nuclei in Solids

Israel Nowik

11.1 Introduction

Nuclear energy level magnetic splitting can be observed in atomic spectra, in electron spin resonance spectra, in nuclear magnetic resonance spectra, in perturbed angular correlation of nuclear γ-ray measurements, in nuclear orientation, and in specific heat measurements. But none of these methods measures this interaction to the variety of details Mössbauer spectroscopy (MS) can obtain. MS measures very accurately the hyperfine interaction energy splitting, at least for two nuclear levels, and yields information frequently unavailable from any of the other methods. This chapter will include the following sections: Sect. 11.2.1. *Magnetic hyperfine interactions in free ions:* since generally the ionic levels are highly degenerate, an effective magnetic hyperfine field cannot characterize the splitting of the nuclear levels. The general magnetic hyperfine Hamiltonian is presented. Section 11.2.2. *Hyperfine fields in nd ($n = 3, 4, 5$) and nf ($n = 4, 5$) transition elements in solids:* crystalline fields and molecular electron sharing and free electrons change completely the form of the magnetic hyperfine interaction. Yet without an external magnetic field or exchange field (in magnetically ordered materials), the ionic levels are, still, either degenerate, and thus not describable in terms of an internal field, or singlet states which are non-magnetic. Section 11.2.3. *Influence of external magnetic fields:* if the Zeeman splitting of the electronic levels, which removes completely their degeneracy, is very large in comparison with the hyperfine interaction, the nuclear level splitting can be described in terms of an effective magnetic field (internal field) acting on the nucleus. Section 11.2.4. *Static internal fields in magnetically ordered systems:* in such systems, the exchange interactions remove completely the degeneracy of the ionic levels, and the hyperfine interaction is described in terms of an effective magnetic field corresponding to each ionic level. Section 11.2.5.

I. Nowik (✉)
The Racah Institute of Physics, The Hebrew University, 91904 Jerusalem, Israel
e-mail: nowik@vms.huji.ac.il

M. Kalvius and P. Kienle (eds.), *The Rudolf Mössbauer Story*,
DOI 10.1007/978-3-642-17952-5_11, © Springer-Verlag Berlin Heidelberg 2012

Dynamics of internal fields in magnetically ordered systems: thermal relaxation among ionic levels leads to relaxation among the hyperfine interaction levels, and if this relaxation rate is very fast relative to the hyperfine interaction splitting, then in paramagnetic systems the magnetic interaction is averaged out to zero, while in magnetically ordered systems, a thermal average of the various ionic effective fields is observed. When the relaxation rate is low, one observes unusual spectra, which can yield the relaxation rate. In small magnetically ordered particles (single domain particles), the total magnetization of the particle is thermally fluctuating, and the rate of this fluctuation can be obtained from the MS spectral shapes. These phenomena will be discussed in detail in the chapter of this book written by Jochen Litterst in chapter 12.

Section 11.2.6. *Internal magnetic fields at phase transitions:* magnetic transition at T_c, transitions from ferromagnetism to antiferromagnetism, spin reorientation, crystallographic transition, and metal–insulator transition can all be observed experimentally, since in all these cases, changes occur in the internal fields. Section 11.2.7. *Internal fields for the following unusual spin structures*: helical, incommensurate with lattice, spin glasses, spin density waves, and magneto-superconductors. The unique characteristics of the internal fields in such systems as observed by MS. Section 11.2.8. *Transferred hyperfine fields*: in any diamagnetic ion exposed to an exchange field from neighboring magnetic ions, its nucleus experiences an internal magnetic field, caused by the exchange polarization of inner electronic closed shells. In metallic systems, the conduction electron spin density in the nuclei produces an internal magnetic field. Section 11.2.9. *Internal fields in thin layers*: on surfaces, thin layers, interfaces, the internal fields display different orientations and absolute values.

Each of the above sections will include experimental results of unique cases (using a diversity of probe nuclei), stressing the information, which could not be obtained by any other method.

11.2 Internal Fields in Magnetic Solids

Hyperfine interactions were already observed in 1881 by A.A. Michelson. The phenomenon was explained in terms of quantum mechanics by W. Pauli in 1924, and since 1958 was extensively studied using Mössbauer spectroscopy. An extended review of the theory of magnetic hyperfine interactions can be found in the books of Abragam and Bleany, and Abragam [1, 2], and the basics in any book dealing with Mössbauer spectroscopy [3].

The general Hamiltonian of the interaction of the magnetic nuclear moment $g_n\mu_n I$ with all the electrons in its environment is given by:

$$\mathcal{H} = -g_n\mu_n I \left\{ 2\mu_B \sum \left(r_i^{-3}\right) l_i + 2\mu_B \sum \left(r_i^{-3}\right) \left(3r_i \left(r_i \cdot s_i\right) / r_i^2 - s_i\right) \right.$$

$$\left. + 2\mu_B \sum \left(8\pi/3\right) s_i \delta(r_i) \right\}. \tag{11.1}$$

Here the sums extend to all the electrons, each located at a distance r_i from the nucleus. The three sums in the parentheses represent the contributions of the electrons with angular momentum l_i, the dipolar contribution of the electron spins s_i, the angular matrix elements of N and the contribution (Fermi contact term) of those electrons with finite probability to be present in the nucleus (s and relativistic $p_{1/2}$ electrons). We may write this Hamiltonian in the form:

$$\mathcal{H} = -g_n \mu_n 2\mu_B < r^{-3} > I \cdot N, \tag{11.2}$$

where $< r^{-3} >$ is the average of r^{-3} over the electronic radial wave functions. This expression looks formally like a nucleus exposed to a "magnetic field" $2\mu_B < r^{-3} > N$ However N is an electronic vector operator. A non-degenerate atomic state will have a single expectation value $< N >$ for N which then can be called the internal magnetic field, $H_{int} = 2\mu_B < r^{-3} >< N >$ (or effective field, H_{eff}) acting on the nucleus due to that state. Such a situation can occur only if a strong external magnetic field, or in magnetic solids the exchange fields, removes all atomic degeneracies.

11.2.1   The Magnetic Hyperfine Interactions in Free Atoms or Ions

The matrix elements of the operator N appearing in (11.2) in an atomic state, well defined by its total angular momentum J, angular momentum L and total spin S, are proportional to those of J. Thus the Hamiltonian in (11.2) can be simplified to the form: $\mathcal{H} = a(J \cdot I)$, and if $F = J + I$, then the energy states of the hyperfine structure are given by $E_F = 1/2a(F(F + 1) - L(L + 1) - S(S + 1))$ and each F state is still $(2F + 1)$ fold degenerate, clearly the hyperfine interaction splitting does not resemble that of a magnetic field acting on the nucleus. In atoms or ions with closed electronic shells, in the ground state, all electrons are paired and both L and S are zero and thus there are no magnetic hyperfine interactions. For a single electron with a wave function characterized by the quantum numbers $|nlsj >$, the expression for "a" is given by:

$$a_j = (16/3)\pi \mu_B g_n \mu_n |\psi_s(0)|^2 \quad \text{for} \quad l = 0, \tag{11.3}$$

and

$$a_j = 2\mu_B g_n \mu_n < nl|r^{-3}|nl > l(l + 1)/j(j + 1) \quad \text{for} \quad l \neq 0.$$

11.2.2   The Magnetic Hyperfine Interactions in nd (n = 3, 4, 5), and nf (n = 4, 5) Elements in Solids

The electrons of nd transition elements in solids are exposed to the influence of their neighboring ions, by their electrostatic potentials and/or electron sharing, or even forming electron conduction bands. This influence may cause the nd ions to display, under various conditions, diamagnetic behavior, or a variety of paramagnetic behavior, which below a certain temperature (T_M) may order magnetically, and in the case of metallic systems, local or itinerant magnetic order may appear. The conditions for all these phenomena are discussed in several books dealing with Mössbauer spectroscopy [3] or magnetic resonance [1]. *3d Mössbauer isotope probes:* in 3d compounds, generally the d electrons are well-shielded by outer electrons, from too strong influence of the ligands, and the ions display local magnetic moments, which lead also to magnetic fields acting on the nuclei of these ions. It turns out that in most cases the orbital angular momentum of these ions is quenched [1] and the observed hyperfine fields [4] are due to polarization of inner closed shell s electrons, by the total spin of the 3d shell. These hyperfine fields are negative (opposite to the direction of the ionic magnetic moment) [5]. Since the isotopes in the 3d series suitable for Mössbauer spectroscopy are very limited ([57]Fe and [61]Ni), only [57]Fe has been used widely for 3d compounds research. It has been also used as a magnetic probe (the observed hyperfine field is due to exchange induced magnetism in the Fe 3d shell) or as a diamagnetic probe (the observed hyperfine field is due to transferred fields, namely direct polarization of the iron inner closed shell s electrons, or penetration of polarized conduction electrons into the Fe nucleus), in many non-containing iron, magnetic compounds of elements from the whole periodic table.

4d and 5d Mössbauer isotope probes: Among the 4d elements only the isotope [99]Ru is suitable for Mössbauer studies; however, it is inconvenient experimentally and very little work has been reported using this isotope, though Ru is one of the magnetic elements contained in the so-called "superconducting-magnetic Ruthenates" [6]. The magnetic information obtained by Mössbauer spectroscopy is generally done by convenient dilute Mössbauer probes like [57]Fe or [119]Sn [7]. Most, though not all, 4d element compounds, due to the less screened, large extension of the 4d electronic wave functions of these compounds, do not order magnetically. In the 5d series, most elements have at least one isotope suitable for Mössbauer studies. However, the 5d electron wave functions are even more extended and no compound of these elements is magnetically ordered due to the 5d element. They participate in many magnetic compounds containing 3d or nf elements, and the study of their Mössbauer spectra reveals the magnetic properties of the compound through the transferred magnetic hyperfine field to be discussed later.

4f and 5f Mössbauer isotope probes: All 4f elements except Ce have at least one isotope suitable for Mössbauer studies. Some of the isotopes like [149]Sm, [151]Eu, [155]Gd, [161]Dy, [166]Er, [169]Tm, [170]Yb, and [171]Yb have made an extensive contribution to the understanding of magnetic and even superconducting rare earth materials. Because the 4f electrons are well-shielded by outer electrons, they fully contribute to the ionic magnetic moment, though sometimes this is reduced by the effect of local crystalline electric fields [1]. In many of their compounds, they display local magnetic moments and order magnetically at low temperatures. The 5f elements have several Mössbauer isotopes ([232]Th, [231]Pa, [234]U, [236]U, [238]U, [239]Pu, [240]Pu, and [243]Am) but only [237]Np is convenient enough for systematic studies. The 5f electrons are more exposed than the 4f electrons to the influence of the neighboring ligands, and thus are less localized, tend to hybridize, and display more complicated magnetism, and at lower temperatures.

11.2.3 Influence of External Magnetic Fields

When an external magnetic field is applied to a diamagnetic compound Mössbauer absorber, the absorbing nuclei will experience only that external field (H_{ext}), and will lead to direct Zeeman splitting of the nuclear levels. If the absorber is a single crystal with only one Mössbauer nucleus per unit cell, one can apply the field in various directions relative to the major axis of the electric field gradient (EFG) acting on this nucleus. The relative intensities of the various absorption lines will change according to the orientation of H_{ext}, the EFG, and the γ-ray direction. In a powder sample, the angle between H_{ext} and the EFG is averaged out on all possible angles, and thus the spectrum depends only on the angle between H_{ext} and the γ-ray. An example of such a measurement using ^{57}Fe is shown in Fig. 11.1. A theoretical analysis of the spectrum [8] with a least-square fit yields the field, H_{int}, acting on the nucleus, and its value equals H_{ext}, confirming the diamagnetism in this system.

For paramagnetic compounds H_{int} is different from H_{ext} due to the polarization of the non-closed shell electrons around the probing nucleus, which produce a

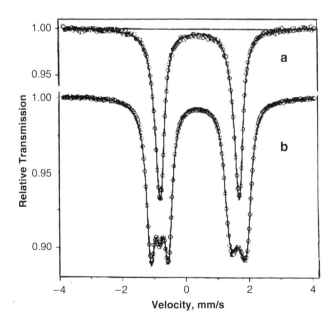

Fig. 11.1 Mössbauer spectra of decamethylferrocene [8], in $H_{ext} = 0.0$ (**a**) and $H_{ext} = 2.36$ T (**b**). The *solid curves* represent a theoretical least-square fit to the spectra, yielding H_{int} equal to H_{ext}

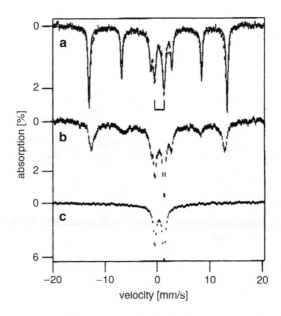

Fig. 11.2 Mössbauer spectra of a paramagnetic divalent iron compound at 4.2 K, in external magnetic fields of, 0.25 T (**a**), 0.15 T (**b**), and 0.05 T (**c**). The induced orbital internal field for the sextet in **a** (82 T) is the largest positive field ever observed for iron (Fig. 4 in [14]). The *solid line* simulation in spectrum **a**, including the presence of the central doublet, has been obtained in terms of the general spin Hamiltonian and containing weak exchange coupling between two ferrous sites

hyperfine field H_{eff} and thus $H_{\text{int}} = H_{\text{ext}} + H_{\text{eff}}$. The two fields H_{ext} and H_{eff} may have the same sign (generally in most nf elements) or opposite sign (generally in nd elements). Applying external fields to magnetically ordered systems, discussed later, can immediately reveal the nature of the magnetic order, for ferromagnets, H_{int} increases or decreases, for ferrimagnets in one of the subspectra the hyperfine field increases, for the other it decreases. For antiferromagnets, a line broadening of the absorption lines is generally observed due to a spread in angles between H_{eff} and H_{ext}, extending the limits $H_{\text{eff}} + H_{\text{ext}}$ to $H_{\text{eff}} - H_{\text{ext}}$ [9–13].

In Fig. 11.2 one observes the spectrum of an unusual case of a divalent iron compound, in which a large positive hyperfine field is induced [14]. It arises from the orbital contribution to the hyperfine field produced by the 3d electrons (first sum in 11.1).

11.2.4 Static Internal Fields in Magnetically Ordered Systems

In any system which orders magnetically below a certain temperature (T_M), the polarized ionic magnetic moments, both the orbital and spin contributions, add to produce a well-defined value for H_{int} (under the condition that the averaging time of the values of H_{int} of all ionic quantum states is much shorter than the inverse nuclear moment Larmor frequency in these fields [15]). The majority of the Mössbauer studies (ME) of magnetic materials deal with the temperature dependence of H_{int}. These are equivalent to measuring the temperature dependence of the local magnetic moment of a single ion in the crystal, and yield frequently much information missed by macroscopic magnetization measurements. The first studies of such internal fields were performed with one of the most common magnetic materials, iron metal, a ferromagnet, in which the internal magnetic field turned out to be negative ($-334\,kOe$ at room temperature) [4], in agreement with theoretical calculations claiming that the field arises from the Fermi contact term by polarized inner closed s shell electrons by the 3d shell total spin [16]. The next was $\alpha - Fe_2O_3$ (rust), an antiferromagnet ($H_{int} = -550\,kOe$) [17], which exhibits combined quadrupole and magnetic interactions, and Fe_3O_4, a ferrimagnet (magnetite, the naturally found bulk magnet) [18], which contains two magnetic sublattices. The ME studies observe simultaneously the behavior of both sublattices. In Fig. 11.3 are displayed the ^{57}Fe spectra of these three magnetic systems mentioned above.

It was mentioned that the measurement of the temperature dependence of the internal field acting on a Mössbauer nucleus is equivalent to the measurement of the local magnetic moment of the ion of that nucleus. Many measurements of this kind have been performed over the last 50 years. Accurate measurements of the temperature dependence of the internal field, for iron metal in [19], may yield even critical exponents at the magnetic phase transition temperature, Fig. 11.4 from [20]. Figure 11.5 displays the temperature dependence of the internal field acting on dilute

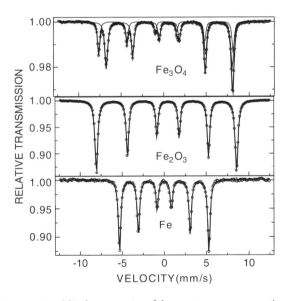

Fig. 11.3 Room temperature Mössbauer spectra of the most common magnetic materials

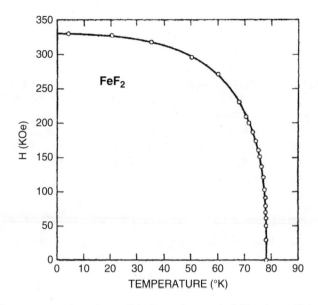

Fig. 11.4 The temperature dependence of the internal magnetic field acting on Fe in FeF$_2$ [20]

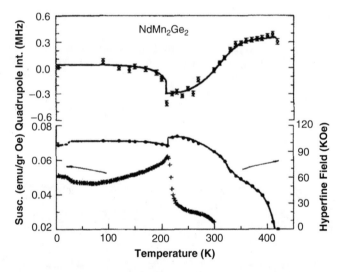

Fig. 11.5 The temperature dependence of the internal magnetic field acting on dilute Fe in NdMn$_2$Ge$_2$ [21]

[57]Fe in NdMn$_2$Ge$_2$. *Four* different magnetic phase transitions are observed. The one at the highest temperature is due to antiferromagnetic ordering of the Mn sublattice, the second one down is a transition of the Mn to a ferromagnetic state, the third one down is due to spin reorientation, and the one at the lowest temperature is due to the

magnetic ordering of the Nd sublattice. All these transitions are observed through the temperature dependence of the transferred magnetic hyperfine fields, discussed later, acting on the nucleus of a *non-magnetic* iron probe [21].

The studies of the temperature dependence of the internal magnetic fields in magnetically ordered systems, powders, and single crystals, under external magnetic fields [9–13], and/or external high pressure, have been practiced by many researchers [22–24]. Many review articles have been published during the years discussing both conventional Mössbauer and nuclear resonant scattering of synchrotron radiation results [23, 24]. Johnson and Thomas and collaborators [25, 26] have studied magnetic low dimensional systems, single crystals, and powders, in external magnetic fields, and were able to derive the values and properties (such as critical exponents) of the various phase transitions.

11.2.4.1   Internal Fields in nf Elements

The many available Mössbauer isotopes of 4f elements enabled extended research of magnetic internal fields in rare earth compounds. Already in 1968, extensive reviews [27, 28] were presented on both static and dynamic magnetic hyperfine interactions in 4f elements. The convenient isotopes of ^{151}Eu, ^{161}Dy, and ^{169}Tm, and also ^{155}Gd, ^{166}Er, and ^{170}Yb have widely contributed to our understanding of magnetism in 4f metals, intermetallics, and other ionic compounds. One of the earliest ^{169}Tm Mössbauer spectra [29] contributing to the understanding of magnetic order in rare earth metals is shown in Fig. 11.6. The results were explained in terms of a magnetic lattice incommensurate with the crystallographic lattice, and in full agreement with neutron diffraction results claiming longitudinal sinusoidal variation of magnetic moments. Thirty years later an attempt was made to interpret the same spectra in terms of spin relaxation theory [30].

Mössbauer studies of the large variety of rare earth isotopes have contributed immensely to the understanding of 4f magnetism. In many metallic compounds, the rare earth ions behave like free ions, where the electric crystalline fields are screened by conduction electrons. In many other compounds, the magnetism can be explained in terms of simple exchange and crystalline field theory. A unique 4f ion is Eu^{3+} [31] in which the ground state has a total angular momentum $J = 0$ and thus is supposed to be diamagnetic. However, the magnetic excited states are not far from the ground state and may mix by crystalline fields and exchange fields, leading to the observation of magnetic hyperfine interactions for the $^{151}Eu^{3+}$ Mössbauer isotope when the ion is located in a magnetic lattice. In these cases, the internal field experienced by the Eu nucleus is proportional to the exchange field acting on the Eu^{3+} ion [32,33]. Thus Mössbauer measurements of ^{151}Eu in $Eu_3Fe_{5-x}Ga_xO_{12}$, shown in Fig. 11.7, enabled one to determine the neighboring iron ions responsible to the exchange field, and its size, acting on the Eu ions in Europium iron garnet ($Eu_3Fe_5O_{12}$) [34].

Actinides: The only isotope, which yields much information about the magnetism in 5f element compounds is ^{237}Np. Because of the need of special facilities for the

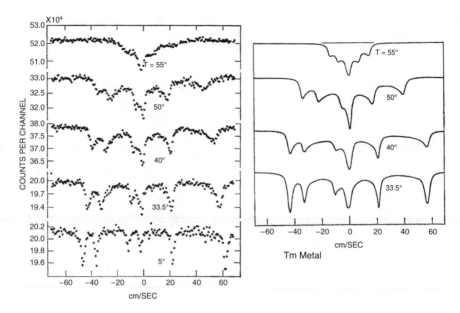

Fig. 11.6 Peculiar Mössbauer spectra of Thulium metal at various temperatures below the magnetic ordering (T_N) temperature. On the right are shown theoretical spectra, from [29]

use of both the radioactive source and the Np absorbers, only a few research groups have been able to contribute to this wide field. Dunlap and Kalvius and collaborators have made the main impact on the field, with also many extensive review articles [35–37]. The studies of Np compounds as a function of temperature and external pressure have stressed the differences between the 5f and 4f elements. Unlike the 4f elements, which are mostly trivalent (except for Ce and Tb which can be also tetravalent, and Sm, Eu, and Yb which can also be divalent), the actinides can appear with many valencies. Np appears as Np^{3+}, Np^{4+}, Np^{5+}, or Np^{6+}, differing in isomer shifts, quadrupole interactions, and magnetic hyperfine fields, and thus displaying a variety of phenomena. The measured temperature and/or pressure dependence of the internal fields in Np compounds yields the magnetic ordering temperature and its dependence on pressure namely on the Np–Np interatomic distances. Though the 5f electrons tend to hybridize, one finds in many cases good linear correlations between the internal magnetic fields and local magnetic moments of the Np ions and obvious correlations between the magnetic ordering temperature and interatomic Np–Np distances. In some cases, the Np ion behaves, though not often, like an ion with well-localized 5f electrons. In Fig. 11.8, one observes typical spectra of ^{237}Np under pressure, for the most studied compound, $NpAl_2$ [38]. The spectra were analyzed in terms of crystalline fields and exchange interactions, similarly to the 4f cases discussed before.

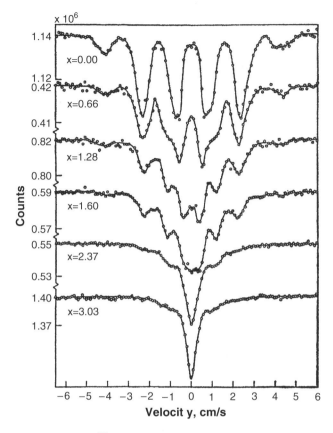

Fig. 11.7 Mössbauer spectra of ^{151}Eu nuclei in $EuFe_{5-x}Ga_xO_{12}$ [34]

11.2.5   Dynamics of Internal Fields in Magnetically Ordered Systems

Dynamic phenomena in Mössbauer spectra of paramagnetic systems have been observed at quite an early stage [28] and are discussed in detail in [15]. Spin relaxation phenomena in magnetically ordered systems are rare because of the strong exchange interactions between the magnetic ions which lead to fast relaxation and consequently to the observation of well-defined internal fields. The first observation of this phenomenon was reported in [39]. Though the spectra looked very unusual, they were analyzed in terms of relaxation between two magnetic levels of unequal populations and yield the relaxation rates and the magnetic exchange splitting of the ionic ground state. In Fig. 11.9, the Mössbauer spectra are shown. Since then, many cases were reported for both 4f and 5f ions, in particular close to the magnetic ordering temperature. In the case of nanoparticles (single domain) [40], one observes similar phenomena. However, in this case it is

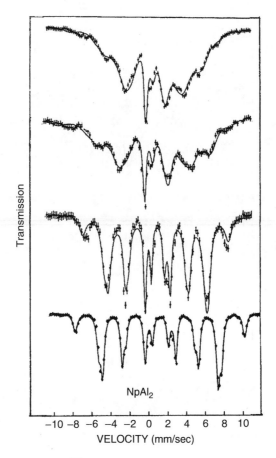

Fig. 11.8 Mössbauer spectra of ^{237}Np in NpAl$_2$ under pressure, from bottom to top: 0, 3.6, 7.2, and 9.0 GPa [38]

a macroscopic phenomenon; in each particle, the magnetization in low anisotropy fields may jump from one easy axis to the opposite direction. If there are exchange interactions among the particles, the population of the two directions may not be equal and it resembles ferromagnetic relaxation. The same phenomena may appear in amorphous systems [41] and in quasicrystals [42].

11.2.6   Internal Magnetic Fields at Phase Transitions

In Figs. 11.4 and 11.5, it was shown how the temperature dependence of the internal fields reveals magnetic phase transitions and spin reorientations. We show here the behavior of the unique system R1$_{1-x}$R2$_x$Fe$_2$, where R1 and R2 are two rare earth elements. These ferrimagnetic systems have high Curie temperatures, strong

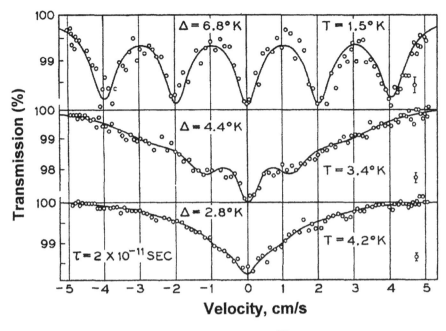

Fig. 11.9 Mössbauer spectra of ferromagnetic relaxation in ^{166}ErFeO$_3$ [39]

magnetism, high magnetic anisotropy, and high magnetosrictive coefficients. The ability to choose R1, R2, and x to obtain very low magnetic anisotropy makes them useful for practical purposes. Mössbauer spectroscopy has made an important contribution by finding the whole phase diagrams of the direction of the magnetic easy axis as a function of R1 and R2 combinations in the x-temperature plane. The different directions of the easy axis display completely different Mössbauer spectra, Fig. 11.10 [43].

The same phenomena can be observed by Mössbauer spectra of ^{170}Yb, using sources of ^{170}TmFe$_2$ and ^{170}Tm$_{0.8}$Ho$_{0.2}$Fe$_2$ with an enriched ^{170}Yb metal foil absorber [44].

Another example in which the Mössbauer spectra reveal the easy magnetization axis is the spectra of ^{155}Gd in Gd metal and alloys [45], Fig. 11.11. In this case and in other Gd cases [46], the spectrum strongly depends on the angle between the electric field gradient major axis (crystallographic c-axis in rare earth metals) and the magnetic field. The spectrum yields directly this angle.

11.2.7   Internal Fields in Unusual Spin Structures

The many available magnetic materials display a large variety of magnetic spin structures. Beside the common magnetic structures mentioned in Fig. 11.3, one

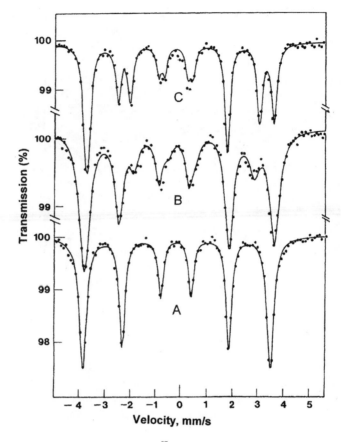

Fig. 11.10 From [43]. Mössbauer spectra of ^{57}Fe in $Ho_{1-x}Er_xFe_2$. (**a**) a one six line spectrum is for $x = 0.2$ for which the easy axis of magnetization is [100]. (**b**) a two six line spectrum with intensity ratio 3:1 for which the easy magnetization axis is [111] for $x = 0.7$ at 4.2 K. C, a two six line spectrum of equal intensity corresponds to the easy axis [110] for $x = 0.35$ at 4.2 K

finds canted antiferromagnets [10–13], helical [45], incommensurate with lattice [29], spin glasses [12, 13], and conduction electron spin density wave structures, all observable by Mössbauer spectroscopy. In addition, one finds magnetic order in some superconducting materials discussed below.

Since 1962, Mössbauer spectroscopy has contributed to the study of conventional superconducting materials, but with relatively low impact. However, in materials containing magnetic ions (Chevrel phases doped with magnetic ions and the RRh_4B_4 systems, R = rare earth [47]), the contribution of Mössbauer spectroscopy has considerably increased. Mössbauer spectroscopy has made even a larger contribution to the research of magnetism in the high T_c superconductors of the $RBa_2Cu_3O_{7-\delta}$ family. It revealed [48], the high temperature antiferromagnetic order of the Cu (2) in oxygen poor systems. In such systems, the magnetism of the Cu is

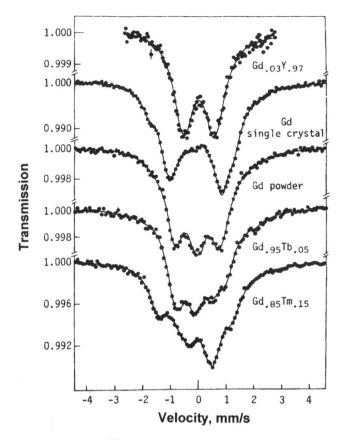

Fig. 11.11 Mössbauer spectra of ^{155}Gd in Gadolinium metal alloys [45]

most easily observed by a ^{57}Fe Mössbauer probe [49]. In the Ruthenium magneto-superconducting materials, several isotopes reveal the magnetic order, including ^{99}Ru [6], ^{57}Fe, and ^{119}Sn [7]. In Fig. 11.12, one observes the magnetic hyperfine structure of ^{99}Ru, below and above the superconducting transition temperature. They look identical.

The recent discovery of materials, in which superconductivity emerges when doping a spin density wave magnetic iron mother compound, with electrons or holes, and thereby suppressing the magnetic order, has triggered extensive research by many methods, including Mössbauer spectroscopy [50]. This was of particular interest since, unlike in all previous cases, the iron (and As), Fig. 11.13, are active participants in the superconductivity phenomenon.

The discovery that $EuFe_2(As_{0.7}P_{0.3})_2$ undergoes a superconducting transition at $T_c = 27$ K, followed by a ferromagnetic ordering of the Eu sublattice at 20 K [52],

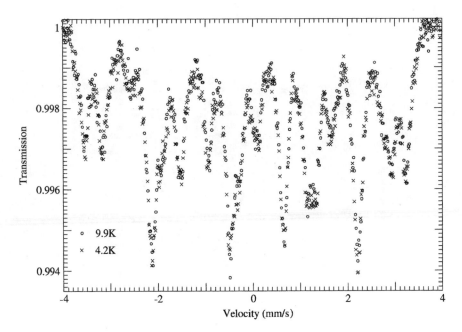

Fig. 11.12 Mössbauer spectra of ^{99}Ru in RuSr$_2$GdCu$_2$O$_8$ at 4.2 K (below the superconducting transition temperature, and at 9.9 K (above the superconducting transition temperature) [6]

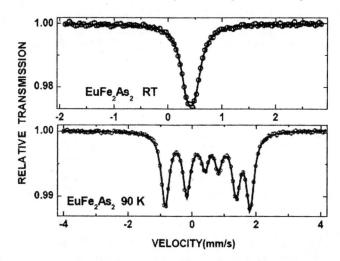

Fig. 11.13 Spin density wave ^{57}Fe Mössbauer spectra of EuFe$_2$As$_2$ at RT and 90 K [51]

encouraged an extensive study of the full EuFe$_2$(As$_{1-x}$P$_x$)$_2$ system by magnetometry and Mössbauer spectroscopy, using both ^{57}Fe and ^{151}Eu isotopes [53].

It was concluded that the Eu sublattice is magnetically ordered for all x values at 16–27 K, Fig. 11.14. The magnetic states of the Fe sublattice can be divided into

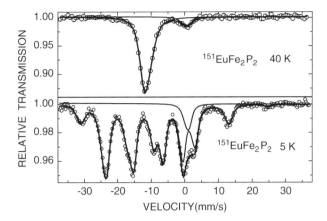

Fig. 11.14 ^{151}Eu Mössbauer spectra of EuFe$_2$P$_2$, below (5 K), and above (40 K) the ferromagnetic ordering temperature of the Eu sublattice (T_M = ∼30 K). The hyperfine field is along the c-axis [53]

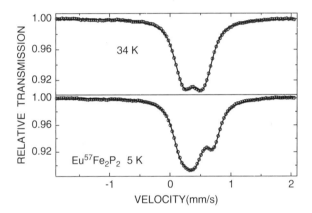

Fig. 11.15 ^{57}Fe Mössbauer spectra of EuFe$_2$P$_2$, below (5 K) and above (34 K) the ferromagnetic ordering temperature of the Eu sublattice (T_M ≈ 30 K). The transferred magnetic hyperfine field at 5 K is along the c-axis [53]

three regions. (1) In the range $0 < x < 0.2$, the Fe ions exhibit a spin density wave behavior, where the magnetic transition decreases sharply from ∼200 K for EuFe$_2$As$_2(x = 0)$, to zero for $x = 0.2$. (2) For $0.2 < x < 0.5$, the samples are superconducting at $T_c = 25$–30 K and superconductivity is confined to the Fe–As layers, and (3) for $x > 0.5$, neither Fe magnetism nor superconductivity is observed. However, in the regions where the iron sublattice does not display any magnetic order of its own, Fig. 11.15, and even in the superconducting region, Fig. 11.16, transferred hyperfine fields (∼1 T) from the Europium ferromagnetic sublattice are observed in the ^{57}Fe Mössbauer spectra.

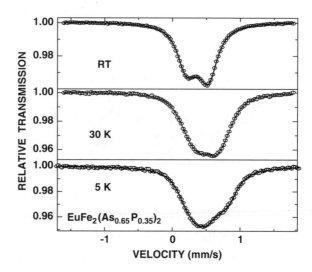

Fig. 11.16 ^{57}Fe spectra of EuFe$_2$(As$_{0.65}$P$_{0.35}$)$_2$ at 5.0 K, in the superconducting state of the FeAs layers, at 30 K above $T_M \approx 27$ K of Eu and at 297 K [53]

11.2.8 Transferred Hyperfine Fields

Nuclei in diamagnetic ions do not experience any internal magnetic field unless the ions are located in solids, which are magnetically ordered due to other paramagnetic ions [54, 55]. The so-called transferred hyperfine fields are produced through spin polarization of the closed shells of the diamagnetic ion by direct exchange interactions with neighboring magnetically ordered spins, by super-exchange through intermediate ligands, and/or by the presence of polarized conduction electrons in the nuclei. The first observations of transferred hyperfine fields were those of diamagnetic isotopes (of Sn, Sb, Au, and Ir) in iron, cobalt, and nickel and other metallic systems. Later, super-transferred fields were observed in various oxides and other compounds. The study of the observed internal fields shed light on the magnetic structure and behavior of the magnetic ions present in the systems. The study of mixed systems, such as Ni$_{1+2x}$Fe$_{2-3x}$ ^{121}Sb$_x$O$_4$ [56] or ^{119}Sn in YFe$_{5-x}$Ga$_x$O$_{12}$ shown in Fig. 11.17, enabled one to determine the magnetic behavior [58] and exact numerical contribution [57] of the neighboring iron ions to the transferred hyperfine field. As mentioned above, the transferred hyperfine fields are due to exchange polarization of closed s shells of non-magnetic ions by direct exchange, super-exchange, or through conduction electron polarization. Recently it was observed [53], Fig. 11.16, that the last mechanism holds even when the material is in its superconducting state.

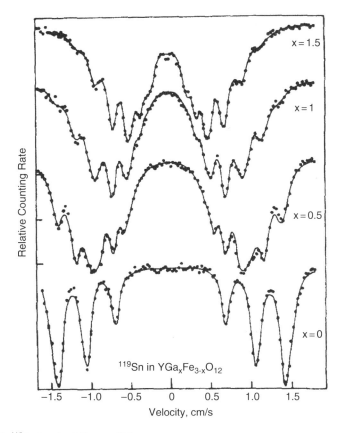

Fig. 11.17 ^{119}Sn in mixed Garnets [57]

11.2.9 Internal Fields in Thin Layers

Mössbauer spectra of any system with somewhat two-dimensional characteristics, such as thin films [59], surfaces [60], and interfaces [61], will display internal fields somewhat different from those in the bulk material. Surfaces are generally studied by conversion electron Mössbauer spectroscopy [62].

11.3 Summary

It was shown that Mössbauer spectroscopy studies of internal magnetic fields has widely contributed to the understanding of magnetism in solids; in insulators, metallic and even superconducting systems. It has also contributed to the understanding and development of a variety of magnetic devices of useful applications, such as magnetic ribbons and wires [63], hard magnetic materials [64], or magnetostictive devices based on the studied RFe_2 materials [43].

References

1. A. Abragam, B. Bleany, *Electron Paramagnetic Resonance of Transition Ions* (Clarendon, Oxford, 1970)
2. A. Abragam, *Principles of Nuclear Magnetism* (Clarendon, Oxford, 1961)
3. An extensive list of books on Mössbauer spectroscopy can be found in http://www.mossbauer. org/books.html
4. S.S. Hanna, J. Heberle, C. Littlejohn, G.J. Perlow, R.S. Preston, D.H. Vincent, Phys. Rev. Lett. **4**, 177 (1960)
5. S.S. Hanna, J. Heberle, G.J. Perlow, R.S. Preston, D.H. Vincent, Phys. Rev. Lett. **4**, 513 (1960)
6. D. Coffey, N. DeMarco, B. Dabrowski, S. Kolesnik, S. Toorongian, M. Haka, Phys. Rev. **B77**, 214412 (2008)
7. I. Felner, E. Galstyan, R.H. Herber, I. Nowik, Phys. Rev. B **70**, 094504 (2004)
8. I. Nowik, R.H. Herber, Eur. J. Inorg. Chem.**24**, 5069 (2006)
9. R.W. Grant, in *Mössbauer Spectroscopy*, vol. 5, ed. by U. Gonser. Topics in Applied Physics (Springer, Berlin, 1975), p. 97
10. C.E. Johnson, Hyperfine Interact. **49**, 19 (1989); **72**, 15 (1992); **90**, 27 (1994), J. Phys. D – Appl. Phys.,**29**, 2266 (1996)
11. Q.A. Pankhurst, R.J. Pollard, in *Mössbauer Spectroscopy Applied to Magnetism and Material Science*, vol. 1, ed. by G.J. Long, F. Grandjean (Plenum, NY, 1993)
12. T.E. Cranshow, G. Lonworth, in *Mössbauer Spectroscopy Applied to Inorganic Chemistry*, vol. 1, ed. by G.J. Long (Plenum, NY, 1984)
13. R.E. Vandenberghe, E. De Grave, in *Mössbauer Spectroscopy Applied to Inorganic Chemistry*, vol. 3, ed. by G.J. Long, F. Grandjean (Plenum, NY, 1989)
14. H. Andres, E.L. Bominaar, J.M. Smith, N.A. Eckert, P.L. Holland, E. Münck, J. Am. Chem. Soc **124**, 3012 (2002)
15. F.J. Litterst, in his chapter in the present book
16. R.E. Watson, A.J. Freeman, Phys. Rev. **123**, 2027 (1961)
17. O.C. Kistner, A.W. Sunyar, Phys. Rev. Lett. **4**, 412 (1960)
18. S.K. Banergee, W. O'Reilly, C.E. Johnson, J. Appl. Phys. **38**, 1289 (1967)
19. R.S. Preston, S.S. Hanna, J. Heberle, Phys. Rev. **128**, 2207 (1962)
20. G.K. Wertheim, Phys. Rev. **161**, 478 (1967)
21. I. Nowik, Y. Levi, I. Felner, E.R. Bauminger, J. Magn. Magn. Mater. **147**, 373 (1995)
22. M.P. Pasternak, R.D. Taylor, in *Mössbauer Spectroscopy Applied to Magnetism and Material Science*, vol. 2, ed. by G.J. Long, F. Grandjean (Plenum, NY, 1996)
23. G. Wortmann, K. Rupprecht, H. Giefers, Hyperfine Interact. **144/145**, 103 (2002)
24. I.S. Lyubutin, A.G. Gavriliuk, Physics **52**, 989 (2009)
25. C.E. Johnson, in *Mössbauer Spectroscopy Applied to Inorganic Chemistry*, vol. 1, ed. by G.J. Long (Plenum, NY, 1984)
26. M.F. Thomas, in *Mössbauer Spectroscopy Applied to Inorganic Chemistry*, vol. 2, ed. by G.J. Long (Plenum, NY, 1987)
27. S. Ofer, I. Nowik, S.G. Cohen, in *Chemical applications of Mossbauer spectroscopy*, ed. by V.I. Goldanskii, R.H. Herber (Academic, NY, 1968)
28. H.H. Wickman, G.K. Wertheim, in *Chemical applications of Mossbauer spectroscopy*, ed. by V.I. Goldanskii, R.H. Herber (Academic, NY, 1968)
29. R.L. Cohen, Phys. Rev. **169**, 432 (1968)

30. B.B. Triplett, N.S. Dixon, L.S. Fritz, Y. Mahmud, Hyperfine Interact. **72**, 97 (1992)
31. J.H. Van Vleck, J. Appl. Phys. **39**, 365 (1968)
32. G. Gilat, I. Nowik, Phys. Rev. **130**, 1361 (1963)
33. I. Nowik, I. Felner, M. Seh, M. Rakavy, D.I. Paul, J. Magn. Magn. Mater. **30**, 295 (1983)
34. I. Nowik, S. Ofer, Phys. Rev. **153**, 409 (1967)
35. B.D. Dunlap, G.M. Kalvius, in *Handbook on the physics and Chemistry of the Actinides*, ed. by A.J. Freeman, G.H. Lander (Elsevier, BV, 1983), pp. 329–434
36. B.D. Dunlap, G.M. Kalvius, J. Nucl. Mater. **166**, 5 (1989)
37. B.D. Dunlap, in *Mössbauer Spectroscopy Applied to Inorganic Chemistry*, vol. 2, ed. by G.J. Long (Plenum, NY, 1987)
38. G.M. Kalvius, W. Potzel, S. Zwirner, J. Gal, I. Nowik, J. Alloys Compd. **213/214**, 138 (1994)
39. I. Nowik, H.H. Wickman, Phys. Rev. Lett. **17**, 949 (1966)
40. S.J. Campbell, H. Gleiter, in *Mössbauer Spectroscopy Applied to Magnetism and Material Science*, vol. 1, ed. by G.J. Long, F. Grandjean (Plenum, NY, 1993)
41. Q.A. Pankhurst, in *Mössbauer Spectroscopy Applied to Magnetism and Material Science*, vol. 2, ed. by G.J. Long, F. Grandjean (Plenum, NY, 1996)
42. Z.M. Stadnik, in *Mössbauer Spectroscopy Applied to Magnetism and Material Science*, vol. 2, ed. by G.J. Long, F. Grandjean (Plenum, NY, 1996)
43. U. Atzmony, M.P. Dariel, E.R. Bauminger, D. Lebenbaum, I. Nowik, S. Ofer, Phys. Rev. B7, 4220 (1973)
44. R. Yanovsky, E.R. Bauminger, D. Levron, I. Nowik and S. Ofer, Sol. State Comm. **17**, 1511 (1975).
45. E.R. Bauminger, A. Diamant, I. Felner, I. Nowik, S. Ofer, Phys. Rev. Lett **34**, 962 (1975)
46. G. Czjzek, in"*Mössbauer Spectroscopy Applied to Magnetism and Materials Science*, vol. 1, ed. by G.J. Long, F. Grandjean (Plenum, NY, 1993)
47. B.D. Dunlap, C.W. Kimball, Hyperfine Interact. **49**, 187 (1989)
48. I. Nowik, M. Kowitt, I. Felner, E.R. Bauminger, Phys. Rev. **B38**, 76677 (1988)
49. I. Felner, I. Nowik, Supercond. Sci. Technol. **8**, 121 (1995)
50. I. Nowik, I. Felner, Physica C **469**, 485 (2009)
51. I. Nowik, I. Felner, Z. Ren, Z.A. Xu, G.H. Cao, J. Phys., Conf. Ser. **217**, 012121 (2010)
52. Z. Ren, Q. Tao, S. Jiang, C. Feng, C. Wang, J. Dai, G. Cao, Z. Xu, Phys. Rev. Lett. **102**, 137002 (2009)
53. I. Nowik , I. Felner, Z. Ren, G. H. Cao, Z. A. Xu, J. of Physics, Condens. Matter, 23, 065701 (2011).
54. F. Granjean, in *Mössbauer Spectroscopy Applied to Inorganic Chemistry*, vol. 2, ed. by G.J. Long (Plenum, NY, 1987)
55. I. Nowik, in *The Proceedings of the International Conference on the Mössbauer Effect*, Cracow, 25–30 August 1975
56. S.L. Ruby, B.J. Evans, S.S. Hafner, Solid State Commun. **6**, 277 (1968)
57. D. Lebenbaum, I. Nowik, E.R. Bauminger, S. Ofer, Solid State Commun. **9**, 1885 (1971)
58. I. Nowik, E.R. Bauminger, J. Hess, A. Mustachi, S. Ofer, Phys. Lett. **34A**, 155 (1971)
59. W. Meisel, in *Mössbauer Spectroscopy Applied to Magnetism and Material Science*, vol. 2, ed. by G.J. Long, F. Grandjean (Plenum, NY, 1996)
60. G. Bayreuther, Hyperfine Interact. **47**, 237 (1989)
61. Ch. Sauer, in *Mössbauer Spectroscopy Applied to Magnetism and Material Science*, vol. 2, ed. by G.J. Long, F. Grandjean (Plenum, NY, 1996)
62. G. Principi, in *Mössbauer Spectroscopy Applied to Magnetism and Material Science*, vol. 1, ed. by G.J. Long, F. Grandjean (Plenum, NY, 1993)
63. Q.A. Pankhurst, in *Mössbauer Spectroscopy Applied to Magnetism and Material Science*, vol. 2, ed. by G.J. Long, F. Grandjean (Plenum, NY, 1996)
64. G.A. Pringle, G.J. Long, in *Mössbauer Spectroscopy Applied to Inorganic Chemistry*, vol. 3, ed. by G.J. Long, F. Grandjean (Plenum, NY, 1989)

Chapter 12
Time-Dependent Effects in Mössbauer Spectra

F. Jochen Litterst

12.1 Introduction

From very early Mössbauer experiments it became apparent that fluctuating hyperfine interactions play a major role for the specific shape of spectra. Depending on the typical time scales of fluctuations, complex spectral line shapes deviating from the common Lorentzian line profiles are found, commonly termed as "relaxation spectra." Proper analysis yields information on the dynamics of the surrounding of the resonant nuclei. This may originate, e.g., from fluctuations of electronic spins related to magnons, coupling to phonons, nearby spins, or also from fluctuations of the neighboring atomic arrangement due to diffusion effects.

Even in a paramagnetic regime, one may observe magnetic hyperfine patterns if spin–lattice and spin–spin relaxation are slow enough to provide a nearly static electronic environment of the Mössbauer nucleus as experienced during a time comparable to nuclear Larmor precession under the locally acting magnetic field. Slow paramagnetic relaxation patterns have been detected in numerous cases, especially in systems when Mössbauer atoms carrying a nearly pure spin magnetic moment are embedded in low concentration in a non-magnetic solid (insulators, like oxides, glasses, and biological systems) thus preventing short spin–spin relaxation times [1–4]. One pioneering paper revealing relaxation spectra in a concentrated spin system is by Nowik and Wickman [5] who to our knowledge have for the first time described relaxation patterns in a concentrated spin system, $ErFeO_3$.[1] In fact, it is surprising to find slowing down of relaxation in a magnetically ordered system where exchange interactions commonly yield fast spin fluctuations. As we

---

[1] See also book Chap. 11 by Israel Nowik.

F.J. Litterst (✉)
IPKM, Technical University Braunschweig, 38106 Braunschweig, Germany
e-mail: j.litterst@tu-bs.de

M. Kalvius and P. Kienle (eds.), *The Rudolf Mössbauer Story*,
DOI 10.1007/978-3-642-17952-5_12, © Springer-Verlag Berlin Heidelberg 2012

will demonstrate below, such cases may occur for specific situations of the electronic ground state splitting in the crystalline electric field.

Early studies of diffusion effects have been described, e.g., by Elliot et al. [6] and Mullen and Knauer [7]. There followed the classical papers by Ruby, Zabransky and Flinn [8, 9] on nuclear motion in viscous surrounding taking into consideration not only the effects of translational but also of rotational diffusion.

The basis for theoretical treatment of relaxation spectra is strongly connected with the instruments developed earlier for NMR [10] with the complication that ground and excited nuclear spin state multiplets are involved connected via multi-pole radiation transitions. Early models of spin relaxation are those of Afanas'ev and Kagan [11], Wegener [12], Van der Woude and Dekker [13], Wickman et al. [4], Bradford and Marshal [14], and Blume [15], to name only a few. Reviews can, e.g., be found in the book of Goldanskii and Herber by Wickman and Wertheim [16] and later by Dattagupta [17]. Theories on the effects of nuclear motion go back to the famous paper by Singwi and Sjölander [18]. A review has been given by Dattagupta [19].

In the following, we will consider various scenarios for relaxation phenomena accessible to Mössbauer spectroscopy. After introducing some fundamental concepts in Sect. 12.2, we will give experimental examples related to fluctuations of the local electronic charge state and their qualitative interpretation in Sect. 12.3. The most frequently met situation of fluctuations of the electronic spin state in magnetic systems will be presented in Sect. 12.4. This section is in close relation to the book Chap. 11. All discussions will be presented without entering details of the sometimes intricate theoretical formalisms.

Section 12.5 deals with the wide field of structural dynamics as far as information can be drawn via hyperfine interactions. We can, however, only touch motional dynamics with its beautiful applications mainly in biology.

We will concentrate here on standard Mössbauer absorber experiments. An overview of time-dependent effects as a consequence of nuclear decay in source experiments (so-called "after-effects") may be found in a review by Friedt and Danon [20].

12.2 General Considerations

We start out by giving a brief introduction to the kind of fluctuations which may affect Mössbauer spectra and to the underlying conditions, together with a short outline of the theoretical concepts.

For the treatment of the interaction of a nucleus with its surrounding, one may start out from an ab initio many body point of view which under modern computer capacities may appear promising, yet being hampered by unreasonably many parameters determining the dynamics in a solid which finally cannot be compared with experiment. In nearly all practical cases, a more pragmatic way is chosen. One tries to use a stochastic approach in which the nuclear probe, or better, a still to be defined "sub-system" comprising the nucleus, is subject to the stochastic fluctuations of its direct surrounding ("interaction" part of Hamiltonian) which in turn are driven by the outside "bath". The proper separation of "sub-system," "interaction zone," and "bath" characterized by Hamiltonians H_p, H_{int}, and H_b will be different from case to case (see, e.g., [21]). The simplest case occurs if we only consider the nucleus as a probe subject to time average hyperfine interaction. The time-dependent part of hyperfine interaction is responsible for the "interaction." The thermal bath is made up from the electrons around the nucleus and the rest of the crystal. In other situations, it may be advantageous to include the electronic systems around the nucleus to the probe "sub-system." For spin–lattice or spin–spin relaxation acting, e.g., on ferrous or ferric iron, the "probe" is the iron nucleus with its hyperfine interactions plus its atomic electrons subject to an average crystalline electric field and eventually exchange fields. The "interaction" part comes from fluctuations of the crystal field and exchange. Still larger entities are adequate for describing superparamagnetic fluctuations of magnetization of small magnetic particles where a complete cluster comprising many magnetically coupled Mössbauer atoms serves as a "sub-system." Here the stochastic "interaction" is transferred by thermal fluctuations of the particle magnetization in an anisotropy potential.

Sometimes difficult is a suitable choice of the "interaction" Hamiltonian describing the coupling between "bath" and "sub-system" which consequently contains correlation functions from the bath and the time-dependent part of the "sub-system" (e.g., Larmor precession). In most practical cases, the variation of the "sub-system" occurs much slower than the decay time of the bath correlations, i.e., the interaction part of the Hamiltonian is only determined by the bath correlations. These are typically characterized by one exponential decay time, assumed independent on frequency. Note that the relaxation times imposed on the "sub-system" are not to be confused with the correlation times of the "bath." Mostly, many fluctuations of the bath are necessary to accumulate enough action inducing a change in the "sub-system." There are in fact a couple of examples where the condition for this so-called "white-noise approximation" is not met. This occurs typically when the separation of "sub-system" and "bath" cannot be clearly made (e.g., for spin–spin relaxation, see later) and where the energy related to H_{int} is not much smaller than that related to H_b.

Although it is desirable to determine the bath correlation times, this is usually precluded since the experiment will only yield the relaxation time of the "sub-system" and a calculation of the bath correlation time involves less well-defined details on the interaction strength. The relaxation mechanisms, however (e.g., Korringa relaxation involving conduction electrons, spin–lattice processes of Raman- or Orbach-type, or spin–spin relaxation related to the same spins forming the "sub-system"), can often be distinguished from their typical temperature dependence. Under many realistic conditions, it is useful to consider so-called jump models. The "system" is conceived as a variety of electronic states the nucleus may experience (e.g., having different electron densities due to different charge states, different electric field gradients due to population of different orbital states, etc.). The electronic state and consequently also the hyperfine interaction at the nucleus may then vary statistically in a jump-like manner. Several models of this type can be found in the literature (see for example the reviews [17, 22]). Their phenomenological use is relatively easy: one has to set-up a relaxation matrix connecting the various possible usually thermally differently populated electronic states with their static hyperfine interactions and define the possible transition paths between them. Using a Liouville superoperator formalism [23], often allows us to derive even analytic solutions for the spectral shape which gives a better transparency to the influence of the determining parameters than purely numeric treatment.

In general, the spectral shape is given by

$$\phi(\omega) = -\text{Im} \sum_{\alpha} \langle \boldsymbol{\eta}_{\alpha} \boldsymbol{j}(\kappa)[A]^{-1} \boldsymbol{\eta}_{\alpha} \boldsymbol{j}^{+}(\kappa) \rangle \tag{12.1}$$

with

$$A = \omega E - L_{\text{A}} - M(\omega) + i^{\Gamma}\!/_{2} E \tag{12.2}$$

L_{A} is the Liouville operator for the transitions caused by the hyperfine interaction in the various states of the "sub-system," $j(\kappa)$ is the current operator describing the nuclear transition probabilities, and $\boldsymbol{\eta}_{\alpha}$ is the γ-ray polarization vector. $M(\omega)$ is the relaxation operator made up from the transition probabilities between the various possible states of the "sub-system." Γ is half width of the resonance and ω stands for the energy (taking $\hbar = 1$), or Doppler velocity. E is the unity operator. The summation runs over all γ-transitions between the excited and ground state Mössbauer levels.

For most practical cases, simplifications are possible. Let us take only two electronic states 1 and 2 that may cause two different hyperfine interactions at the Mössbauer nucleus. For simplicity, we assume a magnetic dipolar interaction of opposite strength for the two states and for both the same electric quadrupole interaction which should be collinear to the magnetic interaction. (In this way, we have a situation where the hyperfine Hamiltonians are already diagonal in a basis set of the nuclear spin wave vector components I_z with quantization axis along the direction of magnetic field and the main axis of electric field gradient.) This situation is actually met for an electronic Kramers doublet with an effective spin $S_{\text{eff}} = 1/2$ exposed to an external magnetic field or to an exchange field. Due to the Zeeman splitting of the Kramers doublet in magnetic field, the thermal population of the two Kramers states will be different ($p(1)$ and $p(2)$) and in general be controlled by Boltzmann statistics. The spectral shape for this case can be straightforwardly calculated:

$$\phi(\omega) \sim \sum_{m_e m_g} \left| \langle m_e | j^{+} | m_g \rangle \right|^2 f_{m_e m_g}(\omega). \tag{12.3}$$

The first term gives the transition probability squared between the exited state level with $I_z = m_e$ and the ground state level $I_z = m_g$, which is determined from the Clebsch–Gordan coefficients and the appropriate polarization factors. The actual shape is determined by

$$f_{m_e m_g}(\omega) = -\mathrm{Im}\left[\frac{\Omega_1 - \langle \varpi \rangle + i\gamma}{(\Omega - \Omega_1)(\Omega_1 - \Omega_2)} + \frac{\Omega_2 - \langle \varpi \rangle + i\gamma}{(\Omega - \Omega_2)(\Omega_2 - \Omega_1)}\right] \quad (12.4)$$

with

$$\Omega_{1,2} = \frac{1}{2}(\omega_1 + \omega_2 - i\gamma) \pm \frac{1}{2}\sqrt{(\omega_1 - \omega_2)^2 - \gamma^2 + 2i(p(1) - p(2))(\omega_1 - \omega_2)\gamma}. \quad (12.5)$$

Here, the following abbreviations are used:

$$\Omega = \omega + i^{\Gamma}/_2$$

$$\omega_1 = \omega_{m_e m_g}(1); \omega_2 = \omega_{m_e m_g}(2)$$

$$\langle \omega \rangle = p(1)\omega_1 + p(2)\omega_2; \langle \varpi \rangle = p(2)\omega_1 + p(1)\omega_2$$

$$\gamma = \gamma_{12} + \gamma_{21}. \quad (12.6)$$

γ_{12} and γ_{21} are the transition rates between the electronic levels 1 and 2 and reverse, respectively. They are connected by detailed balance. The doubly indexed ω_{eg} are the hyperfine transition energies between excited and ground state levels. These are the well-known results for two-level systems as presented, e.g., by Wickman and Wertheim [16]. At this point, we will not follow the algebraic solution of the above given equations for the limiting cases of slow and fast relaxation. We rather would like to present examples which can be treated with the same formalism, yet can be understood also intuitively.

Note that the fluctuations between different states of the "sub-system" may involve non-commuting Hamiltonians, e.g., when the direction of the locally acting magnetic field is changing. The solution is then more intricate. An extreme case is that of strong-collision with randomly fluctuating directions of the quantization axis of the stochastic states. An analytical solution has been presented by Dattagupta [17].

12.3   Fluctuations of Ionic Charge State and Symmetry

How and under which conditions do relaxation effects show up at all? As a general rule, line broadenings and deviations from Lorentzian shape may become apparent only when the inverse relaxation time (=rate γ) is larger than inverse nuclear lifetime, i.e., nuclear linewidth Γ. Lorentzians of natural width will, however, also occur in the opposite limit of fast fluctuations, when the relaxation rate becomes large compared to the differences in hyperfine energies for the various possible states of the "sub-system." Typical examples are situations of ions with well-defined yet different ionic charge states. One famous case is Eu_3S_4 containing Eu^{2+} and Eu^{3+}[24]. At 80 K, the ^{151}Eu spectra reveal well-separated narrow resonances for both valence states (see Fig. 12.8 left, in book Chap. 7), i.e., one line with an isomer shift S_2 typical for Eu^{2+} and one with an isomer shift S_3 for Eu^{3+}. At room temperature, only one resonance of an averaged charge state is found. In between these temperatures, broad spectra of non-Lorentzian shape indicate relaxation effects caused by hopping of the extra charge of Eu^{2+}. At low temperatures, the electron hopping occurs with a rate of less than $10^7 \, s^{-1}$ which is small compared to nuclear line width. At high temperature, the rate is fast compared to the difference in hyperfine frequencies $|S_2 - S_3|/\hbar$ corresponding to about $10^{11} \, s^{-1}$. This is the well-known case of "motional averaging" or "motional narrowing." For the description of the spectra at intermediate temperatures, a simple two-level model was used with populations of the charge states according to the stoichiometry, i.e., $Eu^{2+}/Eu^{3+} = 1/2$ and allowing stochastic jumps between the hyperfine interactions (i.e., here only the isomer shifts) of the two states. From the varying spectral shapes, one could derive the temperature-dependence electron hopping and its activation energy.

There are numerous examples for electron hopping in comparable, so-called mixed valence compounds. Well-localized valence conditions are often met in oxides and minerals containing cations of different valence. At low temperatures, these compounds are insulating; however, when increasing the temperature, they reveal semiconducting and sometimes even metallic behavior related to an activated charge hopping. Often these compounds comprise chain or planar units giving rise also to peculiar magnetic behavior connected with the charge correlations along these low-dimensional units which is of interest by itself.

In Fig. 12.1, we show spectra of Fe in Ilvait [25], a silicate mineral, with composition $CaFe_2^{2+}Fe^{3+}[Si_2O_7/O/OH]$. Ilvait has a double chain structure containing Fe^{2+} and Fe^{3+} in octahedral oxygen coordinations (8d sites) of slightly different symmetries in alternating sequence. These octahedra are edge sharing.

An additional site of Fe^{2+} (4c) is bridging the double chain (see Fig. 12.2), yet is not forming extended ribbons as the 8d sites. The spectra consist of three subspectra with different isomer shifts and quadrupole interactions connected to the three types of sites (see bar diagram in Fig. 12.1). Hopping of the extra 3d electrons of Fe^{2+} along the chains leads not only to fluctuating charge states but also to a simultaneously varying quadrupole interaction. The changes caused by the electron hopping are mainly seen from the changes in the central structure of the spectra being the signal of iron on the 8d sites. The increase of hopping frequency can directly be traced from the temperature dependence of spectra. One may also see that the 4c sites appear to be practically unaffected by the hopping since these sites are not part of the double chain ribbons.

The used relaxation model included electron hopping with jump-like changes of charge state and simultaneous change of electric field gradient and is very similar to that given in (12.3)–(12.6). The temperature dependence of the hopping rates can be described by an Arrhenius law with an activation energy of $\sim 0.11 \, eV$ [25].

Such studies give, firstly, information on the preferential occupation of certain lattice sites and, secondly, about the preferred paths of electron hopping. The derived

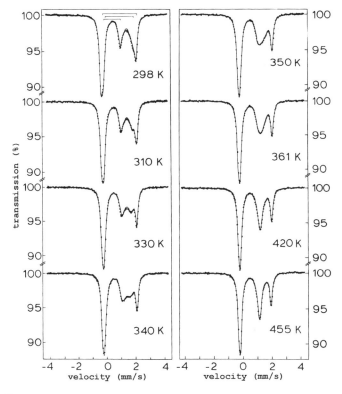

Fig. 12.1 ^{57}Fe absorption patterns of Ilvait (after [25]). *Bars* in 298 K pattern indicate quadrupole doublets for Fe^{2+}(4c) (*upper bar*), Fe^{2+}(8d) (*middle bar*), and Fe^{3+}(8d) (*lower bar*) sites

hopping rates are directly reflecting the local charge fluctuations, information which is complementary to macroscopically averaged transport behavior.

Charge fluctuations may also be traced in metallic systems. Examples are intermediate valence systems like EuCu$_2$Si$_2$ which are relatives of heavy Fermion and Kondo systems [27–31]. In contrast to the above-mentioned mixed-valent oxides we have here a scenario of fast charge fluctuations, i.e., the spectra reveal a strongly changing isomer shift with changing temperature, yet always an averaged charge state is represented (see Fig. 12.8 right, in Chap. 7). This type of studies, especially when combined with photoelectron spectroscopy, has its relevance in revealing details of the band structure of this important class of correlated electron systems.

Intermediate valence also occurs in intermetallic compounds containing Ce or Yb. Unfortunately, there is no resonance available for studying Ce compounds, yet for Yb the resonances of ^{170}Yb and ^{174}Yb appear suitable. Unfortunately, they are of minor usefulness due to the smallness of isomer shifts upon changing the valence state. Effects related to the electronic 4f quadrupolar moment in YbCu$_2$Si$_2$ have been reported in [32].

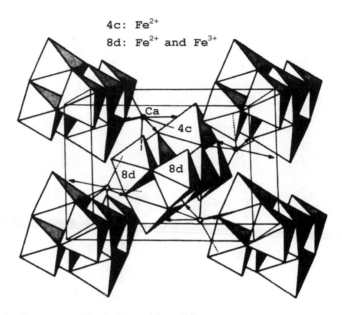

Fig. 12.2 Chain structure of Ilvait. Adopted from [26]

Fast repopulation of crystalline electric levels typically leads to the observation of temperature dependent thermally averaged electric field gradients, i.e., one is in the very fast relaxation regime with sharp resonance lines. Relaxation effects are observed especially in cases where spontaneous changes occur in crystalline electric field. Jahn–Teller distortions may lead either to a static distribution of field gradients due to various local symmetries or to a dynamic hopping between the different Jahn–Teller distorted states (see, e.g., [33, 34]). Stochastic jump models describing field gradient fluctuations have been presented in [35–37]. A proper treatment of the dynamic system including vibronic states, however, turns out to be very complicated in most cases. Recently, the subject has gained renewed actuality by the observation of orbital ordering in various spinels and perovskite materials which may exhibit colossal magnetoresistance and/or multiferroic properties.

A highly developed field of still growing interest is connected with so-called spin-crossover compounds. These are molecules containing transition metal cations whose spin state (low spin, intermediate, or high spin) may be influenced by various parameters, like temperature, applied magnetic fields, or irradiation with light. At low temperatures, one may obtain very long-lived metastable states which can be depopulated and repopulated by irradiation with light of proper optical wavelengths. Some of these compounds are, therefore, suitable for optical switching and storage and thus of interest for potential optical information technological applications. For studying spin-crossover iron compounds, Mössbauer effect is the appropriate tool. The different spin states are characterized by typical isomer shifts and quadrupole interactions. Time-dependent effects play an essential role in the interpretation of data. Reviews are given in [38, 39].

12.4 Spin Relaxation

12.4.1 Relaxation Spectra in Paramagnets and Magnetically Ordered States

In the simplest case, a magnetic hyperfine splitting may be caused by an electronic Kramers doublet with effective spin $S_{eff} = 1/2$ and equal population of both spin sub-states with an effective $S_z = \pm 1/2$. This is the situation represented by (12.3)–(12.6).

A full hyperfine splitting appears only when the relaxation time becomes longer than nuclear lifetime (this is the case of static paramagnetic hyperfine patterns). On the other hand, if the relaxation time is appreciably shorter than Larmor precession time, the magnetic splitting will collapse to a non-magnetic pattern (revealing for sure the other hyperfine interactions, i.e., isomer shift and quadrupole interaction). For relaxation times between both limiting cases, one will observe complex spectral patterns.

In the case of non-equal thermal population of the spin states (due to a Zeeman splitting in an applied magnetic field or due to an exchange field in a magnetically ordered regime), one will find in the fast limit sharp magnetic patterns with splittings representing the average local magnetization, i.e., in a paramagnet in applied field the splitting will follow a Brillouin function. In analogy, the magnetic hyperfine splitting below a magnetic ordering transition will reflect the sublattice magnetization at the Mössbauer atom as commonly observed in ferro- and antiferromagnets (see book Chap. 11). The situation becomes more complicated if relaxation times are already long enough above the magnetic transition, so that the change from slow paramagnetic relaxation to magnetic order can be traced less clearly. An example is shown in Fig. 12.3 for amorphous DyAg where the magnetic Curie temperature is around 18 K, yet the ^{161}Dy hyperfine patterns are already developing below about 100 K [40]. Only a slight additional line broadening is observed below about 20 K related to a distribution of exchange field and electric field gradients in the ferromagnetic state.

Cubic crystalline DyAg is an antiferromagnet with a Néel temperature of $T_N \approx$ 60 K. In contrast to the amorphous compound, one finds an onset of magnetic hyperfine splitting directly below T_N; however, the spectral shape also is clearly revealing a slowing down of relaxation upon decreasing temperature (Fig. 12.4). The relaxation rates are at least an order of magnitude faster when compared to the amorphous sample. The reason for this different behavior can be found in the different electronic ground states. For 4f electron systems, the magnetic ground state is given by the Hund's rules. The crystalline electric field can be treated in the "weak" limit, since the 4f electrons are well-shielded against the fields from neighboring ion charges. For Dy, we expect a Hund's ground state with a total angular momentum $J = 15/2$. In fact for both amorphous and cubic DyAg, the hyperfine field at lowest

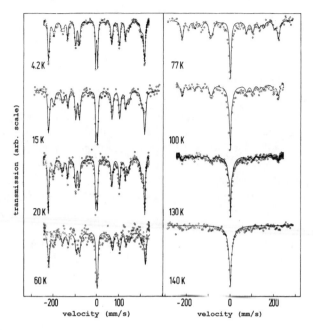

Fig. 12.3 ^{161}Dy absorption spectra of amorphous DyAg (ferromagnetic with $T_c \approx 18$ K). Adopted from [40]

temperatures is typical for a wave function dominated by the maximum angular momentum of $J_z = 15/2$ as expected in a strong exchange field.

The faster relaxation in the crystalline compound with increasing temperature is related to the decrease of influence of exchange field on the cubic ground state Γ_8. In amorphous DyAg, the ground state is at all temperatures dominated by a crystal field term of axial symmetry leaving a Kramers doublet with $J_z = \pm 15/2$ which is split by the exchange field. This situation only permits relaxation via higher crystalline electric field states explaining the slower relaxation compared with the crystalline compound.

Whereas 4f electrons in rare-earth systems are usually well-localized, the situation in actinide systems with their 5f electrons is more complex. For 5f systems, hybridization effects play a stronger role and the crystalline electric field should rather be treated as one of "intermediate" strength. Nevertheless, Neptunium systems with more localized moments may – though with some caution – be discussed with electronic ground states determined from Hund's rules.

In the following example, we discuss the spectra of ^{237}Np in NpCu$_4$Al$_8$. As pointed out in book Chap. 6, ^{237}Np has a $5/2 \rightarrow 5/2$ nuclear transition of 59.6 keV. From neutron scattering on NpCu$_4$Al$_8$, the absence of magnetic order is demonstrated down to $T = 4$ K. Below about 45 K, however, one can clearly resolve magnetic hyperfine patterns (Fig. 12.5) indicating slow paramagnetic relaxation [41]. From isomer shift, a trivalent ionic state is assigned to Np. In contrast,

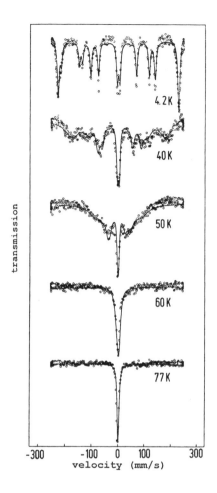

Fig. 12.4 [161]Dy absorption spectra of crystalline DyAg (antiferromagnetic with $T_N \approx 60$ K). Adopted from [40]

$NpCo_2Si_2$ is an antiferromagnet with a magnetic moment on Np. Here, Np is tetravalent. In both cases, the crystal field has tetragonal symmetry.

An apparent argument for the difference in magnetic behavior of the two intermetallics is their difference in electronic ground state. Whereas Np^{4+} is a Kramers ion with a magnetic doublet momentum ground state even under distorted symmetries, the situation is completely different for the non-Kramers ion Np^{3+}. Under cubic symmetry, the ground state would be a triplet state which is split under tetragonal symmetry to a singlet and a lower lying doublet. In fact, the wave functions of this doublet are mainly consisting of total angular momentum states with $|J_z = \pm 3>$ whereas the singlet contains $|J_z = +2>$ and $|J_z = -2>$. This means that relaxation between the doublet states is inhibited by the momentum difference whereas relaxation may occur to the non-magnetic singlet state. This, however, becomes less occupied with decreasing temperature, i.e., spin relaxation is getting static for low temperatures as observed. The fits to the experimental data in Fig. 12.5 have been performed with a model involving relaxation between the low

Fig. 12.5 Paramagnetic relaxation spectra of ^{237}Np in NpCu$_4$Al$_8$ (after [41])

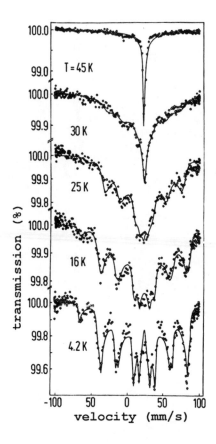

lying doublet and the singlet lying about 25 K above. The derived relaxation rates range from less than 10^{10} s^{-1} below about 16 K to about 3×10^{11} s^{-1} at 45 K. The major influence on the varying shape of spectra comes from the changing population between the magnetic doublet and the non-magnetic singlet. The effective magnetic hyperfine splitting at lowest temperatures corresponds to a field of 330 T, a value which is in good agreement with the expected moment of 1.5 μ_B for the ground state doublet.

We should not leave this section without commenting on the many reports on relaxational effects close to magnetic transitions. Indeed one has to expect and also has observed clear cases of critical slowing down near phase transitions (for a review, see Hohenemser et al. [42]). Yet, also a warning may be given since it is principally difficult to separate homogeneous (i.e., by relaxation) and inhomogeneous line broadening caused by random static distributions of hyperfine fields. This only can be achieved with careful systematic comparison of various spectra over a wide temperature range. Critical broadenings often get blurred or in contrary mimicked by inhomogeneities of samples, e.g., by a local distribution of ordering temperatures or short-range order. Therefore, in this respect proper care is needed.

12.4.2   Superparamagnetic Relaxation

The magnetism of small particles (for a general review, see [43]) is one of the most intensely studied fields via Mössbauer spectroscopy. For an early review, see [44]. Small particles play essential roles in soils, in mineralogy [45], in archeology [46], and nowadays in all areas of materials science. The widespread interest in nanophysics is still expanding rapidly and especially iron based oxides of various composition and structure are in use for medical and pharmaceutical purposes, for catalysis, storage media, etc. Publications under applied aspects are in the dozens per year. Mostly the data analysis is purely phenomenological using the Mössbauer spectroscopy as finger print method without a rigorous and tiresome adjustment using complex dynamic models. This is understandable and mostly justified under the aspect that Mössbauer spectroscopy serves as an easy tool for discriminating various compositional components in often inhomogeneous material and also for giving a rough picture of magnetic behavior which in many cases can be considered as sufficient.

For iron containing particles with sub-micrometer dimensions, one can easily estimate that, despite of magnetic ordering temperatures of the corresponding bulk materials on the order of hundreds of degrees, the overall magnetic behavior of an ensemble of particles will be "superparamagnetic," since the direction of moment of each particle may fluctuate between the easy axes fast on the time scale of measurement (i.e., nuclear Larmor precession time in the locally acting field). "Blocking" of these fluctuations will only be observed as soon as thermal energy becomes small compared to anisotropy energy and eventually an additional Zeeman energy if a magnetic field is applied. Rough analysis may usually be made using standard hopping models for magnetization fluctuations between different minima of the anisotropy potential, in simplest case between moments up and down. The temperature dependence of the fluctuation rates will mostly follow an Arrhenius law as expected from the Néel–Brown model for superparamagnetism [47, 48] of ferromagnetic particles. From this, one may derive an activation energy giving a rough measure of the anisotropy barrier heights. Whereas this simplified treatment neglects particle–particle interactions this has to be taken into account especially when nanoparticles tend to agglomerate. For a realistic treatment also details of the symmetry of the anisotropy play a role. Also not only overbarrier fluctuations may lead to dynamic patterns but also collective excitations with smaller angular deviations from the easy axes. Extensive and thorough studies of various well-defined particle ensembles have been performed in the group of Mørup (see [49–51] and references given there). Figure 12.6 presents a set of spectra of hematite nanoparticles [50]. At low temperatures only the typical six line pattern of magnetically ordered hematite is found. Starting at about 50 K, one observes the development of a single line in the center of the spectrum representing particles with already fast fluctuations of magnetization leading to an average zero hyperfine magnetic field. This "superparamagnetic" pattern is finally dominating at high temperatures. The only gradual development of the non-magnetic line with

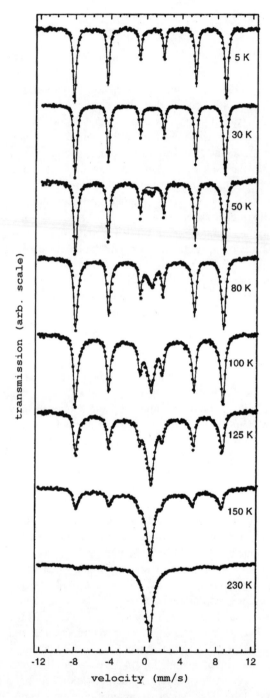

Fig. 12.6 ^{57}Fe absorption spectra of small hematite particles (after [50])

rising temperature is mainly caused by inherent distributions in particle volume and anisotropy density constants, i.e., blocking takes place over a range of temperatures. As "Mössbauer blocking temperature," one commonly defines the temperature where magnetic and superparamagnetic response have equal spectral weight (in the example shown in Fig. 12.6, it is 143 ± 5 K [50]). Actually, different blocking temperatures will be observed when different experimental methods with different typical time scales are applied. E.g., dc magnetization is probing on a static scale, ac susceptibility typically in the 10 ms, Mössbauer effect in the range of ns and shorter, corresponding to the Larmor precession times. Consequently, Mössbauer blocking temperatures are considerably higher than those determined from susceptibility.

In the quoted example of hematite particles, systematic studies of various particle sizes and shapes allow a reliable determination of anisotropy and also changes in the pre-exponential factor of the Néel–Brown formula for the fluctuation rates [50].

The role of various relaxation mechanisms for small particle magnetism influencing the dynamics of hyperfine interactions has been described in [52].

A basic problem in the analysis of the magnetic spectra of small particles is how to separate inhomogeneous broadening from relaxational broadening. Except in the rare examples of studies on monodisperse particles, additional complications arise due to the distribution of relaxation frequencies caused by a distribution in particle size and/or anisotropy.

12.4.3   Examples of Non-white Noise Relaxation

We had pointed out in Sect. 12.2 that within the "white noise approximation" many bath fluctuations are needed to induce a change in the magnetic sub-system. A consequence when this approximation is violated is that a frequency dependence of the relaxation operator needs to be taken into account. Usually, this is not needed (i.e., the "white noise approximation" holds). An exceptional example is $Cs_2NaYbCl_6$ as already described in Chap. 8. Indications came from the ^{170}Yb paramagnetic hyperfine patterns which revealed a slowing down of spin–spin relaxation, however, with a too big magnetic hyperfine parameter and deviations in reproducing the lineshape when using standard relaxation fitting procedures [53,54]. Via defolding experimental data of extremely good statistics, it was possible to derive the frequency dependence of a relaxation function [55]. Its real and imaginary parts reveal a bimodal frequency dependence as shown in Fig. 12.7.

Another case where a non-white noise approach turned out to be necessary for a proper analysis of the hyperfine spectra is presented in Fig. 12.8 [56]. Dilute ^{160}Dy in Thorium metal shows slowing down of spin relaxation with the conduction electron bath via the Korringa mechanism. In the case of the nuclear $I_e = 2 \rightarrow I_g = 0$ E2 transition of ^{160}Dy and an isolated Kramers doublet with an effective spin $S = 1/2$, the isotropic hyperfine coupling between nuclear spin I and S

$$\mathbf{H} = A\mathbf{I} \cdot \mathbf{S} \qquad (12.7)$$

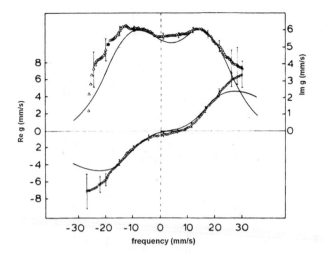

Fig. 12.7 Frequency dependence of the imaginary part (upper curve, right scale) and the real part (lower curve, left scale) of the relaxation function in $Cs_2NaYbCl_6$. The *lines* represent the sum over two Gaussians centered at $\omega_1 = -(3/2)A$ and $\omega_2 = A$. After [55]

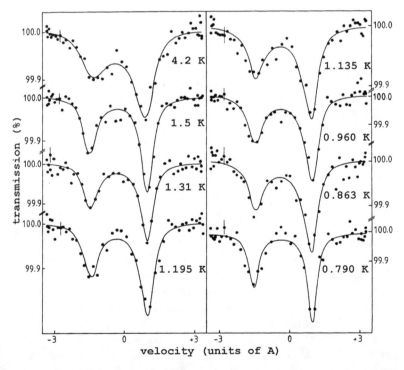

Fig. 12.8 Absorption spectra of dilute ^{160}Dy in Thorium near the superconducting transition at $T_c = 1.27\,K$. After [56]

Fig. 12.9 Development of superconducting gap below T_c derived from relaxation patterns in Fig. 12.8. After [56]

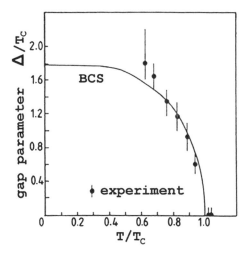

leads to two transitions from total angular momentum states with $F_e = I_e + S = 5/2$ and $3/2$ to the unsplit ground state $I_g = 0$. Whereas line width is continuously decreasing when coming from higher temperature, a broadening is found close to the superconducting transition temperature of Dy metal at $T_c = 1.27$ K. At still lower temperatures, this broadening vanishes. The reason for the additional line width around T_c is related to the build-up of density of states near the Fermi surface by the formation of the BCS state. In NMR, this effect is well-known as the Hebel–Slichter anomaly [57].

In the present case, it was possible to derive the relaxation function in the framework of the BCS model. It is strongly frequency dependent with respect to the resonance frequencies where it is peaked. In contrast to the normally conducting state, it is strongly asymmetric with respect to the absorption frequencies below the superconducting transition. An analysis of the spectra using a frequency-independent relaxation rate would yield erroneous results. The observed slowing down of relaxation below T_c is directly related to the decreasing availability of thermally excited quasiparticles due to the opening superconducting gap. From the fit to the experimental data, it was possible to derive the temperature-dependent development of the BCS gap (Fig. 12.9).

12.5 Diffusion

Diffusion processes are playing an essential role especially in material science. Changes in mechanical and electrical properties upon annealing or chemical treatment are mostly influenced by formation of mobile vacancies, interstitials, or dislocations which become mobile at elevated temperatures. Mössbauer probe atoms may directly participate in the diffusion process or may simply be spectators of the changing surrounding related to diffusion.

Examples of the first type can be of metallurgical relevance. Iron diffusion in several metallic alloys under various conditions has been described [58–60]. From line broadening, one can derive diffusion constants, hopping rates, and their activation energies. Especially informative are studies on single crystals where also the directional properties of the elementary diffusion processes can be characterized. In many cases, it is possible to imply so-called cage hopping, i.e., spatially restricted hopping processes under defined geometrical conditions. For hopping rates faster than the inverse nuclear lifetime, the spectra reveal a typical separation into a narrow elastic line and a broad quasielastic line with a width proportional to the hopping rate. For high enough rates, this quasielastic line will be lost in the non-resonant background. The quasielastic line has its origin in random phase shifts of the γ wave train during the absorption process due to nuclear jump motion. Various models for various jump geometries have been described also including additional correlated changes in hyperfine interactions [61–65].

Diffusion of hydrogen in metals has been a thoroughly studied subject revealing quasielastic lines [66–68]. Figure 12.10 shows spectra of iron in niobium hydride [67]. Spectra in the right diagram have been recorded with a velocity sweep of $1.5 \, \text{mm s}^{-1}$. The major contribution is a quadrupole doublet. At negative velocities, there appears an impurity signal which turns out to be temperature independent. The left diagram shows spectra recorded with a ten times larger velocity. It becomes apparent that a quasielastic line is hidden under the central part of the spectra. At 223 K, the surroundings of the iron appear static. When rising the temperature, diffusion of hydrogen along interstitial sites of the cubic lattice causes positional shifts of the iron atoms by about 0.02 nm which occur randomly in time leading to the broad line seen in the spectra between 227 K and 312 K. This line disappears into the background around 350 K where hydrogen diffusion has become fast. A cage model including correlated jumps of the electric field gradient allowed deriving mean times of residence of hydrogen in an interstitial site. The activation energy of 0.24 eV is in good agreement with NMR and neutron data on similar samples.

Hydrogen diffusion in tantalum metal has been studied using the [181]Ta resonance [69]. When changing the temperature, the diffusion of hydrogen leads to a change of isomer shift due to the changed electron density close to Ta. There are also observed changes with hydrogen concentration since the probability of having a hydrogen close to a [181]Ta nucleus is changing. In this case, both linewidth and isomer shift reflect the diffusion process.

High temperature cation diffusion via vacancies in slightly off-stoichiometric magnetite has been described in [70]. Such studies are helpful for a better understanding of high temperature solid state reactions.

The detection of broad liquid-like lines in addition to sharp elastic lines in biological matter caused a stir in the 1980s [71–74]. In Fig. 12.11, the development of the broad quasi-elastic line in ferritin, an iron-storage protein, is shown [71]. The crystalline material is containing water. In the frozen state at 258 K (spectrum b in Fig. 12.11) a static hyperfine pattern of the protein is recorded. At 278 K

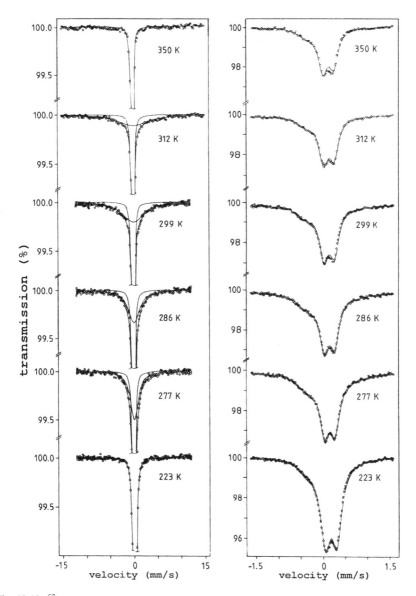

Fig. 12.10 ^{57}Fe absorption spectra of Fe in NbH$_{0.78}$ (left side taken with high velocity sweep demonstrating the development of a quasielastic line). After [67]

(spectrum a), the molten water induces localized diffusional motion leading to the quasielastic line. A preliminary description was based on a hopping model which in detail is certainly insufficient. In biological systems, one rather has to consider a micro-Brownian movement of overdamped oscillations of conformational units. Major highly successful investigations on these subjects have been performed by the

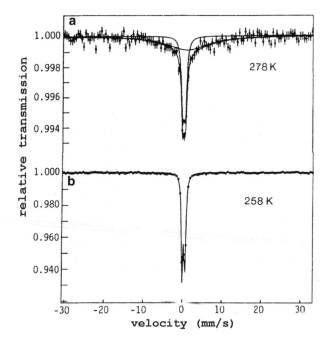

Fig. 12.11 Appearance of a quasielastic line in ferritin (spectrum a) due to micro Brownian movement (after [70])

group of Parak (see [72, 74–76], and references given therein). For a recent review on these experiments and concepts with broad biological and medical relevance, we refer to his review [77].

12.6 Conclusion

Studies of time-dependent hyperfine interactions have proven to be highly powerful in all fields of application of Mössbauer effect, ranging from basic science to applied research and development. Its peculiar aspects are the sensitivity as local probe with its typical time window. Most surplus of information is gained from its complementarity to other local methods using other nuclei or particles at different substitutional or interstitial positions, with different frequency dependence like NMR, PAC, and μSR [22], and bulk methods like dc magnetization and ac susceptibility.

Future highly hopeful developments are envisaged by the nowadays widely accessible possibilities of synchrotron methods of Mössbauer effect opening still further ranging horizons (see book Chap. 17 and 18).

References

1. G.K. Wertheim, J.P. Remeika, Proc. Colloq. Ampere **13**, 147 (1965) and Phys. Lett. **10**, 14 (1964)
2. C.R. Kurkjian, D.N.E. Buchanan, Phys. Chem. Glasses **5**, 63 (1964)
3. H.H. Wickman, G. Wertheim, Phys. Rev. **148**, 211 (1966)
4. H.H. Wickman, M.P. Klein, D.A. Shirley, Phys. Rev. **152**, 345 (1966)
5. I. Nowik, H.H. Wickman, Phys. Rev. Lett. **17**, 942 (1966)
6. J.A. Elliot, E. Hall, D.S.t.P. Bunburry, Proc. Phys. Soc. London **89**, 595 (1965)
7. J.G. Mullen, R.C. Knauer, in Mössbauer Effect Methodology, vol. 5, ed. by I.J. Gruverman (Plenum Press, New York, 1969), p. 197
8. P.A. Flinn, B.J. Zabransky, S.L. Ruby, J. Phys. **37**(C6), 739 (1976)
9. S.L. Ruby, B.J. Zabranski, P.A. Flinn, J. Phys. **37**(C6), 745 (1976)
10. A. Abragam, B. Bleaney, in Electron Paramagnetic Resonance of Transition Ions (Clarendon, Oxford, 1970)
11. A.M. Afanas'ev, Y.u. Kagan, Sov. Phys. JETP **18**, 1139 (1964)
12. H. Wegener, Z. Physik **186**, 498 (1965)
13. F. Van der Woude, A.J. Dekker, Phys. Status Solidi **9**, 775 (1965)
14. E. Bradford, W. Marshal, Proc. Phys. Soc. **87**, 731 (1966)
15. M. Blume, Phys. Rev. Lett. **14**, 96 (1965) and **18**, 305 (1967)
16. H.H. Wickman, G.K. Wertheim, in Chemical Applications of Mössbauer Spectroscopy, ed. by V.I. Goldanskii, R.H. Herber (Academic, New York, 1968), p. 548
17. S. Dattagupta, Hyperfine Interact. **11**, 77 (1981)
18. K.S. Singwi, A. Sjölander, Phys. Rev. **120**, 1093 (1960)
19. S. Dattagupta, Phys. Rev. **B12**, 47 (1975)
20. J.M. Friedt, J. Danon, Atomic Energy Rev. **18**, 893 (1980)
21. H. Wegener, Proc. Int. Conf. Mossb. Spectr. **2**, ed. by A.Z. Hrynkiewic, J.A. Sawicki (Powielarnia, AGH, Cracow, 1975), p. 189
22. S. Dattagupta, Hyperfine Interact.**49**, 253 (1989)
23. F. Hartmann-Boutron, J. Physique **40**, 57 (1979)
24. O. Berkooz, M. Malamud, S. Shtrikman, Solid State Commun. **6**, 185 (1968)
25. F.J. Litterst, G. Amthauer, Phys. Chem. Minerals **10**, 250 (1984)
26. N.V. Belov, V.I. Mokeeva, Trudy Inst. Krist. Akad. Nauk SSSR **9**, 89 (1954)
27. I. Nowik, E.V. Sampathkumaran, G. Kaindl, Solid State Commun. **55**, 721 (1985)
28. I. Nowik, M. Campagna, G.K. Wertheim, Phys. Rev. Lett. **38**, 43 (1977)
29. E.R. Bauminger, D. Froindlich, I. Nowik, S. Ofer, I. Felner, I. Mayer, Phys. Rev. Lett. **30**, 1053 (1973)
30. J. Röhler, D. Wohlleben, G. Kaindl, H. Balster, Phys. Rev. Lett.**49**, 65 (1982)
31. I. Nowik, I. Felner, Hyperfine Interact. **28**, 959 (1986)
32. K. Tomala, D. Weschenfelder, G. Czjzek, E. Holland-Moritz, J. Magn. Magn. Mater. **89**, 143 (1990)
33. F.S. Ham, Phys. Rev. **155**, 170 (1967)
34. J.R. Regnard, U. Dürr, J. Phys. **40**, 997 (1979)
35. M. Blume, J.A. Tjon, Phys. Rev. **165**, 446 (1968)
36. J.A. Tjon, M. Blume, Phys. Rev. **165**, 466 (1968)
37. F. Hartmann-Boutron, J. Phys. **29**, 47 (1968)
38. P. Gütlich, H.A. Goodwin, Top. Curr. Chem. **233**, 1 (2004)
39. P. Gütlich, A. Hauser, H. Spiering, Angew. Chem. **106**, 2109 (1994)
40. G.M. Kalvius, F.J. Litterst, L. Asch, J. Chappert, Hyperfine Interact. **24–26**, 709 (1985)
41. J. Gal, H. Pinto, S. Fredo, M. Melamud, H. Shaked, R. Caciuffo, F.J. Litterst, L. Asch, W. Potzel, G.M. Kalvius, J. Magn. Magn. Mater. **50**, L123 (1985)
42. C. Hohenemser, N. Rosov, A. Kleinhammes, Hyperfine Interact. **49**, 267 (1989)

43. I.S. Jacobs, C.P. Bean, in Magnetism, vol. 3, ed. by G.T. Rado, H. Suhl (Academic, London, 1965), p. 271
44. D.W. Collins, J.T. Dehen, L.N. Mulay, in Mössbauer Effect Methodology, vol. 3, ed. by I.J. Gruverman (Plenum, New York, 1968), p. 103
45. F.E. Wagner, U. Wagner, Hyperfine Interact. **154**, 35 (2004)
46. F.E. Wagner, A. Kyek, Hyperfine Interact. **154**, 5 (2004)
47. L. Néel, Ann. Geophys. **5**, 99 (1949)
48. W.F. Brown, Phys. Rev. **130**, 1677 (1963)
49. F. Bødker, M.F. Hansen, C.B. Koch, K. Lefmann, S. Mørup, Phys. Rev. **B61**, 6826 (2000)
50. F. Bødker, S. Mørup, Europhys. Lett. **52**, 217 (2000)
51. S. Mørup, M.F. Hansen. C. Frandsen, in Comprehensive Nanoscience and Technology, vol. 1, ed.by D.L. Andrews, G.D. Scholes, G.P.Wiederrecht, (Academic Press, Oxford, 2011), p. 437
52. K.K.P. Srivastava, J. Phys., Condens. Matter **15**, 549 (2003)
53. S. Dattagupta, G.K. Shenoy, B.D. Dunlap, L. Asch, Phys. Rev. **B16**, 3813 (1977)
54. G.K. Shenoy, B.D. Dunlap, S. Dattagupta, L. Asch, Phys. Rev. Lett. **37**, 539 (1976)
55. A.M. Afanas'ev, E.V. Onishchenko, L. Asch, G.M. Kalvius, Phys. Rev. Lett. **40**, 816 (1978)
56. W. Wagner, F.J. Litterst, V.D. Gorobchenko, A.M. Afanas'ev, G.M. Kalvius, J. Phys. F: Metal Phys. **11**, 959 (1981)
57. L.C. Hebel, C.P. Slichter, Phys. Rev. **113**, 1504 (1959)
58. W. Petry, G. Vogl, W. Mansel, Phys. Rev. Lett. **45**, 1862 (1980)
59. G. Vogl, B. Sepiol, Acta Metal. Mater. **42**, 3175 (1994)
60. R. Feldwisch, B. Sepiol, G. Vogl, Acta Metal. Mater. **43**,2033 (1995)
61. F.J. Litterst, A.M. Afanasev, P.A. Aleksandrov, V.D. Gorobchenko, Solid State Commun. **45**, 963 (1983)
62. F.J. Litterst, V.D. Gorobchenko, Phys. Status Solidi b**113**, K135 (1982)
63. F.J. Litterst, V.D. Gorobchenko, G.M. Kalvius, Hyperfine Interact. **14**,21 (1983)
64. V.D. Gorobchenko, F.J. Litterst, Phys. Status Solidi b**127**, 351 (1985)
65. F.J. Litterst, Hyperfine Interact. **27**, 123 (1986)
66. F.E. Wagner, F. Pröbst, R. Wordel, M. Zelger, F.J. Litterst, J. Less-Common Metals **103**, 135 (1984)
67. R. Wordel, F.J. Litterst, F.E. Wagner, J. Phys. F: Metal Phys. **15**, 2525 (1985)
68. M. Zelger, R. Wordel, F.E. Wagner, F.J. Litterst, Hyperfine Interact. **28**, 1009 (1986)
69. A. Heidemann, G. Kaindl, D. Salomon, H. Wipf, G. Wortmann, Phys. Rev. Lett.**36**, 213 (1976)
70. K.D. Becker, V.v. Wurmb, F.J. Litterst, J. Phys. Chem. Solids**54**, 923 (1993)
71. S.G. Cohen, E.R. Bauminger, I. Nowik, S. Ofer, J. Yariv, Phys. Rev. Lett. **46**, 1244 (1981)
72. K.H. Mayo, F. Parak, R.L. Mössbauer, Phys. Lett. **82A**, 468 (1981)
73. I. Nowik, S.G. Cohen, E.R. Bauminger, S. Ofer, Phys. Rev. Lett. **50**, 1528 (1983)
74. E.W. Knapp, S.F. Fischer, F. Parak, J. Chem. Phys. **78**, 4701 (1983)
75. V.E. Prusakov, J. Steyer, F.G. Parak, Biophys. J. **68**, 2524 (1995)
76. F. Parak, M. Fischer, G.U. Nienhaus, J. Mol. Liquids **42**, 145 (1989)
77. F.G. Parak, K. Achterhold, S. Croci, M. Schmidt, J. Biol. Phys. **33**, 371 (2007)

Chapter 13
Mössbauer Spectroscopy of Biological Systems

Eckard Münck and Emile L. Bominaar

13.1 Introduction

[57]Fe Mössbauer spectroscopy had an exceptional impact on the development of iron-based metallobiochemistry (we might call it [57]Fe-based metallobiochemistry) and its offspring, bioinorganic chemistry. Right from the outset, the new technique has shed considerable light on the nature of various biological problems. In particular, the application of Mössbauer spectroscopy has led to the discovery of previously unknown clusters as well as cluster assemblies that has stimulated research into new biochemical avenues. Moreover, the application of Mössbauer spectroscopy has made possible the characterization of many intermediates in the catalytic cycle of enzymes that sustain life. Some of the intermediates studied have life times of only a few milliseconds and had to be trapped by rapid quench techniques. For these endeavors, the researchers could draw on parallel developments in biological electron paramagnetic resonance (EPR) spectroscopy and, as we shall see, EPR has been a productive complement to biological Mössbauer spectroscopy all the way.

Mössbauer spectroscopy has many attributes that makes it so powerful for the study of biological systems. Two of these are particularly noteworthy. First, the technique sees all iron in the sample, regardless of its oxidation and spin state. Thus, iron cannot "hide". Second, at 4.2 K virtually all protein iron sites have practically the same Debye–Waller factor ($f \approx 0.8$), which allows one to determine the relative amount of distinct iron species and, with some effort, the total amount of Fe/protein. To give an example: in 1978, the biochemical literature indicated that the MoFe protein of nitrogenase (the enzyme that converts dinitrogen to ammonia) contains anywhere between 14 and 36 Fe atoms (and 2 Mo). After studying and analyzing

E. Münck (✉) · E.L. Bominaar
Department of Chemistry, Carnegie Mellon University, 4400 Fifth Avenue,
Pittsburgh PA 15213 USA
e-mail: emunck@cmu.edu

M. Kalvius and P. Kienle (eds.), *The Rudolf Mössbauer Story*,
DOI 10.1007/978-3-642-17952-5__13, © Springer-Verlag Berlin Heidelberg 2012

the Mössbauer spectra of the MoFe protein for three years, one of us (E.M.) and R. Zimmermann deduced in 1978 that the MoFe protein contained 30 ± 2 Fe atoms (The authors lived to see the X-ray structure in 1992: the protein contains indeed exactly 30 Fe).

We do not have the space to review even only the major accomplishments in biological Mössbauer spectroscopy. Therefore, we confine our presentation to a few examples which, however, will emphasize the particular strength of the technique. The reader will recognize that it was Mössbauer spectroscopy that made the described discoveries possible. Without Rudolf Mössbauer's discovery, iron-based biochemistry would be substantially less mature than it is today.

13.2 Spin Hamiltonian Analysis

Iron binds to proteins through amino acids that act as donors of nitrogen (histidine), oxygen (glutamic acid, aspartic acid, and tyrosine), and sulfur (cysteine and methionine). Often it is part of a prosthetic group (e.g., a porphyrin as in hemoglobin) which is incorporated into the protein matrix. Protein-bound iron centers function, for example, as catalytic sites (in nitrogenases, oxygenases, hydrogenases, oxidases, radical generators...), sulfur donors, and as electron storage sites (ferredoxins) in a broad range of biochemical processes. In ferritin, as many as 4,500 iron atoms (Fe^{3+}) are encapsuled for iron storage.

$Fe^{2+}(3d^6)$ and $Fe^{3+}(3d^5)$ are the most common oxidation states encountered in biological systems, in mononuclear sites as well as in polynuclear clusters. $Fe^{4+}(3d^4)$ has been observed in reactive intermediates in the reaction cycles of oxygen activating enzymes, while lower oxidation states have been identified in hydrogenases in an organometallic coordination site (CO and CN^-). Owing to their partially filled 3d shells, iron centers can carry electronic spin: $Fe^{2+} S = \underline{0}, 1, \underline{2}$; $Fe^{3+} S = 1/2, 3/2, \underline{5/2}$; $Fe^{4+} S = 0, \underline{1}, 2$, where the most common biological spin states have been underscored. The ligands, in general, remove the orbital degeneracy, leading to a ground state with only spin degeneracy ($M_S = -S$, $-S+1, \ldots, +S$). The spin degeneracy is (partly) removed by magnetic interactions such as spin-orbit coupling of the 3d electrons and Zeeman interactions with an external field, B, (first term of (13.1)). As the crystal-field splitting of the 3d orbitals is typically considerably larger than the spin-orbit coupling, the effect of the latter on the energies of the magnetic sublevels can often be expressed, using second-order perturbation theory, by the bilinear zero-field splitting operator (second term of (13.1)). The third, fourth, and last term of (13.1) represent the magnetic hyperfine, the nuclear Zeeman, and electric quadrupole interactions, respectively. These terms are assembled in the celebrated spin Hamiltonian that is widely used in the simulation of the Mössbauer spectra of mononuclear Fe complexes.

$$\hat{\mathcal{H}}_{\text{spinH}} = \beta B \cdot g \cdot \hat{S} + \hat{S} \cdot D \cdot \hat{S} + \hat{S} \cdot A \cdot \hat{I} - g_n \beta_n B \cdot \hat{I} + \hat{I} \cdot P \cdot \hat{I} \qquad (13.1)$$

For $S = 0$, (13.1) simplifies to the sum of the nuclear Zeeman and quadrupole interactions. For $S \geq 1$, all terms are present. In many cases of biological interest, the various tensors of (13.1) have, not surprisingly given the asymmetric protein environments, non-collinear principal axes systems.

In biological systems, iron frequently occurs as a constituent of polynuclear clusters, in which case it is covalently linked to other metal sites by small anions, such as sulfides (S^{2-}), oxides (O^{2-}), and, as we shall see (Fig. 13.1e–g), sulfur atoms from cysteinyl residues. The spin Hamiltonian then is a sum of operators for the individual metal sites (first term of (13.2)) and coupling terms between them (second term of (13.2)),

$$\hat{\mathcal{H}}_{\text{spinH}} = \sum_i \hat{\mathcal{H}}_{\text{spinH},i} + \sum_{i<j} J_{ij}\hat{\mathbf{S}}_i \cdot \hat{\mathbf{S}}_j, \tag{13.2}$$

where the J_{ij} are, in most cases, isotropic exchange coupling constants. (In two protein-bound iron clusters, antisymmetric exchange was found to be important.) Application of a small (≈ 50 gauss) magnetic field is sufficient to split the magnetic sublevels of the electronic system to the extent that mixing of the electronic states by hyperfine interactions can be ignored. Under this condition, finding solutions to (13.1) and (13.2) is simplified by diagonalizing the electronic spin Hamiltonian, consisting of the first two terms of (13.1) and, if present, the second term of (13.2), followed by diagonalization of the modified nuclear Hamiltonian obtained from (13.1) by replacing the spin operator $\hat{\mathbf{S}}$ in the magnetic hyperfine term with the spin expectation values, $< \hat{\mathbf{S}} >$. For slow spin relaxation, each thermally accessible level produces its own Mössbauer spectrum; for fast relaxation, one spectrum is observed, and $< \hat{\mathbf{S}} >$ has to be replaced by its thermal average, $< \hat{\mathbf{S}} >_{\text{th}}$ [1, 2]. The case of intermediate relaxation requires a more complex treatment which, however, is rarely needed for extracting the spin Hamiltonian parameters of biological iron sites. Equations (13.1) and (13.2) have proven to be extremely successful for providing accurate simulations of some thousand Mössbauer spectra. Many reviews have been published that specifically address biological Mössbauer spectroscopy, among them treatments by Lang [1], Münck [2], Trautwein et al. [3], and Krebs and Bollinger [4].

Frequently, one encounters iron-containing clusters in an $S = 1/2$ ground state. In this case, one uses an effective Hamiltonian in the coupled representation,

$$\hat{\mathcal{H}}_{S=1/2} = \beta B \cdot \mathbf{g} \cdot \hat{\mathbf{S}} + \sum_i \hat{\mathbf{S}} \cdot \mathbf{A}_{\text{eff},i} \cdot \hat{\mathbf{I}}_i - g_n\beta_n B \cdot \hat{\mathbf{I}} + \hat{\mathbf{I}} \cdot \mathbf{P} \cdot \hat{\mathbf{I}}, \tag{13.3}$$

in which $\hat{\mathbf{S}}$ is the total spin of the system. The $\mathbf{A}_{\text{eff},i}$ are now effective A tensors, which are linked to the local $\mathbf{A}_i$ by $\mathbf{A}_{\text{eff},i} = K_i \mathbf{A}_i$, where the K_i are spin projection factors. For a dinuclear $Fe^{3+}Fe^{2+}$ or $Fe^{4+}Fe^{3+}$ complex with antiferromagnetically coupled ($J > 0$) high-spin sites ($S_i = 2$ for Fe^{2+} and Fe^{4+} and $S_i = 5/2$ for Fe^{3+}) these factors are $K_{Fe(3+)} = +7/3$ and $K_{Fe(2+),Fe(4+)} = -4/3$, the minus sign indicating that the local $S_i = 2$ spin is antiparallel to the total spin (see Sect. 13.4).

Mössbauer characterization in biological systems may serve a number of purposes. To lists a few: (i) identification of iron sites in proteins, establishing their nuclearity, and quantifying the amount of iron, (ii) determination of the oxidation/spin state of the iron site(s), (iii) identification of reactive intermediates, which elude characterization by conventional structural methods such as X-ray crystallography, (iv) assessment of protein dynamics, and (v) characterization of electronic features, such as the spin ordering and valence distribution in clusters. All these applications rest on the notion that the spin Hamiltonian parameters contain information about the molecular structure and the electronic state of the iron complex. Quantum chemistry, in particular density functional theory (DFT), has evolved into a useful tool for extracting this information [5, 6].

13.3   Sulfite Reductase: The Discovery of Coupled Chromophores

Sulfite reductase (SiR) is an important enzyme that catalyzes the six-electron reductions of sulfite to sulfide and of nitrite to ammonia. By 1980, biochemical, EPR, and optical data had shown that the enzyme contained a heme (a rare variety called siroheme) and an Fe_4S_4 iron–sulfur cluster. Fe_4S_4 clusters (Fig. 13.1d) were discovered in the 1960s by EPR and they have been characterized by many techniques, including X-ray crystallography [7]. With a few exceptions, the Fe_4S_4 cluster is anchored to the protein matrix by coordination to four cysteinyl sulfurs (see Fig. 13.8 below). Nature employs Fe_4S_4 clusters ubiquitously in electron

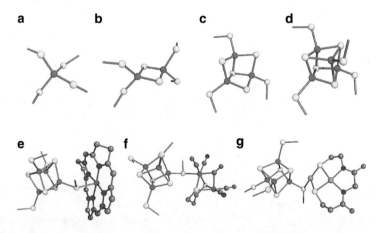

Fig. 13.1 Examples of iron–sulfur clusters and their assemblies with other cofactors. All Fe_4S_4 clusters are coordinated by four cysteinyl sulfurs. In (e–g), the two subassemblies are linked by a cysteinyl residue. Color code: grey (C), blue (N), brown (O), yellow (S), red (Fe), and green (Ni). Hydrogen atoms and side chains of the siroheme are not shown

transfer reactions and elsewhere. They have been studied with many techniques in four oxidation states designated by $[Fe_4S_4]^z$, where $z = 0$, $1+$, $2+$, and $3+$ gives the charge of the cluster's core. Most frequently, the cluster is observed in the $2+$ state ($2Fe^{3+}$, $2Fe^{2+}$, and $4S^{2-}$). In this state, the cluster has spin $S = 0$, a result of antiferromagnetic and double exchange coupling between the iron ions. A typical zero-field Mössbauer spectrum of the $2+$ state consists of one doublet with $\Delta E_Q \approx 1$ mm/s and $\delta = 0.45 \pm 0.02$ mm/s; δ is a very reliable indicator of the core oxidation state. The lowest excited states in the exchange coupling ladder are at energies $> 200\, \text{cm}^{-1}$.

In the resting state of SiR, the cluster is in the $2+$ state ($\Delta E_Q = 1.00$ mm/s and $\delta = 0.45 \pm 0.01$ mm/s at 100 K), while the siroheme is high-spin Fe^{3+} as witnessed by its low-temperature EPR signal ($g_\perp \approx 6$, $g_z = 1.98$) which is typical of the $M_S = \pm 1/2$ ground Kramers doublet of a high-spin Fe^{3+} (the $M_S = \pm 3/2$ state of SiR is at 16 cm^{-1}). Significantly, and as anticipated, the EPR signal is quantified to 1 spin/siroheme. The 4.2 K Mössbauer spectrum was thus expected to exhibit paramagnetic hyperfine structure for the siroheme (readily simulated given the EPR information) and a quadrupole doublet for the $[Fe_4S_4]^{2+}$ cluster. The 50 mT spectrum of Fig. 13.2 shows indeed the typical heme spectrum (left panel, solid line) but the spectrum of the $[Fe_4S_4]^{2+}$ cluster exhibited, surprisingly, paramagnetic hyperfine structure! The cluster not only exhibited magnetic features,

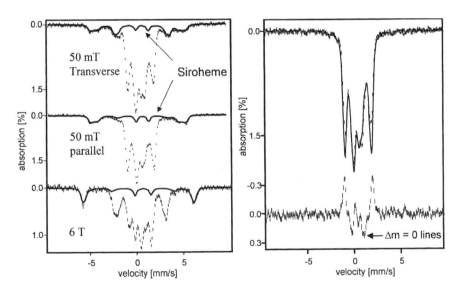

Fig. 13.2 *Left*: Mössbauer spectra of sulfite reductase recorded in 50 mT magnetic fields applied transverse and parallel to the γ beam, and a 6 T spectrum in parallel field (*bottom*). The *solid line* is a simulation of the siroheme spectrum. *Right panel*: 50 mT transverse-field spectrum of the $[Fe_4S_4]^{2+}$ cluster (after subtraction of the siroheme contribution) showing magnetic hyperfine structure for a nominally diamagnetic cluster, and a simulation assuming that the cluster is part of an $S = 5/2$ system (*top*). The difference "transverse minus parallel" of the 50 mT spectra in the left panel is shown at the bottom

the intensities of its absorption bands changed when the 50 mT applied field was altered from parallel to transverse to the observed γ rays; the downward features in the difference spectrum of Fig. 13.2, right panel, are due to nuclear $\Delta m = 0$ transitions, which are enhanced in transverse field. This observation implied that the $[Fe_4S_4]^{2+}$ cluster must be associated with an EPR signal. However, there was no signal other than that attributed to the siroheme. Moreover, spectra taken in strong fields suggested that the field was mixing a low-lying state at $\approx 16 \, cm^{-1}$ into the cluster's ground state. But the state at $16 \, cm^{-1}$ belonged to the heme! Given the above information, the conclusion was inescapable that the heme and the $[Fe_4S_4]^{2+}$ cluster were covalently linked, sharing the $S = 5/2$ spin [8]. Six years after these Mössbauer results were reported, the X-ray structure of SiR was solved. Not surprisingly (at least for researchers who believe in the power of Mössbauer spectroscopy), the X-ray structure showed an Fe_4S_4 cluster covalently linked to the iron of the siroheme by a sulfur atom of a cysteinyl residue (Fig. 13.1e); for most recent structure, see [9].

Subsequent Mössbauer studies of SiR showed that both the Fe^{3+} of the siroheme and the $[Fe_4S_4]^{2+}$ cluster can be reduced by one electron each, while retaining the coupled conformation. The presence of the coupled units allows SiR to store two electrons at the catalytic center, thereby enabling the enzyme to perform two-electron reductions of the substrate, avoiding hazardous reactive intermediates that could be generated by one-electron reductions.

The discovery of the coupled chromophores in SiR relied on the assertion that a field-dependent (parallel vs. perpendicular field) Mössbauer spectrum implied the presence of an EPR signal. This assertion can be understood as follows: A Kramers doublet is EPR-active if the spin-up and spin-down states for a field along z are connected by the S_+ and S_- operators, such as the $M_S = \pm 1/2$ doublet states of the $S = 5/2$ system. In this doublet, a change in the direction of the applied field, as weak as 5 mT, will reorient the expectation value of the electronic spin, $< \hat{S} >$, and the associated internal magnetic field at the ^{57}Fe nucleus, $\boldsymbol{B}_{int} = - < \hat{S} > \cdot A / g_n \beta_n$. $\boldsymbol{B}_{int}$ generally determines the nuclear quantization axis. Since $\boldsymbol{B}_{int}$ of an EPR-active species depends on the direction of the applied field (relative to the γ rays), the Mössbauer line intensities depend on the orientation of the applied field as well. (The above argument does not work, for practical purposes, if the g-tensor is strongly uniaxial, say for example $g_z >> g_x, g_y$).

The nature of the exchange coupling between the heme iron and the $[Fe_4S_4]^{2+}$ cluster has been analyzed in detail [10]. Briefly, antiferromagnetic exchange mediated by the sulfur bridge between the heme iron and one of the cluster irons, Fe_D, promotes an antiparallel alignment between the $S = 5/2$ spin of the siroheme and the spin of Fe_D. The siroheme thus perturbs the internal coupling of the cluster, admixing an excited triplet state ($S_{cube} = 1$) into the $S_{cube} = 0$ ground state by which the $[Fe_4S_4]^{2+}$ cluster acquires paramagnetic properties. In other words, S_{cube} is not a "good" quantum number anymore. Most importantly, through the coupling, the cube, a non-Kramers system, becomes part of a Kramers system. Analysis of the A-tensors of the cluster [10] indicates a heme-cluster exchange coupling constant $|J| \approx 17 \, cm^{-1}$ (in the $J\hat{S}_1 \cdot \hat{S}_2$ convention).

Since their discovery, assemblies involving $[Fe_4S_4]^{2+}$ clusters have been observed in a number of biological systems. In acetyl CoA synthase, the $[Fe_4S_4]^{2+}$ cluster is covalently linked, again by a cysteinyl sulfur, to a nickel site that is part of a dinuclear nickel complex (Fig. 13.1g) [11, 12]. [FeFe] hydrogenases catalyze the reversible heterolytic cleavage of H_2. Here, the $[Fe_4S_4]^{2+}$ cluster is linked to a very unusual [2Fe] subcluster, an assembly called the H-cluster (Fig. 13.1f) [13,14]. The two irons of $[2Fe]_H$ are ligated by CO and CN^- (a major surprise in modern biochemistry) and bridged by a dithiol. The $[Fe_4S_4]^{2+}$ cluster shuttles electrons between $[2Fe]_H$ and other iron–sulfur clusters. In both CoA synthase and [FeFe] hydrogenases, the covalent linkage between the $[Fe_4S_4]^{2+}$ cluster and the subcluster was deduced from the observation of ^{57}Fe paramagnetic hyperfine structure for all Fe sites of the assembly, and the implied covalent linkage between the subclusters preceded their "visualization" by X-ray crystallography.

13.4 Biomimetic Chemistry: Insights into the Dependence of Reactivity on Structure

One of the most intensely studied iron-containing proteins is soluble methane monooxygenase (sMMO), the enzyme that catalyzes the conversion of methane to methanol. In the course of the catalytic reaction, O_2 is heterolytically cleaved such that one oxygen atom ends up in water while the other is incorporated into the substrate as an OH group. The active site of sMMO, a diiron center [15], traverses during the catalytic cycle a diiron(III) state that is reduced to diiron(II), subsequently forming an O_2 adduct (called P^*) that is turned over to a diiron(III)-peroxo state (P). Subsequently, the oxygen is cleaved and a short-lived ($T_{1/2} \approx 15\,s$) intermediate, Q, is formed. Q was shown to be a bis-μ-oxo bridged diiron(IV) complex by Mössbauer and EXAFS spectroscopy [16]. The high oxidation states of the irons in Q give it the oxidative power to abstract a hydrogen atom from methane. The H atom becomes attached to an $Fe^{IV} = O$ group. (In this section, we employ Roman numerals to designate the oxidation state of iron, as generally used in the chemical literature.) The resultant OH is thought to be captured by the methyl radical in a "rebound" mechanism and finds a new home in CH_3OH.

In parallel with these biochemical studies, synthetic bioinorganic chemists have attempted to design complexes that mimic the structure and function of Q. Since the iron sites in sMMO have $FeO_{4-5}N$ coordinations (carboxylates, oxo- and hydroxo groups, water, histidine) they are high-spin; e.g., $S = 2$ for high-spin Fe^{IV}. In contrast, most biomimetic complexes synthesized to date have, for reasons of chemical stability, predominantly nitrogen donor ligands (e.g., amines and pyridine). The octahedral $FeN_{\geq 4}O_{\leq 2}$ coordination sites in these complexes stabilize Fe^{IV} in the low-spin ($S = 1$) state.

We have recently studied a series of biomimetic $Fe^{IV}Fe^{IV}$ and $Fe^{IV}Fe^{III}$ complexes that yielded novel insights into the dependence of the oxygen transfer and C–H cleaving reactivities of high-valent iron on molecular geometry and electronic structure [17, 18]. These studies would have been unthinkable without the application of Mössbauer spectroscopy in *all* phases of the project. The Fe^{IV}-containing intermediates generated are so reactive that they had to be prepared in organic solvents at temperatures between $-40°C$ and $-80°C$, using optical cuvettes for instantaneous monitoring of their electronic absorption spectra, followed by transfer with precooled syringes into EPR tubes and Mössbauer cups. Despite all precautions, these short-lived species partly decayed (up to 40% of total Fe) into $Fe^{III}Fe^{III}$ (the thermodynamic sink in this chemistry), $Fe^{IV}Fe^{III}$, and multiple mononuclear Fe^{III} species. These contaminants, however, could be identified and removed from the Mössbauer spectra with reasonable accuracy. Identification of these contaminants played an essential role in the design of preparative strategies, illustrating the importance of Mössbauer spectroscopy as an analytical tool. In the following, we outline some of the major findings. In Fig. 13.3, we have assembled cartoons of the core structures of the complexes mentioned. The focus will be on $Fe^{IV}Fe^{III}$ complex **5**.

The starting material is a diiron(III) complex, **1**. It has a structure as shown in Fig. 13.4, except that the position of the terminal oxo group of the left iron is occupied by a water molecule. Addition of hydrogen peroxide to **1** yields diiron(IV) complex **2**. It required 14 preparations and nearly 1,000 simulations of magnetic Mössbauer spectra to unravel its electronic structure. Briefly, **2** is a diiron(IV) complex with inequivalent iron sites. The local spins of both Fe sites, $S_a = 1$ and $S_b = 1$, are *ferromagnetically* coupled to yield a cluster ground state with $S_{cluster} = 2$. The observation of ferromagnetic coupling was a big surprise for a structure with a large Fe–O_{bridge}–Fe angle of $130°$, but could be nicely explained when it was realized that the different ligand fields at both sites produced a crucial pair of orthogonal orbitals on sites *a* and *b* [19]. Site *b* (isomer shift $\delta_b = -0.03$ mm/s) has a terminal oxo ligand, $Fe^{IV}=O$. This was established from the set of local Mössbauer parameters extracted from the parameter set of

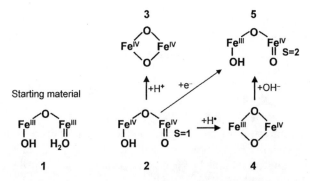

Fig. 13.3 Cartoons of the core structures of the complexes discussed in Sect. 13.4. Each Fe is also ligated to four nitrogens of the TPA ligand (see Fig. 13.4). For **2** and **5**, the spin state of the $Fe^{IV}=O$ site is indicated

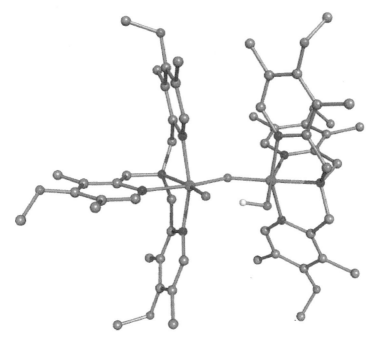

Fig. 13.4 DFT-optimized structure of $Fe^{IV}Fe^{III}$ complex **5**. Coordinated to each iron is a TPA ligand (TPA = tris(3,5-dimethyl-4-methoxylpyridyl-2-methyl)amine). Note that pyridine is coordinated *trans* to the oxo bridge at Fe^{IV}=O while an amine takes the *trans* position at the Fe^{III}–OH site. The DFT structures of **2** and **5** are essentially the same. Color code: grey (C), blue (N), brown (O), and red (Fe); for clarity, only one H atom (white) is shown

the exchange-coupled system and by EXAFS studies of **2** that yielded an Fe–O bond length of 1.65 Å (the DFT structure of Fig. 13.4 yielded 1.66 Å). Fe^{IV} site a ($\delta_a = 0.00$ mm/s) has a terminal hydroxo group [19].

Addition of acid (H^+) to **2** yields, by elimination of H_2O, the diamagnetic $Fe^{IV}Fe^{IV}$ complex **3**, while addition of a hydrogen atom donor (H^+ plus e^-) yields the valence-delocalized $Fe^{IV}Fe^{III}$ complex **4**. Both **3** and **4** have a closed bis-μ-oxo bridged core, unlike the open, singly μ-oxo bridged core structure of **2**. The difference in the core structure has a dramatic effect on the reactivity of the two species (see below).

We now introduce complex **5**, the newest member in this group of high-valent complexes. It is an $Fe^{IV}Fe^{III}$ complex (called **1–OH** in [18]) that was generated either by one-electron reduction of **2** or by the addition of hydroxide to **4**, all at about –70°C. EPR showed a resonance at $g = 2.0$, indicating that the ground state of **5** has cluster spin $S_{cluster} = 1/2$. A Mössbauer spectrum taken at 120 K revealed two quadrupole doublets with $\delta_a = 0.45$ mm/s and $\delta_b = 0.11$ mm/s. The site with δ_a undoubtedly is high-spin ($S_a = 5/2$) Fe^{III}. This assignment was supported by the analysis of the low-temperature spectra, which yielded the magnetic hyperfine

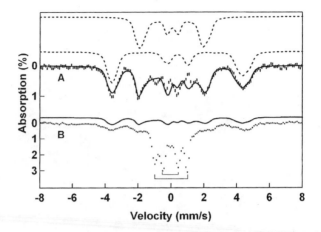

Fig. 13.5 4.2 K Mössbauer spectra of **5** (A) and cryoreduced **2** (B), recorded in a parallel field of 50 mT. For clarity, we have removed from the spectrum of **5** the contributions of an $Fe^{III}Fe^{III}$ decay product (30% of Fe) and a mononuclear Fe^{III} contaminant (10%). The *solid line* drawn through the data of **5** is a spectral simulation, using an $S = 1/2$ spin Hamiltonian (parameters given in ref. [18]) obtained by cryoreduction of **2** at 77 K. The two brackets indicate the doublets of the two Fe^{IV} sites of **2**. The *solid line* drawn above (B) outlines the new feature that appeared upon cryoreduction; it is the same curve that fits **5** of spectrum (A)

coupling constant $A_{eff,iso}/g_n\beta_n = -21.1$ T, a value typical for mononuclear high-spin Fe^{III} complexes of the TPA ligand. DFT calculations and general considerations strongly suggested that the Fe^{III} site was the iron bound to the hydroxo group. The value $\delta_b = 0.11$ mm/s indicated an Fe^{IV} site. Thus, **5** was an exchange-coupled $Fe^{III}Fe^{IV}$ complex with an Fe^{III} spin $S_a = 5/2$ and total spin $S_{cluster} = 1/2$, implying that the Fe^{IV} site must have $S_b = 2$. The 4.2 K Mössbauer spectra of **5** exhibited paramagnetic hyperfine structure and were analyzed in the frame work of (13.1) and (13.2), using an exchange-coupling constant $J = 90$ cm^{-1} determined from EPR [18]. Figure 13.5a shows a simulation of the 4.2 K spectrum recorded in a parallel field of 50 mT; the subspectra of sites a and b are plotted above the data. High-field measurements (up to 8 T) showed that the internal magnetic field, B_{int}, was negative for site a and positive for site b, confirming that the spins of the two sites were coupled by antiferromagnetic exchange.

The observation of a high-spin ($S_b = 2$) configuration suggested that a spin transition of $Fe_b^{IV} = O$ had occurred upon reduction of the neighboring Fe_a–OH site from $S_a = 1$ Fe^{IV} to $S_a = 5/2$ Fe^{III}. A plausible explanation for the spin change of Fe_b would have been that its terminal oxo group had been released in the reduction. To investigate this possibility, we performed a cryoreduction experiment, irradiating **2** with ^{60}Co γ-rays at 77 K, a temperature at which the ligands are locked in the frozen matrix. The irradiation generated a new spectral component with features identical to those observed for **5** (the solid line above spectrum in Fig. 13.5B is the theoretical curve of 5A), leading to the conclusion that the Fe^{IV} site of **5** had retained the terminal oxo group. If it was not a ligand change, what then had caused the spin transition? The observation that the coupling was ferromagnetic in **2** and antiferromagnetic in **5** suggested that exchange interactions between the irons played a decisive role in the spin transition. To investigate this possibility, we performed DFT calculations that gave the following results. The ligand field at site Fe_b gives an $S_b = 1$ ground state that is nearly degenerate ($\approx 200\,cm^{-1}$) with the $S_b = 2$ state, fulfilling a necessary condition for spin transitions. The reduction of Fe_a communicates through the oxo bridge a change in the ligand field at Fe_b that lowers the $S_b = 2$ level slightly below the $S_b = 1$ level. However, by far the largest contribution to the spin transition comes from the strong antiferromagnetic exchange interaction in the $(S_a = 5/2, S_b = 2)$ couple, which lowered the energy of the $|(S_a = 5/2, S_b = 2)\ S = 1/2 >$ state relative to the energy of the $|(S_a = 5/2, S_b = 1)\ S = 3/2 >$ state by $\sim 1{,}600\,cm^{-1}$ (see Fig.13 of [18]). Neither an exchange-driven spin transition nor a spin transition at an Fe^{IV} site had previously been reported in the literature.

The availability of **2** and **5** raised the exciting possibility of measuring the reactivity of an $Fe^{IV} = O$ site of given coordination in different spin states. These experiments demonstrated that the rate of hydrogen atom abstraction for the high-spin state $(S_b = 2)$ of **5** is three-orders of magnitude faster than for the low-spin state $(S_b = 1)$ of **2** [17]. This result is the first experimental evidence for a (perhaps) functional relevance of the high-spin state of $Fe^{IV} = O$ intermediates in oxygen activating enzymes. Ref. [17] also reports a more than million-fold increase in C–H bond cleaving reactivity between the closed core complex **4** and the open core complex. The studies just reported, performed in collaboration with Lawrence Que, Jr, at the University of Minnesota, indicate that biomimetic chemistry has reached a stage where it can develop catalytic complexes with reactivities rivaling or even exceeding those of the natural systems.

13.5   Fe$_3$S$_4$ Clusters, a New Role for Iron–sulfur Clusters

In 1979, an X-ray study was published for a protein from the bacterium *Azotobacter vinelandii* that indicated the presence of two iron–sulfur clusters. One was identified as an Fe_4S_4 cluster and the other was suggested to be of the Fe_2S_2 type (Fig. 13.1b). The presence of an Fe_2S_2 cluster (**X**) seemed odd because biochemical studies had shown that this cluster exhibited an EPR signal at $g = 2.01$ upon *oxidation*, rather than upon reduction as is usually observed (the characteristic $g = 1.94$ EPR signal) for this cluster type. These puzzling results prompted a Mössbauer study of the protein.

The Mössbauer study revealed a novel iron–sulfur cluster, now known to be ubiquitously distributed in nature [20]. This cluster has an Fe_3S_4 core (Fig. 13.1c), essentially a cubane Fe_4S_4 cluster that has lost one Fe. This discovery illustrates how Mössbauer spectroscopy can be used as a tool for structure determination by applying the tricks of the trade. The redox potential of the Fe_4S_4 cluster of the *A. vinelandii* protein was such that **X** could be obtained in both the oxidized and reduced state without changing the diamagnetic 2+ state of the Fe_4S_4 cluster. This condition suggested that one could remove the Fe_4S_4 contribution by taking difference spectra. We consider first the oxidized state, in which **X** exhibited an $S = 1/2$ signal at $g = 2.01$. Recording 4.2 K spectra in a 50 mT field applied parallel and perpendicular to the γ beam and taking the difference "parallel minus perpendicular" revealed *two* sets of nuclear $\Delta m = 0$ lines. (N.B.: the nuclear Zeeman splitting for $B = 50$ mT is too small to leave observable remnants of the Fe_4S_4 cluster in the difference spectrum.) This observation implied the presence of two iron sites, both having isomer shifts characteristic of high-spin ($S = 5/2$) Fe^{III} with tetrahedral sulfur coordination. However, as two $S = 5/2$ spins cannot combine to a cluster spin $S_{cluster} = 1/2$, a site carrying a half-integer spin was apparently missing. To address this issue one-electron reduced **X** was studied. In this oxidation state, the cluster spin of **X** is $S = 2$. The spin quintet was split in zero field with the two lowest sublevels, levels (roughly of $M_S = \pm 2$ parentage) being separated by about 0.3 cm^{-1}. As reduced **X** was in a non-Kramers state, the zero-field spectra consisted of quadrupole doublets that were essentially masked by the absorption of the Fe_4S_4 cluster. However, **X** could be pulled from behind the mask by applying a field of 50 mT that induced (unresolved) magnetic hyperfine interactions by mixing the two lowest spin levels. By taking the difference spectrum "zero-field minus 50 mT", again canceling the contribution of the $[Fe_4S_4]^{2+}$ cluster, a quadrupole doublet pattern appeared with a 2:1 intensity ratio (Fig.3 of [20]). Thus, the cluster had *three* Fe sites. (N.B.: the third site was not revealed in the difference spectrum for the $S = 1/2$ state because the spin of site 3 is oriented perpendicular to the cluster spin, yielding vanishingly small hyperfine interactions. The spectrum of this site had been canceled in the subtraction, just as the spectrum of the Fe_4S_4 cluster.) A revised X-ray structure, containing a cubane Fe_3S_4 cluster like the one shown in Fig. 13.1c, was published in 1988 [21].

Proteins containing Fe_3S_4 clusters are now known to participate in numerous biochemical reactions. The discovery of Fe_3S_4 clusters rapidly was followed by a variety of interesting, and consequential, observations. In the early 1980s, *Desulfovibrio gigas* ferredoxin II (FdII) emerged as the work horse for Fe_3S_4 cluster studies. It was soon discovered that in some proteins, Fe_3S_4 and Fe_4S_4 clusters could be reversibly interconverted, i.e., the Fe_4S_4 cluster contained one labile iron site. H. Beinert, the co-discoverer of iron–sulfur proteins, observed a $g = 2.01$ signal in aconitase, a key enzyme in the citric acid cycle (Krebs cycle). A Mössbauer study of active aconitase revealed an Fe_4S_4 cluster with a labile fourth iron, and this labile site turned out to be the binding site of the substrates; see [2]. (The Fe_3S_4 form of aconitase is inactive and exhibits the $g = 2.01$ EPR signal.) Interestingly, ions of metals like Zn, Cu, Co, Ni, and Mn can be incorporated into the labile site of the Fe_3S_4 cluster to yield MFe_3S_4 cores with interesting magnetic properties.

We wish to discuss briefly the $S_{cluster} = 2$ state of the Fe_3S_4 cluster as the evaluation of its Mössbauer spectra yielded new insights into magnetochemistry. The zero-field spectrum of the FdII cluster exhibited two doublets with 2:1 intensity ratio. The minor doublet had $\Delta E_Q(1) = 0.52$ mm/s and $\delta(1) = 0.32$ mm/s (relative to Fe metal at 298 K); the value of δ is at the upper end of high-spin $Fe^{3+}S_4$

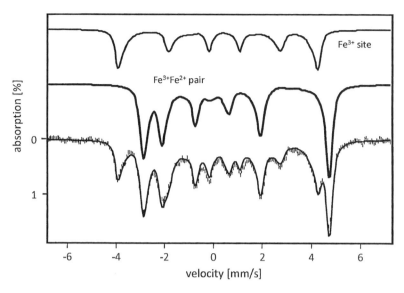

Fig. 13.6 Mössbauer spectrum of *D. gigas* FdII recorded at 1.5 K in a magnetic field applied parallel to the observed γ rays (*bottom*). The *solid line* drawn through the data is a spectral simulations, using an $S = 2$ spin Hamiltonian (parameters are quoted in [24]). Above the data are shown the subspectra of the Fe^{3+} site (*top*) and the delocalized $Fe^{2.5+}Fe^{2.5+}$ pair

sites. The majority doublet, representing two Fe, had $\Delta E_Q(2, 3) = 1.47\,\text{mm/s}$ and $\delta(2, 3) = 0.46\,\text{mm/s}$. As $Fe^{3+}S_4$ and $Fe^{2+}S_4$ sites would have δ values near $0.25\,\text{mm/s}$ and $0.65\,\text{mm/s}$, respectively, the isomer shift of the majority doublet indicated a valence-delocalized pair: $Fe^{2.5+}Fe^{2.5+}$. Figure 13.6 shows a $B = 1\,\text{T}$ Mössbauer spectrum of FdII taken at 1.5 K. The solid line is a spectral simulation for three subsites, using an $S = 2$ spin Hamiltonian. Even in the presence of magnetic hyperfine interactions, the spectra of the $Fe^{2.5+}Fe^{2.5+}$ pair are indistinguishable, which indicates complete valence delocalization.

We struggled for a while with the question of how to describe the delocalized pair in the framework of a spin Hamiltonian. The crucial insight came from papers of Zener [22] and of Anderson and Hasegawa [23], who described a spin-dependent electron delocalization phenomenon called "double exchange". In this mechanism, a delocalized "excess" electron aligns the $S = 5/2$ core spins of the two Fe sites of an $Fe^{3+}Fe^{2+}$ dimer *parallel*. This introduces a coupling term that competes with the antiferromagnetic exchange interaction (J terms in (13.2)). In combination, the two interactions yield for the energies of the spin ladder the expression $E = (J/2)S(S+1) \pm B(S+1/2)$, where B is the double exchange parameter related to inter-iron electron transfer. Because double exchange is proportional to $S + 1/2$, rather than quadratic in S, there is no simple term in the spin Hamiltonian to represent this interaction. However, the interaction can be described by introducing occupation operators (O_A and O_B in [24]) and transfer operators that give non-zero matrix elements between the configurations with the "excess electron on the left or on the right."

Fig. 13.7 Binding of SAM
to the "labile" Fe site of the
Fe_4S_4 cluster in a radical
SAM protein. Ado represents
the adenosine moiety of SAM

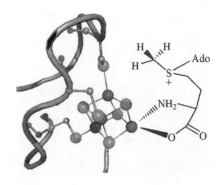

What appeared to be a unique property of aconitase in 1980 turned out to be a characteristic feature of the radical SAM superfamily which, at current count, comprises more than 2,800 enzymes (SAM = S-adenosylmethionine). In these enzymes, SAM binds through a carboxylate and an amino group to the labile iron of the $[Fe_4S_4]^{2+}$ cluster (Fig. 13.7). The catalytic cycle involves the reduction of the cluster by one electron, which subsequently is transferred to the sulfonium sulfur of SAM to promote bond cleavage with generation of a deoxyadenosyl radical. This radical abstracts a hydrogen atom from a substrate which, in turn, initiates a variety of radical-based chemistries [25]. Again, it was Mössbauer spectroscopy that led the way. By means of ^{57}Fe labeling of the labile iron, this site was identified as the locus of SAM binding [26]. These studies prepared the ground for the application of electron nuclear double resonance spectroscopy [27]. An ENDOR study suggested that a direct contact between a sulfide of the iron–sulfur cluster and the sulfonium of SAM may play a central role in the electron transfer and initiation of the radical generation. Radical SAM enzymes are involved in DNA repair, virus protection, biosynthesis, and maturation of the nitrogenase and hydrogenase clusters, incorporation of sulfur into C–H bonds, heme biosynthesis, to name a few, and they may have played an important role in the early forms of life.

13.6   Core Distortion in an All-ferrous Fe_4S_4 Cluster

Perhaps the most complex biological Mössbauer project was the unraveling of the structure of two novel clusters in the molybdenum and iron-containing protein of nitrogenase (MoFe protein), the biological system that catalyzes the $6e^-$ reduction of dinitrogen to ammonia [28]. This protein contains two copies each of two unique clusters, termed the M-center (now known to have the composition $MoFe_7S_9X$, where X = C, N, or O is an as yet unidentified caged light atom) and the Fe_8S_7 P-cluster (two Fe_4S_3 clusters bridged by a hexacoordinate sulfur and two cysteinyl sulfurs, according to present understanding). The M-center got its name from the

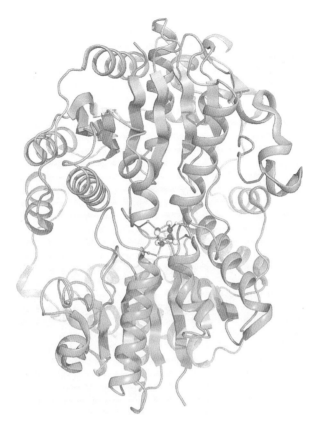

Fig. 13.8 Ribbon diagram of the Fe protein of nitrogenase. The Fe_4S_4 cluster is bound at the interface of two identical protein subunits

observation of a magnetic subcomponent in the Mössbauer spectrum of as-isolated nitrogenase.

The electrons required for the reduction of dinitrogen are transferred to the MoFe protein by the so-called Fe protein, the subject of this section. The Fe protein is a dimer consisting of two identical subunits. Bound at the interface of the subunits is an Fe_4S_4 cluster that is coordinated by four cysteinyl residues, two from each subunit (Fig. 13.8). It is thought that the Fe_4S_4 cluster shuttles between the $1+$ and $2+$ core oxidation states, transferring one electron to the P cluster each time it binds to the MoFe protein. Each binding/transfer process requires hydrolysis of two molecules of adenosine triphosphate (ATP). It came as a surprise when it was found out in 1997 that the $[Fe_4S_4]$ cluster of the Fe protein could be reduced by two electrons, attaining a core oxidation state containing four ferrous ions, $[Fe_4S_4]^0$, that had never been observed for any Fe_4S_4 cluster [29]. This observation raised the intriguing, and as yet unresolved, question of whether the electrons transferred: ATP ratio could be improved from 1:2 to 2:2, increasing the energy efficiency of nitrogenase two-fold. In the following, we wish to discuss some interesting aspects of the all-ferrous cluster.

The spin of the all-ferrous cluster results from exchange coupling between four high-spin Fe^{2+} sites with spins $S_i = 2$, $i = 1$–4. Given that the core of the cluster is reasonably symmetric, one would expect the six exchange-coupling constants, J_{ij}, to have about the same value, yielding a ground state with $S_{cluster} = 0$. The first surprise was the observation of a parallel mode (oscillatory B-field parallel to $\mathbf{B}$) EPR signal at $g_{eff} = 16$, indicating an $S_{cluster} = 4$ ground state. The second surprise was the observation of two quadrupole doublets with 3:1 intensity ratio. In an applied field, the cluster produced a magnetic pattern (Fig. 13.9) for which sites 1–3 were found to have $B_{int} < 0$ while site 4 had $B_{int} > 0$, indicating that $\mathbf{S}_1$, $\mathbf{S}_2$, and $\mathbf{S}_3$ were parallel to $\mathbf{S}_{cluster}$ while $\mathbf{S}_4$ was aligned antiparallel. These results left us facing an enigmatic cluster that exhibited a 3:1 symmetry while being symmetrically bound between two identical protein subunits. The decisive clue to this riddle came 10 years later when R. H. Holm and coworkers prepared a synthetic Fe_4S_4 cluster in the all-ferrous state.

For chemical stability reasons, the Holm cluster was coordinated by four carbene ligands rather than four thiolates as the cluster in the Fe-protein. Quite surprisingly, the carbene-coordinated cluster exhibited essentially the same Mössbauer and EPR spectra as the Fe protein; Fig. 13.9 shows a 4.2 K spectrum. This observation suggested that the 3:1 symmetry was not the result of constraints imposed by the external ligands but reflected an intrinsic distortion of the $[Fe_4S_4]^0$ cluster core. This

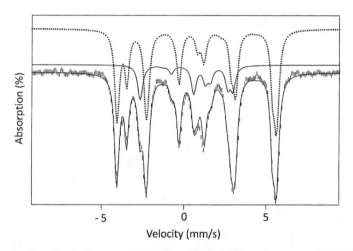

Fig. 13.9 4.2 K Mössbauer spectrum of the carbene-coordinated, all-ferrous Fe_4S_4 cluster. The *solid line* is a spectral simulation, using an $S_{cluster} = 4$ spin Hamiltonian with two types of Fe sites in a 3:1 ratio. Above the spectrum are shown simulations for the subsites. The magnetic splitting of site 4 increases in a strong applied field while that of sites 1–3 (*dotted line*) decreases. The parameters of site 3 are slightly different from those of sites 1 and 2 (see [30]). The nearly degenerate ground doublet of the $S_{cluster} = 4$ system is magnetically uniaxial, $< S_x > \approx < S_y > \approx 0$, $< S_z > = \pm 4$, hence the observation of six-line subspectra

possibility is supported by a theoretical analysis [30]. Assuming that the exchange-coupling constants, J_{ij}, depend on the $Fe_i–S–Fe_j$ bond angles, it was shown that the removal of the degeneracy of the spin states of the symmetric core by distortion lowers both the energy and symmetry of the core (Jahn–Teller distortion). Assuming a linear relationship between the exchange-coupling constants and bond angles, the elastic potential that holds the cluster together can be expressed as a quadratic function of J_{ij} variations. In this way, finding the equilibrium conformations for a spin state reduced to a minimization in the six dimensional J_{ij} space. This theory predicted two possible ground states: an $S_{cluster} = 4$ state with C_{3v} symmetry or an $S_{cluster} = 0$ state with D_{2d} symmetry. Both the spin ordering and core distortion predicted for the $S_{cluster} = 4$ state match those observed in the all-ferrous clusters in the Fe protein and the carbene-ligated synthetic cluster.

13.7 An Example of Whole-cell Mössbauer Spectra

While most biological Mössbauer work has focused on purified iron proteins, a number of groups have reported studies of whole cells and mitochondria. We like to give an, albeit somewhat atypical, example of whole cell work. Rubredoxin (Rd) is a well-studied protein containing an Fe site that is tetrahedrally coordinated by four sulfurs provided by cysteinyl residues. By over-expressing Rd in *E. coli* and growing the bacterium on an ^{57}Fe enriched medium, followed by spinning the cells in a centrifuge into a nylon cup, one can obtain a good Mössbauer absorber. Figure 13.10

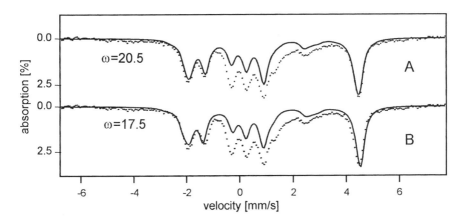

Fig. 13.10 Mössbauer spectrum of whole *E. coli* cells containing a $\sim$100-fold over-expressed ferrous rubredoxin. The *solid curve* in (A) is a simulation, using an $S = 2$ spin Hamiltonian that fits perfectly the spectrum of the purified protein. The central part of the spectrum exhibits absorption from other cell components. The theoretical curve in (B), which clearly deviates from the data, was generated by adopting an $S–Fe–S–C_\beta$ torsion angle, ω, this is only 3° smaller than in (A), demonstrating that Mössbauer spectroscopy can be used to monitor minute changes in the structure

shows a 4.2 K spectrum of the cell paste recorded in a parallel field of 4.0 T (same data in A and B). The solid lines drawn are spin Hamiltonian simulations to the spectrum of the over-expressed high-spin ($S = 2$) ferrous Rd. A comparison with the purified protein (see [31]) showed that the Mössbauer spectra of Rd as purified and in whole cells were indistinguishable, an observation that may not surprise anybody. We are showing the spectrum for the following reason. A combined DFT and crystal-field analysis had shown that the Mössbauer parameters were highly susceptible to changes in the torsion angles ω of the S–Fe–S–C$_\beta$ cysteinyls (Fig.6 of [31]). By implication, these angles had to have a large effect on the spectra, as illustrated in Fig. 13.10. On the basis of this sensitivity we were able to conclude that the torsion angles in the cellular and purified protein must be equal within $2°$.

Mössbauer spectra of an over-expressed protein in whole cells can also be of considerable practical value for biochemical research. Thus, a number of iron-containing proteins lose their iron during purification. (When his students suggested that a protein is "unstable" during purification on a column, I.C. Gunsalus quipped: "There are unstable biochemists but no unstable proteins.") In those instances, biochemists routinely attempt to reconstitute the protein by suitable addition of iron. However, as it is not clear whether the added iron ends up in the depleted binding sites, a comparison of the spectrum of the reconstituted protein with that observed in the whole cells could provide a check on the success of the attempt.

The age of Mössbauer spectroscopy of whole cells has barely begun. We anticipate that in the near future it will be possible to quantify the entire set of iron components of any organelle, provided they can be enriched with ^{57}Fe. Moreover, progress in genetics and biochemistry of iron–sulfur proteins has enabled researchers to study the cluster assembly processes in nitrogenase and hydrogenase systems. Here, the Mössbauer spectroscopists meets chaperone and scaffold proteins involved in the synthesis and maturation of these complex clusters. Undoubtedly, Mössbauer spectroscopy will continue to play a prominent role in all iron-based biological systems, and the glory added by future synchrotron-based approaches can barely be imagined.

Acknowledgements This work was supported by grant EB 001475 from the National Institutes of Health.

References

1. G. Lang, Quart. Rev. Biophys. **3**, 1 (1970)
2. E. Münck, in *The Porphyrins*, vol. IV, ed. by D. Dolphin (Academic, 1978), p. 379
3. A.X. Trautwein, E. Bill, E.L. Bominaar, H. Winkler, Struct. Bonding **78**, 1 (1991)
4. C. Krebs, E.J. Bollinger, Photosynth. Res. **102**, 295 (2009)
5. L. Noodleman, D.A. Case, Adv. Inorg. Chem. **38**, 423 (1992)
6. F. Neese, Curr. Opin. Chem. Biol. **7**, 125 (2003)
7. H. Beinert, R.H. Holm, E. Münck, Science **277**, 653 (1997)
8. J.A. Christner, E. Münck, P.A. Janick, W.M. Siegel, J. Biol. Chem. **256**, 2098 (1981)

9. B.R. Crane, L.M. Siegel, E.D. Getzoff, Biochemistry **36**, 12120 (1997)
10. E.L. Bominaar, Z. Hu, E. Münck, J.-J. Girerd, S.A. Borshch, J. Am. Chem. Soc. **117**, 6976 (1995)
11. J. Xia, Z. Hu, C.V. Popescu, P.A. Lindahl, E. Münck, J. Am. Chem. Soc. **119**, 8301 (1997)
12. C. Darnault, A. Volbeda, E.J. Kim, P. Legrand, X. Vernede, P.A. Lindahl, J.C. Fontecilla-Camps, Nat. Struct. Biol. **10**, 271 (2003)
13. J.W. Peters, W.N. Lanzilotta, B.J. Lemon, L.C. Seefeldt, Science **282**, 1853 (1998)
14. K.K. Surerus, M. Chen, W. van der Zwaan, F.M. Rusnak, M. Kolk, E.C. Duin, S.P.J. Albracht, E. Münck, Biochemistry **33**, 4980 (1994)
15. B.J. Wallar, J.D. Lipscomb, Chem. Rev. **96**, 2625 (1996)
16. L. Shu, J.C. Nesheim, K. Kauffmann, E. Münck, J.D. Lipscomb, L. Que, Science **275**, 515 (1997)
17. G. Xue, R.F. De Hont, E. Münck, L. Que, Nat. Chem. **2**, 400 (2010)
18. R.F. De Hont, G. Xue, M.P. Hendrich, L. Que, E.L. Bominaar, E. Münck, Inorg. Chem. **49**, 8310 (2010)
19. M. Martinho, G. Xue, E.L. Bominaar, E. Münck, L. Que, J. Am. Chem. Soc. **131**, 5823 (2009)
20. M.H. Emptage, T.A. Kent, B.H. Huynh, J. Rawlings, W.H. Orme-Johnson, E. Münck, J. Biol. Chem. **255**, 1793 (1980)
21. G.H. Stout, S. Turley, L.C. Sieker, L.H. Jensen, Proc. Natl. Acad. Sci. U.S.A. **85**, 1020 (1988)
22. C. Zener, Phys. Rev. **82**, 403 (1951)
23. P.W. Anderson, H. Hasegawa, Phys. Rev. **100**, 675 (1955)
24. V. Papaefthymiou, J.-J. Girerd, I. Moura, J.J.G. Moura, E. Münck, J. Am. Chem. Soc. **109**, 4703 (1987)
25. E.M. Shepard, J.B. Broderick, in *Comprehensive Natural Products II Chemistry and Biology*, vol. 8, ed. by L. Mander, H.-W. Lui (Elsevier, Oxford, 2010), pp. 626–661
26. C. Krebs, W.E. Broderick, T.S. Henshaw, J.B. Broderick, B.H. Huynh, J. Am. Chem. Soc. **124**, 912 (2002)
27. C.J. Walsby, D. Ortillo, J. Yang, M.R. Nnyepi, W.E. Broderick, B.M. Hoffman, J.B. Broderick, Inorg. Chem. **44**, 727 (2005)
28. R. Zimmermann, E. Münck, W.J. Brill, V.K. Shah, M.T. Henzl, J. Rawlings, W.H. Orme-Johnson, Biochim. Biophys. Acta **537**, 185 (1978)
29. S.J. Yoo, H.C. Angove, B.K. Burgess, M.P. Hendrich, E. Münck, J. Am. Chem. Soc. **121**, 2534 (1999)
30. M. Chakrabarti, L. Deng, R.H. Holm, E. Münck, E.L. Bominaar, Inorg. Chem. **48**, 2735 (2009)
31. V.V. Vrajmasu, E.L. Bominaar, J. Meyer, E. Münck, Inorg. Chem. **41**, 6358 (2002)

Chapter 14
Relativistic Phenomena Investigated by the Mössbauer Effect

W. Potzel

14.1 Introduction

The Mössbauer effect [1] is sensitive to relativistic phenomena due to the very narrow relative energy distribution Γ/E, where Γ is the natural linewidth and E is the energy of the Mössbauer transition. For example, for the 93.3 keV resonance in ^{67}Zn, this ratio is $\Gamma/E \approx 5 \times 10^{-16}$ and energy shifts as tiny as $\Delta E/E \approx 10^{-18}$ have been measured [2]. Three major aspects will be considered:

(a) Time dilatation due to thermal motion and zero-point motion of the Mössbauer nucleus in a lattice.
(b) Change of the Mössbauer frequency (energy) observed with high-speed centrifuges. Such experiments have been able to give upper limits on a velocity relative to a hypothetical ether (ether-drift experiments). In addition, they provide upper limits on a possible dependence of clock rates on the velocities of clocks relative to distant matter, and they open a new way to experimentally distinguish between Einstein's Special Theory of Relativity and Covariant Ether Theories.
(c) Gravitational redshift (GRS) experiments with γ ray photons in the gravitational field of the earth. GRS experiments are important because they provide an independent test of the Einstein Equivalence Principle, which states that in curved spacetime the acceleration for test bodies and photons is the same. The conclusion is reached that for Mössbauer GRS experiments it is unlikely to obtain a sensitivity better than 10^{-3} for the frequency shift in the gravitational field of the earth because various solid-state effects, which give rise to small center shifts, are difficult to control and thus severely limit the accuracy of a GRS determination. The field equations of gravitation are not tested by GRS experiments.

W. Potzel (✉)
Physics Department, Technical University Munich, 85747 Garching, Germany,
e-mail: walter.potzel@ph.tum.de

M. Kalvius and P. Kienle (eds.), *The Rudolf Mössbauer Story*,
DOI 10.1007/978-3-642-17952-5__14, © Springer-Verlag Berlin Heidelberg 2012

14.2   Time Dilatation and Second-Order Doppler Shift

14.2.1   Thermal Motion of a Mössbauer Nucleus in a Lattice

We consider a Mössbauer experiment where source (S) and absorber (A) are chemically identical but kept at different temperatures T_S and T_A, respectively. The emission and absorption lines will not coincide but will be shifted in frequency (energy). This frequency shift is temperature dependent and is caused by a relativistic effect. In a lattice, a nucleus emitting γ radiation may be considered as a moving clock exhibiting a relativistic time dilatation (reduction in frequency) w.r. to an observer at rest. For a clock moving with velocity v, we have [3,4]

$$t' = \frac{t}{\sqrt{1 - v^2/c^2}} \tag{14.1}$$

or

$$\omega' = \omega \sqrt{1 - v^2/c^2} \approx \omega(1 - v^2/2c^2). \tag{14.2}$$

Thus, the reduction in frequency is given by

$$\Delta\omega = \omega' - \omega \approx -\frac{\omega}{2}\left(\frac{v}{c}\right)^2, \tag{14.3}$$

where c is the speed of light in a vacuum.

Since $\Delta\omega \propto (v/c)^2$ this frequency shift is called second-order Doppler shift (SOD).

In a lattice, the Mössbauer atom is vibrating around its equilibrium position and v^2 has to be replaced by the mean-square velocity $<v^2>$.

Approximating the phonon spectrum of a lattice by the Debye model, the relative energy shift $(\Delta E/E)_{th}$ due to thermal motion is obtained as [5]

$$(\Delta\omega/\omega)_{th} = (\Delta E/E)_{th} = \frac{3k_B}{2Mc^2}[T_S \times f(T_S/\theta_S) - T_A \times f(T_A/\theta_A)] \tag{14.4}$$

where

$$f(T/\theta) = 3\left(\frac{T}{\theta}\right)^3 \times \int_0^{\theta/T} \frac{x^3}{\exp(x) - 1}dx \tag{14.5}$$

and T_S, T_A and θ_S, θ_A are the temperatures and the Debye temperatures of source S and absorber A, respectively.

One might argue that not only $<v^2> \neq 0$ but also the mean-square acceleration $<b^2> \neq 0$. Considering lattice vibrations at room temperature, the nuclei have rms velocities of $\sqrt{<v^2>} \approx 3 \times 10^2$ m/s. Assuming a typical Debye frequency of $\omega = 10^{14}$ Hz, the rms acceleration of the nuclei is calculated as $\sqrt{<b^2>} = \omega\sqrt{<v^2>} \approx 10^{15}$ g, where g is the gravitational acceleration at the surface of the earth. Such accelerations in a lattice are very large by macroscopic standards. Does this lead to an additional relativistic change in energy, similar to a gravitational field (see Sect. 14.4)? The answer is no. Of course, the acceleration is changing with time, changes sign, and the average acceleration $<b> = 0$. More importantly, according to the Special Theory of Relativity (STR), for an ideal clock moving non-uniformly (i.e., there is acceleration present at least for some time) in an inertial frame O, acceleration as such has no effect on the rate of the clock. If this clock in O made a trip with a time-varying velocity $v(t)$ and returned to the starting point, the elapsed time t' indicated by this moving clock would be retarded compared to an identical clock (indicating the time span t), which was kept at rest in O at the starting point. This is a variety of the familiar "twin paradox". The elapsed time t' is given by [3]

$$t' = \int_0^t \sqrt{1 - [v(t)]^2/c^2} \, dt. \tag{14.6}$$

Thus according to the so-called "clock hypothesis" [3,4] advanced by A. Einstein in the development of the STR, clocks having a velocity in an inertial frame S are slowed down by the speed, not by acceleration. In this respect, inertial frames are unique [6]. Concerning lattice vibrations, the relativistic temperature shift can also be derived quantum mechanically [7] by only assuming the equivalence of energy and mass. The Hamilton operator for the lattice energy is given by

$$H_L = \sum_i \frac{\mathbf{p_i}^2}{2M} + V(\mathbf{x_0}, \mathbf{x_1}, ...), \tag{14.7}$$

where $\mathbf{x_i}$ and $\mathbf{p_i} = M \times \mathbf{x_i}$ are the operators for space and momentum coordinates and the sum has to be carried out over all atoms i of the lattice. $V(\mathbf{x_0}, \mathbf{x_1}, ...)$ is the potential energy.

When emitting a γ quantum of frequency ω, the mass M_0 of the emitting nucleus is reduced from $(M + \hbar\omega/c^2)$ to M. This corresponds to a change of the Hamilton operator

$$\Delta H_L = \frac{\mathbf{p_0}^2}{2(M + \hbar\omega/c^2)} - \frac{\mathbf{p_0}^2}{2M} \approx -\frac{\hbar\omega}{c^2 M} \times \frac{\mathbf{p_0}^2}{2M} = -\frac{\hbar\omega}{2c^2} \times \mathbf{v_0}^2. \tag{14.8}$$

Applying ΔH_L to the lattice states, one obtains

$$\Delta\omega = -\frac{\omega}{2c^2} \times <v^2>, \tag{14.9}$$

where $<v^2>$ is no longer the classical time average but the quantum mechanical expectation value. In this way, the question of acceleration of the atom due to lattice vibrations does not enter the derivation. Only $<v^2>$ is important. Equation (14.9) is in full agreement with experiment (see below) demonstrating that Einstein's "clock hypothesis" mentioned above is fully justified. At low temperatures ($T_S <<$ $\theta_S, T_A << \theta_A$) and if source and absorber are kept at about the same temperature

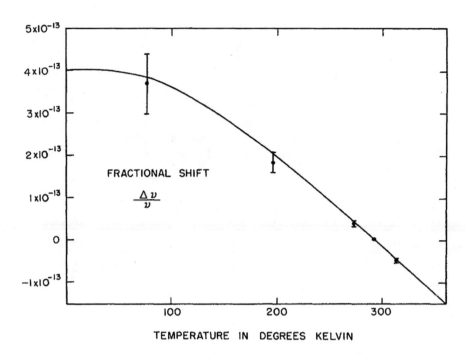

Fig. 14.1 Fractional energy shift of the 14.4 keV Mössbauer transition of ^{57}Fe in an iron-metal absorber versus temperature. The *solid line* is derived from assuming a Debye temperature of 420 K. Reprinted figure with permission from [8]. Copyright (2010) by the American Physical Society

($T_S \approx T_A$), the SOD due to thermal motion is negligibly small (see (14.4)), even if $\theta_S \neq \theta_A$.

The first measurements exhibiting the relativistic temperature shift were carried out with both source and absorber of metallic iron [8] and first results obtained with the 23.8 keV resonance of ^{119}Sn were published only a few weaks later [9]. The shift observed in [8] is shown in Fig. 14.1. The source (^{57}Co$\underline{Fe}$) was kept at $T_S = 295$ K. The temperature T_A of the absorber (metallic iron enriched in ^{57}Fe) was varied between ∼80 and 310 K. The solid line is a fit according to the Debye model with a Debye temperature of 420 K. The total shift between 80 and 310 K is ∼1.5 Γ_{Fe}, where $\Gamma_{Fe} \approx 0.09$ mm/s $\hat{=} 4.3 \times 10^{-9}$ eV is the natural width of the 14.4 keV Mössbauer transition in ^{57}Fe.

More recent examples feature the high-resolution 93.3 keV transition in ^{67}Zn. Various Cu–Zn alloys were investigated between 4.2 and 60 K [10]. Temperature shifts up to ∼40 Γ_{Zn}, where $\Gamma_{Zn} \approx 0.155 \mu$m/s $\hat{=} 4.8 \times 10^{-11}$ eV, were observed. Good agreement was found with the results of model calculations of interatomic force constants based on phonon dispersion relations, which were derived from inelastic neutron scattering data.

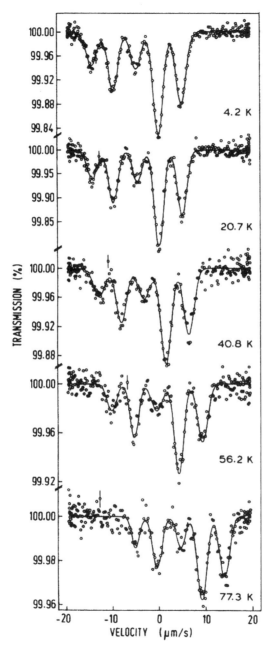

Fig. 14.2 ^{67}Zn Mössbauer absorption spectra recorded at source temperatures between 4.2 and 77.3 K. The source is a ^{67}Ga$\underline{ZnO}$ single crystal with the c axis parallel to the direction of observation of the 93.3 keV γ rays. The absorber is ^{67}ZnO powder (enriched in ^{67}Zn) kept at 4.2 K. Reprinted figure with permission from [11]. Copyright (2010) by the American Physical Society

Another interesting case is ZnO. Figure 14.2 shows absorption spectra obtained with a source of ^{67}GaZnO single crystal at various temperatures between 4.2 and 77.3 K [11]. The absorber was ^{67}ZnO powder (enriched in ^{67}Zn) kept at 4.2 K. The total shift between 4.2 and 77.3 K amounts to ∼60 Γ_{Zn}. A detailed analysis revealed that in ZnO not only the relativistic temperature shift is important. In addition, already at low temperatures, phonon-induced electron transfer from zinc to oxygen is observed [11, 12]. In agreement with theoretical calculations, the shift caused by charge transfer shows a T^4 dependence at low temperatures which partially hides the relativistic effect, and vice versa.

14.2.2 Zero-Point Motion of a Mössbauer Nucleus in a Lattice

According to (14.4), the second-order Doppler shift (SOD) vanishes if source and absorber are chemically identical ($\theta_S = \theta_A$) and both are kept at the same temperature ($T_S = T_A$). If source and absorber are chemically different, an isomer shift [13, 14] is usually observed. In addition, since $\theta_S \neq \theta_A$ also a contribution to the SOD is still present even if source and absorber are at the *same* temperature. This contribution is due to the zero-point motion (energy) of an atom in a crystal lattice and is present even in the limit of absolute zero temperature ($T \to 0$ K). The zero-point motion is a consequence of Heisenberg's uncertainty relation $\Delta x \times \Delta p \geq \hbar/2$ according to which the momentum spread $\Delta p > 0$, even for $T \to 0$, if the uncertainty of the position of an atom, e.g., within a lattice is small.

In the Debye approximation, the relative energy shift $(\Delta E/E)_{ZP}$ due to zero-point motion is given by [5]

$$(\Delta E/E)_{ZP} = \frac{9k_B}{16Mc^2}(\theta_S - \theta_A). \tag{14.10}$$

For example, assuming $\theta_S - \theta_A = 100$ K gives $(\Delta E/E)_{ZP}^{Fe} \approx 2.7 \times 10^{-2}$ mm/s $\approx$ $0.3 \times \Gamma_{Fe}$ for the 14.4 keV resonance in ^{57}Fe, and for the 93.3 keV transition in ^{67}Zn, one obtains $(\Delta E/E)_{ZP}^{Zn} \approx 23.1$ μm/s $\approx 150 \times \Gamma_{Zn}$.

In a Mössbauer experiment, of course, $(\Delta E/E)_{th}$ and $(\Delta E/E)_{ZP}$, can be present at the same time:

$$(\Delta E/E) = \frac{9k_B}{16Mc^2}(\theta_S - \theta_A) + \frac{3k_B}{2Mc^2}\left[T_S \times f(T_S/\theta_S) - T_A \times f(T_A/\theta_A)\right]. \tag{14.11}$$

This shift occurs in addition to the isomer shift [13, 14].

For most Mössbauer isotopes (including ^{57}Fe and ^{119}Sn), the shifts due to temperature and zero-point motion can usually be neglected compared to the isomer shift. For high-resolution isotopes such as ^{67}Zn and ^{181}Ta, however, this is not true. For example, to derive accurate values for the isomer shifts in Zn alloys and Zn compounds, the SOD has to be taken into account (for details, see [12]).

14.3 Experiments with High-Speed Centrifuges

14.3.1 Ether Drift and Relativity Theory

For most physicists before A. Einstein, a hypothetical ether was required for the propagation of electromagnetic signals, analogous to the propagation of sound in air. As C. Møller stated [15], "... when I say *ether* I mean a particular ether, the only ether which I think was treated in a really serious way, namely the absolute ether ..., which cannot be carried along. ... The ether itself is (was) the absolute state of rest – the absolute system of coordinates".

In pre-relativistic absolute ether theory, the Doppler shift depends not only on the relative velocity $\mathbf{u}$ of the source with respect to the observer but also on the *absolute* velocity $\mathbf{v}$ of the observer relative to an *absolute* frame. To second order in the velocities, the frequency ν' emitted in the direction of the unit vector $\mathbf{e}$ is given by [15]

$$\nu' = \nu_0[1 + \mathbf{e} \cdot \mathbf{u}/c + (\mathbf{e} \cdot \mathbf{u})^2/c^2 + \mathbf{v} \cdot \mathbf{u}/c^2], \tag{14.12}$$

where ν_0 is the frequency emitted when the source is at rest in the observer system.

To verify (14.12), several Mössbauer experiments were carried out with rotors rapidly spinning around a vertical axis. In such experiments, various geometrical arrangements have been used, e.g., the source being placed at a certain distance R_S from the center ($R_S = 0$ corresponding to the rotational axis), while the absorber is fixed at the distance R_A, e.g., along the periphery of the rotor. The detectors are usually placed either along the rotor's vertical axis or outside of the rotor setup [16–21].

With respect to the observer in the laboratory system $\mathbf{O}$, both source and absorber are moving with velocities $\mathbf{u}_S$ and $\mathbf{u}_A$, respectively. Thus, we have two equations

$$\nu'_S = \nu_S[1 + \mathbf{e} \cdot \mathbf{u}_S/c + (\mathbf{e} \cdot \mathbf{u}_S)^2/c^2 + \mathbf{v} \cdot \mathbf{u}_S/c^2] \tag{14.13}$$

$$\nu'_A = \nu_A[1 + \mathbf{e} \cdot \mathbf{u}_A/c + (\mathbf{e} \cdot \mathbf{u}_A)^2/c^2 + \mathbf{v} \cdot \mathbf{u}_A/c^2], \tag{14.14}$$

where ν_S and ν_A are the frequencies of a γ ray as seen in the rest systems of the source (S) and absorber (A), respectively, and ν'_S, ν'_A are the frequencies as seen by the observer in the laboratory system $\mathbf{O}$. Further, $\mathbf{v}$ is the absolute velocity of $\mathbf{O}$, and $\mathbf{e}$ is the direction of the γ ray in $\mathbf{O}$. Fulfilling the condition for resonance absorption ($\nu'_S = \nu'_A$), one obtains (to second order in the velocities)

$$\nu_A = \nu_S[1 + \mathbf{e} \cdot \mathbf{u}/c + (\mathbf{e} \cdot \mathbf{u}/c) \times (\mathbf{e} \cdot \mathbf{u}_S/c) + \mathbf{v} \cdot \mathbf{u}/c^2], \tag{14.15}$$

where

$$\mathbf{u} = \mathbf{u}_S - \mathbf{u}_A. \tag{14.16}$$

In a rotor experiment, $\mathbf{u}$ is perpendicular to $\mathbf{e}$ for all rays emitted by S and absorbed by A [15]. Therefore, in *pre-relativistic absolute ether theory*,

$$\nu_A = \nu_S \left[1 + \frac{\mathbf{v} \cdot \mathbf{u}}{c^2} \right], \tag{14.17}$$

where ν_A is the frequency seen in the absorber system and ν_S is the emitted frequency in the system of the source.

The relativistic Doppler effect of the STR is given by either of the following equations (14.18) or (14.19):

$$\nu_A = \nu_S \frac{1 - \mathbf{e} \cdot \mathbf{u}_A / c}{\sqrt{1 - (u_A/c)^2}}, \tag{14.18}$$

where $\mathbf{u}_S = \mathbf{0}$ and $\mathbf{u}_A \neq \mathbf{0}$, i.e., for a source at rest emitting the frequency ν_S and an observer (absorber) moving with velocity $\mathbf{u}_A$; the frequency is emitted in the direction of $\mathbf{e}$; or

$$\nu_A = \nu_S \frac{\sqrt{1 - (u_S/c)^2}}{1 - \mathbf{e} \cdot \mathbf{u}_S / c}. \tag{14.19}$$

Here, $\mathbf{u}_A = \mathbf{0}$ and $\mathbf{u}_S \neq \mathbf{0}$, i.e., the observer (absorber) is at rest and the source moves.

For *any* coordinate system in which both $\mathbf{u}_S \neq \mathbf{0}$ and $\mathbf{u}_A \neq \mathbf{0}$ we have [22]

$$\nu_A = \nu_S \frac{\sqrt{1 - (u_S/c)^2}}{\sqrt{1 - (u_A/c)^2}} \times \frac{1 - \mathbf{e} \cdot \mathbf{u}_A / c}{1 - \mathbf{e} \cdot \mathbf{u}_S / c}. \tag{14.20}$$

Equation (14.20) can be derived by using (14.18) and (14.19) or by using one of them twice and reduces to (14.18) and (14.19) in the two particular coordinate systems, respectively.

In a rotor experiment, the last factor in (14.20) is equal to unity [22], no matter where the source and the absorber are located on the rotor disk. This is due to the fact that the distance between source and absorber remains fixed, and the first-order Doppler shift does not appear. Thus, we get

$$\nu_A = \nu_S \frac{\sqrt{1 - (u_S/c)^2}}{\sqrt{1 - (u_A/c)^2}} \approx \nu_S \left[1 + (u_A^2 - u_S^2)/2c^2 \right] = \nu_S \left[1 + \frac{\Omega^2}{2c^2}(R_A^2 - R_S^2) \right], \tag{14.21}$$

where Ω is the angular velocity of the rotor; R_A and R_S are the radii of the source and absorber orbits, respectively. If $R_A > R_S$, the frequency which the absorber receives is increased. To put it differently, clocks on a rotating disk run the more slowly the larger the radius R. This effect is proportional to $(\Omega^2 \times R^2)$. For $R_A = R_S$, (14.21) gives $\nu_A = \nu_S$, while the ether theory according to (14.17) predicts $\nu_A = \nu_S \left[1 + \frac{\mathbf{v} \cdot \mathbf{u}}{c^2} \right]$, where $|\mathbf{u}|$ depends *linearly* on Ω and R with a coefficient proportional to the ether drift $\mathbf{v}$.

Equation (14.21) can also be derived by taking into account that absorber and source on the rotor can be considered as accelerated systems if $R \neq 0$. As will be shown in Sect. 14.4.3, such accelerations (e.g., on a rotor) are equivalent to *local* gravitational fields [23]. In the gravitational potential of a mass M a length contraction parallel to the direction of the gravitational field as well as a time dilatation occur.

As described in Sect. 14.4.3, we have

$$v' = v\sqrt{1 - \frac{2GM}{c^2 r}} = v\sqrt{1 - \frac{2\phi_G}{c^2}} \approx v(1 - \phi_G/c^2) \qquad (14.22)$$

or

$$\frac{v' - v}{v} = \frac{\Delta v}{v} \approx -\frac{\phi_G}{c^2}, \qquad (14.23)$$

where G is the gravitational constant and $\phi_G = GM/r$ is the Newtonian gravitational potential of the mass M at the distance r [24, 25], with $\phi_G \to 0$ for $r \to \infty$.

For the rotor, the acceleration b at an orbit with radius R is given by $b = \Omega^2 \times R = -\frac{d\phi_{\text{rot}}}{dr}$ with $\phi_{\text{rot}} = -(1/2)\Omega^2 R^2$, where $\phi_{\text{rot}} = 0$ at $R = 0$ and $\phi_{\text{rot}} < 0$ for $R > 0$.

For the frequency change, one obtains $\frac{\Delta v}{v} \approx -\frac{\phi_{\text{rot}}}{c^2}$ or

$$\frac{v_A - v_S}{v_S} = \frac{\Omega^2}{2c^2}(R_A^2 - R_S^2) \qquad (14.24)$$

in agreement with (14.21).

Similar to the curved spacetime in a gravitational field of a mass, the geometry on a rotating disk is also no longer Euclidean [23].

The two derivations of the frequency shift ((14.21) and (14.22)) demonstrate that this relativistic effect is described either by the time dilatation (second-order Doppler shift) or by the influence of an equivalent gravitational potential on clocks. However, both effects must *not* be added! The reason is that the time-dilatation effect and the effect on clocks due to an equivalent potential are related to *different* observers: Time dilatation occurs for an observer in a laboratory external to the rotor, whereas the change of the clock rate due to an equivalent potential is related to an observer involved in the rotation. This is in contrast to a clock on a satellite in the gravitational field of the earth. Viewed from an observer on the surface of the earth, the clock on a satellite orbiting at a certain height h runs faster (see (14.22)) due to the decreased gravitational potential. However, due to its orbital velocity relative to the surface of the earth, the clock is expected to run more slowly (see (14.2)). Here, both effects have to be added since they are related to the *same* observer (on the surface of the earth) and thus lead to a certain height ($h \approx 3,200$ km) above the surface of the earth where both effects cancel each other and the clock in the satellite shows no deviation with respect to a clock at the surface.

14.3.2   Ether-Drift Experiments

Several Mössbauer experiments were performed with high-speed rotors. In all cases, a quadratic dependence of the frequency shift on the angular velocity was observed as expected from the Special Theory of Relativity (STR):

(1) The first experiment [16] was carried out already less than two years after the discovery of the Mössbauer effect. A ^{57}Co source was plated onto the surface of a 0.8 cm diameter iron cylinder in the center of a rotor. A cylindrical shell of Lucite, ~13.3 cm in diameter and concentric with the iron cylinder, held an iron foil enriched in ^{57}Fe. The assembly was rotated at angular velocities up to 500 cycles per second. The results are depicted in Fig. 14.3. A clear dependence of the relative count rate of the transmitted γ photons on Ω^2 was observed in full agreement with (14.21). The absolute ether theory has a term $\mathbf{vu}/\mathbf{c}^2$ (see (14.17)), which is linear in Ω with a coefficient proportional to the ether drift $\mathbf{v}$. No term of this kind was found [15, 17].

(2) In an interesting modification described in [26], source (^{57}Co in a ^{56}Fe matrix) and absorber (natural iron) were placed at opposite ends of a rotor ($R_S = R_A$ in (14.21)). Within the error bars, the Mössbauer absorption was found to be unaffected by rotation.

(3) A technically even more sophisticated experiment has been reported in [20]. A ^{57}Fe metal absorber was placed at $R_A = 9.3$ cm from the axis of an

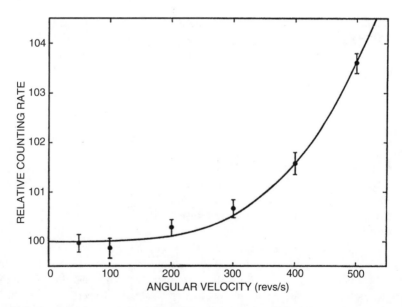

Fig. 14.3 Transmitted count rate of 14.4 keV γ rays as a function of the angular velocity of a rotor. The curve clearly shows a quadratic dependence. Reprinted figure with permission from [16]. Copyright (2010) by the American Physical Society

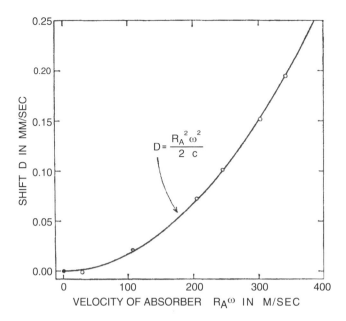

Fig. 14.4 Observed second-order Doppler shift versus the velocity of a ^{57}Fe metal absorber on a high-speed centrifuge. The statistical error corresponds to the radius of the circles. Reprinted figure with permission from [20]. Copyright (2010) by the American Physical Society

ultracentrifuge rotor. A ^{57}Co source plated on an iron foil was mounted on a piezoelectric transducer at the center of the rotor. By applying a voltage to the transducer, the source could be moved relative to the absorber. This made it possible to obtain an accurate velocity calibration and to observe the entire resonance line at values of the angular velocity Ω of up to $\sim$600 revolutions per second. The results are shown in Fig. 14.4. The measured change in frequency of the Mössbauer transition agrees within an experimental error of 1.1% with the prediction of SRT (14.21).

(4) In [18], an experiment is described with high rotor speeds of up to 1,230 cycles/s. A source of ^{57}Co plated on a copper foil and a *single-line* absorber of stainless steel were used. Source and absorber were attached to opposite tips (i.e., $R_A = R_S$) of the rotor. The distance between source and absorber was 8.1 cm. In this experiment, the search for some steady ether drift past the earth included also a term, which takes into account the daily rotation of the earth which – if present – would result in a diurnal periodicity in the count rate. Thus, the results were analyzed taking two different velocities, v and v', into account: v is the magnitude of the component of an ether drift past the earth in a plane perpendicular to the earth's axis of rotation, v' is the magnitude of an ether drift past the apparatus resulting from the laboratory's velocity relative to the center of the earth due to the earth's rotation. The analysis gave $v = (1.6 \pm 2.8)$ m/s and $v' = (2.2 \pm 2.2)$ m/s. The same authors

published another paper in 1965 [27] based on experiments with a high-speed rotor (up to 1,400 cycles/s) and various chemical combinations of ^{57}Co sources and absorber materials and different geometrical arrangements on the rotor disk. Using a source-absorber combination with a suitable inherent frequency difference, the experiment exhibits the feature of a null experiment when the rotor speed is arranged such that the total frequency difference between source and absorber vanishes. In addition, the factor $(\Omega^2/2c^2)$ in (14.21) was extended to $(K \times \Omega^2/2c^2)$, where K has the value unity in SRT. A fit to the data gave $K = 1.021 \pm 0.019$.

The accuracy reached in these experiments has been a remarkable achievement and strongly supports the notion of STR that a distinguished frame of reference does not exist [28].

14.3.3 Dependence of Clock Rates on the Velocity Relative to Distant Matter

In [21], the results of experiments with high-speed centrifuges were given a different interpretation: A high-speed rotor measurement can also be regarded as an experiment to investigate a dependence of clock rates on the velocities of clocks relative to distant matter. If there would be a gravitational-tensor interaction in addition to that provided by a metric tensor, or a vector interaction between matter in the laboratory and distant matter, this interaction would, in general, depend on the relative velocity of the two. Such a dependence would be expected to cause a frequency shift [21]

$$\Delta v/v = 2\gamma(\mathbf{v_S} - \mathbf{v_A}) \times \mathbf{V}/c^2, \qquad (14.25)$$

where γ is an unknown constant and $\mathbf{V}$ is now the laboratory velocity relative to a reference frame in which distant matter has an isotropic velocity distribution. The analysis of the data gave the value (2.2 ± 8.4) m/s for the component of $\mathbf{V}$ lying in the earth's equatorial plane and a value of $\gamma = (1 \pm 4) \times 10^{-5}$.

14.3.4 Distinction Between the Special Theory of Relativity and Covariant Ether Theories

Einstein's *Special Theory of Relativity (STR)* is based on two well-known postulates:

(a) All inertial frames (IFs) are equivalent. The fundamental physical equations are the same, they are form-invariant, in all IFs.
(b) The velocity of light in vacuum, c, is a constant in all IFs.

From these two postulates follow the Lorentz transformations in Minkowski space-time. In particular, an "absolute" inertial frame, distinguished among all other IFs, does not exist. Lorentz transformations between two IFs are fully determined by their relative velocity.

The principle characteristics of *Covariant Ether Theories (CETs)* are the following [29–33]:

(a) Space–time homogeneity, space–time isotropy, the causality principle, and the principles of General Relativity are all valid.
(b) An "absolute" inertial frame K_0 is allowed to exist. As a consequence, the above-mentioned postulates of STR are violated. For example, CETs usually operate with an "absolute" simultaneity of events.
(c) The "absolute" inertial frame K_0, if it exists, is unique.

   A transformation of measured space and time intervals from K_0 to an arbitrary IF K has Lorentz form. Lorentz transformations between two IFs K and K_i always have to proceed via the absolute frame K_0. Nature does not "know" a direct relative velocity between two IFs K and K_i. Nature only "operates" with absolute velocities being applied in the Lorentz transformations.

An important consequence of this transformation rule via the absolute frame K_0, i.e., of two consecutive transformations from the IF K (observer) to K_0 (absolute frame) and then from K_0 to the IF K_i, is a rotation of the coordinate axes of K_i by an angle Φ with respect to the axes of K. The angle Φ is given by [34]

$$\Phi = \left| \frac{\mathbf{u} \times \mathbf{v}}{2c^2} \right|, \tag{14.26}$$

where $\mathbf{u}$ is the velocity of K_i relative to K and $\mathbf{v}$ is the "absolute" velocity of K relative to K_0. If $\mathbf{u}$ and $\mathbf{v}$ are parallel to each other, $\Phi = 0$. In STR, there is only one (direct) transformation between K and K_i and thus there is no rotation of the axes of K_i with respect to those of K. For a high-speed centrifuge with $u = 300$ m/s, $v = 300$ km/s (a typical value for Galaxy objects relative to the cosmic microwave background radiation) we have $\Phi \approx 10^{-9}$ rad which is tiny. To measure Φ, one could think about a change of length: $L \times \Phi \approx 10^{-9}$ m for $L = 1$ m, or a time measurement: $L\Phi/u \approx Lv/c^2 \approx 1$ ps. A very interesting case is the determination of Φ via the measurement of an energy change of a Mössbauer transition in a synchrotron radiation experiment [33] using a high-speed centrifuge, e.g., with a peripheral speed of $u \approx 300$ m/s. The basic idea is that the rotation of the coordinate axes by an angle Φ causes an additional velocity component $u \times \Phi$, and thus an additional Doppler shift

$$\frac{\Delta E}{E} = \frac{u}{c}\Phi \approx \frac{u^2 v}{2c^3}. \tag{14.27}$$

With the numbers given above, $\Delta E / E \approx 5 \times 10^{-16}$ is expected.

In [33], an experiment at an undulator beamline of a synchrotron radiation (SR) source has been proposed. A high-speed rotor carries two targets, an inner target close to the central axis (velocity $u \approx 0$) and an outer target covering the circumference ($R = R_t$) of the rotor. Both targets consist of metal foils containing the isotope ^{57}Fe. Synchrotron radiation pulses monochromatized to a few meV around the Mössbauer transition (14.4 keV) excite the nuclei in both targets. Such a collective nuclear excitation, which is phased in time by the SR pulse and extends over both spatially separated targets, follows the rotation (nuclear lighthouse effect [35]). As a result, the time evolution of the nuclear decay is mapped to an angular scale and can be recorded with a position sensitive detector [36].

The energy difference $\Delta E / E$ between the radiation from both targets is given by [33]

$$\frac{\Delta E}{E} = \frac{u^2}{2c^2} + \frac{u^2 v \cos(\gamma - \alpha)}{4c^3} = \frac{\Delta E_{\text{SOD}}}{E} + \frac{\Delta E_{\text{CET}}}{E}, \qquad (14.28)$$

where $u = \Omega \times R_t$ is the velocity of the outer target at the radius R_t and v is the "absolute" velocity of the laboratory system. The first term in (14.28) describes the well-known second-order Doppler shift (see Sect. 14.2). The second term denotes the additional shift due to CET (see (14.26) and (14.27)); α is the angle between the **x** axis of the laboratory coordinate system and the direction of the synchrotron radiation beam (wave vector $\mathbf{k}_\gamma$) and can be chosen as $\alpha = 0$; γ is the angle between the "absolute" velocity **v** and $\mathbf{k}_\gamma$. Due to the rotation of the earth, the angle γ changes in the range $[0, 2\pi]$ during 24 h. This is also a characteristic feature of the CET effect.

The energy difference $\Delta E / E$ of (14.28) causes a characteristic Quantum Beat (QB) interference pattern. With the numbers assumed above for u and v, the relative energy shifts according to (14.28) are $\dfrac{\Delta E_{\text{CET}}}{E} = 2.5 \times 10^{-16}$, whereas the term $\frac{\Delta E_{\text{SOD}}}{E} = 5 \times 10^{-13}$ which corresponds to an SOD of 0.15 mm/s. Thus, in a QB experiment at an SR facility, the beat pattern due to the SOD of 0.15 mm/s will be *modulated* between $\sim$0.1499 and $\sim$0.1501 mm/s if CET is valid. Calculations show that using the 14.4 keV resonance in ^{57}Fe, such a modulation, if present, would be sufficient for a reliable measurement of this effect predicted by CETs [33].

14.4 Gravitational Redshift

In a typical gravitational redshift (GRS) experiment, the rates of "clocks" at different heights in a gravitational field are compared. Mössbauer transitions in nuclei provide excellent clocks due to the sharply defined frequency of the emitted γ radiation. GRS experiments test predictions of the General Theory of Relativity.

14.4.1   Importance of Gravitational Redshift Experiments

The importance of GRS experiments relies on three aspects:

(a) They are an independent test of the Einstein Equivalence Principle (EEP).
(b) They provide constraints on new (hypothetical) macroscopic gravitational forces which may produce anomalous values for the GRS.
(c) For gravitational theories which *assume* conservation of energy, GRS measurements are a test whether conservation of energy is valid.

The EEP itself is based on three subprinciples [37, 38]:

(a) Local Position Invariance (LPI), which states that there are *no preferred-location effects*. The result of any "local test experiment" is independent of where and when it is performed.
(b) Weak Equivalence Principle (WEP), which requires that the free fall of a "test body" is independent of its internal structure and composition. Sometimes the WEP is also called the *principle of uniqueness of free fall*.
(c) Local Lorentz Invariance (LLI), which states that there are *no preferred-frame effects*. The result of any "local test experiment" is independent of the velocity of the (freely falling) apparatus.

"Local test experiments" are experiments [37, 38] which extend only over small regions of spacetime; self-gravitation can be neglected; external (nongravitational) fields are completely shielded.

"Test bodies" are electrically neutral and so small that self-gravitation as well as coupling to inhomogeneities of the gravitational field may be neglected. Test body trajectories are determined by the free fall of test bodies in a gravitational field.

The EEP states that the test body trajectories are identical to the paths of photons (geodesics) in any local freely falling reference frame (in which gravity is absent locally, a local Lorentz frame.) If the EEP is valid, the gravitational theory has to be a theory with a symmetric metric [37–40].

Within the Einstein Equivalence Principle, GRS experiments point out three aspects:

(a) In gravitational fields, spacetime is curved [41]. As a consequence, to describe GRS consistently, General Relativity is required. Special Relativity with its flat spacetime is not sufficient.
(b) In curved spacetime, the acceleration for test bodies and photons is the same, e.g., on earth the acceleration of photons is also given by $g \approx 9.81$ m/s^2.
(c) GRS experiments test the principle of Local Position Invariance.

The field equations of gravitation, unfortunately, are not tested by GRS experiments.

14.4.2   Gravitational Redshift Implies Curved Spacetime

If two clocks in a gravitational field of a mass M are at rest at their respective places (top: at z_2 and bottom: at z_1), in the hypothetical flat space (Minkowski geometry) their world lines in a (z, t)-diagram are straight lines parallel to the t-axis at $z = z_1$ and $z = z_2$. A γ wave is emitted from the bottom to the top. Let us consider two successive "crests" with a time difference between the two crests of $\Delta t = 1/\nu$, where ν is the frequency of this γ wave. No matter how the γ radiation may be affected by gravity the effect must be the same on both wave crests, since the gravitational field does not change from one time to another. As a consequence, the paths of the two crests are congruent and $\Delta t_{\text{top}} = \Delta t_{\text{bottom}}$, i.e., there is no change in frequency. As we will see below in Sect. 14.5, this is in contradiction to experiment. The conclusion from Minkowski geometry is wrong [41]. The reference frame at rest on earth is not a Lorentz frame. GRS implies curved spacetime.

14.4.3   Changes of Length and Time in a Gravitational Field

A gravitational potential of a massive object M causes length contraction (along the direction of the gravitational field, not perpendicular to it) [23] and time dilatation as compared to an observer far away from the mass M, where its gravitational field can be neglected. Concerning changes of length and time, the following two statements are equivalent [24, 42]:

(1) An observer is at rest at a distance r from a mass M with a gravitational field.
(2) There is no field and the observer has an instantaneous velocity v given by

$$v_e^2 = \frac{2GM}{r}, \tag{14.29}$$

where G is the gravitational constant and v_e is equal to the escape velocity to leave M from the distance r.

Then the formulae of STR can be used:

$$l' = l\sqrt{1 - \frac{v_e^2}{c^2}} = l\sqrt{1 - \frac{2GM}{c^2 r}} = l\sqrt{1 - \frac{2R_s}{r}} = l\sqrt{1 - \frac{2\phi_G}{c^2}} \tag{14.30}$$

and

$$\nu' = \nu\sqrt{1 - \frac{v_e^2}{c^2}} = \nu\sqrt{1 - \frac{2GM}{c^2 r}} = \nu\sqrt{1 - \frac{2R_s}{r}} = \nu\sqrt{1 - \frac{2\phi_G}{c^2}}, \tag{14.31}$$

where $R_s = GM/c^2$ is the Schwarzschild radius and $\phi_G = GM/r$ is the Newtonian gravitational potential [24, 25]. Equations (14.29), (14.30), and (14.31) are valid in the weak-field limit ($R_s \ll r$) and $v_e \ll c$.

Thus, seen from an observer far away from M, lengths as well as frequencies in a gravitational field are reduced by

$$\frac{\delta l}{l} = \frac{l' - l}{l} \approx -\frac{\phi_G}{c^2} \tag{14.32}$$

and

$$\frac{\delta v}{v} = \frac{v' - v}{v} \approx -\frac{\phi_G}{c^2}. \tag{14.33}$$

To derive the difference in the clock rates for observers at $r = r_2$ and $r = r_1$ with $r_2 > r_1$ and $\Delta r = r_2 - r_1$, we use (14.33):

$$\frac{\Delta v}{v} = \left(\frac{\delta v}{v}\right)_{r_2} - \left(\frac{\delta v}{v}\right)_{r_1} = -\left(\frac{\phi_G}{c^2}\right)_{r_2} + \left(\frac{\phi_G}{c^2}\right)_{r_1} \approx \frac{GM}{c^2}\frac{\Delta r}{r_1^2} = \frac{g\Delta r}{c^2}, \tag{14.34}$$

i.e., the clock of the observer at $r_2 > r_1$ runs fast and he "sees" a reduced frequency of a γ ray reaching him from r_1 below.

Often the GRS is derived using energy conservation: If a γ ray of energy $E = h\nu$, i.e., of mass $m = h\nu/c^2$ is climbing in the gravitational field of the earth a short distance Δr, it loses energy according to $\Delta E = h\Delta\nu = m \times g \times \Delta r = (h\nu/c^2)g\Delta r$. Thus,

$$\frac{\Delta v}{v} = \frac{g\Delta r}{c^2}. \tag{14.35}$$

Although (14.35) gives the correct result (14.34), its derivation is not satisfactory because it hides that GRS implies curved spacetime of General Relativity and, as a consequence, the concept of energy is not well defined [43, 44].

In the General Theory of Relativity, also the speed of light is no longer constant. It depends on the curvature of spacetime. Seen from an observer far away from M, the speed of light in a gravitational field is reduced. In addition, because of curved spacetime the speed of light is anisotropic. Perpendicular and parallel to the direction of the gravitational field one obtains [45]:

$$c_\perp = c\left(1 - \frac{R_s}{r}\right)$$
$$c_\parallel = c\left(1 - \frac{2R_s}{r}\right), \tag{14.36}$$

where c is the speed of light far away from M.

At the surface of the earth, $\Delta c = c_\perp - c_\parallel \approx 21$ cm/s. Unfortunately, it is not possible to measure Δc in a Mössbauer experiment. In a Michelson interferometer arranged in such a way that the velocity of light would be measured perpendicular as well as parallel to the direction of the gravitational field, Δc could be determined.

14.5 Gravitational Redshift Experiments Using the Mössbauer Effect

Gravitational redshift (GRS) experiments can be divided into two groups: ordinary and differential redshift measurements. An *ordinary* GRS experiment is described by the change $\Delta\phi_G$ of the gravitational potential between r_{top} and r_{bottom}:

$$\Re = \frac{\Delta\phi_G}{c^2}. \qquad (14.37)$$

Experimental results are usually parameterized as

$$\tilde{\Re} = (1+\alpha)\frac{\Delta\phi_G}{c^2}, \qquad (14.38)$$

where α describes possible deviations from $\Re$ in (14.37) and experiments give upper limits for α.

Differential redshift experiments are sensitive only to changes of α which, for example, might be caused by different chemical compositions (A and B) of the clocks. In such an experiment, one measures

$$\Delta\tilde{\Re} = [\alpha(A) - \alpha(B)] \times \frac{\Delta\phi_G}{c^2} \qquad (14.39)$$

and is interested in upper limits for $[\alpha(A) - \alpha(B)]$.

In this contribution, we will only focus on ordinary GRS experiments. We just mention that the best limit $[\alpha(A) - \alpha(B)] < 4 \times 10^{-5}$ for a differential GRS Mössbauer experiment is given in [46], where two chemically different "clocks", a ^{57}CoRh source and a stainless steel (SS304) absorber, were compared.

14.5.1 GRS Experiments with the 14.4 keV Transition in ^{57}Fe

The discovery that the ^{57}Fe nucleus can absorb 14.4 keV γ rays in a resonance whose width is $\sim 3 \times 10^{-13}$ of the γ-ray energy [47–49] made GRS experiments a practical possibility and was suggested already in 1959.

(1) The first results were published in [50]. A ^{57}CoFe source and an iron foil enriched in ^{57}Fe as absorber were used. The 14.4 keV γ rays passed through an evacuated tube a total difference in height of 12.5 m. Measurements were

also repeated at a difference in height of 3 m. An effect with an uncertainty of about 47% was observed.

(2) Already 1.5 weeks later, a paper was published [51] where the results of [50] were critically discussed concerning two aspects: (a) a possible temperature difference between source and absorber of only 0.6 K would produce an SOD (see Sect. 14.2) as large as the whole effect observed, and (b) a possible frequency difference inherent in a given combination of source and absorber (the first paper on the isomer shift was published two months later [13]) could be responsible for the shift measured in [50]. In [51], a source of ^{57}CoFe and an absorber composed of seven separate units each consisting of enriched ^{57}Fe electroplated onto a beryllium disk were used. The difference in height was 22.5 m, the γ rays propagating in a tube constantly being flushed with helium. The potential difference of a thermocouple with one junction at the source and the other at the main absorber was recorded and used later to correct for the SOD. An additional absorber (similar to the main absorber) and a detector were mounted to see the source from only about 0.9 m away. This monitor system measured the stability of the overall setup. The result obtained in about 10 days of operation was in agreement with the General Theory of Relativity within an accuracy of $\sim$10%: $(\Delta\nu)_{exp}/(\Delta\nu)_{theor} = +(1.05 \pm 0.10)$, where the plus sign indicates that the frequency increases in falling, as expected.

(3) A similar accuracy was achieved in [52]. To reach high sensitivity to small shifts, alternating positive and negative velocities – corresponding to the steepest slopes of the Lorentzian line – were applied to the source by a double moving-coil transducer (double-loudspeaker drive) operated at 50 cycles/s in square-wave mode. A gate was arranged to eliminate counts during the change-over periods between the two velocities of opposite sign. The source was ^{57}Co electroplated on a Fe metal foil. The absorber was an iron-metal foil enriched to 50% in ^{57}Fe. Runs were performed with the source 13.5 and 18.0 m above the absorber, an evacuated tube with Mylar windows being placed in between. The main sources of systematic error were the difficulty of measuring the temperatures and temperature distributions of the source and absorber foils as well as errors connected with the moving-coil transducer. The final result was (0.859 ± 0.085) times the value predicted by the General Theory of Relativity. The chemical (isomer) shift between source and absorber was found to be 4.72×10^{-3} mm/s, i.e., $\sim$10 times larger than the gravitational shift.

(4) An enormous improvement by a factor of $\sim$10 (i.e., an accuracy of 1%) was reached in the experiment described in [53]. This experiment is a great achievement in Mössbauer spectroscopy in terms of experimental skills, data acquisition and evaluation, minimization of statistical and systematic errors as well as achieved accuracy. It is an improvement of the experimental setup described in [51] in three main areas, (a) temperature control of source and absorber, (b) monitor system to eliminate any perturbing effects in the source and its motion, and (c) gating and switching functions of the electronics. Temperature-regulating ovens were used for the source, absorber and monitor units. The time-averaged temperature for a given run could be estimated to

within less than 5 mK, the fluctuations being less than 10 mK. The ^{57}Co activity was electroplated onto an Fe foil enriched in ^{56}Fe to reduce the resonant absorption by ^{57}Fe. The main absorber and the monitor absorber were made from Fe foils enriched in ^{57}Fe cemented onto a Be sheet. Permanent magnets were used to provide a transverse field in the plane of the source and in the plane of the absorbers (parallel to that in the source) to polarize the iron. The difference in height was 22.48 m. A barium titanate transducer provided a velocity modulation $\pm V_M$ (of the source) at the regions of maximum slopes of the absorption line. The transducer was fastened to the piston of a hydraulic (slave) cylinder to enable a calibration motion $\pm V_J$. This hydraulic cylinder was moved by a master cylinder moved in turn by the rotation of a screw. An electronic gate assured precisely equal times for $+V_M$ and $-V_M$. From the counts measured at the four combinations of the source velocities $\pm V_M$ and $\pm V_J$ and from the measured value for V_J, the effective displacement V_D of the line center could be precisely determined. After a thorough evaluation of statistical and systematic errors, the final result was $(\Delta \nu)_{exp}/(\Delta \nu)_{theor} = +(0.9990 \pm 0.0076)$, where the error is statistical only. The estimated limit for the systematic error was ± 0.010, the thermal instability being the major source of systematic errors. For further improvements, it was suggested to use an environment being itself stabilized thermally, e.g., a liquid-He environment (see Sect. 14.5.2). In addition, earth's rotation causes a central acceleration and thus a reduction of the purely gravitational acceleration by about 0.3%. Gravitation and acceleration would occur at the same time, the latter having separately been investigated by high-speed centrifuges (see Sect. 14.3).

14.5.2 GRS Experiments with the 93.3 keV Transition in ^{67}Zn

Already at the early stage of Mössbauer spectroscopy several proposals were made to use the 93.31 keV transition in ^{67}Zn for precision energy-shift measurements. The lifetime of the Mössbauer state is 13.1 μs resulting in a natural linewidth of 0.155 μm/s which is ~600 times smaller than the natural linewidth (0.09 mm/s) of the 14.4keV resonance in ^{57}Fe. Due to the large transition energy, the recoilfree fraction is small, for ^{67}ZnO, e.g., it is only about 2%. However, using carefully annealed ZnO single crystals, the natural linewidth could be observed [54], which opened the door to many applications in high-resolution Mössbauer spectroscopy [12].

(1) The first GRS experiment using the ^{67}Zn resonance is described in [55]. The main aim was the investigation of the dependence of the GRS on the angle ϑ between the propagation direction of the photon and the acceleration of gravity **g**.

In all metric gravitational theories (based on the Einstein Equivalence Principle), the GRS is conservative, i.e., it is only a function of the potential difference and thus independent of the route a photon would take from one

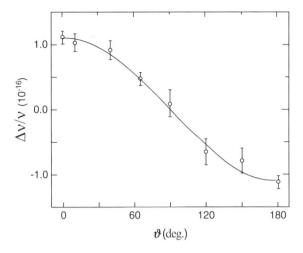

Fig. 14.5 Relative change of the 93.3 keV Mössbauer resonance in ^{67}Zn due to the interaction of the photons with the gravitational field of the earth as a function of the vertical deviation angle ϑ. The *solid line* is a fit according to (14.40). Reprinted from [55], copyright (2010), with permission from Elsevier

potential level to another. Extending (14.34) to tracks not parallel to **g**, the GRS is then given by

$$\frac{\Delta \nu}{\nu} = \frac{g \Delta r}{c^2} \times \cos \vartheta. \qquad (14.40)$$

Equation (14.40) is expected to be true for all gravitational theories, which are characterized by a metric but not necessarily for nonmetric theories as, e.g., the theory discussed in [56]. The apparatus of [55] allowed the angle ϑ to be set to any value between 0 and 180°. The source was ^{67}Ga (decaying with $T_{1/2} = 78.3$ h to ^{67}Zn) in ZnO single crystals. Inside a liquid-He cryostat a ZnO powder reference absorber (enriched in ^{67}Zn) together with a piezoelectric drive moving the source sinusoidally at 130 Hz were fixed at one end of a 1 m long metal tube. At the opposite end was the main absorber, again ZnO powder enriched in ^{67}Zn from the same lot as the reference absorber.

Figure 14.5 shows the results obtained for 8 different angles ϑ. The ϑ dependence can be expanded in a Fourier series,

$$\frac{\Delta \nu}{\nu} = a_0 + \sum_{n=1}^{\infty} a_n \cos(n\vartheta + \varphi_n). \qquad (14.41)$$

Only nonmetric theories of gravitation can give higher-order ($n \geq 2$) contributions. In the actual fit, only the zeroth and first-order terms could be determined with the results: $a_0 = (2 \pm 5) \times 10^{-18}$ and $a_1 = (1.10 \pm 0.06) \times 10^{-16} = (0.99 \pm 0.05) \times g \Delta r/c^2$. The sensitivity was not enough to estimate higher-order terms of (14.41).

(2) About a decade later, another GRS experiment with the ^{67}Zn resonance employ-
    ing improved source production and drive-velocity calibration was carried out
    [2]. As source, again ^{67}GaZnO single crystals were used. After an elaborate
    annealing procedure in oxygen atmosphere [11] the sources showed natural
    linewidth (0.155 μm/s). The basic setup of the GRS measurement was similar
    to the one described in experiment (1) above. In a liquid-He cryostat, two
    Mössbauer experiments were carried out simultaneously, through the reference
    and through the main absorber, both consisting of ^{67}ZnO powder enriched to
    about 88% in ^{67}Zn. To generate Doppler velocities in the sub-μm/s range, a
    piezoelectric transducer was employed. To check and monitor the motion of
    the source, a small permanent magnet was rigidly connected to the source
    holder. Thus, a tiny (a few pV) voltage was induced in the pickup coil of a
    dc-SQUID readout system [57]. The drive system together with the source and
    the reference absorber was fixed at one end of two concentric about 1m long
    aluminum and copper tubes [55]. The temperatures of both absorbers were
    continuously monitored by Ge resistors. The calibration of the drive system
    was obtained via the quadrupole splitting in ZnO [58, 59]. Since both source
    and absorber are split, the complete absorption pattern consists of five lines
    [11]. The line at $\sim$0 μm/s exhibiting the highest intensity originates from two
    absorption channels of source and absorber. In the actual redshift (blueshift)
    experiments, a velocity of $\sim\pm0.35$ μm/s, i.e., the major part of this central line
    was scanned. The GRS was measured over $\sim$1 m and $\sim$0.6 m. Figure 14.6
    displays low-velocity spectra for the determination of the gravitational redshift
    (a) and blueshift (b), respectively. The accuracy achieved in determining the
    line position in the reference spectrum after $\sim$5 days of measuring time was
    $\Delta v = 3$ Å$/s \; \hat{=} \; 1$ cm/year $\hat{=} \; 10^{-3} \times 2\Gamma$, where $\Gamma = 0.155$ μm/s is the natural
    width.

In the redshift (blueshift) experiments [2], a 6% deviation from the expected
result has been noticed, the main reason being systematic errors, which are due
to solid-state effects and thus, unfortunately, appear difficult to control. A variety
of systematic errors was discussed in [2] and dismissed. Two important sources of
error, however, were found to be responsible for the unexpected deviation:

(a) Pressure dependence of the electric field gradient in ZnO. Small pressures
    originating from the contraction of the particles of the ^{67}ZnO powder themselves
    when cooled to cryogenic temperatures cause a change of the asymmetry
    parameter η of the quadrupole tensor and thus an apparent shift of the center
    line when analyzed by a single Lorentzian. In a redshift (blueshift) experiment,
    already a difference of $|\Delta\eta| \approx 0.01$ of the asymmetry parameter between the
    reference and main absorbers could easily explain the 6% deviation mentioned
    above.
(b) The change of η when cooling to low temperatures causes a change of the
    lineshape, which can be different in both absorbers. This again can lead to an
    apparent shift of the central line.

Fig. 14.6 Low-velocity Mössbauer spectra of the 93.3 keV resonance in ^{67}Zn for the determination of the gravitational redshift (**a**) and blueshift (**b**). Source: ^{67}GaZnO single crystal, absorber: ^{67}ZnO powder enriched to ∼88% in ^{67}Zn. Reprinted from [2], copyright (2010), with permission of Springer Science and Business Media

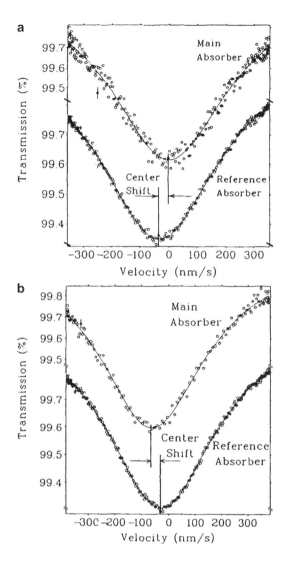

To avoid these two systematic errors, one could think of single-line systems, e.g., ^{67}ZnAl$_2$O$_4$, a spinel compound exhibiting a single absorption line and a relatively large Debye temperature [12]. In this way, an accuracy for the determination of the GRS of 10^{-3} may be reached, still, by far not sufficient to be competitive with other types of measurements (see Sect. 14.6).

A possible exception might be polarization-dependent effects, which might influence the GRS [60]. Mössbauer radiation from transitions between hyperfine-split levels are usually polarized. Thus, the Mössbauer effect might turn out to be a sensitive tool to search for polarization effects changing the GRS, if such effects are present at all. Another exception might be the search for (hypothetical) new forces or for a possible deviation from the r^{-2} behavior of the gravitational force at distances $r \lesssim 1$ m.

14.6   Relativity and Anisotropy of Inertia

Newton and Mach came to different conclusions about the origin of inertia. Newton believed that inertial forces, e.g., the centrifugal force, arise from acceleration with respect to "absolute space". According to Mach's principle [61, 62], the inertial forces observed in an accelerated laboratory are gravitational, having their origin in the distant matter of the whole universe, accelerated relative to the laboratory. It was suggested in [63] that due to the flattened rotating mass distribution in our galaxy, the inertial mass should be spatially anisotropic, i.e., the inertial mass of a particle would be (slightly) different when this particle is accelerated toward or perpendicular to the galactic center. This was checked by a Mössbauer measurement [64] and by two experiments using nuclear magnetic resonance techniques [65, 66]. From the experimental linewidth of the 14.4 keV Mössbauer transition in ^{57}Fe, an upper limit for the mass anisotropy of $(\Delta m/m) < 5 \times 10^{-16}$ was estimated. The magnetic resonance measurements on the ^{7}Li nucleus gave an even stronger upper limit of $(\Delta m/m) < 5 \times 10^{-23}$ [66]. However, it was argued in [67] that these highly accurate null results are in full agreement with the requirements of Mach's principle. These experiments show with high precision that the anisotropy of the inertial mass is universal, i.e., the same for all particles (including photons). Therefore, spatial anisotropy is unobservable locally, and the precise null results mentioned above are expected and in accordance with Mach's principle. In present times, various theoretical ideas and experiments suggested in connection to Mach's principle – in particular, the observation of the dragging of inertial frames (frame-dragging forces, see Sect. 14.7) – are being investigated [68, 69] on the basis of Einstein's principle of equivalence.

14.7   Selected Modern Tests of the General Theory of Relativity

(1) *GRS experiments.* The presently most accurate (7×10^{-9}) GRS experiment was performed – not with photons – but using the interference of quantum-mechanical matter waves (de-Broglie waves) of laser-cooled Cs atoms at different gravitational potential. The measurements were carried out in an atom interferometer in the gravitational field of the earth. In such experiments, the total energy of the Cs atom $E = Mc^2 = h\nu_c$ includes the kinetic, gravitational, internal energies, and most importantly, the rest-mass energy. As a consequence, the Compton frequency ν_c is very high: $\nu_c \approx 3.2 \times 10^{25}$ Hz, as compared to a typical (100 keV) Mössbauer γ ray, the frequency of which

($\sim$2.4 $\times$ 10^{19} Hz) is more than 6 orders of magnitude smaller. The frequency change $\Delta v_c / v_c = g \Delta r / c^2$ (see (14.34)) is derived from the phase difference of the de-Broglie waves accumulated due to the difference of the gravitational potential, taking full advantage of the high value of v_c [70, 71]. Even more sensitive atom interferometers can be expected to be available in the future, which will be able to also test Einstein's field equations [72] (See, however, the critical comments on using atoms as clocks ticking at the Compton frequency in [73, 74]). For Mössbauer spectroscopy, to be competitive with such types of measurements, a relative accuracy of at least 10^{-9} would be required. Even when using single-line sources and absorbers such a precision can probably not be achieved with Mössbauer spectroscopy, because – as mentioned above – solid-state effects like internal stresses in the chemical materials would become dominant and difficult, if not impossible, to control. In particular, possible changes of the hyperfine interactions (e.g., s-electron density at the Mössbauer nucleus) and of the lattice dynamics (zero-point motion) would be crucial.

(2) *Field equations.* (a) Concerning experimental tests of the General Theory of Relativity including the field equations, binaries consisting of two neutron stars (pulsars) have played an important role. Pulsars are also excellent clocks. The discovery of the system (B1913+16) in 1974 [75] was awarded by the 1993 Nobel Prize in Physics for Hulse and Taylor. An even more startling double pulsar, about 2000 light-years from the earth, known by its catalog number J0737-3039 for its coordinates in Canis Major, was discovered in 2003 [76,77]. It consists of two active radio pulsars (neutron stars). Pulsar A rotates every 22 ms, its companion, B, shows a period of 2.7 s. The two members orbit each other in $\sim$145 min with orbital speeds of $\sim$10$^{-3} \times c$ and a separation of $\sim$900,000 km, i.e., $\sim$2 times the size of the earth–moon system. The tight orbits and high orbital speeds make the effects of the General Theory of Relativity very large; for example, the orbital precession is 16.9^o per year, which is $\sim$141,000 times the rate of the precession of Mercury's orbit (perihel precession) around the sun. The gravitational time dilatation (up to $\sim$400 μs), the Shapiro delay ($\sim$100 μs) of signals propagating in the curved spacetime [78, 79] of the pulsar, as well as the masses of the two pulsars have been determined with high precision: 1.3381 $\pm$ 0.0007 and 1.2489 $\pm$ 0.0007 solar masses for the pulsars A and B, respectively. The data also show that the separation of the two pulsars is shrinking by (7.42 $\pm$ 0.09) mm per day leading to a collision in $\sim$85 million years. (b) Another example where the field equations are tested is gravitational effects due to a spinning mass. These effects, often characterized by "dragging of inertial frames", are also called gravitomagnetism, in analogy to the observation in electromagnetism that a spinning charge creates additional effects named magnetism. A pertinent example is the Lense–Thirring effect [80, 81]. A gyroscope placed in an orbit around a rotating star (or the earth) precesses by a (small) amount which is proportional to the angular momentum of the star (earth), in analogy to the precession of a spinning electron if it orbits through a magnetic field. The Lense–Thirring precession (and other

relativistic effects) have been measured by investigating the orbits of two geodesy satellites, LAGEOS I and II [82,83] and by another satellite experiment called Gravity Probe B, which has determined the precession of on-board gyroscopes [84]. The agreement of the LAGEOS II results with the predictions of Einstein's theory is ~99.8%. In the double-pulsar system J0737-3039 mentioned above, the gravitational spin-orbit coupling, which leads to a precession of B's rotation axis in the strong gravitational field of A, has been measured and will also provide stringent tests of the General Theory of Relativity. Gravitomagnetic effects are also important for modeling the X-ray emission near black holes [85].

Up to now, General Theory of Relativity has passed every experimental test with flying colors and at the same time stringent constraints have been set on alternative theories of gravity. Still, theorists have been unable to unify General Theory of Relativity with Quantum Mechanics and modifications (possibly to both theories) might be necessary.

14.8   Conclusions

The high sensitivity of the Mössbauer effect to tiny energy shifts of photons made it possible to investigate basic questions in connection with Einstein's Special Theory of Relativity and General Theory of Relativity. The energy shifts due to lattice vibrations at various temperatures as well as due to the zero-point motion were studied with high precision. In experiments with high-speed centrifuges a convincing upper limit for a velocity v relative to an ether of $v < 5$ m/s could be established. A similar upper limit could be obtained for a dependence of clock rates on the velocities relative to distant matter. The gravitational redshift of γ rays could be determined in a local experiment (on earth) with an accuracy of 1% in full agreement with General Theory of Relativity. Although nowadays such an accuracy can no longer compete with the sensitivity reached by other methods, Mössbauer-type experiments at beam lines of modern synchrotron radiation (SR) facilities received an enormous boost and could well reach an energy resolution sufficient to discriminate between the predictions of Einstein's and other theories, which are being discussed in connection with a possible violation of Lorentz invariance at very high energies. It will be exciting to follow future developments in gravitational physics both in theory and experiment, where SR might play an important role.

Acknowledgements It is a pleasure to thank A.L. Kholmetskii at the Department of Physics, Belarus State University, Minsk, and S. Roth and J. Winter at the Physik-Department E15, Technische Universität München, Garching, for stimulating discussions. This work was supported by funds of the Deutsche Forschungsgemeinschaft DFG (Transregio 27: Neutrinos and Beyond), the Munich Cluster of Excellence (Origin and Structure of the Universe), and the Maier-Leibnitz-Laboratorium (Garching).

References

1. R.L.M. Mössbauer, Z. Physik **151**, 124 (1958)
2. W. Potzel, C. Schäfer, M. Steiner, H. Karzel, W. Schiessl, M. Peter, G.M. Kalvius, T. Katila, E. Ikonen, P. Helistö, J. Hietaniemi, K. Riski, Hyp. Interact. **72**, 197 (1992)
3. A. Einstein, Ann. Physik **17**, 891 (1905)
4. A. Einstein, *The Meaning of Relativity*, 6th edn. (The Electric Book Company, London, 2001)
5. H. Wegener, *Der Mössbauereffekt und seine Anwendung in Physik und Chemie* (Hochschultaschenbücher 2/2a, Bibliographisches Institut Mannheim, 1966)
6. C.W. Sherwin, Phys. Rev. **120**, 17 (1960)
7. B.D. Josephson, Phys. Rev. Lett. **4**, 341 (1960)
8. R.V. Pound, G.A. Rebka, Phys. Rev. Lett. **4**, 274 (1960)
9. A.J.F. Boyle, D.St.P. Bunbury, C. Edwards, H.E. Hall, Proc. Phys. Soc. **76**, 165 (1960)
10. M. Peter, W. Potzel, M. Steiner, C. Schäfer, H. Karzel, W. Schiessl, G.M. Kalvius, U. Gonser, Hyp. Interact. **69**, 493 (1991) and Phys. Rev. B**47**, 753 (1993)
11. C. Schäfer, W. Potzel, W. Adlassnig, P. Pöttig, E. Ikonen, G.M. Kalvius, Phys. Rev. B **37**, 7247 (1988)
12. W. Potzel, in *Mössbauer Spectroscopy Applied to Magnetism and Materials Science*, vol. 1, ed. by G.J. Long, F. Grandjean (Plenum Press, New York, 1993), p. 305
13. O.C. Kistner, A.W. Sunyar, Phys. Rev. Lett. **4**, 412 (1960)
14. G.K. Shenoy, F.E. Wagner (eds.), *Mössbauer Isomer Shifts* (North-Holland, Amsterdam, 1978)
15. C. Møller, Proc. Roy. Soc. **A270**, 306 (1962)
16. H.J. Hay, J.P. Schiffer, T.E. Cranshaw, P.A. Egelstaff, Phys. Rev. Lett. **4**, 165 (1960)
17. M. Ruderfer, Phys. Rev. Lett. **5**, 191 (1960) and erratum, Phys. Rev. Lett. **7**, 361 (1961)
18. D.C. Champeney, G.R. Isaak, A.M. Khan, Phys. Lett. **7**, 241 (1963)
19. D.C. Champeney, G.R. Isaak, A.M. Khan, Nature **198**, 1186 (1963)
20. W. Kündig, Phys. Rev. **129**, 2371 (1963) However, see A.L. Kholmetskii, T. Yarman, O.V. Missevitch, Phys. Scripta **77**, 035302 (2008)
21. K.C. Turner, H.A. Hill, Phys. Rev. **134**, B252 (1964)
22. E.T.P. Lee, S.T. Ma, Proc. Phys. Soc. **79**, 445 (1962)
23. H. Melcher, *Relativitätstheorie in elementarer Darstellung* (Aulis Verlag, Deubner & Co KG, Köln, 1976), Chap. 10.4.
24. H. Melcher, see [23], p. 181
25. B. Schutz, *A first course in General Relativity*, second edition (Cambridge University Press, Cambridge, 2009), p. 186
26. D.C. Champeney, P.B. Moon, Proc. Phys. Soc. **77**, 350 (1961)
27. D.C. Champeney, G.R. Isaak, A.M. Khan, Proc. Phys. Soc. **85**, 583 (1965)
28. W.-D. Schmidt-Ott, Die Naturwissenschaften **52**, 636 (1965)
29. A.L. Kholmetskii, M. Mashlan, B.I. Rogozev, K. Ruebenbauer, D. Zak, Nucl. Instr. Meth. **B108**, 359 (1996)
30. A.L. Kholmetskii, Hyp. Interact. **126**, 411 (2000)
31. A.L. Kholmetskii, Phys. Scripta **55**, 18 (1997)
32. A.L. Kholmetskii, Phys. Scripta **67**, 381 (2003)
33. A.L. Kholmetskii, W. Potzel, R. Röhlsberger, U. van Bürck, E. Gerdau, Hyp. Interact. **156/157**, 9 (2004)
34. C. Møller, *The Theory of Relativity* (Oxford University Press, London, 1972), Chap. 2.8.
35. R. Röhlsberger, T.S. Toellner, W. Sturhahn, K.W. Quast, E.E. Alp, A. Bernhard, E. Burkel, O. Leupold, E. Gerdau, Phys. Rev. Lett. **84**, 1007 (2000)
36. R. Röhlsberger, K.W. Quast, T.S. Toellner, P. Lee, W. Sturhahn, E.E. Alp, E. Burkel, Appl. Phys. Lett. **78**, 2970 (2001)
37. M.P. Haugan, Ann. Phys. (NY) **118**, 156 (1979)
38. C.M. Will, Ann. New York Acad. Sci. **336**, 307 (1980) and Living Rev. Relativity **9**, 3 (2006)
39. C.W. Misner, K.S. Thorne, J.A. Wheeler, *Gravitation* (Freeman, San Francisco, 1973)

40. S. Weinberg, *Gravitation and Cosmology: Principles and Applications of the General Theory of Relativity* (Wiley, New York, 1972)
41. A. Schild, Texas Quart. **3**, 42 (1960); and in: *Evidence for Gravitational Theories*, ed. by C. Møller (Academic Press, New York, 1962); and in: *Relativity Theory and Astrophysics: I. Relativity and Cosmololgy*, ed. by J. Ehlers (Amer. Math. Soc. Providence, RI, 1967), p. 1
42. B. Schutz, see [25], p. 180, 259, and 277.
43. H. Stephani, *Allgemeine Relativitätstheorie*, 2nd edn. (VEB Deutscher Verlag der Wissenschaften, Berlin 1980), Chap. 14.4
44. B. Schutz, see [25], Chap. 4.7
45. H. Melcher, see [23], p. 196
46. H. Vucetich, R.C. Mercader, G. Lozano, G. Mindlin, A.R. López-García, J. Desimoni, Phys. Rev. D **38**, 2930 (1988)
47. R.V. Pound, G.A. Rebka, Phys. Rev. Lett. **3**, 554 (1959)
48. J.P. Schiffer, W. Marshall, Phys. Rev. Lett. **3**, 556 (1959)
49. G. De Pasquali, H. Frauenfelder, S. Margulies, R.N. Peacock, Phys. Rev. Lett. **4**, 71 (1960)
50. T.E. Cranshaw, J.P. Schiffer, A.B. Whitehead, Phys. Rev. Lett. **4**, 163 (1960)
51. R.V. Pound, G.A. Rebka, Phys. Rev. Lett. **4**, 337 (1960)
52. T.E. Cranshaw, J.P. Schiffer, Proc. Phys. Soc. (London) **84**, 245 (1964)
53. R.V. Pound, J.L. Snider, Phys. Rev. Lett. **13**, 539 (1964) and Phys. Rev. **140**, B788 (1965)
54. W. Potzel, A. Forster, G.M. Kalvius, J. Phys. (Paris) **37**, C6-691 (1976)
55. T. Katila, K. Riski, Phys. Lett. **83A**, 51 (1981)
56. P. Kustaanheimo, Phys. Lett. **23**, 75 (1966) and references therein.
57. E. Ikonen, H. Seppä, W. Potzel, C. Schäfer, Rev. Sci. Instrum. **62**, 441 (1991); IEEE Trans. Instrum. Meas. **40**, 196 (1991)
58. G.J. Perlow, W. Potzel, R.M. Kash, H. de Waard, J. de Phys. **35**, C6-197 (1974)
59. W. Potzel, Th. Obenhuber, A. Forster, G.M. Kalvius, Hyp. Int. **12**, 135 (1982)
60. M.D. Gabriel, M.P. Haugan, R.B. Mann, J.H. Palmer, Phys. Rev. D **43**, 308 (1991)
61. E. Mach, *The Science of Mechanics*, trans. by T.J. McCormack (2nd edn., Open Court Publishing Co., 1893)
62. D.W. Sciama, *The Unity of the Universe* (Doubleday, New York, 1959)
63. G. Cocconi, E.E. Salpeter, Nuovo Cimento **10**, 646 (1958); Phys. Rev. Lett. **4**, 176 (1960)
64. C.W. Sherwin, H. Frauenfelder, E.L. Garwin, E. Lüscher, S. Margulies, R.N. Peacock, Phys. Rev. Lett. **4**, 399 (1960)
65. V.W. Hughes, H.G. Robinson, V. Beltrow-Lopez, Phys. Rev. Lett. **4**, 342 (1960)
66. R.W.P. Drever, Phil. Mag. **6**, 683 (1961)
67. R.H. Dicke, Phys. Rev. Lett. **7**, 359 (1961)
68. see [40], pp. 17, 86, 239
69. C.W. Will, *Was Einstein Right?* (Oxford University Press, Oxford 1988), Chaps. 8 and 11
70. H. Müller, A. Peters, S. Chu, Nature **463**, 926 (2010) and Nature **467**, E2 (2010)
71. M.A. Hohensee, S. Chu, A. Peters, H. Müller, Phys. Rev. Lett. **106**, 151102 (2011)
72. S. Dimopoulos, P.W. Graham, J.M. Hogan, M.A. Kasevich, Phys. Rev. D **78**, 042003 (2008)
73. P. Wolf, L. Blanchet, C.J. Bordé, S. Reynand, C. Salomon, C. Cohen-Tannoudji, Nature **467**, E1 (2010) and Class. Quantum Grav. **28**, 145017 (2011)
74. S. Sinha, J. Samuel, Class. Quantum Grav. **28**, 145018 (2011)
75. R.A. Hulse, J.H. Taylor, Astrophys. J. **195**, L51 (1975)
76. M. Kramer, I.H. Stairs, R.N. Manchester, M.A. McLaughlin, A.G. Lyne, R.D. Ferdman, M. Burgay, D.R. Lorimer, A. Possenti, N. D'Amico, J.M. Sarkissian, G.B. Hobbs, J.E. Reynolds, P.C.C. Freire, F. Camilo, Science **314**, 97 (2006) and arXiv: 0609417 [astro-ph] 2006
77. M. Kramer, Sky and Telescope, August 2010, p. 28
78. I.I. Shapiro, Phys. Rev. Lett. **13**, 789 (1964)
79. I.I. Shapiro, M.E. Ash, D.B. Campbell, R.B. Dyce, R.P. Ingalls, R.F. Jurgens, G.H. Pettengill, Phys. Rev. Lett. **26**, 1132 (1971)
80. H. Thirring, Phys. Zeitschr. **19**, 33 (1918)

81. J. Lense, H. Thirring, Phys. Zeitschr. **19**, 156 (1918)
82. I. Ciufolini, E.C. Pavlis, R. Peron, New Astron. **11**, 527 (2006)
83. D.M. Lucchesi, R. Peron, Phys. Rev. Lett. **105**, 231103 (2010)
84. C.W.F. Everitt, D.B. DeBra, B.W. Parkinson, J.P. Turneaure, J.W. Conklin, M.I. Heifetz, G.M. Keiser, A.S. Silbergleit, T. Holmes, J. Kolodziejczak, M. Al-Meshari, J.C. Mester, B. Muhlfelder, V.G. Solomonik, K. Stahl, P.W. Worden, Jr., W. Bencze, S. Buchman, B. Clarke, A. Al-Jadaan, H. Al-Jibreen, J. Li, J.A. Lipa, J.M. Lockhart, B. Al-Suwaidan, M. Taber, S. Wang, Phys. Rev. Lett. **106**, 221101 (2011)
85. L. Brenneman, C. Reynolds, Astrophys. J. **652**, 1028 (2006)

Chapter 15
Extraterrestrial Mössbauer Spectroscopy

Göstar Klingelhöfer

15.1 Introduction

To understand the origin of the Solar system and the origin of Life itself is one of
the longest standing goals of human thought. Our Sun and its planets have formed
out of an interstellar cloud which collapsed due to gravitational forces, forming a
disk shaped so-called protosolar nebula, with the young star in the centre. Such
disk shaped and dust grain containing protosolar nebulae have been observed. One
of them is surrounding the young star Beta pictoris [1, 2]. Silicates, carbon and
metal grains, oxides and sulfides should have been present. One of the important
elements with relatively high abundance is iron. It is believed that simple molecules,
such as water (H_2O), carbon monoxide (CO), and hydrocarbons, were formed
in this protosolar nebula [3]. As we know very well, at least in one case – our
own Solar system – a variety of different objects were formed: planets, asteroids,
and comets. At least on one of these planets, the Earth, life has formed. Today
comets are believed to be remnants of the protosolar nebula, and the Sun and the
planets are processed bodies, whereas asteroids are supposed to be only partially
processed. The process of birth and evolution of our Solar system can be investigated
indirectly by studying all the different members of the planetary system by means
of remote sensing and planetary robotic space missions. One of the key elements
in the evolution of the Solar system, and life itself, is iron. The chemistry of iron
is strongly coupled to the chemistry of abundant elements as hydrogen, oxygen,
and carbon. For instance, the oxidation state of iron in surface rocks of the planets
is an important aspect because according to theoretical studies, iron contained in
a planetary body should be the more oxidized the farther away from the sun this
body has formed. By studying the cosmic history of iron, we have the possibility
of understanding the chemical evolution of matter and life itself. Here, Mössbauer

G. Klingelhöfer (✉)
Institute for Inorganic and Analytic Chemistry, University of Mainz, 55099 Mainz, Germany
e-mail: klingel@mail.uni-mainz.de

M. Kalvius and P. Kienle (eds.), *The Rudolf Mössbauer Story*,
DOI 10.1007/978-3-642-17952-5_15, © Springer-Verlag Berlin Heidelberg 2012

spectroscopy is the obvious tool, because it is a unique method for determining the oxidation state of the element iron, the mineralogical composition of iron containing rocks and their weathering products, meteorites and small grains from solid bodies, directly. This contributes to the understanding of the history and evolution of the planetary surfaces, for instance, the Martian surface, and their atmospheres.

Mössbauer spectroscopy has made decisive contributions to the understanding of the evolution of the Solar system by laboratory studies of meteorites, especially carbonaceous chondrites, metallic Fe–Ni meteorites and Martian meteorites, laboratory studies of Lunar material brought back to Earth by the US-American Apollo missions and the Soviet Union (UdSSR) robotic sample return missions, and laboratory experiments on analogue materials under extraterrestrial conditions like Venus or Mars.

Recently, Mössbauer spectroscopy has been applied for in-situ exploration of the surface of Mars. The miniaturized spectrometer MIMOS II is part of the science payload of the NASA Mars Exploration Rover (MER) mission searching since January 2004 for mineralogical and geological signatures of aqueous processes at the two MER landing sites Gusev Crater and Meridiani Planum, Mars. MIMOS II identified, besides many others, the minerals jarosite, an Fe-sulfate, and goethite, an Fe-hydroxide, both mineralogical markers of aqueous and hydrothermal processes active in the past at both landing sites on Mars. The two spectrometers on MER also identified the mineral hematite, iron sulfates, nanophase oxides, and the rock forming silicates olivine and pyroxene. The scientific results from the Mössbauer instruments contributed significantly to the understanding of the history of water on Mars and the evolution of Mars and its surface in general. In the coming years, new space missions will carry Mössbauer spectrometers to Mars and other solar system bodies to increase our knowledge on the origin and evolution of the solar system.

15.2   Comparative Planetology

Knowledge on how Earth's closest neighbors Venus and Mars evolved climatically into opposite directions after they presumably formed under very similar conditions is of special concern. A planet's climatic history leaves traces on its surface. Mars in particular shows evidence of once abundant water flowing over its surface and thus a warmer climate whereas it appears cold and dry today. On the other hand, Venus has developed, probably due to a 'Greenhouse effect,' to a planet with very high surface temperatures of up to 500–600 K, and high atmospheric pressure of up to 100 bar at the surface, with the atmosphere composed of nearly 100% CO_2. Understanding of the evolution of the terrestrial planets Mars and Venus will help us to understand the future evolution of our planet Earth.

15.2.1 The Planet Venus

At present, there is no direct information about the mineralogy of Venus' surface. Instead, our knowledge of the surface mineralogy is derived from chemical analyses and radar observations of the surface, and thermodynamic calculations and laboratory studies of atmospheric weathering reactions [4]. γ-Ray analyses by five landers (Venera 8, 9, 10 and Vega 1 and 2; all from Soviet Union, now Russia) give elemental abundances for U, Th, and K. X-ray fluorescence (XRF) analyses by three landers (Venera 13, 14, and Vega 2) give elemental abundances for major elements heavier than sodium. All the analytical data are summarized and discussed in [4]. Iron is one of the most abundant elements detected by the XRF analyses of Venus' surface. The theoretical mineral compositions of the three landing sites analyzed by X-ray fluorescence have been calculated by [5, 6]. The normative minerals at the Venera 13, 14, and Vega 2 landing sites include three Fe-bearing phases (fayalite Fe_2SiO_4 dissolved in olivine, ferrosilite $FeSiO_3$ dissolved in pyroxene, and ilmenite $FeTiO_3$). Laboratory studies and theoretical models predict that other Fe-bearing minerals such as magnetite (Fe_3O_4), hematite ($\alpha - Fe_2O_3$), and pyrrhotite (non-stoichiometric Fe sulfide with the general formula $Fe_{1-x}S$, where x is less than or equal to 0.125) may also be present on Venus' surface.

Why is information about Fe-bearing minerals on Venus' surface important? First, Fe-bearing minerals are involved in chemical reactions with C-, O-, H-, and S-bearing gases in Venus' atmosphere. These reactions include oxidation–reduction reactions [4, 7, 8] and production of reduced sulfur gases in the Venus sulfur cycle [9–13]. Thus, determination of the nature and the abundance of Fe bearing minerals in rocks and soil will provide important information about atmospheric weathering processes. Second, percent levels of Fe-bearing sulfides such as pyrite (FeS_2) have been suggested to explain low radar emissivity regions on Venus [14, 15]. However, Laboratory studies, using Mössbauer spectroscopy, indicate a limited life time of those minerals under Venus surface conditions [10, 16]. Other materials, such as volatile compounds of trace metals with high dielectric constants, have been proposed to explain the low radar emissivity [17, 18]. However, it is still important to consider analytical methods to detect and distinguish Fe sulfides such as pyrrhotite and pyrite that may be present at abundances of several mass percent. Finally, knowledge of the nature and abundance of Fe-bearing minerals in basaltic rocks will provide important information about igneous processes inside Venus.

With MB spectroscopy, the Fe minerals present, their relative abundances, and the oxidation state(s) of Fe in the different minerals could be determined directly, which cannot be obtained by XRF spectroscopy. Synthetic MB spectra computed for the normative Fe mineralogies at the Vega 2 and the Venera 14 landing sites [19] indicate that MB spectroscopy can identify the Fe minerals and can distinguish the different Fe mineralogies (see Fig. 15.1). They also show that especially high temperature MB spectra are useful for mineral identification. In-situ MB spectroscopy would be very useful for identifying and measuring the relative abundances of Fe-bearing minerals present on Venus' surface. The harsh environmental conditions on

Fig. 15.1 Calculated room
temperature (300 K)
backscattering Mössbauer
(MB) spectra for the
iron-bearing minerals present
at the Venera 14 and Vega 2
landing sites according to
CIPW normative
mineralogies calculated by
Barsukov et al. [5, 6] from the
XRF elemental analyses
listed in Table 15.1. The
vertical lines at the *top* of the
figure show the positions of
MB peaks for the different
minerals (from [19])

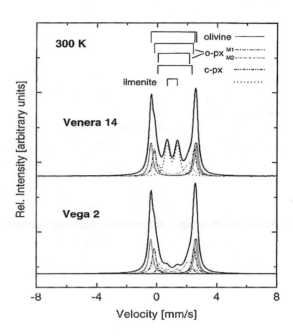

Table 15.1 Mineral composition at three landing sites on Venus, obtained by normative calculations (from [6, 19])

Normative mineral compositions			
Mineral	Venera 13	Venera 14	Vega 2
Orthopyroxene ($En_{75}Fs_{25}$)	–	18.2	25.4[a]
Clinopyroxene ($Wo_{48}En_{36}Fs_{16}$)	–	–	2.5[b]
Diopside ($CaMgSi_2O_6$)	10.2	9.9	–
Olivine ($Fo_{75}Fa_{25}$)	26.6	9.1	13.9[c]
Anorthite ($CaAl_2Si_3O_8$)	24.2	38.6	38.3
Albite ($NaAlSi_3O_8$)	3.0	20.7	18.9
Orthoclase ($KAlSi_3O_8$)	25.0	1.2	0.5
Nepheline ($NaAlSiO_4$)	8.0	–	–
Ilmenite ($FeTiO_3$)	3.0	2.3	0.5
Total	100.0	100.0	100.0

Note: Barsukov et al. [5, 6]
[a] The orthopyroxene is 75 mol% enstatite (En = $MgSiO_3$) and 25 mol% ferrosilite (Fs = $FeSiO_3$)
[b] The clinopyroxene is 48 mol% wollastonite (Wo = $CaSiO_3$), 36 mol% enstatite, and 16 mol% ferrosilite
[c] The olivine is 75 mol% forsterite (Fo = $Mg_2)SiO_4$ and 25 mol% fayalite (Fa = Fe_2SiO_4)

Venus' surface constrain the operation of a spacecraft MB spectrometer on Venus.
Operation of a MB spectrometer on the surface of Venus would probably require
cooling by phase change materials and/or active refrigeration. In particular, variation
of the Debye–Waller factors as a function of temperature should be investigated
for several common Fe minerals for temperatures up to about 800 K. Design of

mechanical equipment for sample acquisition and handling at high temperatures and high pressures for a MB spectrometer is currently under study at the Russian Space Research Institute (IKI), Moscow, for the Venus lander mission Venera-D, planned for year ~2016–2018.

15.2.2 The Planet Mars

Mars is the most Earth like planet, and there are strong indications from space missions, that there has been water present in the past in large amounts. It is also believed that the atmosphere of Mars in the past has been very similar to the Earth atmosphere. This has changed significantly, today Mars is a dry planet with a very thin atmosphere of about 6–7 mbar of mainly CO_2, and the open questions are: (i) why?, (ii) where did the water go to?, and (iii) did life develop on Mars?

Mars is covered by weathered material: weathered rocks, soil, dust, and sediments. The thickness of this regolith layer is not known, but estimations range from meter scale to hundredths of meter scale [20]. These surface materials of weathered rocks and soils on the surface of Mars, and planetary surfaces in general, provide important information and constraints on the nature, timing, and duration of surface processes due to surface–atmosphere interaction. These materials play a significant role in the geologic history of the planet, and in particular in the case of Mars these materials are keys to understanding: (1) the processes that are currently or recently active; (2) the history of erosion and deposition of sediments; (3) the relations between weathered material (e.g., sediments) and igneous rocks; and (4) most importantly changes in environmental conditions through time.

These open questions on the mineralogy on the surface of Mars have been addressed recently by NASA (US Space Agency) and ESA (European Space Agency) space missions in 2003 to the surface of Mars. Part of the scientific payload of both the ESA MarsExpress Beagle-2 mission [21] and the NASA Mars Exploration Rover (MER) mission [22] is our miniaturized Mössbauer spectrometer MIMOS II [23]. It was the first time ever that a Mössbauer spectrometer has been used in situ in space and on extraterrestrial planetary surfaces. For more details and results, see Sect. 15.3.

15.3 Laboratory Studies of Extraterrestrial Materials: Lunar Samples from the Apollo Era

Meteorites have been the first extraterrestrial material available for detailed investigation in terrestrial laboratories. Most meteorites studied in the laboratory are "finds," and for the majority of these samples little is known about the length of time they have resided on Earth, during which they were exposed to terrestrial environments [24]. Even meteorite fragments, which are retrieved relatively quickly (days to weeks after the fall), are still exposed to potentially contaminating terrestrial environments at their landing site or during collection and curation. For example, extensive terrestrial amino acid contamination from the impact site environment has made it difficult to determine whether indigenous amino acids are present in the *Allan Hills* (ALH) 84001 and *Nakhla* Martian meteorites [25, 26], and the amino acids initially observed in the Tagish Lake meteorite appear to be primarily the result of terrestrial contamination from the lake ice meltwater [27]. This has to be considered when analyzing meteorites.

Mössbauer spectroscopy has made decisive contributions to the understanding of the evolution of the Solar system by laboratory studies of meteorites, especially carbonaceous chondrites (see for instance [28–30], metallic Fe–Ni meteorites (e.g., [31, 32]), and Martian meteorites (see e.g., [33, 34]), which are available on Earth. Laboratory studies of Lunar material (see for instance [35–40], which has been brought back to Earth by the US-American Apollo missions in the 1960s to 1970s, and the Soviet Union (UdSSR) robotic sample return missions.

15.3.1   Lunar Samples from the Apollo Era

Previous Mössbauer studies of lunar samples are largely from the early to mid 1970s when lunar samples were being returned from the Moon as a part of the Apollo program (e.g., [35, 37, 41–45]). Fe is one of the major elements in lunar material, and is present in the oxidation states Fe^0 and Fe^{2+} only. Mössbauer spectra show that Fe is present mainly in silicates (pyroxene, olivine, etc.), glass, ilmenite, troilite, and as Fe-metal (see Fig. 15.2). The particle size of Fe-metal is in the Ångstrœm range ($< \sim 40$ Å), exhibiting partially superparamagnetic behavior (Fig. 15.2). The nanophase FeO (np-FeO; fine grained-metal and excess metal) formed during micrometeorite impact by reduction of Fe2+ in silicate and oxide phases as discussed by Housley et al. [46] and Morris et al. [40] (see Fig. 15.3). Since then, more conventional geological techniques (e.g., optical petrography, electron probe microanalysis, and scanning electron microscopy) have dominated mineralogical studies of lunar samples (e.g., [47–49]).

There are still many open scientific questions concerning the evolution of the Moon itself, the system Earth–Moon, and the Solar system, which probably could be answered only by manned missions, giving reasons to think about building lunar bases, where human beings would be permanently present. Also the Moon can act as an ideal platform for different kinds of observatories. The feasibility of all these plans (some of them sound very futuristic) depends strongly on the possibility of producing oxygen on the Moon. Oxygen will be used as a propellant (together with hydrogen), and it is needed for the life support system of a Lunar base [50, 51]. A very promising mineral for oxygen extraction seems to be ilmenite, present in Lunar basalts, soils, and rocks, with a natural abundance of about 5 wt% (in average) [52]. To use ilmenite as a source for oxygen production would need knowledge about locations on the Lunar surface with a high abundance of this mineral. As on Earth, some kind of exploration is required to find these ilmenite rich places. Here

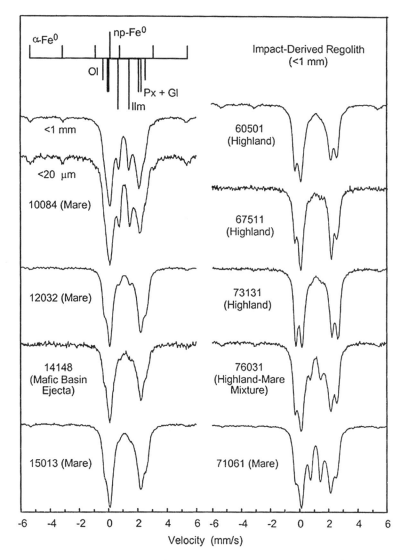

Fig. 15.2 Mössbauer spectra (293 K) for impact-derived lunar regolith. Locations of peaks for individual phases are indicated by the stick diagram. Ilmenite and pyroxene are the dominant crystalline phases in mare samples, and pyroxene and olivine are the dominant crystalline phases in highland samples. Fe^0 is present as $np - Fe^0$ and $\alpha - Fe^0$ metal (from [40])

Mössbauer spectroscopy may play an important role [40,50]. Ilmenite can be easily determined by MBS due to its characteristic pattern (see e.g., [40,52]). A Mössbauer instrument mounted on a Lunar rover would allow us to search for ilmenite very effectively. Also the absolute amount of this mineral could be determined if the MBS will include XRF capabilities, too (see Sect. 15.4). Furthermore, MBS could

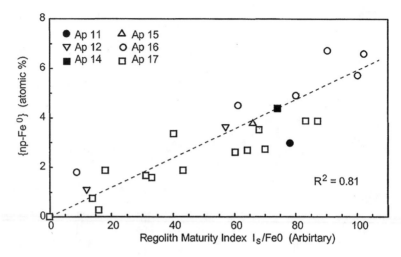

Fig. 15.3 Plot of $\{np - Fe^0\}$ versus the lunar regolith maturity index Is/FeO. The correlation coefficient $R2$ from a linear least square fit constrained to pass through the origin is 0.9. (from [40])

be used to monitor in situ the degree of reduction of ilmenite during the oxygen production process.

15.4   In Situ Exploration of the Martian Surface by Mössbauer Spectroscopy: The Mars Exploration Rover Mission 2003

The major scientific questions in the exploration of our solar system are: "What are the conditions of planet formation and the emergence of life?" and, in particular, "Life and habitability in the Solar System." To achieve these goals, it will be necessary to: (1) explore in situ the surface and subsurface of the solid bodies in the Solar system most likely to host, or have hosted, life and (2) explore the environmental conditions that make life possible. Of particular interest are the terrestrial planets, Earth, Mars, and Venus, which had very similar environmental conditions early in their histories. All three planets had a CO_2 rich atmosphere enabling the presence of liquid water on their surfaces, as well as an abundance of prebiotic organic molecules. Earth, Mars, and Venus could have supported the independent appearance of life.

To answer the question on the presence of significant amounts of water on Mars in the past, NASA designed the Mars Exploration Rover (MER) twin mission according to the main theme: "Follow the Water." Primary goals of the MER missions are: (1) mineralogical and geological characterization of the two landing sites; (2) search for evidence of the presence of water in the past. The instrument payload [22] includes the Mössbauer spectrometer MIMOS II [23]. The scientific basis for landing a Mössbauer spectrometer on Mars is extensively discussed by Knudsen [1, 53], and several unsuccessful attempts have been made to design a small MB instrument [54, 55], until the miniaturized spectrometer MIMOS II was realized and used on the MER mission [23].

The NASA Mars Exploration Rovers (MER) (Fig. 15.4) Spirit and Opportunity landed on the Red Planet in January 2004. Both NASA Mars Exploration Rovers, Spirit and Opportunity [22, 23], carry the miniaturized Mössbauer spectrometer (MIMOS II) as part of their scientific payload.

The scientific objectives of the MIMOS II Mössbauer spectrometer investigation on Mars are (1) to identify its mineralogical composition and (2) to measure the relative abundance of iron-bearing phases (e.g., silicates, oxides, carbonates, phyllosilicates, hydroxyoxides, phosphates, sulfides, and sulfates), (3) to distinguish between magnetically ordered and paramagnetic phases and provide, from measurements at different temperatures, information on the size distribution of magnetic particles, and (4) to measure the distribution of Fe among its oxidation states (e.g., Fe^{2+}, Fe^{3+}, and Fe^{6+}). These data characterize the present state of Martian surface materials and provide constraints on climate history and weathering processes by which the surface evolved to its present state. The MER mission was planned to last for three months on the Martian surface. At the time of writing (March 2011), both rovers and their instruments have spent more than seven years exploring the two landing sites Gusev Crater and Meridiani Planum, Mars, and are still operational. The mission success criteria for the rovers had been to drive more than 600 meters, and the goal for the Mössbauer spectrometers had been to collect spectra from at least three different soil and rock samples at each landing site. To date, the rovers have covered distances of more than 7 km (Spirit) and more than 28 km (Opportunity), respectively. The total amount of scientific targets investigated by the Mössbauer spectrometers exceeds 300, and the total number of integrations exceeds 600. A total of 15 unique Fe-bearing phases were identified up to now: Fe^{2+} in olivine, pyroxene, and ilmenite; Fe^{2+} and Fe^{3+} in magnetite and chromite; Fe^{3+} in nanophase ferric oxide (npOx), hematite, goethite, an Fe–Mg-carbonate, jarosite, an unassigned Fe^{3+} sulfate, and an unassigned Fe^{3+} phase associated with jarosite. Fe^{0} was identified in kamacite, an Fe–Ni alloy, and schreibersite $((Fe, Ni)_3P)$ present in Fe-meteorites. The minerals jarosite [56], the Fe-sulfates, goethite [57, 58], and the Fe–Mg–carbonate [59] are mineralogical markers of aqueous and hydrothermal processes active in the past at both landing sites on Mars. Both Mössbauer spectrometers remain operational and continue to return valuable scientific data [60–64].

15.4.1 The Instrument MIMOS II

The instrument MIMOS II is extremely miniaturized compared to standard laboratory Mössbauer spectrometers and is optimized for low power consumption and high detection efficiency [20, 22, 23, 50, 65, 66].

MIMOS II operates in backscattering geometry, irradiating the samples through a collimator (Fig. 15.5), and detecting the resonantly scattered 14.4 keV Mössbauer radiation, and simultaneously the 6.4 keV Fe-X-ray's (accompanying the resonant absorption internal conversion process) with its four Si-PIN diodes (each of them with sensitive area of 1 cm^2) (see Fig. 15.5; [20, 23, 50, 65]).

All components were selected to withstand high acceleration forces and shocks, temperature variations over the Martian diurnal cycle, and cosmic ray irradiation. Mössbauer measurements can be done during day and night covering the whole diurnal temperature variation between about −100°C (min) and about +10°C (max) of a Martian day [67–70]. To minimize the experiment time, as strong a main ^{57}Co/Rh source as possible was used, with an initial source strength of about 350 mCi at launch. Instrument internal calibration is accomplished by a second, less intense radioactive source mounted on the end of the velocity transducer opposite to the main source and in transmission measurement geometry with a reference sample. For further details, see [23, 71].

The MIMOS II Mössbauer spectrometer sensor head is located at the end of Instrument Deployment Device IDD (see Figs. 15.4 and 15.6). On Mars-Express Beagle-2, an European Space Agency (ESA) mission in 2003, the sensor head was mounted also on a robotic arm integrated to the Position Adjustable Workbench (PAW) instrument assembly [72, 73]. The sensor head shown in Figs. 15.5 and 15.6 carries the electromechanical transducer with the main and reference ^{57}Co/Rh sources and detectors, a contact plate and sensor. The contact plate and sensor are

Fig. 15.4 NASA Mars Exploration Rover artist view (courtesy NASA, JPL, Cornell University, USA). On the front side of the rover the robotic arm, carrying the Mössbauer spectrometer and other instruments, can be seen

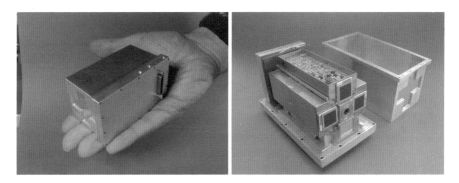

Fig. 15.5 External view of the MIMOS II sensor head without contact plate assembly (*left*); internal view of the sensor head (cover removed) (right), showing the four Si-PIN-diodes (*dark squares*; 1 cm × 1 cm) and the front-end electronics. In the center, the collimator for the MB source can be seen

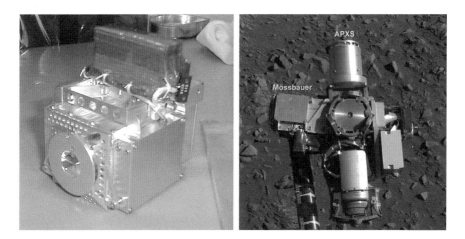

Fig. 15.6 (*Left*) Flight unit of the MIMOS II Mössbauer spectrometer sensor head, with the *circular* contact plate assembly (*front side*). The *circular* opening in the contact plate has a diameter of 15 mm, defining the field of view of the instrument. (*Right*) Contact instruments at the robotic arm (IDD) on spirit rover at Gusev crater, Mars. MIMOS II can be seen at the *left side*

used in conjunction with the IDD to apply a small preload when it places the sensor head, holding it firmly against the target.

Because of the complexity of sample preparation, backscatter measurement geometry is the choice for an in-situ planetary Mössbauer instrument [23]. No sample preparation is required, because the instrument is simply presented to the sample for analysis. Both the 14.41 keV γ-rays and 6.4 keV Fe X-rays are detected simultaneously.

MIMOS II has three temperature sensors, one on the electronics board and two on the sensor head. One temperature sensor in the sensor head is mounted near the internal reference absorber. The other sensor is mounted outside the sensor head at the contact ring assembly to determine temperature of the sample on the Martian surface.

This temperature is used to route the Mössbauer data to the different temperature intervals (maximum of 13, with the temperature width software selectable) assigned in memory.

15.4.2   Gusev Crater Landing Site, Mars

The Gusev Crater landing site, a flat-floored crater with a diameter of 160 km and of Noachian age, is located at about 14.5°S of the equator. Gusev was hypothesized to be the site of a former lake, filled by Ma'adim Vallis, one of the largest valley networks on Mars. The first Mössbauer spectrum ever recorded on the Martian surface was obtained on soil at Spirit's landing site on the plains in Gusev Crater. It shows a basaltic signature dominated by the minerals olivine and pyroxene.

This type of soil and dust were found to be globally distributed on Mars: spectra obtained on soil at Opportunity's landing site in Meridiani Planum are almost identical to those recorded in Gusev Crater.

The collection of spectra obtained on the plains of Gusev Crater and in the Columbia hills reveals various mineralogical signs of weathering. Spectra obtained on the basaltic rocks and soil on the plains show mainly olivine and pyroxene and small amounts of non-stoichiometric magnetite, and only comparably small amounts of weathering. An example of such a rock named *Adirondack*, found close to Spirits landing site, is shown in Fig. 15.7, together with its Mössbauer spectrum. The Fe mineralogy of this ∼50 cm wide rock is composed of the dominating mineral olivine, pyroxene, nanophase ferric oxide (npOx), and a small contribution of non-stoichiometric magnetite. On a similar rock, also in the plains, named *Mazatzal*, a dark surface layer was detected with the Microscopic Imager after the first of two **R**ock **A**brasion **T**ool (RAT) grinding operations. This surface layer was removed except for a remnant in a second grind. Spectra – both 14.4 keV and 6.4 keV – were obtained on the undisturbed surface, on the brushed surface and after grinding. The sequence of spectra shows that npOx is enriched in the surface layer, while olivine is depleted [62, 74, 75]. The thickness of this surface layer was determined to about 10 μm by a Monte Carlo simulation analysis program [74, 75].

The spectra of the rocks in the plains (e.g., Fig. 15.7) are very similar to the spectra obtained on the soil. The ubiquitous presence of olivine in soil suggests that physical rather than chemical weathering processes currently dominate at Gusev crater.

On the contrary, thoroughly weathered rocks were encountered in the Columbia hills, about 2.5 km away from the Spirit landing site. The Mössbauer signature is characteristic of highly altered rocks. Spectra obtained on these samples show larger amounts of nanophase ferric oxides and hematite. In spectra obtained on ∼20 rock samples, the mineral goethite (α-FeOOH) was identified (see Fig. 15.8), a clear

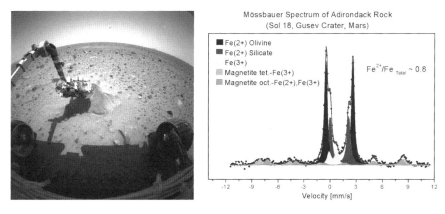

Fig. 15.7 *Left:* robotic arm with MIMOS II positioned on the rock *Adirondack*, as seen by the navigation camera of the rover; *right:* Mössbauer spectrum (14.4 keV; temperature range 220 to 280 K) of the rock Adirondack at Spirit landing side Gusev Crater, plains. The data were taken at the as-is dusty surface (not yet brushed). The spectrum shows an olivine-basalt composition, typical for soil and rocks in Gusev plains, consisting of the minerals olivine, pyroxene, an Fe^{3+} doublet, and non-stoichiometric magnetite

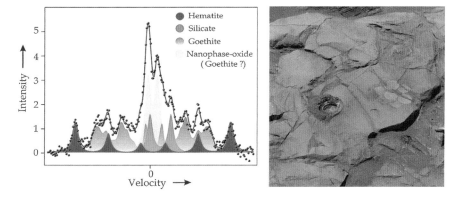

Fig. 15.8 (*Left*) In the Mössbauer spectrum taken in the Columbia Hills at a rock called *Clovis*, the mineral goethite ($\alpha - FeOOH$) could be identified. Goethite is a clear mineralogical evidence for aqueous processes on Mars. (*Right*) The rock *Clovis* is made out of rather soft material as indicated by the electric drill-current when drilling the ~ 1 cm deep hole seen in the picture. Drill fines are of brownish color. The pattern to the right of the drill hole was made by brushing the dust off the surface by using the RAT (courtesy JPL, Cornell University)

indicator for aqueous weathering processes in the Columbia hills in the past. In particular, one of the rocks, named Clovis (Fig. 15.8), contains the highest amount of the Fe oxyhydroxide goethite (GT) of about 40% in area found so far in the Columbia Hills. A detailed analysis of these data indicates a particle size of ~ 10 nm [57]. The rock Clovis also contains a significant amount of hematite. The behavior

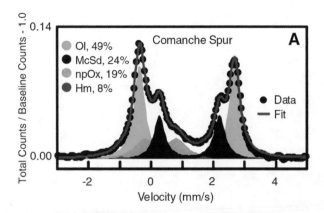

Fig. 15.9 Identification of Fe–Mg–carbonate in the Gusev Crater-Columbia Hills outcrop Comanche Spur Mössbauer spectrum [composite of Horseback and Palomino targets]: Ol = Fe2+ in olivine; McSd = Fe2+ in Mg–Fe carbonate; npOx = Fe3+ in nanophase ferric oxide; and Hm = Fe3+ in hematite (from [59])

of hematite is complex because the temperature of the Morin transition ($\sim$260 K) lies within diurnal temperature variations on Mars.

Decades of speculation about a warmer, wetter Mars climate in the planet's first billion years call upon a denser CO_2-rich atmosphere than at present. Such an atmosphere should have led to the formation of outcrops rich in carbonate minerals, for which no in situ evidence has been found up to now. The Mars Exploration Rover Spirit has now identified outcrops rich in Mg–Fe carbonate (16 to 34 wt%) in the Columbia Hills of Gusev crater (see Fig. 15.9). The Gusev carbonate probably precipitated from carbonate-bearing solutions under hydrothermal conditions at near neutral pH in association with volcanic activity during the Noachian era [59].

15.4.3 Meridiani Planum Landing Site

Opportunity touched down on 24 January 2004 in the eastern portion of the Meridiani Planum landing ellipse in an impact crater 20 m in diameter named Eagle crater. The Meridiani Planum landing site is the top stratum of a layered sequence about 600 m thick that overlies Noachinan cratered terrain. Orbital data indicated the presence of significant amounts of the mineral hematite, an indicator for water activities. The surface of Meridiani Planum explored by the Opportunity rover can be described as a flat plain of sulfur-rich outcrop that is mostly covered by thin superficial deposits of aeolian basaltic sand and dust, and lag deposits of hard Fe-rich spherules (and fragments thereof) that weathered from the outcrop, and small unidentified rock fragments and cobbles, and meteorites. Surface expressions of the outcrop occur at impact craters (e.g., Eagle (Fig. 15.10; [64]), Fram, Endurance,

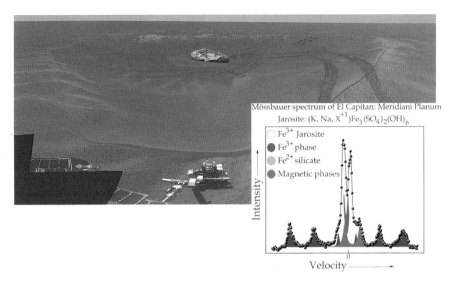

Fig. 15.10 (*Top*) Eagle crater, where the rover Opportunity landed on 24th January 2004. Close to the landing platform (inside the crater), outcrop rocks are found at the crater wall. Picture taken after Opportunity left the crater, investigating with the IDD the soil outside the crater; (*bottom*) MB spectrum of the outcrop rocks, taken on sol 33 (sol = Martian day) of the mission, the mineral Jarosite, an Fe^{3+}-sulfate, could be identified at the Meridiani Planum landing site. It forms only under aqueous conditions at low pH ($< \sim$ 3–4) and is, therefore, a clear mineralogical evidence for aqueous processes on Mars

Erebus, Victoria (see Fig. 15.12) craters), and occasionally in troughs between ripple crests [60, 64, 76].

Mössbauer spectra measured by the Opportunity rover at the Meridiani Planum landing site revealed four mineralogical components in Meridiani Planum at Eagle crater: jarosite- and hematite-rich outcrop (see Fig. 15.10), hematite-rich soil, olivine-bearing basaltic soil, and a variety of rock fragments such as meteorites and impact breccias. Spherules (Fig. 15.11), interpreted to be concretions, are hematite-rich and dispersed throughout the outcrop.

The same Fe-sulfate jarosite containing material was found everywhere along the several (more than 15) kilometers long drive way of opportunity to the south, in particular at craters Eagle, Fram, Endurance, and Victoria, suggesting that the whole area is covered with this sedimentary jarositic, hematite, and Fe-silicate (olivine; pyroxene) containing material. The mineral jarosite $((K, Na)Fe_3(SO_4)_2(OH)_6)$ contains hydroxyl and is thus direct mineralogical evidence for aqueous, acid-sulfate alteration processes on early Mars. Because jarosite is a hydroxide sulfate mineral, its presence at Meridiani Planum is mineralogical evidence for aqueous processes on Mars, probably under acid-sulfate conditions as it forms only at pH-values below $\sim$3–4.

Hematite in the soil is concentrated in spherules and their fragments, which are abundant on nearly all soil surfaces. Several trenches excavated using the rover

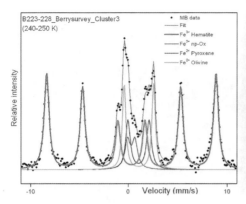

Fig. 15.11 *Left*: spectrum of an accumulation of hematite rich spherules (*Blueberries*) on top of basaltic soil (Sol 223–228 of the mission). The spectrum is dominated by the hematite signal. Estimations based on area ratios (bluerries/soil) and APXS data indicate that the blueberries are composed mainly of hematite. *Right*: MI picture (3 cm × 3 cm) of hematitic spherules (*blueberries*) on basaltic soil at Meridiani Planum (courtesy NASA-JPL, USGS, Cornell University, USA). Besides the four spherules in the center (average diameter ∼4–5 mm/spherule) the "nose" print of the MIMOS contact plate can be seen (outer diameter ∼32 mm; inner diameter (=field of view of the MIMOS instrument) 15 mm; see also Fig. 15.6, *left*)

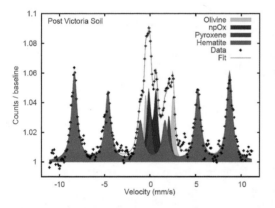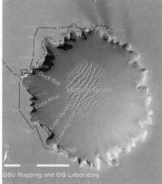

Fig. 15.12 *Left*: spectrum of the soil close to the crater rim where Opportunity entered and exited the crater. The basaltic soil is very high in hematite (probably due to contributions from hematitic spherules). *Right*: ∼750 m diameter (∼75 m deep) eroded impact crater Victoria crater, formed in sulfate-rich sedimentary rocks. Image acquired by the Mars Reconnaissance Orbiter High Resolution Science Experiment camera (Hirise). The *red line* is the drive path of opportunity exploring the crater. Courtesy NASA, JPL, ASU, Cornell University

wheels showed that the subsurface is dominated by basaltic sand, with a much lower abundance of spherules than at the surface. Olivine-bearing basaltic soil is present throughout the region. At several locations along the rover's traverse, sulfate-rich bedrock outcrops are covered by no more than a meter or so of soil.

15.4.4   Victoria Crater

After a journey of more than 950 Sols (1 Sol = 1 Martian day ~24 hours and 37 minutes) or more than 2.5 Earth years (~1.3 Martian years), Opportunity arrived at the 750 m diameter crater Victoria (Fig. 15.12) [64], after traveling across the plains of Meridiani Planum more than 10 km to the south from its landing site in Eagle crater. The rover has explored this eroded impact crater for more than 750 Sols (more than 2 Earth years or 1 Martian year), descending into the crater on Sol 1293 to begin in situ physical and chemical/mineralogical observations. After exiting the crater on Sol 1634, Opportunity performed more imaging and in situ investigations to the southwest of Duck Bay, name of a part of the crater rim, before leaving Victoria heading south towards a 20 km diameter crater named Endeavour.

The outline of Victoria crater is serrated, with sharp and steep promontories separated by rounded alcoves (Fig. 15.12). The crater formed in sulfate-rich sedimentary rocks, and is surrounded by smooth terrain that extends about one crater diameter from the rim [64]. On the crater floor is a dune field. There are no perched ejecta blocks preserved on the smooth terrain around the crater rim, probably planed off by aeolian abrasion. The Mössbauer mineralogy of the sedimentary rocks at the crater rim and inside the crater itself is nearly the same as at Eagle crater landing site and Endurance crater, both about 6–8 km away [64].

The soil close to the crater rim, at the exit point of Opportunity, shows a high hematite content in the Mössbauer spectrum (Fig. 15.12), which can be attributed to the presence of hematite spherules (blueberries).

15.4.5   Meteorites on Mars

Meridiani Planum is the first Iron meteorite discovered on the surface of another planet, at the landing site of the Mars Exploration Rover Opportunity [77]. Its maximum dimension is ~30 cm (Fig. 15.14). Meteorites on the surface of solar system bodies can provide natural experiments for monitoring weathering processes. On Mars, aqueous alteration processes and physical alteration by, e.g., aeolian abrasion may have shaped the surface of the meteorite, which, therefore, has been investigated intensively by the MER instruments. Observations at mid-infrared wavelengths with the Mini-TES (Thermal Emission Spectrometer) instrument indicated the metallic nature of the rock [67]. Observations made with the panoramic camera and the microscopic image revealed that the surface of the rock is covered with pits interpreted as regmaglypts and indicate the presence of a coating on the surface. The Alpha-Particle-X-ray Spectrometer (APXS) and the Mössbauer spectrometer were used to investigate the undisturbed and the brushed surface of the rock. Based on the Ni and Ge contents derived by APXS, Meridiani Planum was classified as an Iron meteorite of the IAB complex. The brushed meteorite surface was found to be enriched in P, S, and Cl in comparison to martian soil (Fig. 15.14).

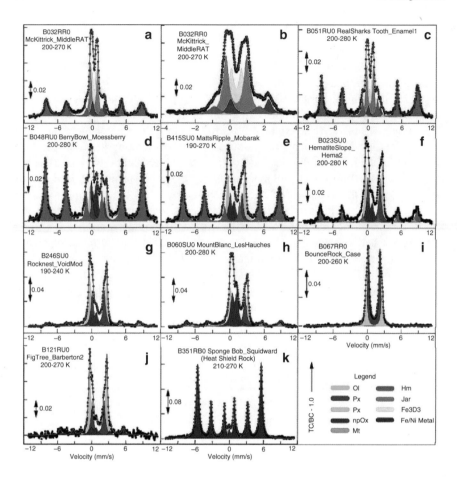

Fig. 15.13 Overview on Meridiani Planum Mössbauer mineralogy. The large variability in mineral composition at this landing site can be seen. Shown are representative Mössbauer spectra. Spectra are the sum over all temperature windows within the indicated temperature ranges. The computed fit and component subspectra (Lorentzian line shapes) from least-squares analyses are shown by the *solid line* and the *solid shapes*, respectively. Full (**a**) and reduced (**b**) velocity Mössbauer spectra for interior Burns outcrop exposed by RAT grinding show that hematite, jarosite, and Fe_3D_3 (acronym for unidentified Fe^{3+} phase; see [61–63, 68, 74, 81]) are the major Fe-bearing phases. Mössbauer spectrum for a rind or crust (**c**) of outcrop material that has an increased contribution of hematite relative to jarosite plus Fe_3D_3. Mössbauer spectra of two soils (**d** and **e**) have high concentrations of hematite. The spectrum (**e**) is typical for hematite lag deposits at ripple crests. The soil spectra (**f**) to (**h**) [68, 82] are basaltic in nature and have olivine, pyroxene, and *nanophase ferric oxide* as major Fe-bearing phases. The soil target named*MountBlanc_LesHauches* (**h**) is considered to be enriched in martian dust. Mössbauer spectra in (**i–k**) are for three single-occurrence rocks: Bounce Rock (**i**) is monomineralic pyroxene. Barberton (**j**) contains kamacite (iron–nickel metal), and heat shield rock is nearly monomineralic kamacite identified as an Fe–Ni meteorite (see below). TC = total counts and BC = baseline counts. Figure taken from Morris et al. [82]

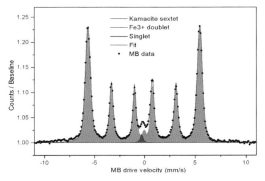

Fig. 15.14 (*Left*) The Mössbauer spectrum of the rock called "Heat Shield rock," clearly shows with high intensity the mineral Kamacite, an Fe–Ni alloy with about 6–7% Ni; (*Right*) the iron–nickel meteorite "Meridiani Planum" (originally called "Heat Shield Rock") at opportunity's landing site, close to the crater Endurance. The meteorite is about 30 cm across. Courtesy NASA, JPL, Cornell University, USA

15.5   Recent Developments

The Mars Exploration Rover mission has demonstrated that Mössbauer spectroscopy is an important tool for the in situ explorationof extraterrestrial bodies and the study of Fe-bearing samples. MIMOS II is part of the scientific payload of future space missions. In comparison to MIMOS II, the new instrument MIMOS IIA, under development for the 2018 ESA-NASA mission to Mars, will use Si-Drift detectors (SDDs) which allows also high-resolution X-ray fluorescence (XRF) analysis in parallel to Mössbauer spectroscopy [78–80]. The new design of the improved version is lighter than the MER instrument. A new ring detector system of four SDDs [79] will replace the four Si-PIN detectors of the current version greatly improving the energy resolution and the area fill factor around the collimator. The SDDs exhibit an energy resolution < 150 eV at a moderate cooling of about –5°C. This results in an increase of the signal to noise ratio (SNR) (or sensitivity) of more than a factor of 10 (Fig. 15.15). In addition to the Mössbauer data, SDD allows the simultaneous acquisition of the X-ray fluorescence spectrum, thus providing data on the sample's elemental composition (Fig. 15.15).

15.6   Summary and Outlook

The miniaturized Mössbauer instruments have proven as part of the NASA Mars Exploration Rover 2003 mission that Mössbauer spectroscopy is a powerful tool for planetary exploration including our planet Earth.

The scientific results from the Mössbauer instruments contributed significantly to the understanding of the history of water on Mars and the evolution of Mars and its

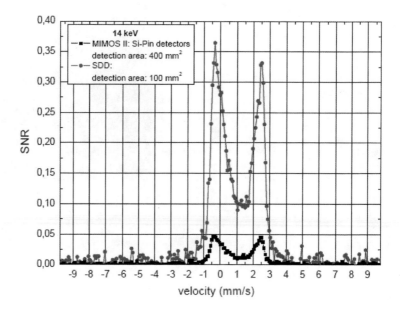

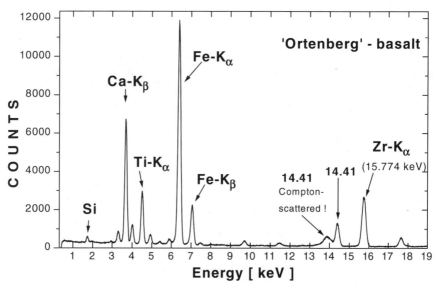

Fig. 15.15 *Top:* comparison of signal to noise ratio of 14.4. keV Mössbauer spectra, taken with a Si-PIN detector system (MER instrument; four diodes; see Fig. 15.5) and with a SDD system (advanced MIMOS instrument; only one diode chip); *Bottom*: X-ray spectrum of a basalt (Ortenberg basalt; see [78, 80], taken with a high resolution Si-drift detector (SDD) system at ambient pressure (1 atm), demonstrating the XRF capability of the "advanced MIMOS" instrument. Excitation source: ^{57}Co/Rh Mössbauer source. Energy resolution: about 170 eV at $+10°$C (SDD temperature)

surface in general. The MER missions with its two Mössbauer spectrometers have changed the picture of Mars as we have seen it before. In the coming years, new space missions will carry Mössbauer spectrometers to Mars and other solar system bodies to increase our knowledge on the origin and evolution of the solar system.

The instrument MIMOS II will be part of the upcoming ESA-NASA space missions to Mars in 2018, and the Russian Space Agency sample return mission Phobos Grunt scheduled for launch in 2011 to visit the Mars moon Phobos.

The miniaturized Mössbauer spectrometer MIMOS II has been used already in several terrestrial applications [66, 71] which would not have been possible before. A number of other possible terrestrial application, e.g., in the field, in industry, and fundamental research, are under consideration. With the new generation of the Mössbauer spectrometer MIMOS IIA, the method itself can be applied to numerous new fields in research, environmental science, planetary science, and many other fields.

References

1. J.M. Knudsen, Hyperfine Interact. **47**, 3–31 (1989)
2. *The Comet Rendezvous Asteroid Flyby Mission. A search for our Beginnings,* National Aeronautics and Space Administration (NASA) and Jet Propulsion Laboratory, Caltech, Pasadena
3. W.M. Irvine, Planet. Rep. **7**(6), 6 (1987)
4. B. Fegley Jr., G. Klingelhöfer, K. Lodders, T. Widemann, in *Venus II*, ed. by S.W. Boucher, D. Hunten, R. Phillips. Geochemistry of Surface–Atmosphere Interactions on Venus (University of Arizona Press, Tucson, 1997), pp. 591–636
5. V.L. Barsukov, Yu. A. Surkov, L.P. Moskalyeva, O.P. Sheglov, V.P. Kharyukova, O.S. Manvelyan, V.G. Perminov, Geokhimiya **19**, 899–919(1982)
6. V.L. Barsukov, Yu.A. Surkov, L.V. Dmitriyev, I L. Khodakovsky, Geochem. Intl. **23**, 53–65 (1986)
7. B. Fegley Jr., G. Klingelhöfer, R.A. Brackett, N. Izenberg, D.T. Kremser, K. Lodders, Icarus **118**, 373–383 (1995)
8. B. Fegley Jr., M.Yu. Zolotov, K. Lodders, Icarus **125**,416–439(1997)
9. B. Fegley Jr., A.H. Treiman, in *Venus and Mars: Atmospheres, Ionospheres, and Solar Wind Interactions*, ed. by J.G. Luhmann, M. Tatrallyay, R.O. Pepin (AGU, Washington, 1992), pp. 7–71
10. B. Fegley Jr., K. Lodders, A.H. Treiman, G. Klingelhöfer, Icarus **115**, 159–180 (1995)
11. B. Fegley Jr., Icarus **128**, 474–479 (1997)
12. Y. Hong, B. Fegley Jr., Ber. Bunsenges. Phys. Chem. **101** 1870–1881 (1997)
13. Y. Hong, B. Fegley Jr., Planet. Space Sci. **46** 683–690 (1998)
14. G.H. Pettengill, P.G. Ford, S. Nozette, Science **217**, 640–642 (1982)
15. G.H. Pettengill, P.G. Ford, B.D. Chapman, J. Geophys. Res. **93**, 14881–14892 (1988)
16. B. Fegley Jr., G. Klingelhöfer, K. Lodders, T. Widemann, in *Venus II*, ed. by S.W. Boucher, D. Hunten, R. Phillips (University of Arizona Press, Tucson, 1997), pp. 591–636
17. R.A. Brackett, B. Fegley Jr., R.E. Arvidson, J. Geophys. Res. Planets **100**, 1553–1563 (1995)
18. G.H. Pettengill, P.G. Ford, R.A. Simpson, Science **272**, 1628–1631 (1996)
19. G. Klingelhöfer, B. Fegley, Icarus **147**, 1–10 (2000)
20. G. Klingelhöfer, B. Fegley Jr., R.V. Morris, E. Kankeleit, P. Held, E. Evlanov, O. Priloutskii, Planet. Space Sci. **44**(11), 1277–1288 (1996)

21. M.R. Sims, C.T. Pillinger, I.P. Wright, G. Morgan, G.W. Fraser, D. Pullan, S. Whitehead, J. Dowson, R.E. Cole, A.A. Wells, L. Richter, H. Kochan, H. Hamacher, A. Johnstone, A.J. Coates, S.C. Peskett, A. Brack, J. Clemmet, R. Slade, N. Phillips, C. Berry, A. Senior, J.S. Lingard, J.C. Underwood, J.C. Zarnecki, M.E. Towner, M. Leese, A. Gambier-Parry, N. Thomas, J.-L. Josset, G. Klingelhoefer, SPIE **3755**, 10–23 (1999)

22. S.W. Squyres, R.E. Arvidson, E.T. Baumgartner, J.F. Bell III, P.R. Christensen, S. Gorevan, K.E. Herkenhoff, G. Klingelhöfer, M.B. Madsen, R.V. Morris, R. Rieder, R.A. Romero, J. Geophys. Res. **108**(E12), 8062 (2003)

23. G. Klingelhöfer, R.V. Morris, B. Bernhardt, D. Rodionov, P.A. de Souza Jr., S.W. Squyres, J. Foh, E. Kankeleit, U. Bonnes, R. Gellert, C. Schröder, S. Linkin, E. Evlanov, B. Zubkov, O. Prilutski, J. Geophys. Res. **108**(E12), 8067–8084 (2003)

24. A.J.T. Jull, in *Meteorites and the Early Solar System II*, ed. by D.S. Lauretta, H.S. McSween (University of Arizona Press, Tucson, 2006), pp. 889–905

25. J.L. Bada, D.P. Glavin, G.D. McDonald, L. Becker, Science **279**, 362–365 (1998)

26. D.P. Glavin, J.L. Bada, K.L.F. Brinton, G.D. McDonald, Proc. Natl. Acad. Sci. U.S.A. **96**, 8835–8838 (1999)

27. G. Kminek, O. Botta, D.P. Glavin, J.L. Bada, Meteorit. Planet. Sci. **37**, 697–701 (2002)

28. T.J. Wdowiak, D.G. Agresti, Nature **311**, 140–142 (1984)

29. M.B. Madsen, S. Moerup, T.V.V. Costa, J.M. Knudsen, M. Olsen, Hyperfine Interact. **41**, 827–830 (1988)

30. R.B. Scorzelli, R.A. Pereira, C.A.C. Perez, A.A.R. Fernandes, Hyperfine Interact. **94**, 2343–2347 (1994)

31. I. Fleischer, C. Schröder, G. Klingelhöfer, J. Zipfel, R.V. Morris, J.W. Ashley, R. Gellert, S. Wehrheim, S. Ebert, Meteorit. Planet. Sci. **46**(1), 21–34 (2011)

32. C. Schröder, D.S. Rodionov, T.J. McCoy, B.L. Joliff, R. Gellert, L.R. Nittler, W.H. Farrand, J.R. Johnson, S.W. Ruff, J.W. Ashley, D.W. Mittlefehldt, K.E. Herkenhoff, I. Fleischer, A.F.S. Haldemann, G. Klingelhöfer, D.W. Ming, R.V. Morris, P.A. de Souza Jr., S.W. Squyres, C. Weitz, A.S. Yen, J. Zipfel, T. Economu, J. Geophys. Res. **113**, E06S22 (2008)

33. M.B. Madsen, M. Olsen, J.M. Knudsen, D. Petersen, L. Vistisen, Lunar Planet. Sci. **XXIII**, 825–826 (1992)

34. C. Schröder, G. Klingelhöfer, W. Tremel, Planet. Space Sci. **52**, 997–1010 (2004)

35. C.L. Herzenberg, D.L. Riley, Science **167**, 683 (1970)

36. H. Fernandez-Moran, S.S. Hafner, M. Ohtsuki, D. Virgo, Science **167**, 626 (1970)

37. P. Gay, G.M. Bancroft, M.G. Brown, Science **167**, 626 (1970)

38. A.H. Muir Jr., R.M. Housley, R.W. Grant, M. Abdel-Gawad, M. Blander, Science **167**, 688 (1970)

39. S.S. Hafner, D. Virgo, Proc. Apollo 11 Lunar Sci. Conf. **3**, 2183 (1970)

40. R.V. Morris, G. Klingelhöfer, R.L. Korotev, T.D. Shelfer, Hyperfine Interact. **117**, 405–432 (1998)

41. T.C. Gibb, R. Greatrex, N.N. Greenwood, M.H. Battey, Proc. 3rd Lunar Planet. Sci. Conf. 2479 (1972)

42. R.M. Housley, M. Blander, M. Abdel-Gawad, R.W. Grant, A.H. Muir Jr., Proc. 1st Lunar Planet. Sci. Conf. 2251 (1970)

43. R.M. Housley, R.W. Grant, A.H. Muir Jr., M. Blander, M. Abdel-Gawad, Proc. 2nd Lunar Planet. Sci. Conf. 2125 (1971)

44. G.P. Huffman, F.C. Schwerer R.M. Fisher, Proc. 5th Lunar Planet. Sci. Conf. 2779 (1974)

45. S. Mitra, *Applied Mössbauer Spectroscopy, Theory and Practice for Geochemists and Archaeologists* (Pergamon, New York, 1992)

46. R.M. Housley, R.W. Grant, M. Abdel-Gawad, Proc. 3rd Lunar Planet. Sci. Conf. 1065 (1972)

47. G. Heiken D.S. McKay, Proc. 5th Lunar Planet. Sci. Conf. 843 (1974)

48. J.J. Papike, S.B. Simon J.C. Laul, Rev. Geophys. Space Phys. **20**, 761 (1982)

49. G.H. Heiken, D.T. Vaniman, B.M. French, *Lunar Sourcebook, A User's Guide to the Moon* (Cambridge Univ. Press, New York, 1991), p. 736

50. G. Klingelhöfer, P. Held, R. Teucher, F. Schlichting, J. Foh, E. Kankeleit, Hyperfine Interact. **95**, 305–339 (1995)
51. W.W. Mendell (ed.), *Lunar Bases and Space Activities of the 21st Century* (Lunar and Planetary Institute, Houston, 1985)
52. M.A. Gibson, C.W. Knudsen, D.J. Brueneman, H. Kanamori, R.O. Ness, L.L. Sharp, D.W. Brekke, C.C. Allen, R.V. Morris, L.P. Keller, D.S. Mckay, Lunar Planet. Sci. **XXIV**, 531 (1993)
53. J.M. Knudsen, M.B. Madsen, M. Olsen, L. Vistisen, C.B. Koch, S. Moerup, E. Kankeleit, G. Klingelhöfer, E.N. Evlanov, V.N. Khromov, L.M. Mukhin, O.F. Prilutskii, B. Zubkov, G.V. Smirnov, J. Juchniewicz, Hyperfine Interact. **68**, 83–94 (1992)
54. J. Galazkha-Friedman, J. Juchniewicz, Martian Mössbauer Spectrometer MarMös, Project Proposal, Space Research Center, Polish Academy of Sciences, February 1989
55. R.V. Morris, D.G. Agresti, T.D. Shelfer, T.J. Wdowiak, Lunar Planet. Sci. **XX**, 721 (1989)
56. G. Klingelhöfer, R.V. Morris, B. Bernhardt, C. Schröder, D.S. Rodionov, P.A. de Souza Jr., A. Yen, R. Gellert, E.N. Evlanov, B. Zubkov, J. Foh, U. Bonnes, E. Kankeleit, P. Gütlich, D.W. Ming, F. Renz, T. Wdowiak, S.W. Squyres, R.E. Arvidson, Science **306**, 1740–1745 (2004)
57. G. Klingelhöfer, E. DeGrave, R.V. Morris, A. Van Alboom, V.G. de Resende, P.A. De Souza Jr., D. Rodionov, C. Schroeder, D.W. Ming, A. Yen, Hyperfine Interact. **166**, 549–554 (2005)
58. R.V. Morris, G. Klingelhöfer, B. Bernhardt, C. Schröder, D.S. Rodionov, P.A. de Souza Jr., A. Yen, R. Gellert, E.N. Evlanov, J. Foh, E. Kankleit, P. Gütlich, D.W. Ming, F. Renz, T. Wdowiak, S.W. Squyres, R.E. Arvidson, Science **305**, 833–836 (2004)
59. R.V. Morris, S.W. Ruff, R. Gellert, D.W. Ming, R.E. Arvidson, B.C. Clark, D.C. Golden, K. Siebach, G. Klingelhöfer, C. Schröder, I. Fleischer, A.S. Yen, S.W. Squyres, Science **329**, 421–424 (2010)
60. C. Schröder, G. Klingelhöfer, R.V. Morris, D.S. Rodionov, I. Fleischer, M. Blumers, Hyperfine Interact. **182**, 149–156 (2008)
61. R.V. Morris, G. Klingelhöfer, C. Schröder, I. Fleischer, D.W. Ming, A.S. Yen, R. Gellert, R.E. Arvidson, D.S. Rodionov, L.S. Crumpler, B.C. Clark, B.A. Cohen, T.J. McCoy, D.W. Mittlefehldt, M.E. Shmidt, P.A. de Souza, S.W. Squyres, J. Geophys. Res. **113**, E12S42 (2008). doi:10.1029/2008JE003201
62. R.V. Morris, G. Klingelhöfer, C. Schröder, D.S. Rodionov, A. Yen, P.A. de Souza Jr., D.W. Ming, T. Wdowiak, R. Gellert, B. Bernhardt, E.N. Evlanov, B. Zubkov, J. Foh, U. Bonnes, E. Kankeleit, P. Gütlich, F. Renz, S.W. Squyres, R.E. Arvidson, J. Geophys. Res. – Planets **111**, E02S13 (2006)
63. G. Klingelhöfer, R.V. Morris, B. Bernhardt, C. Schröder, D.S. Rodionov, P.A. de Souza Jr., A. Yen, R. Gellert, E.N. Evlanov, B. Zubkov, J. Foh, U. Bonnes, E. Kankeleit, P. Gütlich, D.W. Ming, F. Renz, T. Wdowiak, S.W. Squyres, R.E. Arvidson, Science **306**, 1740–1745 (2004)
64. S.W. Squyres, A.H. Knoll, R.E. Arvidson, J.W. Ashley, J.F. Bell III, W.M. Calvin, P.R. Christensen, B.C. Clark, B.A. Cohen, P.A. de Souza Jr., L. Edgar, W.H. Farrand, I. Fleischer, R. Gellert, M.P. Golombek, J. Grant, J. Grotzinger, A. Hayes, K.E. Herkenhoff, J.R. Johnson, B. Jolliff, G. Klingelhöfer, A. Knudson, R. Li, T.J. McCoy, S.M. McLennan, D.W. Ming, D.W. Mittlefehldt, R.V. Morris, J.W. Rice Jr., C. Schröder, R.J. Sullivan, A. Yen, R.A. Yingst, Science **324**, 1058–1061 (2009)
65. G. Klingelhöfer, Hyperfine Interact. **113**, 369–374 (1998)
66. G. Klingelhöfer, in *Industrial Applications of the Mössbauer Effect*, ed. by M. Garcia, J.F. Marco, F. Plazaola (American Institute of Physics, 2005), pp. 369–377
67. J. Bell (ed.), *The Martian Surface* (Cambridge University Press, 2008)
68. R.V. Morris, G. Klingelhöfe, in *The Martian Surface*, ed. by J. Bell (Cambridge University Press, 2008), pp. 339–365
69. H.H. Kieffer, B.M. Jakosky, C.W. Snyder, M.S. Mathews (eds.), in *Mars* (The University of Arizona Press, Tucson, 2008)
70. N. Barlow (ed.), *Mars: An Introduction to its Interior, Surface and Atmosphere* (Cambridge University Press, 2008)
71. P. Gütlich, E. Bill, A.X. Trautwein (ed.), *Mössbauer Spectroscopy in Transition Metal Chemistry* (Springer, Heidelberg, 2011)

72. E.K. Gibson, C.T. Pillinger, I.P. Wright, G.H. Morgan, D. Yau, J.L.C. Stewart, M.R. Leese, I.J. Praine, S. Sheridan, A.D. Morse, S.J. Barber, S. Ebert, F. Goesmann, P. Roll, H. Rosenbauer, M.R. Sims, Lunar Planet. Sci. **35**,1845 (2004)

73. *Space is a Funny Place*, Collin Pillinger, Barnstorm Productions (2007)

74. I. Fleischer, G. Klingelhöfer, C. Schröder, R.V. Morris, M. Hahn, D. Rodionov, R. Gellert, P.A. de Souza, J. Geophys. Res. **113**, E06S21 (2008)

75. I. Fleischer, G. Klingelhöfer, C. Schröder, D. Rodionov, Hyperfine Interact. **186**, 193–198 (2008)

76. G. Klingelhöfer, R.V. Morris, P.A. de Souza Jr., D. Rodionov, C. Schröder, Hyperfine Interact. **170**, 169–177 (2006)

77. C. Schröder, D.S. Rodionov, T.J. McCoy, B.L. Joliff, R. Gellert, L.R. Nittler, W.H. Farrand, J.R. Johnson, S.W. Ruff, J.W. Ashley, D.W. Mittlefehldt, K.E. Herkenhoff, I. Fleischer, A.F.S. Haldemann, G. Klingelhöfer, D.W. Ming, R.V. Morris, P.A. de Souza Jr., S.W. Squyres, C. Weitz, A.S. Yen, J. Zipfel, T. Economu, J. Geophys. Res. **113**, E06S22 (2008)

78. G. Klingelhöfe, in *Mössbauer Spectroscopy in Materials Science*, ed. by M. Miglierini, D. Petridis (Kluwer, 1999), pp. 413–426

79. L. Strüder, P. Lechner, P. Leutenegger, Naturwissenschaften **11**, 539–543 (1998)

80. M. Blumers, B. Bernhardt, P. Lechner, G. Klingelhöfer, C. d'Uston, H. Soltau, L. Strüder, R. Eckhardt, J. Brückner, H. Henkel, J. Girones Lopez, J. Maul, Nucl. Instrum. Meth. A **624**, 360–366 (2010). doi: 10.1016/j.nima.2010.04.007

81. S.W. Squyres, R.E. Arvidson, J.F. Bell III, J. Brückner, N.A. Cabrol, W. Calvin, M.H. Carr, P.R. Christensen, B.C. Clark, L. Crumpler, D.J. Des Marais, C. d'Uston, T. Economu, J. Farmer, W. Farrand, W. Folkner, M. Golombek, S. Gorevan, J.A. Grant, R. Greely, J. Grotzinger, L. Haskin, K.E. Herkenhoff, S. Hviid, J. Johnson, G. Klingelhöfer, A.H. Knoll, G. Landis, M. Lemmon, R. Li, M.B. Madsen, M.C. Malin, S.M. McLennan, H.Y. McSween, D.W. Ming, J. Moersch, R.V. Morris, T. Parker, J.W. Rice Jr., L. Richter, R. Rieder, M. Sims, M. Smith, P. Smith, L.A. Soderblom, R. Sullivan, H. Wänke, T. Wdowiak, M. Wolff, A. Yen, Science **305**, 794–799 (2004)

82. R.V. Morris, G. Klingelhöfer, C. Schröder, D.S. Rodionov, A. Yen, D.W. Ming, P.A. De Souza Jr., T. Wdowiak, I. Fleischer, R. Gellert, B. Bernhardt, U. Bonnes, B.A. Cohen, E.N. Evlanov, J. Foh, P. Gütlich, E. Kankeleit, T. McCoy, D.W. Mittlefehldt, F. Renz, M.E. Schmidt, B. Zubkov, S.W. Squyres, R.E. Arvidson, J. Geophys. Res. Planets **111**, E12S15 (2006)

Chapter 16
Coherent Nuclear Resonance Fluorescence

G.V. Smirnov

16.1 Introduction

For a long time, the impact of ideas with respect to coherency and scattering on the art and science of Mössbauer spectroscopy was not very essential. The main line of development of Mössbauer experiments rested in the frame of absorption spectroscopy. Mössbauer physicists dealt mostly with absorption spectra taken either in traditional transmission experiments or in measurements of the conversion electron yield. For the interpretation and description of these spectra, it was appropriate to use the picture of interaction of γ-quantum with an individual nucleus where the nuclear resonant absorption cross-section was applied. The coherent properties of radiation and those of the interaction mechanism were not explicitly involved in these studies. In the meantime, coherent phenomena with Mössbauer γ-rays were thoroughly investigated, starting soon after Mössbauer's discovery. The present paper is aimed to illuminate this side of the Mössbauer story.

Based on the idea of the inverse Fourier transform, one can regard the time dependence of the spontaneous decay of an isomeric nuclear state as a result of interference of the spectral components of the energy distribution in the excited state. All the frequency components of the nuclear transition current are added coherently in the interference process. This way the Lorentzian distribution of energy in the excited state yields the exponential decay. In the case of hyperfine splitting of nuclear level, the spontaneous decay pattern exhibits beating, which also gives evidence for the coherent superposition in time of the hyperfine components of the nuclear current (the basic result of the time differential PAC spectroscopy). The examples of nuclear de-excitation discussed show clearly how the intrinsic coherence of nuclear radiative transition reveals itself.

G.V. Smirnov (✉)
Russian Research Center "Kurchatov Institute", 123182 Moscow, Russia
e-mail: g.smirnov@gmx.net

M. Kalvius and P. Kienle (eds.), *The Rudolf Mössbauer Story*,
DOI 10.1007/978-3-642-17952-5__16, © Springer-Verlag Berlin Heidelberg 2012

However, Doppler frequency modulation of the emitted radiation prevents it to inherit the properties of coherence of nuclear decay. A remarkable feature of Mössbauer effect is the fact that the recoil-free emission preserves the phase coherence of nuclear transitions in the emitted γ-radiation.

With the recoilless or Mössbauer radiation, an opportunity appeared *to check the coherence of the nuclear resonance fluorescence*. Owing to the Mössbauer effect, the linewidth of incident radiation is comparable to the breadth of the resonance. In this case, as already Heitler concluded [1], absorption and emission cannot be regarded as two subsequent independent processes. The resonance fluorescence has to be considered a single-quantum process and the re-emitted radiation should be coherent with the incident radiation. Yet, resonance fluorescence appeared only as a very minor contribution to nuclear scattering due to strong dominance of the internal conversion in a low energy nuclear isomeric transition. Hence the question as to whether coherence is preserved in nuclear resonance fluorescence sounded academically lofty. Fortunately, however, the advanced methodology presented a more optimistic picture: besides the demonstration of coherence in nuclear fluorescence, it was also shown that the coherence permits the radiative channel to become even dominant in the process of interaction of a γ-ray photon with a nuclear array.

Soon after the discovery of Mössbauer, Black and Moon [2] observed the interference between the nuclear and electronic scattering of γ-ray photons by ^{57}Fe atoms, i.e., the intra-atomic interference. In this way, it was established that the process of resonant scattering by an individual nucleus is fully coherent. This was evidence for the temporal coherence in the nuclear resonant scattering process. The time scale of the preservation of coherence is determined by the collision time of γ-ray photon with the target nucleus. During the collision time with a ^{57}Fe nucleus, more than 10^{11} oscillations with the phase of the nuclear transition current occur. This truly is an impressive scale of the phase memory conservation in nature!

Somewhat later Bernstein and Campbell [3] detected the total external reflection of Mössbauer radiation from an ^{57}Fe mirror. This was the first evidence for spatial coherence in scattering of γ-rays from an ensemble of nuclei. The spatial coherence was finally proven firmly in the observation of Bragg scattering from the nuclear lattice by Black and Duerdoth [4]. Certainly, the energy of an individual γ-quantum is sufficient to excite only a single nucleus. On the other hand, it is obvious that the participation of all nuclei in such scattering events like total reflection or Bragg scattering is obligatory. To combine these seemingly incompatible features of the nuclear resonant scattering, the concept of a delocalized nuclear excitation was put forward by Trammell [5] and by Afanas'ev and Kagan [6]. For the description of such an excitation state, the term "nuclear exciton" was coined by Zaretskii and Lomonosov [7].

In fact, to understand diffraction one has to consider the interference of all scattering paths, where, in each path, the quantum is scattered by an individual nucleus.

That is, one must accept the existence of internuclear interference depending on the phase correlation between the individual paths. This multiple-optical-path quantum interference results in the coherent response of the nuclear system. To account for the collective nuclear response, one has to assume an excitation probability for each nucleus in accordance with the quantum-mechanical principle of superposition of states. In the superposition state, the nuclear excitation is delocalized and thus the incident γ-ray is shared by many nuclei. This is referred to as nuclear exciton. The nuclear exciton is related to the nuclear transition currents and can be perceived as polarization of nuclear media.

It is clear that the currents cannot exist independently of the electromagnetic field of the γ-radiation. The two subsystems must be united into a single physical entity. They interact with each other and exchange energy. Both propagate through the target as a coupled state of nuclear polarization and electromagnetic waves. At the exit of the target they generate a coherent beam of radiation. In following the terminology accepted in optics, one calls the coupled system of nuclear currents and radiation field inside the target a *nuclear exciton polariton*, or simply a *nuclear polariton* [8–11] and [12, 13]. Recoilless nuclear scattering, the process preserving coherence, is responsible for creating and sustaining the nuclear polariton.

Nuclear polaritons can be created in the propagating- and in the standing-wave spatial modes. The temporal and spatial phase relations of nuclear polarization and of the radiation field is existing all over the target. It plays a decisive role in the formation of the wavefield at the exit from target [10]. In later experiments on coherent scattering of γ-radiation, it was demonstrated that collective nuclear scattering can play a crucial role in the whole picture of interactions of γ-ray photons with a nuclear ensemble. In case that a nuclear polariton is created, the nuclear system behaves like a macroscopic resonator whose properties differ quantitatively and qualitatively from those of an individual nucleus. The parameters specific for the individual resonant interaction, like linewidth, lifetime, and conversion ratio undergo drastic changes.

The discovery of nuclear γ-resonance without recoil opened thus a new field of theoretical and experimental optics. That subject covers a wide scope of coherent phenomena with resonant γ-radiation both in the energy and time domains. Since 1960, many scientific groups all over the world invested extensive work in order to advance our knowledge and skills in the new field. It would be difficult, or even impossible, in a short paper to trace this development and to assess all achievements of the scientists in this field. Quite an amount of review papers have already been published. We list only some of them here [14–21]. In the present paper, we would like to familiarize the reader with the most impressive contributions and important results. The aim is to present the essentials of the physics of coherent nuclear resonant scattering and to display some of its applications.

16.2 When and How the Radiative Channel Starts Playing a Key Role

16.2.1 Absorption at Nuclear γ-Resonance from the Point of View of the Optical Theory

A Mössbauer spectroscopist deals mostly with absorption spectra. In the traditional experiment, a diminution of photons is registered in the transmitted beam when the energy of incident γ-radiation is tuned to the nuclear transition energy in the absorber. Being asked what happened with the missing quanta, one usually says: "they are just absorbed by the nuclei". Indeed, to describe the results of a measurement, one need not be bothered by the further fate of the "disappeared" quanta. Meanwhile, it is known that the absorption represents only the first stage of the interaction of the γ-ray with a nucleus. The total process also includes the de-excitation of the nucleus accompanied by emission of secondary radiation (e.g., conversion electrons and secondary γ-ray photons). Actually, it is a process of scattering. In the following, we attempt to describe the transmission of γ-ray photon through the resonant target in terms of optical theory.

The primary quantum when being absorbed by a nucleus excites the transition between the ground and excited nuclear state. The low multipolarity of low energy nuclear transitions causes in the absence of hyperfine interaction the *decay products from a single nucleus*, i.e. secondary γ-ray photon or conversion electron, to be *distributed nearly isotropically in space.*

Let us next consider a plane wave of γ-radiation that transverses perpendicularly a thin slab with infinite lateral extent, composed of resonant nuclei. As said earlier, a nuclear exciton is created when the primary photon is absorbed in an array of nuclei. In accordance to the picture of multipath quantum interference the partial wavelets from all nuclei have to be summed to give the resultant probability wave for γ-ray emission in a chosen direction. With a given geometry of the incident wave, and the target, the *resultant probability wave will be a plane wave propagating in the direction of the primary radiation*, because only in that direction all the scattered wavelets are summed *constructively*. In the approximation where the interaction of nuclei with the primary radiation is solely taken into account, the direct calculation gives the following amplitude of the scattered wave

$$A_s = -i\frac{1}{2}A_i\eta_0(\omega)k\Delta z \quad \text{with} \qquad (16.1)$$

$$\eta_0(\omega) = -\frac{1}{k}\mu\frac{\Gamma/2\hbar}{\omega - \omega_0 - i\Gamma/2\hbar},$$

where k is wave number related to the wavelength of radiation $k = 2\pi/\lambda$, μ is linear nuclear absorption coefficient, proportional to the maximum resonance cross-section σ_0, ω is frequency of incident radiation, A_i is amplitude of the incident wave and Δz is the slab thickness. As seen from the upper line of (16.1), the nuclei arranged in a thin sheet parallel to the primary wavefront produce a resultant probability wave whose phase is retarded by $\pi/2$ ($-i = \exp\{-i\pi/2\}$) with respect to the waves scattered by the individual nuclei in the slab. A wave scattered by a nucleus at the exact resonance energy already lags $\pi/2$ behind the phase of the primary wave (at $\omega - \omega_0 = 0$, $\eta_0(\omega) \propto -i$), as shown in the lower line of (16.1). The retardation of the resultant scattered wave will be then $-\pi$, meaning that it opposes the primary wave. The effect of scattering will be to add a wave that at each point of the medium is out of phase with the primary wave, at resonance $A_s = -\frac{1}{2}A_i\mu\Delta z$. There is thus a *progressive decrease in the amplitude of the transmitted wave due to the destructive interference of the incident and the forward scattered waves*, or, in other words, *absorption*. Using the picture described one can claim that the coherent scattering of γ-ray photon does reveal itself in a transmission measurement, although in an indirect way. Nonetheless, it is worthwhile to emphasize that, *due to the constructive summation* of the wavelets scattered by nuclei in the forward direction, the resultant amplitude is *enlarged proportionally to the slab thickness* Δz. In the following we shall consider means how this quite strong radiative channel can be visualized directly.

16.2.2   Visualization of the Coherent Forward Scattering

16.2.2.1   Transient Scattering Processes

The preceding discussion forms the basis for the most direct way to visualize coherent nuclear scattering. The method is to suddenly stop the primary radiation and then to observe the delayed emission from the target slab. At the interruption of the incident beam, the external excitation of nuclei ceases. Let this moment be the origin of time, $t = 0$. The nuclei now start to de-excite spontaneously. We assume no order in the spatial arrangement of nuclei in the slab. Then a phase correlation between the secondary wavelets emitted by the nuclei in all directions, except the primary one, is absent, meaning the interference conditions are not fulfilled. As a result, the slab shows only a weak spontaneous emission into surrounding space at $t > 0$. The increase of emission probability is expected *only for the radiative channel in the forward direction* where the wavelets from all the nuclei are in phase irrespectively of the arrangement of the nuclei. Hence the angular distribution of γ-ray photons emitted by a slab must be quite different from that emitted by a single nucleus. In contrast to the isotropic emission from a single nucleus, the emission of γ-radiation from the slab must be sharply directed along the k-vector of the primary radiation. This is illustrated in Fig. 16.1.

The amplitude of the forward scattering is at resonance $A_s \sim A_i\mu\Delta z$, i.e., it becomes comparable to the amplitude of incident wave when the target thickness is of the order of the absorption length. Thus the nuclear excitation created in the slab of the appropriate thickness decays (when the shutter is closed) *predominantly via the radiative channel and a γ-ray photon is emitted in the forward direction.*

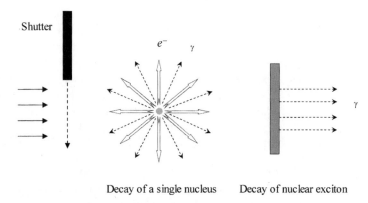

Decay of a single nucleus   Decay of nuclear exciton

Fig. 16.1 Spontaneous decay of nuclear excitation after the primary beam is shut off by closing the shutter

This is certainly correct for times just after the interruption moment, i.e., when the emitted radiation is not yet re-absorbed by nuclei in the slab. During this time, the radiative channel is highly enhanced due to the coherent action of all nuclei, whereas the conversion channel on the contrary is nearly closed. In close approximation, the time dependence of the coherent emission at resonance for $\mu \Delta z \lesssim 1$ is given by

$$I \approx I_0 \exp \left\{ -\frac{t}{t_0} \left(1 + \frac{\mu \Delta z}{4} \right) \right\} , \qquad (16.2)$$

where I_0 is the incident intensity. Re-absorption and re-emission of γ-radiation in the target were here taken into account. The time evolution of the spontaneous coherent emission is shown in Fig. 16.2. At the left-hand side the time dependence of the incident, exciting radiation is represented and at the right-hand side those of the transmitted (until time 0) and forward scattered radiation. The peak intensity of the spontaneously emitted γ-radiation is equal to that in the incident beam. This sharp peak is an evidence of the coherent decay of the nuclear exciton which proceeds mostly via the radiative channel at initial time. The drop of intensity at later times is caused by the resumption of the normal mode of propagation of the emitted γ-rays.

We see from the figure that the method presented does allow to observe the forward scattering and the enhancement of the radiative channel, but only *as a transient phenomenon*. These unique features could indeed be verified in an actual experiment [22], employing an original magnetic shutter for the γ-radiation. With its help, the beam could be shut off within 5–10 ns, i.e., quite fast compared to the mean lifetime of the excited state of ^{57}Fe nuclei. The coherent emission was detected as a sharp flash of γ-radiation in the forward direction similar to that displayed in Fig. 16.2.

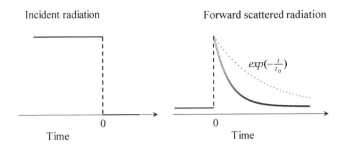

Fig. 16.2 Time dependences of incident and forward scattered radiation. Shutter is closed at time zero

Another way to allow the forward scattering to manifest itself, is to manipulate the interference between the primary and secondary waves as to make their amplitudes add positively. In other words, to transform the normally destructive interference between them into a constructive one. This could be achieved by a sudden change of the relative phase between the waves. The stepwise phase modulation was accomplished successfully by several authors for the ^{67}Zn resonance [23] and for the ^{57}Fe resonance [24], by a stepwise displacement of the source towards the absorber. After the prompt source displacement of $\Delta x = \lambda/2$, both primary and forward scattered waves got in phase resulting in a strong flash of transmitted intensity.

An interesting possibility to observe the coherent enhancement of the radiative channel was found with the help of the fast reversal of the hyperfine magnetic field. When the field is rapidly reversed (much faster than the Larmor precession), the nuclear sublevels appear reversed as well. The energy of the excited nuclei is shifted due to inversion of the levels and the nuclei concerned exhibit free decay. The coherent mechanism results in the directed and enhanced γ-ray emission. The interference between primary and the forward scattered waves leads to quantum beats in the transmitted intensity [25]. This represents a case of *inelastic coherent scattering*. The incoherent channel associated with the spin-flip turns out to be suppressed, as well as the internal conversion channel.

We did not yet consider the time evolution of nuclear excitation inside the target, our attention rather concentrated on the enhanced radiative response of the nuclear array. During a short time of the intensified re-emission of γ-ray photon, the delocalized nuclear excitation could actually be considered as an immobile formation. However, the further development of the temporal response can demonstrate the processes of the second, third, – multiple nuclear resonant scattering in the forward direction. The distributions of nuclear excitation and of γ-ray field strength develop in space and time as the scattering proceeds. This just represents the propagation mode of the nuclear polariton. In solid matter, a nuclear polariton propagating in the forward direction is always created when the nuclear ensemble is excited by a photon, irrespective of the structure of the nuclear ensemble.

16.2.2.2   Pulsed Excitation of Nuclei by Synchrotron Radiation

The best conditions under which the coherent radiative decay of nuclear exciton can be observed are realized by the pulsed excitation of the nuclear ensemble. This approach became possible with the creation of the powerful sources of synchrotron radiation (SR). The idea to apply SR to study nuclear fluorescence was put forward by Mössbauer [26, 27] and Ruby [28]. The first coherent response of a nuclear ensemble to an SR pulse was unequivocally observed by Gerdau et al. [29]. With the pulsed excitation, the processes of absorption and emission of γ-ray photon can be treated as two sequential, temporally decoupled events. The time and spatial distributions of γ quanta emitted after excitation are strictly dictated by the spontaneous development of the nuclear exciton polariton, whose initial phase and amplitude distribution inside the target are determined by the SR pulse. When an SR pulse is spatially a plane wave falling on the target at off-Bragg angle, the nuclear polariton possesses propagating plane wave behavior, resulting in an emission of the γ-ray photon in the forward direction [30]. The first observation of this process was accomplished in [31]. If, however, the excited nuclear ensemble as a whole changes its angular position within the time interval between excitation and emission then the γ-ray photon is emitted in a new direction, fixed by, besides external conditions, the excitation-induced phase memory. This effect was observed in [32] and called the *nuclear lighthouse effect*. It is another convincing example of the temporal and spatial phase preservation in a nuclear exciton polariton. Many other coherent phenomena in the nuclear resonant scattering of SR have been found and are extensively described in the collection of papers [33] devoted to that subject.

Among those phenomena, two types of beating features in the forward scattering intensity deserve special attention. The obvious consequence of hyperfine splitting of the resonance is quantum beating of the forward scattered intensity due to interference of the hyperfine components [34]: the larger the spacing of the spectral components, the higher the beating frequency. In this manner, a hyperfine structure of a Mössbauer spectrum produces its image in the time domain. The less obvious feature is the so called propagation beating. This pattern is specific for the *temporal response of a resonant nuclear target* [30, 35]. Propagation beating just reflects the dynamics of the interaction of the γ-ray photon with the nuclei, and in particular, the multiple process of absorption and re-emission of a quantum in its passage through the nuclear system [10]. For that reason this temporal response is also called *dynamical beating* [21].

We have shown in the preceding paragraphs that the radiative channel can indeed play a decisive role in the interaction of γ-rays with nuclei. However, its dominance in the methods discussed can be observed only within a certain time span as a transient phenomenon, where the enhancement of γ-ray emission is provided by the collective coherent response of nuclear array. In the following, we shall discuss the possibilities to make use of coherent scattering under steady state conditions.

16.2.2.3 Steady State Enhancement of γ-Ray Reemission: Scattering from a Target Vibrated by Ultrasound

In a target vibrated sinusoidally with the appropriate ultrasound frequency the resonance line is split: a set of equidistant satellite resonances appears in the absorption spectrum [36]. The additional resonance lines are associated with the process of absorption of a γ-ray photon simultaneously with creation or annihilation of acoustic phonons whose number in the phonon spectrum is greatly enlarged by the application of ultrasound. One can easily imagine that in the presence of ultrasound, the re-emission of γ-rays from the target can also be accompanied by the analogous processes. The resonance spectrum of the emitted radiation then should consist of a set of equidistant components as well. An energy analysis of the radiation emerging from the target subjected to ultrasound vibrations confirmed these expectations [37]. The same expectations were the basis of an experiment where the enhancement of the radiative channel in the vibrating target was studied [38, 39]. Radiation from a Mössbauer single line source passed through a target of resonant nuclei to which ultrasound could be applied by a piezoelectric crystal. A second resonant absorber was used to measure the spectrum of radiation emerging from the target. The first target was chosen to be thick enough to absorb nearly completely the resonant radiation from the source. This feature of the target is illustrated in Fig. 16.3 where the spectrum of the radiation from the source (Fig. 16.3a) is compared with the spectrum of radiation emerging from the target in the absence of ultrasonic vibrations (Fig. 16.3b). Only small amount of radiation located at the wings of the resonance, could actually pass through the target. The picture changes drastically when ultrasound is applied to the target. The emerging radiation contains sidebands separated from the original resonant frequency by multiples of the ultrasound frequency (Fig. 16.3c). The measured form of spectrum is interpreted in the following way: the nuclei in a vibrating target can be represented as having their ground and excited state energy levels

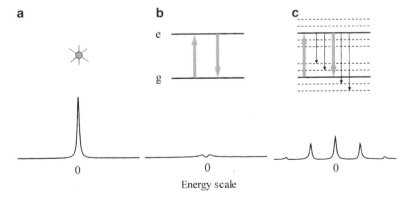

Fig. 16.3 Transmission of γ-radiation through a resonant target under ultrasound vibration

split into two sets of energy levels, with equal spacing $\hbar\omega_{ph}$ due to presence of a sharp peak in the phonon spectrum at angular frequency ω_{ph}. In the ground state, absorber nuclei are distributed between the ground state sublevels, according to the number of quanta of energy of vibration. The allowed transitions involving a change in phonon number of $n = 0, \pm 1, \pm 2...$ are represented in the transition scheme of Fig. 16.3c. Thus, a photon of energy corresponding to the absorber resonance energy can produce forward scattered photons of the unchanged energy and photons having their energy shifted by $\pm n\hbar\omega_{ph}$. Each scattering channel is being realized with a certain probability amplitude. In [38, 39], it is shown that the entire intensity of the radiation scattered into the sidebands essentially exceeds the limit allowed by internal electron conversion. In this manner, a *steady state enhancement of the radiative channel* is achieved via coherent inelastic scattering of γ-radiation from the array of nuclei in the vibrating target. A detailed interpretation of the phenomenon is given in [39].

16.2.3 Nuclear Resonant Bragg Scattering: Peculiar Features

16.2.3.1 Total Reflection

So far we have considered mostly the coherent scattering in the forward direction, while at the same time a nuclear array in a crystal represents for the γ-radiation a resonating three-dimensional grating. Hence, spatial interference in resonant scattering by the individual nuclei must be expected. When the incident wave meets the planes of nuclei at the correct angle for reflection, a reflected wave is built up due to constructive interference. Nuclear resonant Bragg reflection takes place. The reflected wave inside the crystal will also meet the same planes and will be reflected a second time, back into the direction of the primary beam. If regularity of the planes persists over a large volume of the crystal, multiple scattering of radiation forth and back occurs. Mutual interference of the propagating and reflected waves produces a resultant wavefield whose structure is now of standing wave type. The nodal and antinodal planes of the interference wavefield are parallel to the reflecting planes of the nuclear grating, and the wavelength of the standing wave is equal to or fractional of the spacing of the planes. Thus, the standing wave mode of the nuclear exciton polariton is realized under conditions of Bragg diffraction. A detailed account of the dynamical diffraction theory is given by Kagan, Trammell and Hannon in their reviews [19, 20].

   In development of the dynamical diffraction theory, the great surprise for the theoreticians was to find 100% reflectivity of resonant γ-rays from nuclear lattice [40] at Bragg angle. Indeed, high reflectivity looks improbable if one takes into account that the intermediate state decays only to a small percentage by re-emission of γ-radiation. Most of the radiation should be lost into incoherent decay channels via internal conversion and spin-flip processes. Furthermore, losses can be expected from the diffraction process itself, which gives rise to incoherent channels due to isotopic and spin incoherence. In the case of ^{57}Fe, the losses into the incoherent channels should amount to at least 95%. Under these circumstances, it was hard to imagine that for resonance Bragg diffraction of γ-rays high reflectivity might ever be reached.

The main reason of the high reflectivity turned out to be a suppression of all inelastic channels of nuclear scattering. The *effect of suppression of inelastic channels of nuclear reaction* is also called the *Kagan-Afanas'ev effect* [41]. The nature of the suppression is related to the formation of the interference field described above. Due to the nodal-antinodal structure of the wavefield inside the crystal *the amplitude of nuclear excitation is strongly modulated in space.* In case of magnetic dipole transition (i.e., the ^{57}Fe nucleus) the amplitude of the nuclear excitation is proportional to the scalar product of the magnetic vector in the wavefield **H** at the nucleus and the magnetic moment $\boldsymbol{\mu}$ of the induced nuclear transition. If the magnetic vector of the summary field at the position $\mathbf{r}_a$ drops to zero, or if it is at right angle to $\boldsymbol{\mu}$ the excitation amplitude becomes zero. Such conditions can be fulfilled at the reflecting nuclear planes where in this case nuclear absorption vanishes. We shall deal in more detail with suppression effect in a later section.

And now look – we face an obvious logical confusion: the total reflection of γ-radiation is only possible because of the full suppression of nuclear absorption. The latter is caused by forcing the excitation amplitude to zero. One should then conclude that the entire reflection takes place under conditions in which the excitation amplitude becomes zero, which sounds really puzzling. This intriguing question is discussed in the work [42] where high nuclear reflectivity is examined.

In the case of resonant scattering of γ-rays, the wave which was reflected twice is in phase with the primary wave, and their interference is constructive. Therefore, *in the case of resonant scattering there is no primary extinction* as is typical for X-rays scattering. The destructive interference between the primary and twice reflected X-ray waves occurs within the finite angular interval near Bragg angle where the radiation is practically prevented from penetrating into the crystal. In contrast, *the resonant γ-radiation is allowed to penetrate deep into the crystal* when the *Bragg angle is approached*. The question now concerns the entire absorption and reflection of the wavefield over the increased penetration depth. It will be determined by the total number of nuclei $N \sim 1/\sqrt{\alpha}$, and by the field amplitude at the positions of nuclei $\mathbf{H}(\mathbf{r}_a) \sim \sqrt{\alpha}$ with α being the deviation from Bragg angle. When absorption is considered, the contributions of all nuclei are summed *incoherently*. The entire absorption then turns out to be given by

$$\text{entire absorption} \sim N \times |\mathbf{H}(\mathbf{r}_a)|^2 \sim \left(1/\sqrt{\alpha}\right)\alpha = \sqrt{\alpha}$$

and, hence, is diminishing proportionally to $\sqrt{\alpha}$. At the exact Bragg position, the entire absorption completely vanishes. When diffraction is considered, the contributions of all nuclei are summed *coherently*. The reflection is then the square of the summary scattering amplitude

$$\text{Reflection} \sim |N \times \mathbf{H}(\mathbf{r}_a)|^2 \sim \left|(1/\sqrt{\alpha})\sqrt{\alpha}\right|^2 = 1$$

In spite of the vanishing wavefield on approach to the Bragg position, reflection does not vanish. In contrast, it becomes 100% at the exact Bragg position. Thus, we see that *the coherent scattering of all nuclei is able to maintain diffraction, even though the field amplitude at the place of each individual nucleus vanishes.* We see further how the coherent enhancement leads to the absolute domination of the radiative channel.

16.2.3.2 Resonance Broadening

An amazing peculiarity of nuclear Bragg reflection of Mössbauer γ-rays is the existence of a combined energy and angular dependent response of a crystal containing the nuclear array. Landscape of a Bragg reflection as two dimensional function in the vicinity of the nuclear resonance and of the Bragg angle is displayed in Fig. 16.4. At the exact Bragg angle and at resonance, total reflectivity is reached. The high reflectivity still remains in a large angular interval near the resonance and in a large energy range close to Bragg angle. One may notice that a particular reflectivity is reached in the two symmetrically situated energy-angular domains, framed by the negative–negative/positive–positive energy and angular coordinates. Such a symmetry is due to varying conditions of the interference of the waves scattered by nuclei and is related to the fact that the real part of the nuclear amplitude is an odd function of the energy deviation from resonance. As may be seen from Fig. 16.4, the width of the resonance increases rapidly in approaching Bragg angle. In the very vicinity of the Bragg angle the resonance line covers the energy range of about 100 natural widths of nuclear level. Resonance broadening of that order was obtained experimentally (see, e.g., [9, 43]) and provides a strong manifestation of the coherent enhancement of the radiative channel.

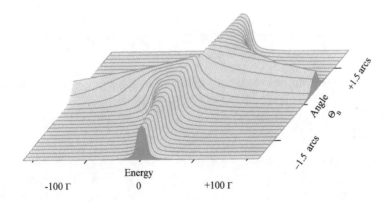

Fig. 16.4 Energy and angular dependence of nuclear resonance Bragg scattering

For the interpretation of the large energy broadening of the reflectivity next to the Bragg angle, the following argument may be used. As outlined in the preceding paragraph, the resonant γ-radiation is allowed to penetrate deep into the crystal when the Bragg angle is approached. The nearer to the Bragg angle radiation falls onto the crystal, the more lattice planes scatter in phase. But the more lattice planes contribute constructively to the diffraction, the smaller the scattering amplitude may be in order to maintain a given level of reflectivity. Therefore, the deviation from resonance where that reflectivity level can be reached gets larger. As a result, the energy broadening becomes more significant the closer the Bragg angle is approached. The observed energy broadening may be interpreted as being *due to an enlarged phase conservation* in the vicinity of the Bragg angle.

16.2.3.3 Interference of Nuclear and Electronic Scattering

A γ-ray photon is scattered recoilless by a crystal through either nuclear resonant or Rayleigh electronic scattering processes. A wonderful thing is that even though Rayleigh scattering is an atomic process which occurs instantaneously and Mössbauer scattering is a nuclear process which covers a large time interval, 1.4×10^{-7} s in the case of ^{57}Fe, the processes may interfere with each other [2,3]. The necessary and sufficient conditions for interference between two or more processes are that *they transform an initial state into the same final state and that the measurement does not permit one to decide which process has caused the transformation*. The processes are called coherent in this case.

Let us assume in a Gedanken Experiment that the source nucleus is formed at $t_0 = 0$. It may emit a γ-ray at $t = t_1 > 0$ which then is scattered by a target instantaneously through Rayleigh scattering. Or the source nucleus may emit a γ-ray at $t = t'$ with $0 < t' < t_1$, which then is resonantly absorbed by a nucleus in the target at the same time t' and re-emitted at time $t = t_1$. Since it is impossible in such an experiment to determine whether the scattering has taken place through a fast or a slow process, the criteria for the interference mentioned above are fulfilled.

The result of that interference is determined by the phase relationship between the amplitudes of Rayleigh electronic and of nuclear resonant scattering. Within the typical energy range for nuclear resonance, the electronic scattering is represented by a phase-lag of π in the scattered radiation, while the phase-lag in nuclear channel is varying from 0 to π in going through resonance. Besides, the probability amplitude for a γ-ray resonance scattering in this range has Lorentzian shape. Thus, below the resonance the electronic and nuclear channels interfere destructively, whereas above the resonance they interfere constructively. Due to this interference between the two processes, an asymmetric, dispersion-like shape of resonance line is produced (see, e.g., the review [15]). In a polyatomic crystal, the group of atoms constituting a unit cell contribute to the summary amplitude. The phases of the contributing waves are included in the structure factors for specific lattice types which are in general complex numbers and specified by a particular reflection and the atomic positions in the unit cell of a crystal. To illuminate the sensitivity of

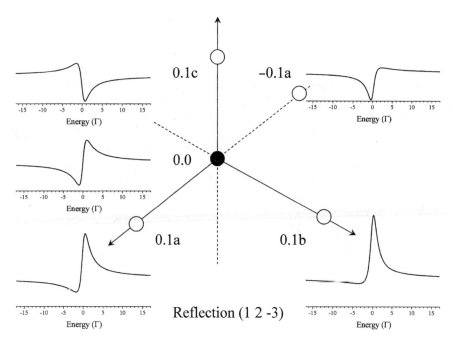

Fig. 16.5 Interference of nuclear resonant and Rayleigh electronic scattering of γ-ray photon for different mutual positions of nucleus and electronic charge in a crystal unit cell; a, b, c are dimensions of an orthorhombic unit cell

the nuclear–electronic interference to the mutual positions of a nucleus and of an electronic charge within a unit cell we have simulated the interference patterns for a chosen reflection and the five different positions of the electronic charge with respect to the nucleus. The result is shown in Fig. 16.5. The displacements from the origin, where the nucleus situated, are given on the lengths scale of the unit cell sides. The drastic change of the pattern is seen in the transition from one case to another not only in contrast the dispersion shape of the intensity line but also in its asymmetry. For example, the dark–light intensity sequence is changed to a light–dark sequence in the transition from case "0.1a" to "0.1c". The idea to employ the nuclear–electronic interference for solving the phase problem of crystal structure determination was put forward and developed in [44–46].

In the presence of a hyperfine splitting of nuclear levels, the nuclear–electronic interference is also sensitive to the strength and orientation of the internal magnetic and electric fields. The unusual shapes and transformation of the interference pattern are shown in Fig. 16.6, where the calculated Mössbauer spectra are represented in the case of the Bragg diffraction of γ-rays from a single crystal of $^{57}FeBO_3$, reflection (444), for the two orthogonal orientations of magnetic fields on nuclei. Peaks at resonance energies appear when the nuclear amplitude dominates over the electronic one, while dips indicate dominance of the electronic amplitude. Such types of spectra were investigated in [47].

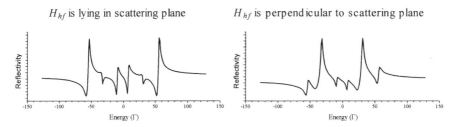

Fig. 16.6 Interference of nuclear resonant and Rayleigh electronic scattering channels for different orientations of the magnetic field on nuclei

16.2.4 Suppression Effect

16.2.4.1 Anomalous Transmission

The total reflectivity of γ-rays from a resonant crystal, discussed above, is a forceful manifestation of the suppression effect (SE), which allows γ-radiation to penetrate deep into a resonating crystal. The nuclear exciton polariton exhibits standing wave behavior in this case. An anomalous transmission of γ-radiation through a thick absorber is another impressive appearance of the SE. This phenomenon can be observed well in the so-called Laue geometry of diffraction, where the reflected beams emerge from the back side of a crystal as shown in Fig. 16.7. The atomic planes are assumed to be perpendicular to the crystal surface. The incident beam gives rise to an energy flow within the crystal, propagating in this case normal to the surface and parallel to the crystal planes toward the back surface of the crystal. On emergence, this gives rise to the Laue transmitted (LT) and Laue reflected (LR) beams. A standing wave structure of nuclear polariton in the direction perpendicular to the reflecting planes is established inside the crystal due to multiple scattering of radiation. Such a state of nuclear polariton is an eigensolution of the relevant Maxwell wave equations. In contrast to the Bragg geometry of reflection, two different types wavefields are produced in Laue reflection. While the one, we call it WF1, is weakly coupled to the nuclei, the other (WF2) is coupled strongly to the nuclei, much more than in usual transmission. The energy splits equally between the two fields in the symmetric geometry of scattering. Therefore, under the conditions of full suppression, only half of the incident radiation can penetrate anomalously through the crystal. As the crystal is rotated into the range of the Bragg angle, the nuclear planes are brought into the proper position for nuclear reflection. The reflection occurs over a certain angular interval where the transmitted intensity maxima are detected in both the LT and LR beams, see Fig. 16.7. Since in the off Bragg propagation, there is no intensity behind the crystal because of its large thickness, the observed *intensity peaks reveal an anomalous transmission of γ-radiation through the crystal*. The energy analysis of the transmitted radiation (right-hand side of Fig. 16.7) fully confirms the anomalous transmission of the resonant quanta: a prominent line of the resonant radiation is observed behind the crystal when the latter is set exactly at the Bragg angle. But

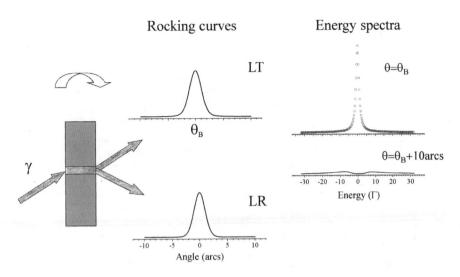

Fig. 16.7 Anomalous transmission of γ-radiation through a resonant crystal

only the far wings of the Lorentzian distribution of the incident radiation are seen when the crystal is tilted off Bragg angle just by 10 arc seconds. Pictures of this type were observed in [48].

A direct observation of the suppression effect was accomplished in [49], where a sharp drop in conversion electron yield from an ^{57}Fe single crystal was measured as soon as the standing wave mode of the nuclear polariton had been built up in the crystal.

There are many peculiarities of the Kagan–Afanas'ev effect which render it different from the analogous effect in X-ray optics,- the Borrmann effect. The most striking peculiarity is an independence of the SE on the thermal oscillations of the atoms. This property is related to the existence of a long-lived nuclear polariton state: during the lifetime of nuclear excitation γ-radiation sees nuclei in their average, i.e., equilibrium positions, right at the place where the excitation amplitude vanishes. The relevant experiment is described in [50]. Another important feature of the SE is the possibility to convert almost the entire incident resonant radiation into the anomalously transmitting state (WF1), using the strong polarization dependence of nuclear resonant scattering [51]. One more distinction of the SE will be considered in the next paragraph.

16.2.4.2 Pendellösung in Energy and Time Domains

The stronger the deviation from nuclear resonance the smaller the absorbing ability of the nuclei causing an increase of the probability for a γ-ray photon to appear in the WF2 state at the exit of crystal. Therefore, at the wings of resonance one can expect to observe interference of the two photon states WF1 and WF2. An

Laue reflection

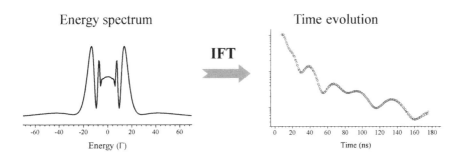

Fig. 16.8 Pendelösung in the energy and time domains of a Laue reflected beam. ITF Inverse Fourier Transform

example of such an interference phenomenon is shown in Fig. 16.8. At the left hand panel, the spectrum of the Laue reflected γ-ray beam is displayed. It is assumed that the incident radiation has a white spectrum and falls onto the crystal in the vicinity of a Bragg angle. The interference pattern exhibits pronounced oscillations symmetrically on either side of nuclear resonance. In our example, the first oscillation appears at a distance from resonance of about $\pm 5\Gamma$, where the wave WF2 starts to penetrate through the crystal. The oscillation amplitude is rising until deviation from the resonance reaches $\pm 15\Gamma$, where the amplitudes of the fields WF1 and WF2 become comparable. Further on, the oscillation amplitude decreases. The progressive increase of the oscillation period in the pattern is an evidence of a rapid decrease of the phase difference between the interfering waves.

Nuclear resonance fluorescence manifests itself not only in the energy domain but in time domain as well. The two representations are mutually related by Fourier transformations (FT). The direct FT allows to find the spectral composition from a time-dependent resonance signal. Backward, the inverse Fourier transform (IFT) provides the transition from the energy to the time representation of resonance response. The IFT of the spectral composition shown in the left hand panel of Fig. 16.8 thus leads to an oscillatory interference pattern in the time domain representing the temporal beating of radiation propagating through the crystal in Laue reflection geometry. The beat period corresponds rigidly to the spacing of the fringes in the energy spectrum. The phenomena of interference represented in Fig. 16.8, *i.e the Pendellösung in the energy and time domains*, are characteristic for and exclusively present in resonance scattering.

An experimental verification most adequate for the IFT procedure has become possible only with the advent of the pulsed synchrotron radiation (SR) sources. The duration of SR pulse is much shorter than the life time of nuclear isomers, and consequently SR exhibits a white spectrum in the range of nuclear γ-resonance.

16.2.5   Speed Up Effect

The first aim of application of SR for studying nuclear resonance fluorescence was to answer the fundamental question about the lifetime the collective nuclear excitation. The broadening of a nuclear level, associated to the coherent collective excitation leads to the increase of the decay probability of the excited nuclear state. As a result, one should expect a *speed-up* of γ-ray photon emission, directly related to the shortening of the lifetime of the collective excited state.

While determining the true lifetime of nuclear excitation one should escape the other reasons of the resonance broadening, e.g., inhomogeneous distribution of the resonance energy over the sample, fluctuation of magnetic fields at nuclei or simply resonance self-absorption of γ-radiation, which can also result in an accelerated decay of the coherent signal. Obviously these reasons can easily hide the effect under investigation. The most pure conditions for observation of the *speed-up effect* are fulfilled in the Bragg reflection from a homogeneous nuclear grating, where the suppression effect excludes also the resonance self-absorption.

The nuclear resonance lines in the vicinity of a Bragg angle are shown in the left-hand panel of Fig. 16.9. The combined energy and angular dependent response of the nuclear array, displayed in Fig. 16.4, is integrated over the symmetric angular limits $\theta_B \pm 3$ arcs and $\theta_B \pm 0.3$ arcs. The breadth of nuclear resonance is increasing from $\sim 8\Gamma$ to $\sim 24\Gamma$ when the angular region centered at the Bragg angle is narrowed. The broadened lines of the collective nuclear resonance are no longer of the Lorentzian form. The resonance response of an isolated nucleus (the narrowest line) is also shown for comparison. The coherent broadening effect is evident: the nearer to a Bragg angle beam is falling on the nuclear grating, the wider the breadth of the resonance. Since the partial width of the conversion channel stays unchanged, the partial width of the radiative channel (in our example that of ^{57}Fe) increases by about 80 and 240! times correspondingly.

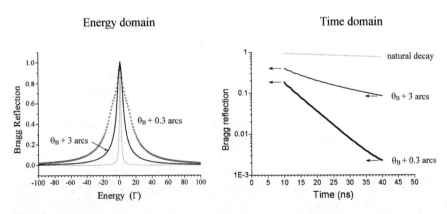

Fig. 16.9 Collective nuclear resonance at Bragg scattering exhibiting large broadening of the resonance linewidth and significant speed-up of the nuclear de-excitation

The time distributions of γ-ray photons emitted by the nuclear array under similar conditions are displayed on the right-hand panel of Fig. 16.9 in semi-log scale. The uppermost straight line represents the natural decay of a ^{57}Fe nucleus, the excited-state energy distribution of which is described by the Lorentzian function. Since that is not the case under diffraction conditions, the time distributions for diffraction appear as complex nonexponential decay functions. During the time interval from 10 to 40 ns, where the intensity of the emitted radiation falls down significantly, we can approximate the decay evolution by exponential functions and estimate the effective lifetime parameters. As the incident radiation concentrates at the Bragg angle, the lifetime of nuclear excitation drastically decreases from 141.2 ns first to $\sim$17 ns and then to $\sim$6 ns. This is a truly amazing effect of quantum phase coherence!

One can of course, make use of the uncertainty principle to explain the decrease of the lifetime by broadening of the resonance. But let us try to find the physical reason of the *speed-up* effect directly in the real time representation. The picture looks as follows. The nuclei of a certain layer in target excited initially by the SR flash are then exposed to the coherent delayed resonant radiation coming from the upstream nuclear layers, i.e., the *decay of these nuclei is driven by the delayed coherent radiation*. Just this coupling of the coherent radiation with the nuclei stimulates at the initial stage γ-ray photon emission. In other words, the coupling induces decay of the excited state. The physical nature of this process is similar to that of "Dicke superradiance". But, in contrast to the Dicke case we have here dealt mostly with a single photon excitation and while describing a single photon scattering event we regard in fact the coherent superposition in time of the probability amplitudes of different scattering channels.

Both a huge broadening of the resonance line and a drastic speed-up of the nuclear de-excitation being impressive manifestations of the collective nuclear resonance were convincingly presented experimentally in the works [9, 52, 53].

16.2.6 On Synchrotron Mössbauer Source

This paragraph represents an example of how a Spiral Model can be applied to describe the evolution of our knowledge about nature. The concepts of coherence for Mössbauer radiation were initially developed in the energy domain, the usual mode in Mössbauer spectroscopy. Then "on the coherence wave" a trip was made from energy to time domain where the nuclear gamma resonance exhibits new complementary features. Presently we, continue moving along an evolutionary spiral, came to the energy domain again but at the upper spiral arc.

The pulsed excitation of a nuclear array by an SR flash lets the nuclei generate and spontaneously emit their own, delayed γ-radiation. In a coherent processes like the forward or Bragg scattering, the radiation is emitted by nuclei without recoil. Thus, the excited nuclear array could play a role of a perfect Mössbauer source, if the enormous problem of a huge non-resonant background would not exist. Even after maximal possible monochromatization of the primary radiation down to $\sim$1 meV, the non-resonant background absolutely dominates. Fortunately the solution of this problem can be found by making use of the so called pure nuclear Bragg reflections [54]. The strong polarization dependence of nuclear scattering allows to suppress completely the non-resonant background and extract resonant γ-radiation. However, the mentioned property of nuclear scattering can be employed only in the presence of hyperfine magnetic or quadrupole interaction, and just under these conditions the nuclear array works as a multi line radiator. Nevertheless a unique opportunity arises, in the case of a combined magnetic dipole and electric quadrupole splitting of nuclear level, where a single line spectrum can be obtained. In particular, for ^{57}Fe it is attainable in an Iron Borate crystal near the Néel point [55]. One deals here with a quite complicated case of multispace interference where 48 paths in geometric, energy and spin spaces are involved in building up the interference pattern of γ ray emission.

The first source of Synchrotron Mössbauer (SM) radiation from ^{57}Fe nuclei has been developed at ESRF in 1997 [56]. After almost 10 years with the use of principally the same scheme, an SR-based ^{57}Fe- Mössbauer spectrometer with a γ-ray focusing device was constructed and set in operation in SPring8 [57]. Nowadays, a permanently working station of SM source in line with focusing mirrors is operating also on the beam line ID18 at ESRF. The test experiments lead to the following characteristics of 14.41 keV γ-radiation generated in the ESRF SM source: the intensity $\sim$40,000 ph/s, the resonant content 99.9%, resonance width $\sim$15 neV, polarization of about 99%, beam collimation 4$\times$ 8 arcs, the beam cross-section after focusing on a target $\sim$10$\times$16 μm [58].

As the first example of application of the SM source, the Mössbauer spectra measured with hematite and magnetite at high pressure are displayed in Fig. 16.10. Each spectrum was taken during 10 min only. Thanks to the high intensity and low background (pay attention on the resonant effect value), spectral details are perfectly apparent.

In addition to the studies at high pressures, the SM source can be used in many other scientific and technological applications where the collimated, polarized,

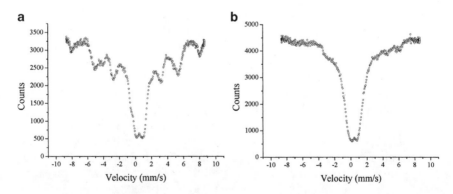

Fig. 16.10 Mössbauer spectra of enriched (**a**) Fe_2O_3, and (**b**) Fe_3O_4 samples taken at the pressure 70 GPa in a diamond anvil cell with the ^{57}Fe SM source in ESRF during 10 min each

highly intense, focused and 100% ^{57}Fe Mössbauer radiation is needed. The SM source is a remarkable application of γ-ray phase coherence where the entire knowledge accumulated up to now in γ-ray optics is employed.

References

1. W. Heitler, *The Quantum Theory of Radiation*, 3rd edn. (Oxford University Press, Oxford, 1954)
2. P.J. Black, P.B. Moon, Nature **188**, 841 (1960)
3. S. Bernstein, E.C. Campbell, Phys. Rev. **132**, 1625 (1963)
4. P.J. Black, I.P. Duerdoth, Proc. Phys. Soc. **84**, 169 (1964)
5. G.T. Trammell, *Proc. Intern. Atomic Energy Agency Symp. on Chemical Effects of Nuclear Transformations*, Vol 1, p75, 1961
6. A.M. Afanas'ev, Yu. Kagan, JETP Lett. **2**, 81 (1965)
7. D.F. Zaretskij, V.V. Lomonosov, JETP **21**, 243 (1965)
8. M. Haas, V. Hizhnyakov, E. Realo, J. Jõgi, Phys. Stat. Sol. (b) **149**, 283 (1988)
9. Yu.V. Schvyd'ko, G.V. Smirnov, J. Phys. Condens. Matt. **1**, 10563 (1989)
10. G.V. Smirnov, Hyperfine Interact. **123/124**, 31 (1999)
11. G.V. Smirnov, U. van Bürck, W. Potzel, P. Schindelmann, S.L. Popov, E. Gerdau, Yu.V. Shvydko, H.D. Rüter, O. Leupold, Phys. Rev. A **53**, 171 (2005)
12. G.V. Smirnov, U. van Bürck, J. Arthur, G.S. Brown, A.I. Chumakov, A.Q.R. Baron, W. Petry, S.L. Ruby, Phys. Rev. A **76**, 043811 (2007)
13. M. Haas, Phys. Lett. A **361**, 391 (2007)
14. V.A. Belyakov, Sov. Phys.-Usp. **18**, 267 (1975)
15. G.V. Smirnov, Hyperfine Interact. **27**, 203 (1986)
16. E. Gerdau, U. van Bürck, in *Resonant Anomalous X-Ray Scattering*, ed. by G. Materlik, C.J. Sparks, K. Fisher (Elsevier, Amsterdam, 1994), p. 589
17. G.V. Smirnov, A.I. Chumakov", in *Resonant Anomalous X-Ray Scattering*, ed. by G. Materlik, C.J. Sparks, K. Fisher (Elsevier, Amsterdam, 1994), p. 609
18. G.V. Smirnov, Hyperfine Interact. **97/98**, 551 (1996)
19. Yu. Kagan, Hyperfine Interact. **123/124**, 83 (1999)
20. J.P. Hannon, G.T. Trammell, Hyperfine Interact. **123/124**, 127 (1999)
21. U. van Bürck, Hyperfine Interact. **123/124**, 483 (1999)
22. Yu.V. Shvyd'ko, G.V. Smirnov, S.L. Popov, T. Hertrich, JETP Lett. **53**, 69 (1991)
23. E. Ikonen, P. Helistö, T. Katila, K. Riski, Phys. Rev. A **32**, 2298 (1985)
24. R. Koch, E. Realo, K. Rebane, J. Jógi, in *Proc. Intern. Conf. on the Applications of the Mössbauer Effect*, 1983
25. Yu.V. Shvyd'ko, S.L. Popov, G.V. Smirnov, J. Phys. Condens. Matter **5**, 1557 (1993)
26. R.L. Mössbauer, Naturwissenschaften **60**, 493 (1973)
27. R.L. Mössbauer, in *Anomalous Scattering*, ed. by S. Ramaseshan, S.C. Abrahams (Munksgaard, Copenhagen, 1975), p. 463
28. S.L. Ruby, J. Phys. (Paris) C **6**, 209 (1974)
29. E. Gerdau, R. Rüffer, H. Winkler, W. Tolksdorf, C.P. Klages, J.P. Hannon, Phys. Rev. Lett. **54**, 835 (1985)
30. Yu. Kagan, A.M. Afanas'ev, V.G. Kohn, J. Phys. C Solid State Phys. **12**, 615 (1979); Erratum: J. Phys. C Solid State Phys. **12**, 615 (1979)
31. J.B. Hastings, D.P. Siddons, U. van Bürck, R. Hollatz, U. Bergmann, Phys. Rev. Lett. **66**, 770 (1991)
32. R. Röhlsberger, T.S. Toellner, W. Sturhahn, K.W. Quast, E.E. Alp, A. Bernhard, E. Burkel, O. Leupold, E. Gerdau, Phys. Rev. Lett. **84**, 1007 (2000)

33. E. Gerdau, H. de Waard (eds.) *Nuclear Resonant Scattering of Synchrotron Radiation*, Baltzer Science Publishers, in Hyperfine Interact. **123–125** (1999/2000)
34. T. Trammell, J.P. Hannon, Phys. Rev. B **18**, 165 (1978)
35. U. van Bürck, D.P. Siddons, J.P. Hastings, U. Bergmann, R. Hollatz, Phys. Rev. B **46**, 6207 (1992)
36. T.E. Cranshaw, P. Reivari, Proc. Phys. Soc. **90**, 1059 (1967)
37. J. Asher, T.E. Cranshaw, D.A. O'Connor, J. Phys. A Math. Nucl. Gen. **7**, 410 (1974)
38. S.L. Popov, G.V. Smirnov, Yu.V. Shvyd'ko, Hyperfine Interact. **58**, 2463 (1990)
39. Yu.V. Shvyd'ko, G.V. Smirnov, J. Phys. Condens. Matter **4**, 2663 (1992)
40. Yu. Kagan, A.M. Afanas'ev, I.P. Perstnev, JETP **27**, 819 (1968)
41. A.M. Afanas'ev, Yu. Kagan, JETP **21**, 215 (1965)
42. H.J. Maurus, U. van Bürck, G.V. Smirnov, R.L. Mössbauer, J. Phys. C Solid State Phys. **17**, 1991 (1984)
43. U. van Bürck, G.V. Smirnov, R.L. Mössbauer, H.J. Maurus, N.A. Semioschkina, J. Phys. C Solid State Phys. **13**, 4511 (1980)
44. P.J. Black, Nature **206**, 47 (1965)
45. F. Parak, R.L. Mössbauer, U. Biebl, H. Formanek, W. Hoppe, Z. Physik **244**, 456 (1971)
46. R.L. Mössbauer, in *Proc. 18th Ampére Congr. on Magnetic Resonance and Related Phenomena*, vol.1, 1974
47. A.N. Artem'ev, I.P. Perstnev, V.V. Sklyarevskii, G.V. Smirnov, E.P. Stepanov, JETP **37**, 136 (1973)
48. U. van Bürck, G.V. Smirnov, H.J. Maurus, R.L. Mössbauer, J. Phys. C Solid State Phys. **19**, 2557 (1986)
49. G.V. Smirnov, A.I. Chumakov, JETP **62**, 673 (1985)
50. G.V. Smirnov, Yu.V. Shvyd'ko, U. van Bürck, R.L. Mössbauer, Phys. Stat. Sol. b **134**, 465 (1986)
51. G.V. Smirnov, U. van Bürck, R.L. Mössbauer, J. Phys. C Solid State Phys. **21**, 5835 (1988)
52. U. van Bürck, R.L. Mössbauer, E. Gerdau, R. Rüffer, R. Hollatz, G.V. Smirnov, J.P. Hannon, Phys. Rev. Lett. **59**, 355 (1987)
53. A.I. Chumakov, G.V. Smirnov, A.Q.R. Baron, J. Arthur, D.E. Brown, S.L. Ruby, G.S. Brown, N.N. Salashchenko, Phys. Rev. Lett. **71**, 2489 (1993)
54. G.V. Smirnov, V.V. Sklyarevskii, R.A. Voskanyan, A.N. Artemev, JETP Lett. **9**, 123 (1969)
55. G.V. Smirnov, Hyperfine Interact. **125**, 91 (2000)
56. G.V. Smirnov, U. van Bürck, A.I. Chumakov, A.Q.R. Baron, R. Rüffer, Phys. Rev. B **55**, 5811 (1997)
57. T. Mitsui, N. Hirao, Y. Ohishi, R. Masuda, Y. Nakamura, H. Enoki, K. Sakaki, M. Seto, J. Synchrotron Rad. **16**, 723 (2009)
58. V.B. Potapkin, A.I. Chumakov, R. Rüffer, G.V. Smirnov, S.L. Popov, L. Dubrovinsky, C. McCammon, XXVI European Crystallographic Meeting, Darmstadt, Germany, 27–29 August 2010

Chapter 17
Nuclear Resonance Scattering of Synchrotron Radiation as a Unique Electronic, Structural, and Thermodynamic Probe

**E. Ercan Alp, Wolfgang Sturhahn, Thomas S. Toellner,
Jiyong Zhao, and Bogdan M. Leu**

17.1 Introduction

Discovery of Mössbauer effect [1] in a nuclear transition was a remarkable development. It revealed how long-lived nuclear states with relatively low energies in the kiloelectron volt (keV) region can be excited without recoil. This new effect had a unique feature involving a coupling between nuclear physics and solid-state physics, both in terms of physics and sociology. Physics coupling originates from the fact that recoilless emission and absorption or resonance is only possible if the requirement that nuclei have to be bound in a lattice with quantized vibrational states is fulfilled, and that the finite electron density on the nucleus couples to nuclear degrees of freedom leading to hyperfine interactions. Thus, Mössbauer spectroscopy allows peering into solid-state effects using unique nuclear transitions. Sociological aspects of this coupling had been equally startling and fruitful. The interaction between diverse scientific communities, who learned to use Mössbauer spectroscopy proved to be very valuable. For example, biologists, geologists, chemists, physicists, materials scientists, and archeologists, all sharing a common spectroscopic technique, also learned to appreciate the beauty and intricacies of each other's fields. As a laboratory-based technique, Mössbauer spectroscopy matured by the end of the 1970s. Further exciting developments took place when accelerator-based techniques were employed, like synchrotron radiation or "in-beam" Mössbauer experiments with implanted radioactive ions. More recently, two Mössbauer spectrometers on the surface of the Mars kept the technique vibrant and viable up until present time.

In this chapter, we will look into some of the unique aspects of nuclear resonance excited with synchrotron radiation as a probe of condensed matter, including

E.E. Alp (✉) · W. Sturhahn · T.S. Toellner · J. Zhao · B.M. Leu
Advanced Photon Source, Argonne National Laboratory, Argonne, Il 60439, USA
e-mail: Eea@aps.anl.gov

M. Kalvius and P. Kienle (eds.), *The Rudolf Mössbauer Story*,
DOI 10.1007/978-3-642-17952-5__17, © Springer-Verlag Berlin Heidelberg 2012

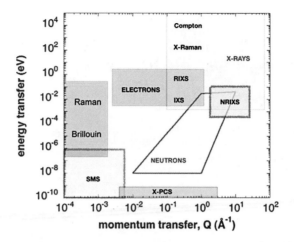

Fig. 17.1 The energy and momentum transfer relationship for several probes like photons, electrons, and neutrons. SMS technique, performed in forward scattering conditions, essentially imparts negligible momentum transfer, but it can distinguish hyperfine interactions with nanoelectron volt energies. NRIXS, on the other hand, imparts all of the momentum the photon has, so the momentum transfer range shown here covers different isotopes, without implying continuous tunability. The region *top-right*, labeled X-rays, indicate the possibility for other types of inelastic scattering, like momentum-resolved high-resolution (meV) inelastic X-ray scattering for phonon dispersion measurements, IXS, resonant inelastic X-ray scattering with 100 meV resolution for electronic excitations, RIXS, X-ray Raman scattering for near edge X-ray absorption spectroscopy of low-energy transitions under 1 keV using hard X-rays, and Compton scattering as an extreme form of inelastic X-ray scattering. X-PCS refers to X-ray photon correlation spectroscopy or speckle spectroscopy, a term inherited from laser spectroscopy

magnetism, valence, vibrations, and lattice dynamics, and review the development of nuclear resonance inelastic X-ray scattering (NRIXS) and synchrotron Mössbauer spectroscopy (SMS). However, to place these two techniques into some perspective with respect to other methods that yield related information, we display our version of a frequently used map of momentum and energy transfer diagram in Fig. 17.1. Here, various probes like electrons, neutrons, or light, i.e., Brillouin or Raman, and relatively newer forms of X-ray scattering are placed according to their range of energy and momentum transfer taking place during the measurements. Accordingly, NRIXS is a method that needs to be considered as a complementary probe to inelastic neutron and X-ray scattering, while SMS occupies a unique space due to its sensitivity to magnetism, structural deformations, valence, and spin states.

17.2 Synchrotron Radiation and Mössbauer Spectroscopy

For almost three decades after its discovery, Mössbauer spectroscopy was practiced using suitable radioactive parents decaying into the isotope of interest. For example,

^{57}Co with a half-life of 270 days decays into ^{57}Fe with an electron capture, populating two energy levels at 136.6 and 14.4 keV, the latter being very suitable for Mössbauer spectroscopic purposes.

Starting from 1974, upon a suggestion of S.L. Ruby at Argonne National Laboratory [2], several groups around the world started research programs to make use of synchrotron radiation to replace the radioactive parent isotope as a source. These studies gave their first fruits when a group led by E. Gerdau of Hamburg University made the first unambiguous observation of the excitation of the first energy level above the ground state in ^{57}Fe [3]. Since then, more than a dozen of these transitions have been observed, including Ta, Tm, Kr, Fe, Eu, Sm, Sn, Dy, Hg, I, K, Te, Sb, Ni, and Ge, given in the order of resonance energy (see Table 17.1).

For synchrotron-based studies, a strong radiation source is needed to have access to X-rays within the range of nuclear transition energies. Moreover, suitable diffractive and reflective optics to scan the photon energy around a particular nuclear resonance must be developed for each isotope separately. In this sense, a "good" radiation source can be defined as the one that provides sufficient number of photons in the energy range of interest of practical Mössbauer isotopes, namely, 6–100 keV. The incident radiation should arrive in a manner that is suitable for distinguishing nuclear resonant photons away from non-resonant radiation. This requires, first, a pulsed source with a sufficient time gap between each pulse commensurate with the nuclear lifetimes, typically from a few tens of nanoseconds to a microsecond. Second, this source should be tunable enough to cover a number of Mössbauer transitions in the energy range where the recoil-free fractions are still substantial. Finally, the radiation source should be bright enough to populate the excited states in sufficient numbers so that experiments can be accomplished in a reasonable time

Table 17.1 Mössbauer transitions observed by synchrotron radiation given with their transition energies, half-lives, and first reported observations

Isotope	Energy (keV)	Half-life (ns)	First published reference
^{181}Ta	6.2155	6,050	A.I. Chumakov, et al., Phys. Rev. Lett. **75**, 16384 (1995)
^{169}Tm	8.4013	4	W. Sturhahn, et al., Europhys. Lett. **14**, 821 (1991)
^{83}Kr	9.4035	147	D.E. Johnson, et al., Phys. Rev. B **51**, 7909 (1995)
^{57}Fe	14.4125	97.8	E. Gerdau, et al., Phys. Rev. Lett. **54**, 835 (1985)
^{151}Eu	21.5414	9.7	O. Leupold, et al., Europhys. Lett. **35**, 671 (1996)
^{149}Sm	22.496	7.1	A. Barla, et al., Phys. Rev. Lett. **92**, 066401 (2004)
^{119}Sn	23.8794	17.8	E.E. Alp, et al., Phys. Rev. Lett. **70**, 3351 (1993)
^{161}Dy	25.6514	28.2	A.I. Chumakov, et al., Phys. Rev. B **63**, 172301 (2001)
^{201}Hg	26.2738	0.63	D. Ishikawa, et al., Phys. Rev. B **72**, 140301 (2005)
^{129}I	27.770	16.8	Unpublished
^{40}K	29.834	4.25	M. Seto, et al., Hyperfine Interact. **141**, 99 (2002)
^{125}Te	35.460	1.48	H.C. Wille, et al., Europhys. Lett.**91**, 62001 (2010)
^{121}Sb	37.129	4.53	H.C. Wille, et al., Europhys. Lett. **74**, 170 (2006)
^{61}Ni	67.419	5.1	T. Roth, et al., Phys. Rev. B **71**, 140401 (2005)
^{73}Ge	68.752	1.86	M. Seto, et al., Phys. Rev. Lett. **102**, 217602 (2009)

period, like from a few minutes to a few days. Such pulsed X-ray sources are only available in the form of synchrotron radiation emitted from high-energy electron storage rings equipped with special devices called undulators. Currently, there are at least five different storage rings that fit this definition in the USA (the Advanced Photon Source, APS, Argonne, Illinois), France (European Synchrotron Radiation Facility, ESRF, Grenoble), Germany (Hasylab, PETRA-III, Hamburg), and Japan (Super Photon ring at 8 GeV, SPring-8, Hyogo and the National Laboratory for High Energy Physics, Accumulator Ring, KEK-AR, Tsukuba, Ibaraki).

17.3 Nuclear Resonance and X-ray Scattering

The scattering of X-rays from charged particles like electrons is described classically by the Thomson cross-section:

$$\sigma_T = \frac{8\pi}{3} r_o^2.$$

Here, r_o is the classical electron radius. The charge scattering from nuclei, on the other hand, will be reduced by a factor of 10^{-8} due to the mass difference between electrons and nuclei. However, this argument does not apply to the special case of a resonance between nuclear energy levels and the incident electromagnetic radiation. In such a case, nuclear resonant scattering cross-section is given by

$$\sigma_N = \frac{\lambda^2}{2\pi} \frac{1}{1+\alpha} \frac{2I_e + 1}{2I_g + 1}.$$

Here, λ is the wavelength of the photon, α is the internal conversion coefficient indicating the branching ratio between photon release and electron conversion, possibly leading to nuclear induced electronic fluorescence if the transition energy is higher than the relevant absorption edge or electron binding energy, and I_e and I_g are the spins of the excited and ground states of the nucleus. A comparison between σ_T and σ_N indicates that the latter can be two or three order of magnitude bigger, albeit over a very narrow energy range. In other words, nuclear resonance can be considered an excellent oscillator with a very high quality factor. This also may explain the reason for a rather late discovery of SMS because despite this favorable cross-section, the product of scattering cross-section with energy bandwidth remains overwhelmingly in favor of electronic scattering. Therein also lies the clue that distinguishes nuclear resonance from the electronic scattering background: it is necessary to use an alternative approach than energy discrimination, namely time discrimination.

A typical experimental set-up for time-resolved synchrotron-based Mössbauer experiments is given in Fig. 17.2. The details of optics may vary for each facility, but essentially remains the same for time-resolved experiments. However, there are also

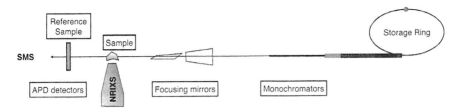

Fig. 17.2 A schematic layout for synchrotron Mössbauer spectroscopy and nuclear resonant inelastic X-ray scattering experiments. While the two detectors, SMS and NRIXS, can work simultaneously, they cannot record data continuously, because, for NRIXS, resonance is only excited for a very short period during the scans

other alternative approaches for using synchrotron radiation to record Mössbauer spectrum in the energy domain that may or may not involve time discrimination. They will not be discussed here. These methods include heating $FeBO_3$ crystal near Néel temperature to generate a single absorption line [4] and [5], a heterodyne [6], stroboscopic [7], or interferometric measurements [8]. Another interesting approach was reported recently by using a single-line analyzer mounted on a Mössbauer velocity drive [9]. A comparison between several of these methods has recently been given [10].

The two types of experiments we will discuss here share almost exactly the same optics. They are synchrotron Mössbauer spectroscopy, SMS [11], and nuclear resonant inelastic scattering, NRIXS [12] and [13]. The former is also called nuclear forward scattering, NFS [14], and the latter nuclear resonant vibrational spectroscopy, NRVS [15], or simply nuclear inelastic scattering, NIS [16].

Apart from the nomenclature, however, it should be noted that the two experiments mentioned, SMS and NRIXS, give very different information. SMS provides the possibility to measure hyperfine interactions leading to isomer shift, quadrupole splitting, and magnetic hyperfine field and its direction. NRIXS, on the other hand, measures lattice dynamics related quantities like recoil-free fraction or Lamb–Mössbauer factor, f_{LM}, phonon density of states, speed of sound, and all the thermodynamic quantities that are associated with the density of states, like vibrational specific heat, kinetic energy, Helmholtz free energy, force constants, and directional dependence of vibrational amplitudes in single crystals [17]. SMS experiments are carried out in the forward direction, while the NRIXS experiments are performed in scattering geometry. Thus, sample preparation and external conditions that can be applied depend somewhat on the geometry of the experimental method used. For example, thick samples are not an issue for NRIXS, while for SMS sample thickness plays a major role for a successful outcome.

Examples are provided in Figs. 17.3 and 17.4 to explain the SMS and NRIXS methods.

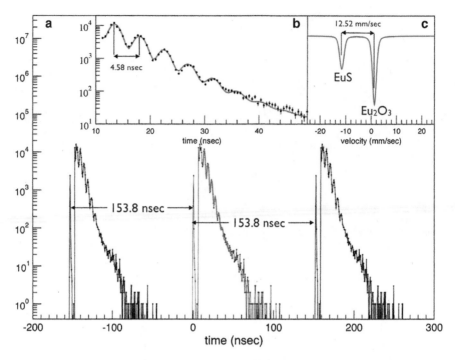

Fig. 17.3 (a) Typical data sequence in an SMS experiment: a pulse arriving to initiate nuclear absorption, followed by a natural or accelerated decay, and repeating itself with electron bunch frequency. This particular data is used to measure the isomer shift between EuS and Eu_2O_3 at the APS facility, where the electron bunches are separated by 153.8 ns, repeating itself with a rate of 270.7 kHz. The pulses at the origin are suppressed electronically in a logic gate to avoid the overload of the signal processing system, and they are shown here only partially. The APD detectors also take time to recover, and hence first 10 ns of the decay data are usually not accessible to the experimenters. (b) A theoretical fit to the data using CONUSS program. (c) A simulation in the energy domain based on the experimental fit shown in (b)

17.4   Synchrotron Mössbauer Spectroscopy

The proposal to use synchrotron radiation to replace parent isotopes [2] had captured the imagination of many experimentalists [3] because of several attractive features, including access to Mössbauer transitions without suitable parent isotope, like ^{40}K with 29.83 keV [18] or ^{73}Ge with 68.752 keV [19]. The unique polarized and collimated nature of high-brightness synchrotron radiation enables experiments under extreme conditions of pressure and temperature, which will be discussed later in detail.

To demonstrate the salient features of SMS, we make use of an experiment designed to establish Mössbauer parameters for EuS and Eu_2O_3 for further measurements of change in the isomer shift and phonon density of states of Eu-compounds as a function of pressure [20]. The data shown in Fig. 17.3 is recorded after X-rays

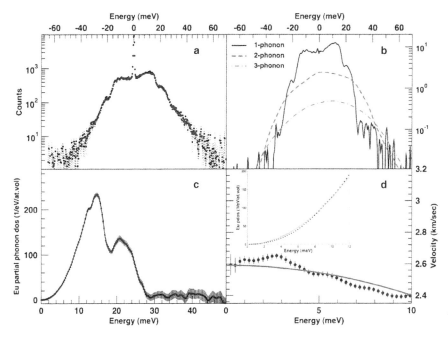

Fig. 17.4 (**a**) Phonon excitation probability of Eu_2O_3, measured at the 21.5414 keV [151]Eu Mössbauer resonance with 0.8 meV energy resolution. The central peak is the elastic contribution, while the negative and positive energies are the result of phonon annihilation and creation, respectively. (**b**) The decoupling of single and multi-phonon contributions, as described in [41]. (**c**) The single Eu-partial phonon density of states extracted and normalized to three degrees of freedom. (**d**) Sound velocity extracted from the low-energy part of the phonon density of states

pass through the sample and the reference. In Fig. 17.3a, a segment of events taking place between bunches are depicted. At the APS, the standard operation involves 24 bunch filling, where the bunch-to-bunch distance is 153.8 ns, repeating itself with approximately 6.5 MHz. The nuclear decay signal between bunches will be accumulated and added to give the spectrum shown in Fig. 17.3b. The data can be analyzed in time domain using an established simulation and fitting program. We made use of CONUSS, which is described in detail [21], to obtain the best fit, as well as to simulate a corresponding energy spectrum, as given in Fig. 17.3c. The energy difference due to the isomer shift between EuS (measured value of -11.49 ± 0.4 mm/s) and Eu_2O_3 (literature value of $+1.024$ mm/s) manifests itself as time beats with 4.58 ns period. For [151]Eu, the conversion factor for 1 mm/s Doppler shift is 17.3743 MHz or 71.8544 neV according to the Mössbauer effect data center compilation [22]. The results obtained here are consistent with a previous measurement [23].

The advantages of SMS can be summarized as follows:

1. High signal-to-noise ratio: there is an overwhelming advantage to SMS compared to conventional measurements in the laboratory due to the time-discrimination.

The entire decay signal arriving in the time period between the electron bunches registers with almost negligible background. Thus, even under most unfavorable conditions of dilution [24], micrometer-size samples, monolayer coating, and under extreme pressures (2 Mb) or temperatures (3,000 K), the SMS experiment will yield a good result [25].

2. The collimated nature of X-ray from the synchrotron renders microfocusing easy, thus creating a Mössbauer beam with 10μm or smaller dimensions which enables experiments inside the diamond anvil cell. This is sufficiently small for measurements above 1 Mbar pressure. Also, the small size of the X-ray beam eliminates the issues related to pressure gradients that beset previous Mössbauer spectroscopy measurements with point sources.

3. The measurements can be easily combined with in situ X-ray diffraction to ascertain the crystallographic structure and to capture structural phase transitions [26].

4. For absolute measurements of isomer shift using time-domain data, higher precision becomes possible as the theoretical fits to the data have more "pinch" points to match than the traditional spectra.

5. The duration of SMS experiments are typically less than an hour, fast enough to look systematically across a wide pressure, temperature, or composition range with sufficient frequency. With some advances in instrumentation, like the use of mechanical shutters, it is conceivable that the experiment times can be reduced to a few seconds or less [27].

6. Another important feature of the SMS experiments is related to the purity of the signal. Thus, even under very extreme conditions of dilution, and sample being surrounded by gasket material, a pressure medium and pressure marker, the only signal that comes between the successive synchrotron pulses is related to the Mössbauer isotope.

7. Hence, details of relaxation, field distributions, or thickness variations can all be addressed in the time domain. On the contrary, laboratory-based Mössbauer experiments do not have this level of selectivity or sensitivity.

A closer look at the scientific activities in the last decade reveals that SMS technique has been frequently adopted for high-pressure studies using, ^{119}Sn [28], ^{151}Eu [29], and ^{161}Dy [30] isotopes, in addition to ^{57}Fe and ^{83}Kr [31].

17.5 Mössbauer Spectroscopy as a Tool for Lattice Dynamics in the Synchrotron Era

Vibrational characteristics of solids, liquids, and gases have been an important tool in understanding the basic nature of the chemical bonds, as well as their collective behavior as atoms and molecules aggregate in liquid or solid form. Vibrational spectra provide important clues about local symmetry, the strength of forces acting on a particular atom of interest, the effect of ligands on the overall dynamics, and possibly chemical and biological activity of complex systems such as enzymes.

Many of the vibrational spectroscopic techniques have been developed in the twentieth century using light that covers a wide range of the electromagnetic spectrum: from far infrared ($\lambda = 15$–$1,000\,\mu$m), to visible ($\lambda = 390$–700 nm), and up to X-rays ($\lambda = 1$–0.01 nm). Techniques like FTIR (Fourier transform infrared) and Raman spectroscopy rely on coupling between the probing light and the spontaneous electric dipole moment generated by a particular vibrational mode. Since the momentum of long-wavelength photons is limited, the dispersion of vibrational frequencies cannot be observed, except for a very limited range near the Brillouin zone center using Brillouin light scattering. Thus, after a theoretical formulation of Weinstock [32], inelastic neutron scattering was demonstrated experimentally by Brockhouse [33] and almost immediately it was understood to be the method of choice to measure phonon dispersion relations, which led to an eventual Nobel Prize in 1994. Momentum dependence of the phonon frequency is highly specific information, providing a good metric for interatomic interactions in the solid and liquid state, which can be computed theoretically using empirical or first principles calculations.

The range and resolution of each technique differ depending on the instruments and sources used. In Fig. 17.1, we provide a modern version showing the energy and momentum transfer in several prominent experiments used for dynamical information.

Lattice dynamics, as directly related to recoil-free fraction, has always been the subject of Mössbauer spectroscopy studies. Almost all Mössbauer conferences include a special section on lattice dynamics studies. However, the so-called "f-factor" measurements which require diligent work on temperature dependence of the area under the resonance peaks reveal one meaningful parameter: average atomic displacement at a given temperature, through which many issues were then addressed. Measurement of the details of the vibrational spectrum remained in the domain of neutron scattering, and to a limited extent, of Brillouin or Raman spectroscopy.

However, from very early on, the narrow line width of Mössbauer resonances inspired scientists to measure the details of lattice vibrations [34,35]. Unfortunately, many attempts to tune the radiation emitted from the radioactive sources proved to be ineffective and the idea never became a practical tool. All this has changed in 1995 when it was demonstrated that it is possible to use tunable crystal monochromators to record an inelastic spectrum and extract phonon density of states [12, 13]. Since then, this field has expanded and several review articles have been written [36–38]. It is our intention in this chapter to articulate the unique aspects of vibrational spectroscopy application of nuclear resonant scattering with synchrotron radiation.

17.6   Why Another Vibrational Spectroscopy?

The dynamical behaviors of atoms are at the core of understanding the electronic, thermal, and elastic properties of materials and living organisms alike. Many of the elastic and thermal properties of materials depend on the distribution of occupational states in various energy levels, as the atoms vibrate around their equilibrium positions. Well-known quantities like Young's modulus of elasticity, Grüneisen parameter, compressibility, specific heat, and thermal conductivity are a reflection of force constants acting between atoms. Also, modern computational methods like density functional theory, DFT, and many variants of it can make many predictions about the forces holding materials together, but they all need to be verified using vibrational spectroscopy. Thus a number of techniques have been developed to study these properties. Among the photon-based probes, we can readily count Raman, Brillouin, and infrared spectroscopy. However, some of these techniques have limited applicability and suffer from the fact that photon energies are very low, and can carry very small momentum. A proper probe would have to cover effectively first and preferably higher Brillouin zones to account for forces acting at different length scales.

Optical techniques like Raman and infrared spectroscopy work well when there is an optical mode with the right polarization to couple to the incident light. Since the incident light is of very low energy compared to X-rays, there is not enough momentum carried by the probing radiation. Hence, while the center-of-zone modes can be measured, their dispersive behavior inside the Brillouin zone cannot be observed. Also, these techniques are not element or isotope selective. The coupling strength is related to the electric dipole generated by a particular mode, and cannot be related to the absolute displacement. Another requirement is related to the transparency of the samples. Not only that many samples are not transparent, some samples may lose their transparency during temperature or pressure changes related to phase transitions.

The nuclear resonance-based method to be described here overcomes many of these limitations. First of all, it is element selective. Second, it is isotope selective, providing yet another window of opportunity when there are multiple sites. Third, the incident X-rays carry large momentum, sufficient to integrate over the entire Brillouin zone. Fourth, the peak intensities measured are directly related to the displacement of the targeted isotope. Finally, for single crystals or oriented samples, the nuclear resonant probes are direction sensitive. It will be difficult to highlight all of these advantages here, but a few examples will be provided.

We can highlight some of the relevant features of the NRIXS technique with an example shown in Fig. 17.4. Here, powder Eu_2O_3 is used as the sample. The synchrotron beam was monochromatized with a combination of C (111) double crystal monochromator and a nested pair of Si (4 4 0) and Si (15 11 3) reflections. This combination gave an overall resolution of 0.8 meV, with a flux of 0.8 GHz. A detailed description of high-resolution, tunable crystal monochromators was given earlier [39]. A silicon avalanche photodiode detector, APD [40], is used to detect the 21.541 keV scattered radiation as a function of energy. Energy tuning is accomplished by the rotation of the outer Si (4 4 0) crystal by 1.464 microradians and the inner Si (15 11 3) crystal by a total of 82.44 microradians for a scan of ±50 meV, in 300 steps while integrating the nuclear delayed signal between 15 and 120 ns time window. Figure 17.4a shows an integrated scan, which may be obtained by averaging typically 10–50 s per data point. This data is the result of all phonon excitations and needs to be "treated" to decouple multi-phonon contributions as described earlier. The details of data analysis have been described several times in the past [12,21,41] and will not be repeated here in detail, except the results, given in Fig. 17.4b.

The phonon density of states, dos, $g(E)$ is obtained from the single phonon excitation probability, $S_1(E)$ according to

$$g(E) = \frac{E}{E_R} \tanh(\beta E/2)[S_1(E) + S_1(-E)],$$

where E is the relative energy of the incident photons with respect to the nuclear resonance energy, E_R is the recoil energy of the resonant nucleus, and β is the reduced temperature (E/kT). (Fig. 17.4c).

The phonon dos is an important quantity in thermodynamics because many intrinsic material parameters like vibrational specific heat, entropy, and internal energy are additive functions of $g(E)$ in the harmonic approximation and hence they can be derived after the measurement of phonon dos is completed. A computer program, PHOENIX, is used to evaluate the data and extract the above-mentioned thermodynamic quantities [21].

Another important quantity that can be derived from $g(E)$ is the Debye sound velocity, V_D, which is extracted from the lower region of the phonon density of states.

$$V_D = \left[\frac{M}{2\pi^2 \rho \hbar^3} \frac{E^2}{g(E)}\right]^{1/3},$$

where M is the mass of the resonant nucleus and ρ is the density. In the inset of Fig. 17.4d, a parabolic fit to the density of states is shown. The curvature obtained with this fit depends on the lower and the upper limits of the data used. Hence, a procedure is developed to obtain this curvature with different upper bounds, and a fit is made to extrapolate the Debye sound velocity (Fig. 17.4d). This quantity can be used to obtain compression, V_P, and shear, V_S, sound velocities, provided that independent knowledge of bulk modulus, B, and density, ρ, is available under the conditions of each experiment, based on the following relationships [19]

$$\frac{3}{V_D^3} = \frac{1}{V_P^3} + \frac{2}{V_S^3}, \quad \text{and} \quad \frac{B}{\rho} = V_P^2 - \frac{4}{3}V_S^2.$$

For the particular example we are using here for Eu_2O_3, we can use 147 GPa as the bulk modulus, and 7.42 g/cm^3 as the density to extract a Debye sound velocity of 2,591 m/s, and 5,181 m/s as the compression, and 2,296 m/s as the shear sound velocity. In addition, if V_P is available from inelastic X-ray/neutron scattering, combining with density, one can extract bulk modulus, as it was shown recently [42].

One common misperception about the nuclear resonance vibrational spectroscopy is that it only measures the properties of the nuclear resonant isotope. In reality, for low-frequency vibrations, it has been demonstrated that all atoms move in unison, thus some of the most relevant properties like velocity of sound, which is extracted from the low-energy region of the density of states, is a reflection of the entire material and not of only the nuclear resonant element [43]. This extension of the application range actually makes this approach all the more relevant to a large class of materials. Furthermore, at least for molecular crystals, like porphyrins or cubanes, through the use of density functional theory, DFT, which is validated by NRIXS or NRVS, it is now possible to solve dynamical matrices and obtain a complete set of eigenvectors. For each mode, we can now determine the direction and amplitude of each atomic movement. Thus, even the data is relayed to us by iron, or other nuclear resonant isotope, the motion of all atoms have been determined.

17.7 Geophysical Applications of NRIXS

The role of high-pressure techniques in geophysical studies cannot be overemphasized. For many of the geophysical issues, one resorts to creating experimental conditions similar to Earth's mantle, outer core, transition zone, and, if possible, the inner core. Since iron is abundant in all strata, which is composed of iron-containing oxides, silicates and metallic alloys, the nuclear resonance NRIXS and SMS experiments provide unique information over a large scientific field. Today, it is possible to reach pressures above 3.7 Mb, and temperatures of 5,700 K [44] using diamond-anvil cell, DAC, and laser heating. The methodology of using Mössbauer nuclei as a probe of vibrational density of states has been successfully incorporated to DAC, making it possible to measure the pressure dependence of Debye sound velocity up to 200 GPa (2 Mb) and, when complemented with independent measurements of bulk moduli, the shear and compression components can be distinguished [19]. This approach also opened the possibility of characterizing opaque samples like iron-rich spinels and perovskites [45]. Since the technique is isotope selective via the Mössbauer nuclei, typical problems associated with sample environment leading to deteriorating signal-to-background ratio are readily resolved. By using focused X-ray beam and tunable high-resolution crystal monochromators with 1 meV resolution, the partial phonon density of states is recorded for iron and a variety of iron alloys containing hydrogen [46], carbon [47], oxygen [48] sulfur [49], silicon and nickel [50]. Studies carried out up to 1,700 K under pressure led to re-evaluation of Birch's law [25], observation of formation of post-perovskite Fe_2O_3 upon heating under pressure [51], and spin changes in magnesiowüstites [52] and perovskites [53].

17.8   Biological Applications of NRIXS

The scientific applications of [57]Fe-based NRIXS in life science-related fields are mostly exploited in bio-inorganic chemistry, which is focused in identifying and characterizing model compounds as bio-mimetic metal complexes [54], and the geometric and electronic nature of the intermediate and final states in enzymatic [55–57] materials. The technique is called nuclear resonant vibrational spectroscopy (NRVS) when applied in this field, to underscore the intended application and the physics origins [15,58].

NRIXS is unique because it is a new tool providing vibrational information that neither Raman nor FTIR spectroscopy can deliver, namely, element and isotope selectivity, no restrictions by the optical selections rules, access to low-frequency modes below $100 \, cm^{-1}$, down to $8 \, cm^{-1}$, complete representation of all modes involving the nuclear resonant isotope [59–62], and access to anisotropic information in heme-based protein mimics [63] and model compounds like porphyrins [64,65]. The ability of NRVS to determine the frequency and absolute amplitude of vibrations and its access to cooperative motion of the molecular groups with energy below $200 \, cm^{-1} (\sim 25 \, meV)$ makes it very attractive for scientists who would like to calculate and validate the structural models of initial, intermediate, and final states of enzymes like *hydrogenase* and *nitrogenase* [66,67].

Results from the normal mode analysis can be compared against a calculation based on a refined set of force constants to assign experimental modes using the modes of the bare heme molecule as a basis set. Availability of single crystal data where in-plane and out-of-plane modes can be positively identified provides a solid confirmation of the DFT or force constant computations. An example is given in Fig. 17.5. The compound is five-coordinated iron tetraphenyl porphyrin with NO ligand. Modes at 74 and $128 \, cm^{-1}$ involve strong coupling between ligand vibrations and the delocalized motion of the porphyrin core. In heme proteins, these modes may provide a channel for communicating the ligand state to the protein side chains. At frequencies below $150 \, cm^{-1}$, the iron VDOS is dominated by out-of-plane porphyrin modes, modes due to collective motion of the peripheral phenyl rings, and acoustic modes. Several porphyrin in-plane modes, such as v_{42}, v_{50}, and v_{53}, have been positively identified for the first time because they are Raman-inactive vibrations [68,69].

17.9   Materials Science Applications of SMS and NRIXS

There are several aspects of SMS and NRIXS that make them particularly suitable for materials science and condensed matter physics experiments. One of them is the ability to work with surfaces, due to extreme collimation and relatively high flux, which is an inherent property of synchrotron radiation. As a result, magnetism of buried surfaces of iron [70, 71], iron Pt thin films [72], and iron oxide [73] have

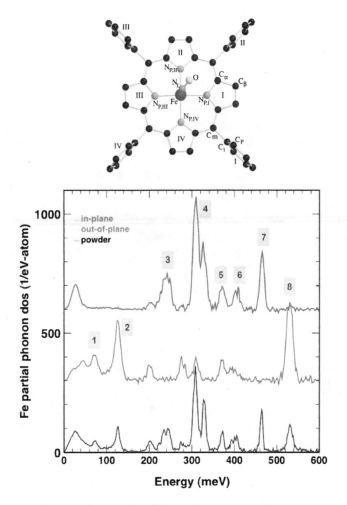

Fig. 17.5 Phonon densities of states of five-fold coordinated Fe tetraphenyl porphyrin–NO (shown above) in powder and single crystal form [68]. The mode assignments are the result of force-constant calculations according to a scheme described earlier [86] are as follows: (1) doming mode of the porphyrin core, (2) pyrrole tilting motion, (3) in-plane porphyric core motion, (4) in-plane motion of the two pyrrole rings, (5) unassigned mode, (6) asymmetric Fe–pyrrole stretch, which in the presence of an off-axis NO ligand removes the two-fold degeneracy, causing two distinct vibrations at 400 and 410 cm^{-1}, (7) in-plane displacement of the iron with a Fe–NO character, which is mixed with tilt and stretch vibrations between iron and nitrogen, and (8) out-of-plane Fe–N stretch and torsional mixed mode, inferred as the line width exceeds the instrumental resolution of 7 cm^{-1}

been studied. Using NRIXS, dynamics of monolayers of iron [71], nanograins [75], and nanoclusters of iron with an effective coverage of less than a monolayer have been studied [76, 77].

Materials science and solid-state physics experiments address a wide field of interest. For example, the changes in the electronic structure of oxides under high-pressure [78, 79] affect the dynamical behaviour such as high-energy phonon confinement in multilayer films [80]. Temperature and pressure dependence of phonons in newly discovered materials like iron pnictides were studied with NRIXS [81]. Complex phases like in Fe–Cr alloys have been the subject of investigations using either SMS or NRIXS and it was shown that relevant thermodynamic quantities can be extracted without the necessity of using empirical parameters. A thorough study of complex alloys with combined NRIXS and theoretical ab initio approach may provide an understanding of lattice dynamics in a wide variety of disordered systems [82]. Study of effects of composition, temperature, and magnetism on phonons in Fe–V alloys reveals how the interatomic forces depend on electronic screening [83].

In nanoscience applications, it is clear that there are serious modifications to the density of states as the particle size is reduced below 50 nm. These measurements can be used in quantifying the phase segregation and alloy decomposition between surface and the interior of nanoclusters and the role it plays in catalytic selectivity and activity in heterogeneous catalysis. Some of these questions can be addressed quantitatively with NRIXS, which is the only direct method for nanosized particles with less than one monolayer surface coverage [76, 77, 80].

More recently, the interplay between spin dynamics and interlayer coupling has also been addressed using SMS [84, 85].

Acknowledgements Many colleagues have over the years contributed to this field. In particular, we would like to acknowledge Brent Fultz, Timothy Sage, Robert Scheidt, Stephen Cramer, Beatriz Roldan-Cuenya, Werner Keune, Jung-Fu Lin, Ho-Kwan Mao, Jennifer Jackson, Jie Li, Guoyen Shen, Michael Hu, Caroline L'abbé, Ralf Röhlsberger, Michael Lerche, Hasan Yavaş, Ahmet Alataş, and Gopal Shenoy who have generously shared their knowledge, skills, software, students, and samples with us.

Use of the Advanced Photon Source, an Office of Science User Facility operated for the U.S. Department of Energy (DOE) Office of Science by Argonne National Laboratory, was supported by the U.S. DOE under Contract No. DE-AC02-06CH11357. The National Science Foundation provided generous support for the NRVS experiments under the contract no PHY-0545787, and for high-pressure experiments by COMPRES under NSF Cooperative Agreement EAR 06–49658. Argonne National Laboratory is operated by The University of Chicago under contract with the U.S. Department of Energy, Office of Science.

References

1. R.L. Mössbauer, Z. Phys. **151**, 124 (1958); Naturwiss. **45**, 538 (1958)
2. S.L. Ruby, J. Phys. (Paris) **35-C6**, 209 (1974)

3. E. Gerdau, R. Rüffer, H. Winkler, W. Tolksdorf, C.P. Klages, J.P. Hannon, Phys. Rev. Lett. **54**, 835 (1985)
4. G.V. Smirnov, U. van Bürck, A.I. Chumakov, A.Q.R. Baron, R. Rüffer, Phys. Rev. B **55**, 5811 (1997)
5. T. Mitsui, M. Seto, N. Hirao, Y. Ohishi, Y. Kobayashi, S. Higashitaniguchi, R. Masuda, Jpn. J. Appl. Phys. **46**, L382 (2007)
6. R. Coussement, J. Odeurs, C. L'abbe, G. Neyens, Hyperfine Interact. **125**, 113 (2000)
7. R. Callens, R. Coussement, C. L'abbe, S. Nasu, L. Vyvey, T. Yamada, Y. Yoda, J. Odeurs, Phys. Rev. B **65**, 180404 (2002)
8. R. Callens, C. L'Abbe, J. Meersschaut, I. Serdons, W. Sturhahn, T. Toellner, Phys. Rev. B **72**, 081402 (2005)
9. M. Seto, R. Masuda, S. Higashitaniguchi, S. Kitao, Y. Kobayashi, C. Inaba, T. Mitsui, Y. Yoda, Phys. Rev. Lett. **102**, 217602 (2009)
10. N. Planckaert, R. Callens, J. Demeter, B. Laenens, J. Meersschaut, W. Sturhahn, S. Kharlamova, K. Temst, A. Vantomme, Appl. Phys. Lett. **94**, 224104 (2009)
11. E.E. Alp, W. Sturhahn, T. Toellner, Nucl. Instrum. Meth. Phys. Res. B **97**, 526 (1995)
12. W. Sturhahn, T.S. Toellner, E.E. Alp, X. Zhang, M. Ando, Y. Yoda, S. Kikuta, M. Seto, C.W. Kimball, B. Dabrowski, Phys. Rev. Lett. **74**, 3832 (1995)
13. M. Seto, Y. Yoda, S. Kikuta, X.W. Zhang, M. Ando, Phys. Rev. Lett. **74**, 3828 (1995)
14. J. B. Hastings, D. P. Siddons, U. van Bürck, R. Hollatz, U. Bergmann, Phys. Rev. Lett, **66**, 770 (1992)
15. J.T. Sage, C. Paxton, G.R.A. Wyllie, W. Sturhahn, S.M. Durbin, P.M. Champion, E.E. Alp, W.R. Scheidt, J. Phys: Condens. Matter **13**, 7707 (2001)
16. R. Rüffer, A.I. Chumakov, Hyperfine Interact. **128**, 255 (2000)
17. E.E. Alp, W. Sturhahn, T. Toellner, J. Phys: Condens. Matter **13**, 7645 (2001)
18. M. Seto, S. Kitao, Y. Kobayashi, R. Haruki, T. Mitsui, Y. Yoda, X.W. Zhang, Yu. Maeda, Phys. Rev. Lett. **84**, 566 (2000)
19. H.K. Mao, J. Xu, V.V. Struzhkin, J. Shu, R.J. Hemley, W. Sturhahn, M.Y. Hu, E.E. Alp, L. Vocadlo, D. Alfè, G.D. Price, M. J. Gillan, M. Schwoerer-Böhning, D. Häusermann, P. Eng, G. Shen, H. Giefers, R. Lübbers, G. Wortmann, Science **292**, 914 (2001)
20. J.Y. Zhao (2011), *unpublished*
21. W. Sturhahn, Hyperfine Interact. **125** (2000)
22. MEDC_Eu (2010), http://www.moesbauer.org/151Eu.html
23. O. Leupold, J. Pollmann, E. Gerdau, H.D. Rüter, G. Faigel, M. Tegze, G. Bortel, R. Rüffer, A.I. Chumakov, A.Q.R. Baron Europhys. Lett. **35**, 671 (1996)
24. J.M. Jackson, W. Sturhahn, O. Tschauner, M. Lerche, Y. Fei, Geophys. Res. Lett. **36**, L17312 (2009)
25. J.F. Lin, W. Sturhahn, J. Zhao, G. Shen, H-K. Mao, R.J. Hemley, Science **308**, 1892 (2005)
26. L.L. Gao, B. Chen, M. Lerche, E.E. Alp, W. Sturhahn, J.Y. Zhao, H. Yavas, J. Li, J. Synch. Rad. **16**, 714–722 (2009)
27. T.S. Toellner, E.E. Alp, T.R. Graber, W. Henning, S.D. Shastri, G. Shenoy, W. Sturhahn, J. Synch. Rad. **18** (2011), in print
28. H. Giefers, E.A. Tanis, S.P. Rudin, C. Graeff, X. Ke, C. Chen, M.F. Nicol, M. Pravica, W. Pravica, J.Y. Zhao, A. Alatas, M. Lerche, W. Sturhahn, E.E. Alp. Phys. Rev. Lett. **98**, 245502 (2007)
29. R. Lübbers, M. Pleines, H. J. Hesse, G. Wortmann, H.F. Gruenstudel, R. Rüffer, O. Leupold, J. Zukrowski, Hyperfine Interact. **120–121**, 49 (1999)
30. E. Tanis, MSc thesis, University of Nevada, Las Vegas, 2010, unpublished
31. J.Y. Zhao, T.S. Toellner, M.Y. Hu, W. Sturhahn, E.E. Alp, G.Y. Shen, H.K. Mao, Rev. Scientific Instrum. **73**, 1608 (2002)
32. R. Weinstock, Phys. Rev. **65**, 1 (1944)
33. B.N. Brockhouse, A.T. Stewart, Phys. Rev. 100, 756 (1955)
34. W.M. Visscher, Ann. Phys. **9**, 194 (1960)
35. K.S. Singwi, A. Sjölander, Phys. Rev. **120**, 1093 (1960)

36. A.I. Chumakov, W. Sturhahn, Hyperfine Interact. **123**, 781–808 (1999)
37. J.T. Sage, S.M. Durbin, W. Sturhahn, D.C. Wharton, P.M. Champion, P. Hession, J. Sutter, E.E. Alp, Phys. Rev. Lett. **86**, 4966 (2001)
38. W. Sturhahn, J.M. Jackson, Geol. Soc. Am. **421**, 157 (2007)
39. T.S. Toellner, Hyperfine Interact., 125, 3 (2000)
40. S. Kishimoto, Rev. Sci. Instr. 63, 824 (1992)
41. M.Y. Hu, W. Sturhahn, T.S. Toellner, P.M. Hession, J.P. Sutter, E.E. Alp, Nucl. Instr. Meth. A **428**, 551 (1999)
42. B.M. Leu, A. Alatas, H. Sinn, E.E. Alp, A.H. Said, H. Yavaş, J. Zhao, J.T. Sage, W. Sturhahn, J. Chem. Phys. **132**, 085103 (2010)
43. M.Y. Hu, W. Sturhahn, T.S. Toellner, P.D. Mannheim, D.E. Brown, J. Zhao, E.E. Alp, Phys. Rev. B **67**, 094304 (2003)
44. S. Tateno, K. Hirose, Y. Ohishi, Y. Tatsumi, Science **330**, 359 (2010)
45. J.M. Jackson, E.A. Hamecher, W. Sturhahn, Eur. J. Mineral. **21**, 551 (2009)
46. W.L. Mao, W. Sturhahn, D.L. Heinz, H-K. Mao, J. Shu, R.J. Hemley, Geophys. Res. Lett. **31**, L15618 (2004)
47. L. Gao, B. Chen, J. Wang, E.E. Alp, J. Zhao, M. Lerche, W. Sturhahn. H.P. Scott, F. Huang, Y. Ding, S.V. Sinogeikin, C.C. Lundstrom, J.D. Bass, J. Li, Geophys. Res. Lett. **35**, L17306 (2008)
48. V.V. Struzhkin, H.-K. Mao, J. Hu, M. Schwoerer-Böhning, J. Shu, R.J. Hemley, W. Sturhahn, M.Y. Hu, E.E. Alp, P. Eng, G. Shen, Phys. Rev. Lett. **87**, 255501 (2001)
49. H. Kobayashi, T. Kamimura, D. Alfè, W. Sturhahn, J. Zhao, E.E. Alp, Phys. Rev. Lett. **93**, 195503 (2004)
50. J.F. Lin, V.V. Struzhkin, W. Sturhahn, E. Huang, J. Zhao, M.Y. Hu, E.E. Alp, H.-K. Mao, N. Boctor, R. J. Hemley, Geophys. Res. Lett. **30**, 2112 (2003)
51. S.H. Shim, A. Bengtson, D. Morgan, W. Sturhahn, K. Catalli, J. Zhao, M. Lerche, V. Prakapenka, Proc. Natl. Acad. Sci. U.S.A. **106**, 5508–5512 (2009)
52. J. Li, W. Sturhahn, J.M. Jackson, V.V. Struzhkin, J.F. Lin, J. Zhao, H.K. Mao, G. Shen, Phys. Chem. Minerals **33**, 575 (2006)
53. W.L. Mao, H.K. Mao, J. Shu, Y. Fei, R.J. Hemley, Y. Meng, G. Shen, V.B. Prakapenka, A.J. Campbell, D.L. Heinz, W. Sturhahn, J. Zhao, Geochim. Cosmochim. Acta **70**, A389 (2006)
54. W.R. Scheidt, G.R.A. Wyllie J. Inorg. Biochem. **96**, 51 (2003)
55. Y. Xiao, H. Wang, S.J. George, M.C. Smith, M.W.W. Adams, F.E. Jenney Jr., W. Sturhahn, E.E. Alp, J. Zhao, Y. Yoda, A. Dey, E.I. Solomon, S.P. Cramer, J. Am. Chem. Soc. **127**, 14596 (2005)
56. Y. Xiao, K. Fisher, M.C. Smith, W.E. Newton, D.A. Case, S.J. George, H. Wang, W. Sturhahn, E.E. Alp, J. Zhao, Y. Yoda, S.P. Cramer, J. Am. Chem. Soc. **128**, 7608 (2006)
57. Y. Xiao, M.L. Tan, T. Ichiye, H. Wang, Y. Guo, M.C. Smith, J. Meyer, W. Sturhahn, E.E. Alp, J. Zhao, Y. Yoda, S.P. Cramer, Biochemistry **47**, 6612 (2008)
58. J.T. Sage, A. Barabanschikov, M. Kubo, P.M. Champion, J. Zhao, W. Sturhahn, E.E. Alp, Int. Conf. Raman Spectrosc. **1267**, 645 (2010)
59. W.R. Scheidt, S.M. Durbin, J.T. Sage, J. Inorg. Biochem. 99, 60 (2005)
60. K.L. Adams, S. Tsoi, J. Yan, S.M. Durbin, A.K. Ramdas, W.A. Cramer, W. Sturhahn, E.E. Alp, C. Schulz, J. Phys. Chem. B 110, 530 (2006)
61. B.K. Rai, S.M. Durbin, E.W. Prohofsky, J.T. Sage, M.K. Ellison, A. Roth, W.R. Scheidt, W. Sturhahn, E.E. Alp, J. Am. Chem. Soc., **125**, 6927 (2003)
62. B.M. Leu, M.Z. Zgierski, G.R.A. Wyllie, W.R. Scheidt, W. Sturhahn, E.E. Alp, S.M. Durbin, J.T. Sage, J. Am. Chem. Soc. **126**, 4211 (2004)
63. J.W. Pavlik, A. Barabanschikov, A.G. Oliver, E.E. Alp, W. Sturhahn, J. Zhao, J.T. Sage, W.R. Scheidt, Angew. Chem. Int. Ed. **49**, 4400 (2010)
64. N.J. Silvernail, A. Barabanschikov, J.W. Pavlik, B.C. Noll, J. Zhao, E.E. Alp, W. Sturhahn, J.T. Sage, W.R. Scheidt, J. Am. Chem. Soc. **129**, 2200 (2007)

65. V. Starovoitova, G.R.A. Wyllie, W.R. Scheidt, W. Sturhahn, E.E. Alp, S.M. Durbin J. Phys. Chem. B **112**, 12656 (2008)
66. S. Cramer, Y. Xiao, H. Wang, Y. Guo, M. Smith, Hyperfine Interact. **170**, 47 (2006)
67. T. Petrenko, S. DeBeer George, N. Aliaga-Alcalde, E. Bill, B. Mienert, Y. Xiao, Y. Guo, W. Sturhahn, S P. Cramer, K. Wieghardt, F. Neese, J. Am. Chem. Soc. **129**, 11053 (2007)
68. B.K. Rai, S.M. Durbin, E.W. Prohofsky, J.T. Sage, G.R.A. Wyllie, W.R. Scheidt, W. Sturhahn, E.E. Alp, Biophys. J. **82**, 2951 (2002)
69. W. Sturhahn, J. Zhao, J.T. Sage, W.R. Scheidt, T. Petrenko, W. Sturhahn, F. Neese, Hyperfine Interact. **175**, 165 (2007)
70. C. L'abbé, J. Meersschaut, W. Sturhahn, J.S. Jiang, T.S. Toellner, E.E. Alp, S.D. Bader, Phys. Rev. Lett. **93**, 037201 (2004)
71. N. Planckaert, C. L'abbé, B. Croonenborghs, B. Laenens, R. Callens, A. Vantomme, J. Meersschaut, Phys. Rev. B **78**, 144424 (2008)
72. S. Couet, M. Sternik, B. Laenens, A. Siegel, K. Parlinski, N. Planckaert, F. Gröstlinger, A.I. Chumakov, R. Rüffer, B. Seipol, K. Tempst, A. Vantome, Phys. Rev. B **82**, 094109 (2010)
73. S. Couet, S. Schlage, K. Rüffer, R. Stankov, S. Diederich, T. Laenens, R. Röhlsberger, Phys. Rev. Lett. **103**, 097201 (2009)
74. S. Stankov, R. Röhlsberger, T. Ślęzak, M. Sladecek, B. Sepiol, G. Vogl, A.I. Chumakov, R. Rüffer, N. Spiridis, J. Łażewski, K. Parliński, J. Korecki, Phys. Rev. Lett. **99**, 185501 (2007)
75. S. Stankov, Y.Z. Yue, M. Miglierini, B. Sepiol, I. Sergueev, A.I. Chumakov, L. Hu, P. Svec, R. Rüffer, Phys. Rev. Lett. **100** 235503 (2008)
76. B. Roldan Cuenya, L.K. Ono, J.R. Croy, A. Naitabdi, H. Heinrich, J. Zhao, E.E. Alp, W. Sturhahn, W. Keune, Appl. Phys. Lett. **95**, 143103 (2009)
77. B. Roldan Cuenya, A. Naitabdi, J. Croy, W. Sturhahn, J.Y. Zhao, E.E. Alp, R. Meyer, D. Sudfeld, E. Schuster, W. Keune, Phys. Rev. B **76**, 195422 (2007)
78. A.G. Gavriliuk, J.F. Lin, I. S. Lyubutin, V.V. Struzhkin, JETP Lett. **84**, 161 (2006)
79. J.F. Lin, H. Watson, G. Vankó, E.E. Alp, V.B. Prakapenka, P. Dera, V.V. Struzhkin, A. Kubo, J. Zhao, C. McCammon, W. J. Evans, Nat. Geosci. **1**, 688 (2008)
80. B. Roldan Cuenya, W. Keune, R. Peters, E. Schuster, B. Sahoo, U. von Horsten, W. Sturhahn, J. Zhao, T.S. Toellner, E.E. Alp, S.D. Bader, Phys. Rev. B **77**, 165410 (2008)
81. O. Delaire, M.S. Lucas, A.M. dos Santos, A. Subedi, A. S. Sefat, M.A. McGuire, L. Mauger, J. A. Muñoz, C.A. Tulk, Y. Xiao, M. Somayazulu, J.Y. Zhao, W. Sturhahn, E.E. Alp, D.J. Singh, B.C. Sales, D. Mandrus, T. Egami, Phys. Rev. B **81**, 094504 (2010)
82. S.M. Dubiel, J. Cieslak, W. Sturhahn, M. Sternik, P. Piekarz, S. Stankov, K. Parlinski, Phys. Rev. Lett. **104**, 155503 (2010)
83. M.S. Lucas, J.A. Muñoz, O. Delaire, N.D. Markovskiy, M.B. Stone, D.L. Abernathy, I. Halevy, L. Mauger, J.B. Keith, M.L. Winterrose, Y. Xiao, M. Lerche, B. Fultz, Phys. Rev. B **82**, 144306 (2010)
84. B. Laenens, N. Planckaert, J. Demeter, M. Trekels, C. L'abbé, C. Strohm, R. Rüffer, K. Temst, A. Vantomme, J. Meersschaut, Phys. Rev. B **82**, 104421 (2010)
85. N. Planckaert, C. L'abbé, B. Croonenborghs, R. Callens, B. Laenens, A. Vantomme, J. Meersschaut, Phys. Rev. B **78**, 144424 (2008)
86. T.G. Spiro, X.-Y. Li, in*Biological Applications of Raman Spectroscopy*, ed. by T.G. Spiro (Wiley, New York, 1988), pp. 1–37

Part III
Future Developments
of the Mössbauer effect

Future-oriented perspectives and opportunities in using advanced synchrotron radiation sources for the study of non-equilibrium phenomena with the Mössbauer effect are elaborated. The most challenging performance requirement for such Mössbauer studies of non-equilibrium systems is, first, to collect information, preferably with a single X-ray pulse from the radiation source when the system dwells in the non-equilibrium (excited) state and second, to probe mesoscopic structures with a spatial resolution of nanometers. In a further contribution, the field of neutrino Mössbauer spectroscopy using mono-energetic anti-neutrinos from bound β^--decay with all its fascination and problems is discussed.

Chapter 18
Dreams with Synchrotron Radiation

G.K. Shenoy

18.1 Introduction

This is the Golden Jubilee year for the Mössbauer Effect Nobel Prize [1]. Since its discovery, the measurements of hyperfine interactions, recoilless fraction (the so-called Lamb–Mössbauer factor), and the second-order Doppler shift obtained from Mössbauer spectra have shed light on the magnetic, structural, and dynamical properties of matter. The applications of this tool have contributed to every area of science and technology including nuclear physics, condensed-matter physics, general physics, chemical physics, materials science, earth science, planetary science, environmental science, life science, and medical science. This volume addresses many of these applications in detail.

In 1962, Seppi and Boehm [2] suggested that resonant gamma-rays from radioactive nuclear decays be replaced with X-ray sources to excite the Mössbauer resonance. This was followed by a concrete proposal to use synchrotron radiation by Ruby in 1974 [3]. The ultra-high collimation and brightness of synchrotron radiation beams compared to those of radioactive gamma-ray sources provided new opportunities as demonstrated by the first experiment in 1985 [4], followed by a suite of new techniques. Three productive techniques have evolved over the years, all of which use time-differential measurements, complementing the traditional Mössbauer spectroscopy performed in the energy domain. They are:

1. Elastic nuclear scattering or the nuclear forward scattering (NFS) to measure hyperfine spectra [5],
2. Nuclear resonant inelastic X-ray scattering (NRIXS) to probe atomic and molecular vibrational dynamics [6, 7]

G.K. Shenoy (✉)
Advanced Photon Source, Argonne National Laboratory, Argonne, Il 60439, USA
e-mail: gks@aps.anl.gov

M. Kalvius and P. Kienle (eds.), *The Rudolf Mössbauer Story*,
DOI 10.1007/978-3-642-17952-5_18, © Springer-Verlag Berlin Heidelberg 2012

3. Synchrotron-radiation-based quasi-elastic nuclear forward scattering (QNFS) to address diffusional problems [8, 9]

These techniques are generally used in the field of condensed-matter physics, materials science, chemical physics, geoscience, and biology. Today the workhorse capabilities for performing NFS, NRIXS, and QNFS experiments are available at the "Third-Generation" electron storage-ring facilities , namely, the Advanced Photon Source (APS) at Argonne, the European Synchrotron Radiation Source (ESRF) at Grenoble, the Super Photon Ring-8 (SPring-8) in Hyogo Prefecture, Japan, and PETRA III facility, Deutsches Elektronen-Synchrotron DESY at Hamburg. The sources of radiation at these facilities are undulators, which enhance the brilliance of the spontaneous radiation beam by many orders of magnitude compared to those of traditional radioactive sources. A large number of Mössbauer resonances have been successfully measured at the storage-ring facilities (see the listing in book chap. 17). Excellent reviews on the current state of development of the field are available [10, 11].

During the next decade, it is expected that new types of radiation sources with unprecedented collimation, brightness, and pulse structure will extend and revolutionize the applications of the nuclear resonance spectroscopy even further. These include ultra-high brightness storage-ring-based sources, X-ray free-electron laser (XFEL) sources such as EuroXFEL under construction in Hamburg, LCLS at Stanford and future seeded-XFEL sources. In this chapter, we will briefly present the characteristics of these new sources and discuss new areas of potential scientific impact from nuclear resonance spectroscopy.

18.2 Future Synchrotron Radiation Sources

18.2.1 Storage Rings

The average brilliance of a synchrotron radiation beam, which is a measure of the X-ray flux density per unit phase space volume, is the most important quantity for nuclear resonance experiments [12, 13]. The phase space volume is a product of the X-ray beam divergence and the X-ray beam size, both in the x- and y-directions. The vertical divergence of the X-ray beams from undulator sources (APS, ESRF, and SPring-8) is adequately matched with the angular acceptance of high-resolution optics for 5 keV to 15 keV radiations. On the other hand, these sources are somewhat marginal in performance for ultra-high spatial resolution experiments. The new storage ring PETRA III (DESY, Hamburg) with a circumference of 2300 m will provide undulator beams many times more brilliant [14] than the operating facilities, resulting from a smaller source size. There are early considerations to build a high-energy ultra-brilliant X-ray facility in India with a multi-bend achromat storage-ring lattice that will further reduce the source size by another order of magnitude. This ring planned by the Saha Institute of Nuclear Physics (SINP), Kolkata, with a circumference of $\sim$800 m will be capable of delivering "diffraction-limited" X-ray beams in the 1 Å range making it an ideal future source to perform spatially resolved NFS measurements.

The storage-ring-based undulator sources have an intrinsic limit in delivering quality beams suitable for nuclear resonance experiments (APS, ESRF, SPring-8, PETRA III, SINP). An important measure of source performance for the nuclear resonance experiments is the flux per meV band width at 14 keV, which is in the range of about $10^9 - 10^{11}$ photons/meV/s. Further improvements will be incremental and mostly governed by the development of X-ray optics, detectors, and timing techniques.

18.2.2 X-ray Free Electron Laser (XFEL) Sources

The next major enhancement in delivering super bright beams requires a new paradigm. The linear accelerators (linacs) fit this bill. When an ultra-relativistic, high-electron-density, low-emittance bunch of electrons from a linac propagates through a long undulator, its interaction with the emitted radiation produces microbunches of electrons separated by the wavelength of emitted radiation. This process, referred to as self-amplified spontaneous emission (SASE), in an XFEL was first demonstrated in the long-wavelength range [12, 13]. The gains made in the peak brilliance by SASE-XFELs are proportional to the second power of the total number of coherent electrons in the microbunches. This is many orders of magnitude higher than that from any spontaneous undulator sources (placed in a storage-ring or a linac) where the flux is linearly proportional to the number of electrons in the bunch. Many new technological innovations, such as the ultra-low emittance photocathode electron-gun ($x-y$ emittance of 10^{-2} nm rad) delivering ultra-short (100–200 fs) bunches with high electron density ($>10^9$) with an energy spread of less than 10^{-4} and long undulators with over 10^4 periods are required to realize SASE-XFEL. A hard X-ray SASE-XFEL is now operating in the United States (LCLS at SLAC, Stanford) [15], while facilities in Germany (European XFEL Laboratory at DESY, Hamburg) and Japan (SPring-8 XFEL, Harima City) will begin operation in 2014 and 2011, respectively. A comprehensive discussion on SASE-XFEL [16] and its potential use in nuclear resonance studies has been presented earlier [17].

The X-ray beam from a SASE-XFEL source is fully coherent in the transverse direction. However, it lacks temporal (longitudinal) coherence due to the noisy startup process and is similar to a pulse from a storage ring. This can be remedied by introducing a "seed" in the FEL process. There are many seeding schemes based on a cascade process [18] or an oscillator [19] or self-seeding [20]. The output from a seeded XFEL (SXFEL) will be proportional to the second power of the number of electrons in a bunch since all of them are coherent. The output will also be an amplified reproduction of the seed with transform-limited properties. For example, the 10–15 keV pulse from a SXFEL could have ∼1 fs pulse width (and 0.6 eV energy band width) or a 6–12 meV energy band width (with 10–20 fs pulse width), thus packing all the X-rays either in a narrow time window or in a few meV energy band.

Table 18.1 A comparison of important characteristics of the X-ray beams from future XFEL and storage-ring sources nuclear resonance scattering experiments

Properties	SASE-XFEL (14 kev)	SXFEL-SEEDED (14 kev)	Third Generation Facilities
Energy (GeV)	14.0–18.0	14.0–18.0	6.0–18.0
Undulator length(m)	80–100	80–100	5–25
Average brilliance (ph/s/0.1%BW/mrad^2/mm^2)	$1–4\times10^{24}$	$10^{25}–10^{26}$	$10^{20}–10^{22}$
Beam divergence (—rad)	1–2	1	v2–20
Average flux (ph/s/meV)	3×10^{14}	3×10^{16}	$10^9–10^{11}$
Bunch length (fs) (energy band width)	100–200	50,000–100,000 (12–24 meV)	~1,000,000
Repetition rate (Hz)	10–120	$10^2–10^5$	$3–6\times10^6$
Photons/pulse/meV	$10^{10}–10^{12}$	$10^{10}–10^{14}$	$10^3–10^5$

The later scheme of starting with a 10 meV beam from a SXFEL undulator will be of great advantage for nuclear resonance experiments. In Table 18.1, we make a comparison of anticipated properties of the X-ray beam from XFELs important for the nuclear resonance experiments, and compare these with those from best storage-rings. The true experimental demonstration of the self-seeding scheme in the hard X-ray range is planned for early 2011 at LCLS [20].

The number of photons delivered per bunch in a given energy band width (see Table 18.1) is nearly 10 orders of magnitude larger from an XFEL compared to that from a storage-ring source. This will allow one to collect an excellent statistical quality nuclear resonance spectrum using the photons from a single X-ray pulse making the XFELs most suitable for nonequilibrium studies of systems; a research area that is hard to address with storage-ring based sources. In addition, the ultra low divergence of the XFEL beam and nearly 100% transverse coherence will allow spatially resolved studies with perhaps 1–5 nm resolution. There is also a significant difference between an XFEL and a storage ring source in their time structure. The storage-ring sources have a typical single electron bunch repetition rate of 200–400 kHz. For ideal nuclear resonance experiments, a fill-pattern with 20–30 electron bunches with a X-ray pulse-to-pulse separation of about 150–200 ns is chosen to match with the lifetime of the ^{57}Fe Mössbauer resonance state. The repetition rate of X-ray pulses in the currently planned XFELs is only 10–120 Hz. Future sources (EuroXFEL) will have much higher repetition rates of 10–100 kHz with an appropriate realization of superconducting radiofrequency cavities. It is of some interest to note that the signal-to-noise ratio is inversely proportional to the square root of pulse repetition rate, and hence in general XFEL is a more favorable source.

In principle, the repetition rate of the pulses from a SXFEL can be tailored to the needs of any experiment by controlled management of superconducting rf and beam switching between beamlines. The average current in XFELs is small (~0.1 mA)

compared to storage rings, and the peak power, although considered high, can be controlled through the manipulation of the electron beam and the "seed".

18.2.3 Return of Classical Mössbauer Energy Spectroscopy at Synchrotron Sources

Almost all the NFS experiments performed at the storage-ring sources use a highly monochromatic X-ray beam to excite the nuclear resonance and an APD detector to measure the time-differential hyperfine spectrum (a beat pattern) in the forward direction [5].[1] The accompanying prompt X-ray intensity overloads the APD detectors even with the use of ultra-small band width of the excitation pulse (e.g., $\sim$1 meV resolution at 14.4 keV for ^{57}Fe). X-ray absorbers and thick APDs are routinely added to mitigate the problem of background prompt intensity, often sacrificing more than an order of magnitude in experimental capability. Recently, a new method that suppresses the prompt radiation completely has been demonstrated [21]. It involves the use of a dynamic shutter located after the resonant sample, thus dispensing with the high-resolution monochromator. This approach will allow one to perform time-integrated classical Mössbauer energy spectroscopy since no time resolution would be needed in such an experiment. A single-line resonant scatterer or absorber, such as potassium ferrocyanide, is placed on a Doppler velocity transducer and an ion chamber as the detector. If it is advantageous to perform the experiment in the time-differential mode currently used at the synchrotron beamlines, the APD detector will handle the enhanced resonant photon in the absence of the prompt excitation pulse. Figure 18.1 shows the schematic layout of the concept. When a scaled up version of this method is fully implemented, one can realize a resonant source intensity that is equivalent to many Curies of ^{57}Co in the case of ^{57}Fe resonant measurements. Such a source will generate linearly polarized X-rays with 5 neV band width at resonant energy of 14.4 keV with an unprecedented intensity of $\sim$10^{11} ph/s or 10^8 ph/pulse. Furthermore, the source will be nearly background-free and, with the use of well-designed X-ray optics will deliver a highly focused resonant beam with a spot size well below a micrometer. This will usher in a new era in performing classical Mössbauer energy spectroscopy. One can begin to dream of experiments that have been hitherto hard to do in the new experiments requiring ultra-high collimation, as well as those requiring space and time resolution.

The method is suited for both high- and low-repetition-rate sources, namely the storage-ring sources and SXFEL delivering very high instantaneous pulse intensities. However, longer separation between the excitation (prompt) pulses at both these sources will have many advantages; the primary one is to achieve a source line-shape that will have minor distortions from a Lorentzian to simplify the data analysis. The ability of the shutter to suppress the enormous intensity of

---

[1] See also book chaps. 16 and 17.

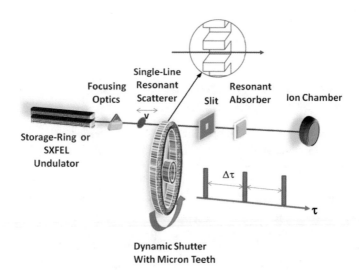

Fig. 18.1 Conceptual layout to perform classical Mössbauer energy spectroscopy at synchrotron sources. The shutter opens over $\Delta\tau$ (few times the life-time of the resonant excited state) to allow the resonant fluorescence in the forward direction to reach the detector. Details are in [21]

Table 18.2 Estimated 14.4 keV flux in a focused spot within a natural energy window of 5 neV

Resonant photon flux	Storage-ring facilities	SASE-XFEL	Seeded XFEL
ph/s/μm^2/5 neV (20% focusing efficiency)	10^4–10^6	2×10^8	2×10^{10}
ph/s/20 nm^2/5 neV (5% focusing efficiency)	3×10^3–10^5	4×10^7	4×10^9
ph/pulse/20 nm^2/5 neV (5% focusing efficiency)	$\sim10^2$	10^4–10^6	10^4–10^8

The SXFEL pulse length is typically 10–20 fs.

both prompt and electronic scattering at a SEXFEL would lead to new experiments to investigate nonequilibrium processes in a pump-probe mode.

Many of the ^{57}Fe nuclear resonance experiments at storage-ring facilities are now performed with a focused beam illuminating an area of $\sim 5 \times 5\,\mu$m^2 or larger. This compares well with the 14.4 keV flux from a 100 mCi ^{57}Co source illuminating a square cm area of absorber located 10 cm away. Typically, a K-B mirror pair or a compound refractive lens has been employed to focus the beams at storage-ring facilities. With an increase in the average flux by many orders of magnitude as described above, a focus spot of 20×20 nm^2 will still contain as many photons as in current experiments, and our ability to perform experiments with a spatial resolution down to a few nm^2 cannot be ruled out. In order to emphasize this point, we present in Table 18.2 the estimated photon flux in a focused beam of 20 nm^2.

Can this new approach be extended to high-energy Mössbauer transitions? The storage-ring sources can routinely deliver very large flux of radiation up to 100 keV, and this capability will improve with the advent of superconducting undulators that are on the horizon. The ability to deliver 30–40 keV SXFEL pulses is well conceived using the self-seeding schemes [20] and will be feasible in a decade. It must be pointed out that at energies above 30 keV, single-crystal monochromators have two unpleasant features that hinder the delivery of suitable highly monochromatic X-ray beams (<5 meV) to perform time-differential measurements. Firstly, at a given reflection, the width of the transmitted energy band increases with the X-ray energy, and secondly the angular acceptance of the reflection decreases being inversely proportional to the value of the incident energy. Hence, there is a need to develop time-integrated energy spectroscopy for the high-energy resonances at synchrotron sources. It should also be noted that, with the exception of a few E1 transitions, the higher energy nuclear transitions have shorter lifetimes generally in the ns range. Can one fully suppress the prompt pulse and open the window to observe nuclear resonant fluorescence from excited states with shorter half-life? The ultra-fast suppression of the high-energy prompt pulse poses another challenge in the mechanical design of shutter blades. The micromachining technology for the high aspect ratio blade structures required for this application must also deliver blade surfaces with nearly optical surface quality. One must confront this major technology issue in the next decade. An alternate approach is now being conceived and developed in order to separate the prompt and resonant fluorescence X-rays. This uses MEMS-based spatial light modulators (SLMs), in the form of torsional micro-mirrors [22]. This scheme has the potential to provide a resolution of better than 100 ps at high X-ray energies.

With continued effort, it should be possible to perform Mössbauer time-integrated energy spectroscopy and, in some instances, the time-differential beat pattern measurements with all the well-known high-energy resonances using synchrotron radiation sources in the next decade.

18.3   Potential Applications of New Generation Sources

The most obvious impact of the new generation of sources depends on the intensity of the source. Instead of the current approach of collecting an energy spectrum or a time- differential beat pattern over hours or days using either a radioactive source or a storage-ring source, the dynamic shutter approach (Sect. 18.2.3) will allow a complete experiment in a few seconds at a storage-ring source or simply with a single pulse from a SXFEL source. This would lead to a very new paradigm in both the method of measurement and the type of research problems that can be performed with these new generation sources. While the potential to extend ongoing studies in various fields using this new tool will be limited only by our imagination, it should be pointed out that the big area of impact will be in understanding matter at a nanoscopic and mesoscopic level, a subject that has been discussed in some detail elsewhere [17]. Radiation damage of the samples under investigation could arise from their being exposed to intense X-ray pulses from these sources. This could be a limitation to this approach. The situation can be mitigated by decreasing the band width of resonant excitation pulse using high-resolution X-ray monochromators that are currently available. This is a too-little appreciated capability of the nuclear resonance tools compared to all the other X-ray tools.

Most of the above mentioned measurements on physical, chemical, and bio-logical systems are to understand their equilibrium properties (Born–Openheimer approximation) which are within the realm of equilibrium statistics. On the other hand, there is a pervasiveness of nonequilibrium systems and processes continually occurring in nature and all around us in real systems (often having aperiodic structures). The dynamics of structure and spins in these real systems often comprising nanostructures and interfaces, or those showing growth of bacteria, or performing photosynthesis for energy storage are main targets of future scientific endeavor. One of the underlying themes of research in the next decades is to characterize and control matter away from equilibrium with a temporal resolution that is governed by the fundamental time scales of functional correlations in matter. These systems are driven away from equilibrium by providing energy to the system using a pump, and then probing the evolution of the system in both time and space. There are two classes of pump–probe studies:

1. Cyclical in which the system relaxes back to its original state, and
2. Non-cyclical in which the system reaches a new equilibrium (or dissipative) state.

The most challenging performance requirement for the Mössbauer studies of non-equilibrium systems is to measure and analyze the evolution of the hyperfine structure following the excitation of the system. The cyclical systems may be probed using the pulses from a storage–ring where enough statistics have been collected from repeated pump-probe cycles. The time-integrated or time-differential spectra are collected as a function of delay time between the pump pulse and the probe. These measurements will yield new information when the interaction energy associated with the relaxation process being investigated is comparable to the hyperfine interaction energy and larger than the natural energy resolution of the resonance. The deconvolution of the measured spectra will require special binning algorithms since the probe has its own characteristic time. Probing with a single X-ray pulse from the SXFEL source is suited for both cyclical and noncyclical processes [17].

A full knowledge of temporal evolution of far-from-equilibrium processes is essential to understand the functions in complex materials. The characteristic times of chemical reactions, structural transformations, and dynamics of spins is typically in the μs, ns to ps domains. Some of the phenomena of interest would include responses to shock-wave propagation, domain excitation, onset of photon-excited ordered magnetism, spin crossover processes, and photo-induced phase transitions. All of these are amenable to Mössbauer spectroscopic measurements. The next decades will be an exciting period to develop new sources of X-rays with tailored properties to perform Mössbauer resonance studies. This will also lead to the next-generation of experiments and tools to investigate real systems in real environments representing both equilibrium and far-from-equilibrium states. The biggest challenge will be to extract the primary process variables using new algorithms to analyze data banks of unprecedented volume.

18.4   Exotic Mössbauer Resonances and Cosmology

There are a few exotic Mössbauer resonances which have the potential to provide super high frequency resolution with a line width in the peV (0.24 kHz) to zeV (0.24 Hz) range. Why should one pursue these? Their attractive properties as a frequency standards and nuclear clocks open the possibility to address one of the most interesting questions in fundamental physics: What are the most stringent tests of General Relativity [23]? Are the fundamental constants of nature actually constant? Recent astrophysical measurements have hinted at possible fundamental constant variation over cosmological time and distance [24]. They predict a fractional time variation in the fine-structure constant of $1 \times 10^{-19}\,\mathrm{year}^{-1}$ making the exotic Mössbauer resonances the most attractive probes.

The big ship that launched the Mössbauer Effect soon after its discovery in the arena of general relativity was the measurement of gravitational red shift, i.e., the change in the energy of a photon as it moves through a gravitational field. In addition, to its inherent simplicity, this application of the Mössbauer Effect made its way into physics folklore and into undergraduate textbooks and pre-graduate laboratories. The measured change amounted to 2.5 parts in 10^{15}, in close agreement with theoretical predictions, making this the most celebrated verification of general relativity [25, 26]. The precision of the red shift measurement has been improved over the years, for example, with the ^{67}Zn Mössbauer resonance. We refer the reader to chap. 14 of this volume and to an excellent review of the entire topic presented by Potzel et al. [27]. In both ^{57}Fe and ^{67}Zn experiments, the Mössbauer clock appears to run more slowly than an identical clock located in a region of lower gravitational potential. The references mentioned above present a very critical analysis on what needs to be done to improve the sensitivity of these "absolute" clocks in the future. Discussed are in particular, the very serious issues of eliminating the systematic errors and residual static and dynamic hyperfine effects in the measurements.

An alternative approach is a so-called "null" experiment . In this, instead of comparing the same Mössbauer clock located in two differing gravitational potentials, one compares two clocks in different chemical environments. These clocks are located at one point where the gravitational potential varies with time. Because of its eccentricity, the Sun gravitational potential at the laboratory will have a sinusoidal variation with the period of one year. The same seasonal periodicity but with a different phase will then appear as a yearly fluctuation of the difference of shift in different chemical environments [28, 29]. The measurement of the influence of changing gravitational field performed over a period of 8 years set a lower limit on the sensitivity to 3.8×10^{-5}.

Can we improve any further either the "absolute" or the "null" measurements discussed above? A glimmer of hope lies in using higher sensitivity exotic Mössbauer transitions. In Table 18.3, we list the dream team of exotic resonances along with their main resonant properties. In the context of this section, the Q value of the resonance is the most important factor and the dream team provides 3–6 orders of magnitude enhancement over ^{67}Zn. There are serious adversities in this path toward higher Q. For example, the interactions of the magnetic moments of

Table 18.3 Dream team of exotic Mössbauer resonances

Nuclide	Abundance (%)	E_0	I_g–I_e	$T_{1/2}$	Multi-polarity	Line width Γ	$Q = \Gamma/E_0$	α	σ_0 (cm^2)
^{57}Fe	Stable (2.14)	14.4 keV	1/2–3/2	99 ns	M1	4.6 neV	3.3×10^{-13}	8	2.6×10^{-18}
^{181}Ta	Stable (100)	6.214 keV	7/2–9/2	6,800 ns	E1	67.0 peV	1.2×14^{-14}	82	9.5×10^{-19}
^{67}Zn	Stable (4.11)	93.32 keV	5/2–1/2	9,150 ns	E2	50.0 peV	5.4×10^{-16}	1	5.0×10^{-20}
^{45}Sc	Stable (100)	12.4 keV	7/2–3/2	318 ms	M2	1.46 feV	1.18×10^{-19}	400	2.0×10^{-20}
^{109}Ag	Stable (49.6)	88.0 keV	1/2–7/2	39.6 s	E3	11.5 aeV	1.31×10^{-22}	20	6.0×10^{-20}
^{107}Ag	Stable (51.4)	93.1 keV	1/2–7/2	44.3 s	E3	10.3 aeV	1.11×10^{-22}	20	5.4×10^{-20}
^{235}U	7.1×10^8 year	73 eV	7/2–1/2	1.56 ks	E3	292 zeV	4.0×10^{-21}	$\sim 10^{17}$	1.2×10^{-26}
^{103}Rh	Stable (100)	39.8 keV	1/2–7/2	3.37 ks	E3	135 zeV	3.39×10^{-24}	1350	4.5×10^{-21}
^{229}Th	7340 year	7.6 eV	5/2–3/2	26 ks	M1	17.5 zeV	2.30×10^{-21}	$\sim 10^3$	2.1×10^{-14}

The ^{57}Fe, ^{181}Ta and ^{67}Zn resonances are included for comparison. Conventional notations are used.

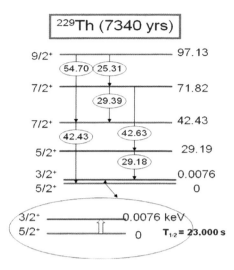

Fig. 18.2 The nuclear decay scheme of ^{229}Th [33]. The inset shows the potential resonance transition

neighboring nuclei or the interactions between the magnetic moments of nuclei and conduction electrons can amount to about 10^{-12} eV, a big term compared to natural line widths. The effect of such broadening mechanisms in the context of resonances in Ag isotopes is well articulated by Davydov [30]. The 39.8 keV resonance in ^{105}Rh (excited by bremsstrahlung) has also received considerable attention [31, 32]. There is evidence for an anisotropy of the gravitational field from these measurements.

Although a low-lying nuclear excitation was predicted in ^{229}Th for many years, only recently has there been a more definitive assignment of energy to this first excited state [33] at 7.6 ± 0.5 eV (see Fig. 18.2), with an estimated natural line width in the range of 17.5 zeV (1 zeV 10^{-21} eV). To put the numbers in perspective, the ^{229}Th resonance will produce a gravitational red shift of about half the line width by having a separation between the clocks by mere 10 μm. This should be compared with the height of Jefferson Tower at Harvard University [25, 26] of 22.6 m, which produced a red shift of only a few thousandths of the ^{57}Fe line width! An attempt to perform this nuclear excitation by pumping the 7.6 eV state with a UV beam from a storage-ring source (Advanced Light Source at Berkeley) is planned [34]. It requires a sequential measurement in which the pump is turned off during the observation of the decay of excited state. Feeding the resonance state by populating the second excited state of 29.19 keV (with ps lifetime) from a storage-ring undulator holds some promise, since the line width of this state matches the band width of undulator radiation. On the downside, the materials physics and technology problems of realizing this resonance are immense. Much of the discussion presented in [27, 30] is even more applicable here. However, if these extremely small hyperfine interactions could be verified, this would be in itself a substantial advance in solid state physics. A potential host for observing the 7.6 eV resonance would be ThO_2, which is diamagnetic, transparent to resonant photons, and provides a cubic environment at the Th^{4+} site. It would be necessary to produce a defect-free crystal with even isotopes of Th to avoid all the interactions of nuclear moments and introduce a low concentration of resonant isotope in the lattice. The recipe for observing the resonance exceeds these basic requirements and will make it intriguing and challenging. Nuclear-electronic double-resonance of a laser-cooled single-ion of $^{229}Th^{3+}$ in a radiofrequency trap [35] may be observed as an alternative approach to resonance detection.

Some of the above dreams may be realized in the next 50 years. One will have to draw heavily on the collective expertise of the atomic, solid-state, and nuclear physics communities, and of modern materials science. New fields of research will be opened, solid-state optical nuclear clocks developed, and better physical standards to test the fundamental laws of physics will be established. The exotic Mössbauer resonances discussed in this section are the only frequency standards that have the sensitivity, when fully realized, to test both the current cosmological theories that attempt to unify gravity with the other fundamental forces and to predict the spatial and temporal variations in the fundamental constants [36].

18.5   Entanglement and coherent control of nuclear excitons

The coherent control of nuclear excitons[2] has been a long-time goal in nuclear physics and is a promising application in nuclear quantum optics. An interesting aspect of coherent control was achieved more than a decade ago in nuclear forward scattering (NFS) of X-rays by exploiting the properties of delocalized nuclear excitations in systems of identical particles. The coherent cooperative decay of nuclei excited by monochromatized synchrotron radiation in NFS was switched on and off by changing the direction of the hyperfine field in a magnetic sample [37]. Recently, a new, ingenious method for coherent control and quantum teleportation has been proposed [38, 39]. This involves the spatial separation of field modes entangled by a single photon in a nuclear excitonic decay. The details of the approach depend on successive switching of the direction of the nuclear hyperfine magnetic field that controls the coherent scattering of photons on nuclei such that two signal pulses are generated out of one initial pump pulse. The resulting two time-resolved correlated signal pulses will have different polarizations and energy at 14.4 keV. It is proposed that the spatial separation and extraction of the entangled field modes will require a set of X-ray polarizers and piezoelectric fast steering mirrors [38, 39]. With a proper choice of switching parameters, specific transitions between hyperfine levels can be restored thus controlling the polarization of the emitted X-ray light. In the coming decades, a new set of nuclear quantum optics experiments using NFS will be performed. These ideas could pave the way to a

---

[2]For definition and details see book Chap. 16.

number of applications for quantum cryptography, quantum key distribution, or macroscopical quantum devices that could profit from single-photon entanglement in the X-ray energy regime.

18.6 Conclusions

In this book chapter, we have addressed anticipated developments in synchrotron radiation sources and speculated their impact on new applications of the Mössbauer Effect during the next few decades. The chapter includes a snapshot of the new and exotic Mössbauer resonance candidates with unprecedented Q factor and their use in the fundamental physics. The field of nuclear quantum optics is only beginning, and we expect the NFS experiments to have a major impact in future. The intent of this chapter is to create a broader science interest in seeking resources to perform nuclear resonance experiments at future storage-ring and XFEL sources.

Acknowledgements This chapter is not intended to be a complete bibliographic account and we regret any omission of important works. The author is grateful to many of his colleagues at many institutions for numerous discussions on the topics covered. In particular, I thank E. E. Alp, J. Evers, E. Hudson, Adriana Pálffy, R. Röhlsberger, R. Rüffer, W. Sturhahn, and T. S. Toellner for their contributions to the field. I wish to acknowledge the support from the US Department of Energy, Office of Science, Office of Basic Energy Sciences under Contract no. DE-AC02-06CH11357.

References

1. R.L. Mössbauer, Nobel Lecture, December 11, 1961 in *Nobel Lectures, Physics 1942–1962* (Elsevier, Amsterdam 1964) (See also reprint in this book)
2. E.J. Seppi and F. Boehm, Phys. Rev. **128**, 2334 (1962)
3. S.L. Ruby, J. Phys. **35**, C6-209 (1974)
4. E. Gerdau et al., Phys. Rev. Lett. **54**, 835 (1985)
5. see articles in E. Gerdau, H. de Waard, Nuclear resonant scattering of synchrotron radiation. Hyperfine Interact. **123/124**, 1 (1999); **125**, 1 (1999)
6. E.E. Alp et al., Hyperfine Interact. **144/145**, 3 (2002)
7. E.E. Alp, W. Sturhahn, T. Toellner, J. Zhao, B.M. Leu, Nuclear Resonance Scattering of Synchrotron Radiation as a Unique Electronic, Structural, and Thermodynamic Probe, Chapter 17 of this volume
8. B. Sepiol et al., Phys. Rev. B **57**, 10433 (1998)
9. B. Sepiol et al., Phys. Rev. Lett. **76**, 3220 (1996)
10. See for example: R. Rüffer, Hyperfine Interact. **141/142**, 83 (2002)
11. W. Sturhahn, J. Phys. Condens. Matter **16**, S497 (2004)
12. S.V. Milton et al., Science **292**, 2037 (2001)
13. Ayvazyan et al., Phys. Rev. Lett. **88**, 104802 (2002)
14. H. Franz et al., PETRA III: DESY's new high brilliance third generation synchrotron radiation source. Synchrotron Radiat. News **19**, 25 (2006)
15. J. N. Galayda et al., J. Opt. Soc. Am. B **27**, B106 (2010)

16. R. Röhlsberger, *Nuclear Condensed Matter Physics with Synchrotron Radiation* (Springer, Heidelberg, 2004)
17. G.K. Shenoy and R. Röhlsberger, Hyperfine Interact. **182**, 157 (2008)
18. W.S. Graves et al., in *Proceedings of the Particle Accelerator Conference, IEEE*, vol. 959 (2003)
19. K-Y. Kim, Yu. Shvyd'ko, S. Reiche, Phys. Rev. Lett. **100**, 244802 (2008)
20. G. Geloni, V. Kocharyan, E. Saldin, Evgeni Cost-effective way to enhance the capabilities of the LCLS baseline, arXiv:1008.3036v1 [physics.acc-ph]
21. T.S. Toellner et al., J. Synchrotron Radiat. **18**, 1 (2011)
22. D. Mukhopadhyay et al., Temporal Modulation of X-rays Using MEMS Micromirrors, Proceedings of *The 16th International Conference on Solid-State Sensors, Actuators and Microsystems*, Beijing, China, June 5–9, IEEE Transducers'11, **T3P.141**, pp. 1562–1564 (2011)
23. N. Ashby et al., Phys. Rev. Lett. **98**, 070802 (2007)
24. J.K. Webb et al., Evidence for spatial variation of the structure constant, arXiv:1008.3907v1 [astro-ph.CO] (2010)
25. T.E. Cranshaw, J.P. Schiffer, A.B. Whitehead, Phys. Rev. Lett. **4**, 163 (1960)
26. R.V. Pound, G.A. Rebka Jr., Phys. Rev. Lett. **4**, 337 (1960)
27. W.Potzel et al., Hyperfine Interact. **72**, 195 (1992)
28. H. Vucetich et al., Phys. Rev. D **38**, 2930 (1988)
29. M. De Francia et al., Mössbauer null redshift experiment II, arXiv:gr-qc/9211005v1 (1992)
30. A.V. Davydov, Phys. Atom. Nucl. **70**, 1182 (2007)
31. Y.Cheng, B. Xia, Y-N. Liu, Q-X. Jin, Chin. Phys. Lett. **22**, 2530 (2005)
32. Y.Cheng et al., Hyperfine Interact. **167**, 833 (2006)
33. B.R. Beck et al., Phys. Rev. Lett. **98**, 142501 (2007)
34. W.G. Rellergert et al., Phys. Rev. Lett. **104**, 200802 (2010)
35. E. Peik, Ch. Tamm, Europhys. Lett. **61**, 181 (2003)
36. V.V. Flambaum, Phys. Rev. Lett. **97**, 092502 (2006)
37. Yu.V. Shvyd'ko et al., Phys. Rev. Lett. **77**, 3232 (1996)
38. A. Pálffy, C.H. Keitel, J. Evers, Phys. Rev. Lett. **103**, 017401 (2009)
39. A. Pálffy, J. Evers, J. Mod. Opt. **57**, 1993 (2009)

Chapter 19
Mössbauer Effect with Electron Antineutrinos

W. Potzel and F.E. Wagner

19.1 Introduction

As shortly as one year after the discovery of the Mössbauer effect with γ-ray photons [1] in nuclear transitions, the possibility has been pointed out [2] to observe recoilless resonant emission and detection of electron antineutrinos (Mössbauer antineutrinos). This suggestion was discussed in greater detail [3] in 1983 and has been revived in 2006 [4]. The basic concept is to use electron antineutrinos ($\bar{\nu}_e$) that are emitted – without recoil – in bound-state β-decays and are resonantly captured – again without recoil – in the reverse bound-state process. In various publications [3–6] the conclusion has been reached that the ^{3}H–^{3}He system appears as the "least impractical" [3]. In the source, ^{3}H decays into ^{3}He and emits an $\bar{\nu}_e$ which can be absorbed in a ^{3}He target to generate ^{3}H. The reaction cross section for weak interaction at low energies (MeV range) is very small, typically 10^{-43} cm^2. The resonance cross section for Mössbauer antineutrinos, however, would be about 25 orders of magnitude larger! This makes the Mössbauer effect with $\bar{\nu}_e$ highly attractive.

We will discuss basic questions and relevant conditions concerning an observation of the Mössbauer effect with $\bar{\nu}_e$ as well as some essential requirements for possible experiments. In particular, we will consider lattice expansion and contraction processes which are not present in conventional Mössbauer spectroscopy (with photons) but might considerably reduce the recoilfree fraction. We will critically review magnetic relaxation phenomena in metallic systems and show that the induced homogeneous line broadening cannot be avoided [5, 7, 8], in contradiction to a more optimistic suggestion [6, 9]. Concerning inhomogeneous broadening we will describe the direct influence of the binding energies of the ^{3}H and ^{3}He atoms in the metal matrix on the energy of the antineutrinos – a problem which will

W. Potzel (✉) · F.E. Wagner
Physics Department, Technical University Munich, 85747 Garching, Germany
e-mail: walter.potzel@ph.tum.de; friedrich.wagner@ph.tum.de

M. Kalvius and P. Kienle (eds.), *The Rudolf Mössbauer Story*,
DOI 10.1007/978-3-642-17952-5_19, © Springer-Verlag Berlin Heidelberg 2012

also make the observation of Mössbauer antineutrinos very difficult [8]. We will describe relativistic effects and their connection to second-order Doppler shifts, which again cannot be discarded since they also give rise to inhomogeneous line broadening. Both homogeneous and inhomogeneous line-broadening effects are estimated to be many orders of magnitude larger than the natural width (minimal width) $\Gamma = \hbar/\tau = 1.17 \times 10^{-24}$ eV as calculated on the basis of the lifetime $\tau = 17.81$ y of ^{3}H [8]. We will also briefly consider the rare-earth system ^{163}Ho–^{163}Dy as an alternative. Using Mössbauer antineutrinos, several basic experiments could be performed concerning, for example, the true nature of neutrino oscillations, the precise determination of oscillation parameters, and the gravitational redshift of antineutrinos as particles, in contrast to photons with a (small) rest mass. The observation of Mössbauer (anti)neutrinos will be a very difficult experiment and may turn out to be unsuccessful with the techniques and ideas available at present.

19.2 Bound-State β-Decay and Its Resonant Character

19.2.1 Usual β-Decay

In the usual β-decay, a neutron in a nucleus transforms into a proton and the emitted electron (e^-) and electron-antineutrino ($\bar{\nu}_e$) occupy states in the continuum. This continuum-state β-decay (Cβ) is a three-body process. Thus the e^- and the $\bar{\nu}_e$ show (broad) energy spectra. The maximum $\bar{\nu}_e$-energy is determined by the Q value and is often called end-point energy. The Q value takes into account the rest masses of all partners involved in the interaction.

19.2.2 Bound-State β-Decay

In the bound-state β-decay (Bβ), again a neutron in a nucleus transforms into a proton. The electron, however, is directly emitted into a bound-state atomic orbit [10]. Since this is a two-body process, the emitted $\bar{\nu}_e$ has a fixed energy $E_{\bar{\nu}_e} = Q + B_z - E_R$. The $\bar{\nu}_e$-energy is determined by the Q value, the binding energy B_z of the atomic orbit the electron is emitted into, and by the recoil energy E_R of the atom formed after the decay. This process occurs, e.g., in the bound-state tritium decay.

The reverse process is also possible: an $\bar{\nu}_e$ and an e^- in an atomic orbit are absorbed by the nucleus and a proton is transformed into a neutron. This is also a two-body process. The required energy of the antineutrino is given by $E'_{\bar{\nu}_e} = Q + B_z + E'_R$, where E'_R is the recoil energy of the atom after the transformation of a proton into a neutron. This process occurs, e.g., in the bound-state electron-capture transformation of ^{3}He irradiated by electron antineutrinos.

Since both $E_{\bar{\nu}_e}$ and $E'_{\bar{\nu}_e}$ are well defined, Bβ has a resonant character which, however, is partially destroyed by the recoil occurring during emission and absorption of the $\bar{\nu}_e$. The resonance cross section is given by [10, 11] $\sigma = 4.18 \cdot 10^{-41} \cdot g_0^2 \cdot \varrho(E_{\bar{\nu}_e}^{\text{res}})/ft_{1/2}$ (in units of cm^2) where $g_0 = 4\pi(\hbar/mc)^3|\psi(R)|^2 \approx 4(Z/137)^3$ for low-Z, hydrogen-like wavefunctions ψ being evaluated at the nuclear surface (R is the radius of the nucleus); m is the electron mass, c is the speed of light in a vacuum, and $\varrho(E_{\bar{\nu}_e}^{\text{res}})$ is the resonant spectral density, i.e., the number of antineutrinos in an energy interval of 1 MeV around $E_{\bar{\nu}_e}^{\text{res}}$. For a super-allowed transition, $ft_{1/2} \approx 1,000$ s.

19.2.3 ^{3}H–^{3}He System

The ^{3}H–^{3}He system has been considered as a favourable example [3, 4] for the observation of Mössbauer $\bar{\nu}_e$. The resonance energy is 18.60 keV, $ft_{1/2} = 1,000$ s, and Bβ/C$\beta = 6.9 \cdot 10^{-3}$ with 80 and 20% of the Bβ events proceeding via the atomic ground state and excited atomic states, respectively [10]. In the case of the Mössbauer resonance only transitions into (from) the ground-state atomic orbit are relevant (see below). If gases of ^{3}H and ^{3}He are used at room temperature the profiles of the emission and absorption probabilities are Doppler broadened and both emission as well as absorption of the electron antineutrinos will occur with recoil. Thus the expected resonance cross-section is tiny, $\sigma \approx 1 \cdot 10^{-42}$ cm^2. Clearly, an observation would require very strong sources of ^{3}H and large target (^{3}He) masses and thus make such an experiment virtually impossible. However, making use of the Mössbauer effect of antineutrinos, i.e., using recoilfree resonant antineutrino emission and absorption, would increase the resonance cross-section by many orders of magnitude (see Sect. 19.3). To prevent the recoil and avoid Doppler broadening, ^{3}H as well as ^{3}He have been considered to be imbedded in Nb metal lattices [4, 6] and this case will be discussed as an example.

19.3 Mössbauer Antineutrinos

19.3.1 Resonance Cross Section

In analogy to conventional Mössbauer spectroscopy with photons, the absorption cross section σ_R at resonance for Mössbauer antineutrinos can be written as [1, 3]

$$\sigma_R = 2\pi \left(\frac{\lambda}{2\pi}\right)^2 s^2\alpha^2 f^2\delta \,, \tag{19.1}$$

where λ is the wavelength of the antineutrinos, s takes into account statistical factors (nuclear spins, isotopic abundance, etc.) and is considered to be of the order of unity,

α is the ratio of bound-state to continuum-state β decays ($\alpha \approx 0.005$ for the ^{3}H–^{3}He system), f is the probability that no phonons are excited in the Nb lattice when the $\bar{\nu}_e$ is emitted or captured, and $\delta = \Gamma / \Gamma_{\exp}$, with Γ the natural linewidth of the weak decay and $\Gamma_{\exp}$ the experimental width due to line broadening. In (19.1) it has been assumed that the properties described by s, α, and f are the same in the absorber (target) and the source. If f and δ would be of the order of unity,

$$\sigma_R \approx 1.8 \cdot 10^{-22} \text{cm}^2 \qquad (19.2)$$

for $\bar{\nu}_e$ with an energy of 18.6 keV ($\lambda \approx 0.67 \cdot 10^{-8}$ cm). This would be a huge cross section for all standards of neutrino interactions and makes Mössbauer (anti)neutrinos highly attractive [4, 5, 12].

Equation (19.1) may look surprising because it is of the Breit-Wigner type [13] and does not explicitly reflect the weak interaction. Equation (19.1) is valid, since we are considering a resonance process and since Mössbauer $\bar{\nu}_e$ are characterized by low energies, where λ is much larger than the dimensions of a nucleus. In this limit, the specific properties of the weak interaction come into play only via the natural linewidth Γ, i.e., the lifetime of the resonant state.

In a real experiment, there are many effects which may drastically reduce σ_R. In the next sections we will demonstrate that it will not be possible to observe $\Gamma = h/(2\pi\tau) = 1.17 \cdot 10^{-24}$ eV, $\tau = 17.81$ y being the lifetime of ^{3}H. Magnetic relaxation phenomena in Nb metal as well as the *direct* influence of the variation of the binding energies (due to the random distributions of ^{3}H and ^{3}He in the Nb lattices) on the energy of the $\bar{\nu}_e$ will cause $\Gamma_{\exp} \gg \Gamma$ by many orders of magnitude. In addition, the factor f^2 in (19.1) may be as small as $\sim 10^{-7}$ due to lattice expansion and contraction processes (see Sect. 19.3.2.2) which are not present in conventional Mössbauer spectroscopy with photons.

19.3.2 Phononless Transitions

The probability of phononless transitions with $\bar{\nu}_e$ is reduced by two kinds of lattice-excitation processes:

1. Momentum transfer due to emission/capture of a $\bar{\nu}_e$,
2. Lattice expansion and contraction when the nuclear transformation occurs during which the $\bar{\nu}_e$ is emitted or absorbed. This does not occur in usual Mössbauer spectroscopy with photons [7, 8].

19.3.2.1 Lattice Excitations Due to Momentum Transfer

Considering the momentum transfer, recoilfree emission and absorption can be achieved by imbedding the atoms of the source (^{3}H) and of the target (^{3}He) into solid-state lattices, e.g., into metallic matrices. Recoilfree processes require that the lattice excitations remain unchanged by the emission and the absorption of the $\bar{\nu}_e$. The recoilfree fraction is given by

$$f_r = \exp\left\{-\left(\frac{E}{\hbar c}\right)^2 \cdot \langle x^2\rangle\right\}, \tag{19.3}$$

where E is the transition energy (18.6 keV for the ^{3}H–^{3}He system) and $\langle x^2\rangle$ is the mean-square atomic displacement. The recoilfree fraction is biggest at low temperatures. However, even at very low temperatures, $f_r < 1$, because of the zero-point motion, which itself is a consequence of the Heisenberg uncertainty principle. In the Debye approximation and in the limit of very low temperatures T,

$$f_r(T \to 0) = \exp\left\{-\frac{E^2}{2Mc^2} \cdot \frac{3}{2k_B\theta}\right\}, \tag{19.4}$$

where θ is the effective Debye temperature, k_B is the Boltzmann constant, and $E^2/(2Mc^2)$ is the recoil energy which would be transmitted to a free atom of mass M. For ^{3}H and ^{3}He in a metal matrix, e.g. Nb, effective Debye temperatures up to $\theta \approx 800$ K have been estimated [4, 6]. Thus $f_r(0) \approx 0.27$ and the probability for recoilfree emission and consecutive recoilfree capture of electron antineutrinos is $f_r^2 \approx 0.07$ for $T \to 0$.

19.3.2.2   Lattice Excitations Due to Lattice Expansion and Contraction

When the $\bar{\nu}_e$ is emitted or captured, one chemical element is transformed into another one, and the $\bar{\nu}_e$ itself takes part in the nuclear processes. More specifically, in the ^{3}H–^{3}He system, ^{3}H and ^{3}He are differently bound in the Nb lattice and use different amounts of lattice space. Thus, when the nuclear transformations occur, the lattice will expand or contract. The lattice deformation energies for ^{3}H and ^{3}He in the Nb lattice are $E_L^H = 0.099$ eV and $E_L^{He} = 0.551$ eV, respectively [14]. Assuming again an effective Debye temperature of $\theta \approx 800$ K one can estimate – in analogy to the situation with momentum transfer – that the probability f_L that this lattice deformation will *not* cause lattice excitations is smaller than

$$f_L \approx \exp\left\{-\frac{E_L^{He} - E_L^H}{k_B\theta}\right\} \approx 1 \cdot 10^{-3}. \tag{19.5}$$

We would like to note that this effect is typical for phononless emission and absorption of neutrinos. In the Mössbauer effect with γ rays it is absent because the atomic number does not change.

   Thus, the total probability for phononless emission and consecutive phononless capture of $\bar{\nu}_e$ is

$$f^2 = f_r^2 \cdot f_L^2 \approx 7 \cdot 10^{-8}; \qquad (19.6)$$

this would be a very tiny probability indeed. The validity of (19.5), including the Debye approximation, has to be checked by theoretical calculations using advanced lattice dynamics [15].

19.3.3 Linewidth

19.3.3.1 Natural Linewidth

The lifetime of ^{3}H is $\tau = 17.81$ y. According to the time-energy uncertainty principle, the natural linewidth Γ, i.e., the minimal width $\Gamma = \hbar/\tau = 1.17 \cdot 10^{-24}$ eV. This value is $\sim$7, respectively $\sim$14, orders of magnitude smaller than those of the Mössbauer resonances (with photons) in ^{107}Ag (93.1 keV, $\tau = 63.9$ s), ^{109}Ag (87.7 keV, $\tau = 57.1$ s), and ^{67}Zn (93.3 keV, $\tau = 13.2 \cdot 10^{-6}$ s). With ^{3}H and ^{3}He being imbedded in Nb metal lattices, it has been argued that this narrow linewidth of the ^{3}H decay can indeed be observed experimentally [6, 9]. This, however, cannot be achieved, because of severe line broadening [8, 16–18].

In conventional Mössbauer spectroscopy (with photons), line broadening is due to variations of electric and magnetic hyperfine interactions and to relativistic effects. One can distinguish between *variations in time* (fluctuating hyperfine interactions, often called relaxation) which lead to homogeneous line broadening and *stationary local variations* within the source and the target (absorber) due to an imperfect lattice which cause inhomogeneous broadening. With Mössbauer $\bar{\nu}_e$, an additional aspect (not present in conventional Mössbauer spectroscopy) is of great importance: The binding energies of the atoms in source (^{3}H) and target (^{3}He) *directly* affect the energy of the $\bar{\nu}_e$. It turns out that a variation of the binding energies in an imperfect lattice will most probably cause the largest contribution to inhomogeneous broadening (see Sect. 19.3.3.3).

19.3.3.2 Homogeneous Broadening

Homogeneous broadening is caused by electromagnetic relaxation. For example, spin-spin interactions between nuclear spins of ^{3}H and ^{3}He and with the spins of the nuclei of the metallic lattice lead to fluctuating magnetic fields. Contrary to the notion in [6] and [9], magnetic relaxations are stochastic processes and can not be described by a periodic energy modulation of the magnetic hyperfine interaction. The simplest magnetic relaxation model consists of a three-level system: the groundstate and two excited hyperfine-split states (energy separation $\hbar\Omega_0$) between which transitions (so-called relaxation processes) take place with an average rate Ω. Stochastic processes lead to sudden, irregular transitions between hyperfine-split states originating, e.g., from magnetic spins of many neighbouring nuclei. Thus the wave function of the emitted particle (photon, $\bar{\nu}_e$) is determined by random (in time) frequency changes which lead to line broadening characterized by the time-energy uncertainty relation [7]. As a consequence, the broadened line *cannot* be decomposed into multiple sharp lines, in contradiction to the claim made in [9].

With stochastic processes, three frequency regimes can be distinguished [19]:

1. $\Omega \ll \Omega_0$. For the simple three-level system, two lines separated by $\hbar\Omega_0$ will be observed. The lines will be broadened to an effective experimental linewidth $\Gamma_{\text{exp}} \approx \hbar\Omega$ as suggested by the time-energy uncertainty principle. Only in the limit of very small Ω, will the lines exhibit the natural width. With increasing Ω the lines broaden.
2. $\Omega \approx \Omega_0$. The lines are severely broadened. In fact, as suggested by the time-energy uncertainty principle, the intensity is distributed over a broad pattern which extends roughly over a range given by the total hyperfine splitting $\hbar\Omega_0$.
3. $\Omega \gg \Omega_0$. This is the frequency regime of motional narrowing. The system stays only for a short time (typically $1/\Omega$) in one of the hyperfine-split levels and then stochastically jumps into the other one. Thus an averaging process over the energies of both levels takes place, causing a collapse into one line at the center of the hyperfine-splitting pattern. In the high-frequency limit ($\Omega \to \infty$) the linewidth is again practically natural.

For typical hyperfine splittings due to nuclear spin-spin interaction in metallic lattices, one has $\Omega_0 \approx 10^5\,\text{s}^{-1}$. Typical relaxation times for ^{3}H and ^{3}He are $T_2 \approx 2\,\text{ms}$ in a Pd lattice, and $T_2 \approx 79\,\mu\text{s}$ for NbH [4, 20]. The latter gives a linewidth Γ_{exp} due to homogeneous broadening, $\Gamma_{\text{exp}} \approx 5 \times 10^{-11}\,\text{eV} \approx 4 \times 10^{13}\,\Gamma$. Due to stochastic relaxation processes, homogeneous broadening by $\sim$13 orders of magnitude has to be expected in the ^{3}H–^{3}He system because the stochastic relaxation frequencies are far below the motional-narrowing regime. As a consequence, for the system ^{3}H–^{3}He in Nb metal, $\Gamma_{\text{exp}} \approx 4 \cdot 10^{13}\,\Gamma$. Thus in (19.1), $\delta_h \approx 2.5 \cdot 10^{-14}$ [7, 8], leading to a drastic reduction of σ_R.

19.3.3.3   Inhomogeneous Broadening

In imperfect lattices, inhomogeneous broadening is caused by stationary effects, in particular by impurities, lattice defects, variations in the lattice constant, and other effects which destroy the periodicity of the lattice. This is a critical issue for the Nb system since – as will be described in Sect. 19.4 – the source contains a large amount of ^{3}H but very little ^{3}He whereas the target contains a lot of ^{3}He but practically no ^{3}H [4, 5]. Clearly, the ^{3}H and ^{3}He distributions on the interstitial sites will be random and will destroy lattice periodicity. An additional process due to variations of the zero-point energy will be discussed in Sect. 19.3.3.4. In general, for photon Mössbauer spectroscopy, in the best single crystals, inhomogeneous broadening is of the order of 10^{-13} to $10^{-12}\,\text{eV}$ [17]. This broadening is due to hyperfine interactions only.

The situation with Mössbauer antineutrinos is much more serious and is related to the variation of the binding energies of ^{3}H and ^{3}He in an inhomogeneous metallic lattice. The binding energy E_B *directly* affects the energy of the $\bar{\nu}_e$ in two ways: (a) The difference ΔE_B in binding energies is given to (or taken from) the $\bar{\nu}_e$; (b) when E_B changes, also the Debye temperature, i.e., the vibrational energy E_{vib} of the lattice changes, and also this difference ΔE_{vib} of the vibrational energies directly affects the energy of the $\bar{\nu}_e$. In a perfect lattice, the differences ΔE_B and ΔE_{vib} in the lattice between ^{3}He and ^{3}H [14], which are given to the $\bar{\nu}_e$ when ^{3}H decays into ^{3}He, exactly compensate the corresponding differences needed for the reverse process, when the $\bar{\nu}_e$ is captured and ^{3}He transforms into ^{3}H. In a real experiment, such an exact compensation implies that the chemical bonds of ^{3}H in the source and in the target (absorber) are equal and that the same condition is fulfilled for ^{3}He. However, in the source and target intended to be used in a real experiment (see Sect. 19.4) such a compensation even within an experimental linewidth of $\Gamma_{exp} \approx 5 \times 10^{-11}$ eV (estimated for homogeneous broadening only, see Sect. 19.3.3.2 above) will be extremely unlikely considering the fact that the binding energies per atom are in the eV range [14]. Vibrational energies per atom are typically one to two orders of magnitude smaller (otherwise the lattice would not be stable). Inhomogeneities in real lattices as mentioned above and, in particular, the vastly different amounts of ^{3}H and ^{3}He in the source and in the target, will result in variations of E_B and E_{vib} of the ^{3}H and ^{3}He atoms in source and target much larger than $\Gamma_{exp} \approx 5 \times 10^{-11}$ eV and thus destroy the resonance condition.

Let us estimate the consequences of the variations of the binding energies E_B. In usual Mössbauer spectroscopy with photons, different binding strengths due to inhomogeneities in source and absorber (target) affect the photon energy only via the *change* in the mean-square nuclear charge radius between the groundstate and the excited state of the nucleus. This leads to the *isomer shift*, i.e., a shift of the photon energy typical for *hyperfine interactions* (in the 10^{-7}–10^{-9} eV range) [21]. Thus variations of isomer shifts due to an inhomogeneous lattice cause a line broadening which is also in the neV range. Since, in the nuclear transformations, the $\bar{\nu}_e$ energy is *directly* affected by E_B, one has to expect that the variations of the $\bar{\nu}_e$ energy are much larger than in usual Mössbauer spectroscopy, probably in the 10^{-6} eV regime or even larger. Thus, inhomogeneous line broadening is estimated to give $\delta_I \ll 10^{-12}$ [7,8], probably $\delta_I \approx 10^{-18}$ and can be more serious than homogeneous broadening.

19.3.3.4 Relativistic Effects

Another contribution to line broadening in an imperfect lattice is due to relativistic effects. An atom vibrating around its equilibrium position in a lattice does not only exhibit a mean-square displacement $\langle x^2 \rangle$ but also a mean-square velocity $\langle v^2 \rangle$. According to the Special Theory of Relativity this causes a time-dilatation effect which results in a reduction of frequency (energy) [22, 23]:

$$\Delta\omega = \omega - \omega' = -v^2\omega/(2c^2). \tag{19.7}$$

Since this reduction is proportional to $(v/c)^2$ it is often called second-order Doppler shift (SOD).

In usual Mössbauer spectroscopy, within the Debye model the energy shift (SOD) between source (s) at temperature T_s and target (t) at temperature T_t is given by [24]

$$\Delta E = \hbar \cdot \Delta\omega = \frac{9k_{\mathrm{B}}E}{16Mc^2}(\theta_s - \theta_t) + \frac{3k_{\mathrm{B}}E}{2Mc^2}\left[T_s \cdot f(T_s/\theta_s) - T_t \cdot f(T_t/\theta_t)\right] \tag{19.8}$$

with the Debye integral

$$f(T/\theta) = 3\left(\frac{T}{\theta}\right)^3 \cdot \int_0^{\theta/T} \frac{x^3}{\exp(x) - 1}\,\mathrm{d}x, \tag{19.9}$$

and M denoting the mass of the Mössbauer nucleus.

The second-order Doppler shift of (19.8) can also be derived quantum mechanically, assuming only the equivalence of energy and mass [24, 25]. Equation (19.8) then reads

$$\Delta E = \hbar \cdot \Delta\omega = \frac{9k_{\mathrm{B}}}{16}\frac{\Delta M}{M}(\theta_s - \theta_t) + \frac{3k_{\mathrm{B}}}{2}\frac{\Delta M}{M}\left[T_s \cdot f(T_s/\theta_s) - T_t \cdot f(T_t/\theta_t)\right], \tag{19.10}$$

where $\Delta M = E/c^2$.

When emitting a γ quantum, the mass of the emitting nucleus is reduced by ΔM from $(M + E/c^2)$ to M. As a consequence, in the source the frequencies of the lattice vibrations where M is involved, are increased after γ emission. This increase in lattice energy of the Mössbauer atom is taken from the γ ray whose energy is thus reduced accordingly. This effect is tiny due to the factor $\Delta M/M$, which is typically 10^{-6}. In the target (absorber), the reverse process takes place, i.e., after absorbing the γ ray, the mass of the Mössbauer nucleus is increased and the lattice energy of the Mössbauer atom is reduced. Thus, if source and target are made from the same material ($\theta_s = \theta_t$) and are at the same temperature ($T_s = T_t$), the influence of the SOD cancels (see (19.8) and (19.10)).

In general, at low temperatures the temperature-dependent term in (19.8) can be neglected, even more so if source and target are at about the same temperature (e.g., in a liquid-He bath at 4.2 K). However, even in the low-temperature limit, the first term in (19.8) and (19.10) which is caused by the zero-point vibrations, cannot be neglected.

For Mössbauer $\bar{v}_e$, the situation is more complex because emission and absorption are always connected to transformations of one chemical element into another one. Therefore the vibrational energy of the lattice will change not only because of the mass of the emitting or absorbing atom changes, but also because the bonding forces change when one element transforms into another. The latter is not a relativistic effect and has already been described in the previous Sect. 19.3.3.3. Here we restrict the discussion to the (much smaller) effect which is due to the change of the mass of the Mössbauer nucleus. Considering again the whole sequence of emission and absorption, one reaches the conclusion [5] that in a Mössbauer $\bar{\nu}_e$ experiment it is *not* required that the chemical bonds (i.e. the Debye temperatures) of ^{3}H and ^{3}He in the metal matrix have to be the same. The Mössbauer resonance condition can be fulfilled if the chemical bond for ^{3}H is the same in source and target, and if the same condition is valid for ^{3}He. For Mössbauer $\bar{\nu}_e$, it is difficult to satisfy this condition because the surroundings of ^{3}H (and those of ^{3}He) including not only nearest but also more distant neighbours in source and target should be as similar as possible. This is critical, since, as already mentioned in Sect. 19.3.3.3, source and target contain vastly different amounts of ^{3}H and ^{3}He [4, 5] which may lead to a difference in the effective Debye temperatures for ^{3}H (and ^{3}He) between source and target.

Another critical issue is that a *variation* of the binding energies E_B of ^{3}H and ^{3}He in an imperfect lattice of both source or target will result in a variation of the effective Debye temperatures and thus also in a *variation* of the zero-point energies. Taking again only the relativistic effect (due to the change of the mass of the Mössbauer nucleus) into account we arrive at the following estimate: If the effective Debye temperature varies by only $\Delta\theta = 1$ K, $(\Delta E/E) = 9k_B\Delta\theta/(16Mc^2) \approx 2 \times 10^{-14}$, which corresponds to a lineshift of 3×10^{14} times the natural width Γ [7]. Thus, this broadening effect is expected to give $\delta_{SOD} \approx 3 \cdot 10^{-15}$ comparable to or even smaller than δ_h (see Sect. 19.3.3.2). In other words, the broadening effect δ_{SOD} is important but less serious than δ_I, the latter being due to variations of E_B and E_{vib} as discussed in Sect. 19.3.3.3.

Additionally, with a natural width $\Gamma = \hbar/\tau = 1.17 \cdot 10^{-24}$ eV, the red (blue) shift due to the gravitational field of the earth would be extremely important. The gravitational redshift is given by

$$\delta E/E = gh/c^2, \tag{19.11}$$

where g is the gravitational acceleration on earth and h is the difference in height. Thus for the ^{3}H–^{3}He system, a shift by $\delta E = \Gamma$ corresponds to a difference in height of $h_\Gamma = \frac{\Gamma}{E} \cdot \frac{c^2}{g} \approx 5.8 \cdot 10^{-13}$ m, much smaller than the diameter of an atom! An experiment with finite dimensions of source and target (absorber) would thus severely suffer from line broadening due to gravitation. If, however, the neutrino resonance is already significantly broadened due to the other effects described above, the influence of gravity would be much less important. For instance, if we assume $\Gamma/\Gamma_{exp} \approx 2.5 \cdot 10^{-14}$ as a typical value for the line broadening, one obtains $h_{\Gamma_{exp}} \approx 23$ m. Line broadening due to gravitation would thus be negligible in such a more realistic case.

19.4 Fundamental Difficulties: Alternative Systems

Taking into account the basic considerations of Sect. 19.3, there exist mainly four serious difficulties to observe Mössbauer $\bar{\nu}_e$ in the system ^{3}H–^{3}He imbedded in Nb metal:

1. The probability of phonon-less emission and detection might be very low ($f^2 \approx 7 \cdot 10^{-8}$) in the system ^{3}H–^{3}He, largely due to the expansion and contraction of the Nb lattice.
2. Homogeneous line broadening ($\delta_h \lesssim 2.5 \cdot 10^{-14}$) alone may lead to an experimental linewidth of $\Gamma_{\mathrm{exp}} \gtrsim 4 \cdot 10^{13} \Gamma$.
3. Inhomogeneous line broadening due to the random distribution of ^{3}H and ^{3}He in the Nb lattice may lead to a line broadening of $\sim 10^{18} \Gamma$, i.e., $\delta_I \approx 10^{-18}$. This is mainly caused by the fact that inhomogeneities in an imperfect lattice *directly* influence the energy of the $\bar{\nu}_e$.
4. Variations of the binding energies of ^{3}H in the source and of ^{3}He in the target will cause variations in the zero-point energy and thus lead to an inhomogeneous line broadening ($\delta_{\mathrm{SOD}} \approx 3 \cdot 10^{-15}$) of $> 10^{14} \Gamma$.

In addition, there will be many technological difficulties [26] which have not been considered in the present article. Altogether it is not certain that the system ^{3}H–^{3}He, imbedded in Nb metal, will work. Possible alternatives can be searched for in two directions: different host lattices and different systems exhibiting the bound-state β-decay. Instead of Nb metal, graphite and related substances like graphfoil might be favourable lattices. The ^{12}C nucleus in its groundstate does not have a magnetic moment. Thus magnetic relaxation phenomena between the nuclei of the ^{3}H and ^{3}He atoms and the ^{12}C nuclei of the graphite lattice can be avoided. Still, it has to be investigated how ^{3}H and ^{3}He are bound in this lattice. In particular, it has to be assured that the ^{3}He atoms formed after the ^{3}H decay occupy the same type of (interstitial) lattice site as the ^{3}He atoms in the target and that the lattice changes accompanying the elemental transformations are small.

The recoilfree fraction due to momentum transfer for ^{3}He could be small because ^{3}He might be expected to be loosely bound in graphite. ^{3}H might be bound more strongly. As mentioned earlier (see Sect. 19.3.3.4), the chemical bonds of ^{3}H and ^{3}He are not required to be the same in the lattice; however, the bond of ^{3}H (and of ^{3}He) has to be the same in source and target to avoid destruction of the resonance by SOD and variations of the SOD.

Concerning alternative systems, ^{163}Ho–^{163}Dy was considered as the next-best case [3]. There are three major advantages:

1. The Q value of 2.6 keV (thus the $\bar{\nu}_e$ energy) is very low, the mass of the nuclei is large. As a consequence, the recoilfree fraction is expected to be $f_r \approx 1$.
2. Due to the similar chemical behaviour of the rare earths, the lattice deformation energies of ^{163}Ho and ^{163}Dy can be expected to be similar, which could lead to a probability of phononless transitions f larger by up to seven orders of magnitude than for the ^{3}H–^{3}He system.

3. Due to the large mass, the relativistic effects mentioned in Sect. 19.3.3.4 will be smaller, typically by a factor of ~ 50.

The main disadvantage will be the large magnetic moments due to the 4f electrons of the Rare-Earth atoms. Electronic magnetic moments are typically three orders of magnitude larger than nuclear moments. Thus, line broadening effects due to magnetic relaxation phenomena will again be decisive for a successful observation of Mössbauer $\bar{\nu}_e$. Profound technological knowledge concerning the fabrication of high-purity materials, preferably as single crystals, in kg quantities will be required [3]. Further feasibility studies should be undertaken.

19.5   Principle of Experimental Setup

The system which has been discussed most thoroughly is ^{3}H–^{3}He [3, 4]. As described in Sects. 19.2–19.4, it is essential that ^{3}H and ^{3}He occupy the same type of (regular or interstitial) lattice site. According to [4], the tetragonal interstitial site (TIS) in Nb metal would be a favourable candidate. ^{3}H can be loaded into Nb metal with all of the ^{3}H atoms occupying the TIS.

Usually, it is much more difficult to introduce ^{3}He into metals. For Nb metal, however, the "tritium trick" can be applied [4]: After introducing ^{3}H to form Nb^3H_x, some of the tritium decays into ^{3}He, $\sim 2/3$ of which stay at the same TIS as ^{3}H ($\sim 1/3$ occupy octahedral interstitial sites) if the Nb metal is kept below 200 K. After some accumulation time (typically 200 days), the remaining ^{3}H in the metal has to be removed or isotopically exchanged by deuterium. According to [4], after an accumulation time of 200 days, the atomic fraction ^{3}He/Nb is ~ 0.03. After desorption of ^{3}H, a huge fraction of the ^{3}He atoms still occupy interstitial sites if the temperature of the sample is kept below 200 K.

A transmission experiment like in conventional Mössbauer spectroscopy including a Doppler drive to move the source (or absorber) and a detector does not look feasible, in particular because the neutrinos transmitted through the target (absorber) would have to be recorded by a neutrino detector comparable in efficiency with a photon detector. Such a device does not exist because the weak interaction determins the detection process for neutrinos. Rather than performing a transmission experiment one could determine the β-activity of ^{3}H generated in the ^{3}He target after a measuring time t_M. In such an experiment, t_M starts at that moment when the ^{3}H source and the ^{3}He target are arranged in their fixed positions. For $t_M = 65\,d \approx 0.01\tau$ (with $\tau = 17.81$ y), a ^{3}H source of 1 kCi, a ^{3}He-target mass of 100 mg and a base line (separation betweeen source and target) of 5 cm, a β-activity of the generated ^{3}H in the ^{3}He target of ~ 40 decays per day could be expected if *only* homogeneous broadening ($\Gamma_{exp} \approx 5 \cdot 10^{-11}$ eV, see Sect. 19.3.3.2) would be present and *no* phonons would be generated by lattice expansion and contraction, i.e., $f^2 = f_r^2 = 0.07$ (see Sect. 19.3.2.2). This would correspond to an effective resonance cross section of $\sim 3 \cdot 10^{-37}$ cm^2 which is still more than 5 orders of magnitude larger than the resonance cross section without Mössbauer effect (see Sect. 19.2). To determine the produced ^{3}H activity by counting its radioactive decays would be very difficult because of the low energy of the emitted electrons. Therefore extraction of ^{3}H and ^{3}He from the target may be unavoidable. One could then imagine to apply accelerator mass spectroscopy to detect the ^{3}H. The separation of ^{3}H and ^{3}He should be easy when both atoms are fully ionized. One could also think of using ^{3}H-NMR and record the intensity of the NMR signal. With all methods, the main difficulty will be the low concentration of ^{3}H. Still, for all standards of neutrino interactions, a count rate of 40 decays per day (as mentioned above) would be large but most probably unattainable because of the fundamental difficulties discussed in Sect. 19.4.

19.6 Interesting Experiments

If Mössbauer $\bar{\nu}_e$ could be observed successfully, several basic questions and interesting experiments in particle physics could be addressed [7]:

1. The neutrino mass eigenstates $|\nu_1\rangle$, $|\nu_2\rangle$, $|\nu_3\rangle$ with masses m_1, m_2, m_3, respectively, do not coincide with the flavour eigenstates $|\nu_e\rangle$, $|\nu_\mu\rangle$, $|\nu_\tau\rangle$, which in turn can be expressed by a coherent superposition of the mass eigenstates:

$$|\nu_\alpha\rangle = \sum_{k=1}^{3} U_{\alpha k} |\nu_k\rangle \qquad (19.12)$$

with $\alpha = e, \mu, \tau$ and

$$|\nu_k\rangle = \sum_{\alpha} U_{\alpha k}^{*} |\nu_\alpha\rangle. \qquad (19.13)$$

Here U is the so-called (unitary) mixing matrix and $|\nu_k\rangle$ are stationary mass eigenstates with the time dependence

$$|\nu_k(t)\rangle = e^{-iE_k t}|\nu_k\rangle, \qquad (19.14)$$

where E_k is the energy of $|\nu_k\rangle$. Thus, a pure flavour state $|\nu_\alpha\rangle$ at $t = 0$, develops with time into

$$|\nu(t)\rangle = \sum_{k=1}^{3} U_{\alpha k} e^{-iE_k t} |\nu_k\rangle = \sum_{k,\beta} U_{\alpha k} U_{\beta k}^{*} e^{-iE_k t}|\nu_\beta\rangle \qquad (19.15)$$

which is a superposition of the flavour eigenstates $|\nu_e\rangle$, $|\nu_\mu\rangle$, and $|\nu_\tau\rangle$.

This means that ν_e, ν_μ, and ν_τ change their identity when they propagate in space and time [27–35].

As an example, let us consider the case of two mass eigenstates ($|\nu_1\rangle$, $|\nu_2\rangle$) and two flavour eigenstates ($|\nu_e\rangle$, $|\nu_\mu\rangle$) only. Then using (19.12), $|\nu_e\rangle$ and $|\nu_\mu\rangle$ can be expressed by $|\nu_1\rangle$ and $|\nu_2\rangle$ according to

$$\begin{pmatrix} \nu_e \\ \nu_\mu \end{pmatrix} = \begin{pmatrix} \cos\Theta & \sin\Theta \\ -\sin\Theta & \cos\Theta \end{pmatrix} \begin{pmatrix} \nu_1 \\ \nu_2 \end{pmatrix}, \tag{19.16}$$

where

$$U = \begin{pmatrix} \cos\Theta & \sin\Theta \\ -\sin\Theta & \cos\Theta \end{pmatrix} \tag{19.17}$$

is the mixing matrix and Θ is the mixing angle. The probability that a ν_e is transformed in flight into a ν_μ can be derived from (19.15) as

$$P_{\nu_e \to \nu_\mu}(L) = \sin^2 2\Theta \cdot \sin^2\left(\frac{\Delta m_{12}^2}{4} \cdot \frac{L}{E}\right) = \sin^2 2\Theta \cdot \sin^2\left(\pi \frac{L}{L_0}\right), \tag{19.18}$$

where L is the distance (base line) between neutrino source and target and

$$L_0 = 2.48 \frac{E}{|\Delta m_{12}^2|} \tag{19.19}$$

is the oscillation length with $|\Delta m_{12}^2| = |m_1^2 - m_2^2|$. L_0 is given in meters, when the neutrino energy E and the mass-squared difference $|\Delta m_{12}^2|$ are in units of MeV and eV^2, respectively. For a given $|\Delta m_{12}^2|$, the oscillation length L_0 is short if E is small, which is the case for Mössbauer (anti)neutrinos. If $\Theta = 0$, $P_{\nu_e \to \nu_\mu} = 0$ and no mixing would occur. If $\Theta = 45°$, $\sin^2 2\Theta = 1$ and the mixing is maximal, i.e., within the distance L_0 a ν_e will completely transform into a ν_μ and then back to a ν_e again.

In Nature, there are (at least) three mass eigenstates with [34] $\Delta m_{12}^2 \approx 7.6 \cdot 10^{-5}$ eV^2, $\Theta_{12} \approx 34°$; $\Delta m_{23}^2 \approx 2.5 \cdot 10^{-3}$ eV^2, $\Theta_{23} \simeq 45°$; $\Theta_{13} < 13°$, the exact value being still unknown.

Mössbauer $\bar{\nu}_e$ are characterized by a very sharp energy distribution. In connection with the time-energy uncertainty relation $\Delta t \cdot \Delta E \geq \hbar$, according to which it should take a long time span Δt for a system consisting of a superposition of mass eigenstates to change if its uncertainty in energy ΔE is small (as in the case of Mössbauer $\bar{\nu}_e$), the question has been asked: Do Mössbauer $\bar{\nu}_e$ oscillate? Considering the evolution of the neutrino state in *time only*, neutrino oscillations are characterized as a non-stationary phenomenon, and Mössbauer $\bar{\nu}_e$ would not oscillate because of their extremely narrow energy distribution. An evolution of the neutrino wave function in *space and time*, however, would make oscillations possible in both the non-stationary and also in the stationary (Mössbauer $\bar{\nu}_e$) case. Thus the investigation of Mössbauer $\bar{\nu}_e$ could lead to a better understanding of the true nature of neutrino oscillations [29–35]. For further details, see [36–38].

2. Due to the low energy of $E = 18.6\,\text{keV}$ (or of $2.6\,\text{keV}$ only for the ^{163}Ho–^{163}Dy system), Mössbauer-antineutrino oscillations would allow us to use ultra-short base lines (see (19.19)) to determine oscillation parameters. For example, for the determination of the still unknown mixing angle Θ_{13}, a base line of only $\sim 10\,\text{m}$ (instead of $\sim 1{,}500\,\text{m}$ as required for reactor neutrinos [39–41]) would be sufficient. In addition, very small uncertainties for Θ_{13} and accurate measurements of $\Delta m_{12}^2 = m_1^2 - m_2^2$ and $\Delta m_{31}^2 = m_3^2 - m_1^2$ could become possible [42].

3. At present, only upper limits for the neutrino masses are known. In addition, the mass hierarchy, i.e., the question which of the three mass eigenstates has the lowest mass, could not yet be determined. As shown in [43,44], Mössbauer $\bar{\nu}_e$ could settle this question using a base line of $\sim 300\,\text{m}$. In such an experiment, the superposition of two oscillations with different frequencies can be observed: a low-frequency (so-called solar neutrino) oscillation driven by $\Delta m_{12}^2 = m_1^2 - m_2^2 \approx 7.6 \cdot 10^{-5}\,\text{eV}^2$ and a high-frequency (so-called atmospheric neutrino) oscillation driven by $\Delta m_{32}^2 = m_3^2 - m_2^2 \approx 2.5 \cdot 10^{-3}\,\text{eV}^2$. In the case of the normal mass hierarchy ($m_1 < m_2 \ll m_3$) the phase of the atmospheric-neutrino oscillation advances, whereas it is retarded for the inverted hierarchy ($m_3 \ll m_1 < m_2$), by a measurable amount for every solar-neutrino oscillation [43,44].

4. Oscillating Mössbauer $\bar{\nu}_e$ could be used to search for the conversion $\bar{\nu}_e \rightarrow \nu_{\text{sterile}}$ [45] involving additional mass eigenstates. Sterile neutrinos (ν_{sterile}) would not show the weak interaction and therefore such a conversion would manifest itself by the disappearance of $\bar{\nu}_e$. The results of the LSND (Liquid Scintillator Neutrino Detector) experiment [46] are consistent with a hypothetical $\Delta m^2 \approx 1\,\text{eV}^2$ [4]. For such a value, the oscillation length would only be $\sim 5\,\text{cm}$! This would require ultra-short base lines, which would be difficult to realize except with Mössbauer neutrinos.

5. In contrast to photons, $\bar{\nu}_e$ are particles with a (small) rest mass and, in principle, could behave differently from photons in a gravitational field. However, gravitational redshift $\bar{\nu}_e$ measurements in the gravitational field of the earth could only be performed if an experimental linewidth of $\Gamma_{\text{exp}} \approx 10^{-10}\,\text{eV}$ (or smaller) could be reached [7]. Otherwise the required difference in height between ^{3}H source and ^{3}He target would have to be too large (see Sect. 19.3.3.4).

19.7 Conclusions

The system ^{3}H (source) and ^{3}He (target) has been considered for a possible observation of recoilfree resonant (Mössbauer) emission and absorption of electron antineutrinos. Several basic aspects have been discussed. The experiment is very challenging. In particular, we have pointed out that – contrary to the claims of [9] – it will not be possible to reach the natural linewidth ($\Gamma = \hbar/\tau = 1.17 \cdot 10^{-24}\,\text{eV}$)

because already homogeneous broadening by itself would result in an experimental linewidth of $\Gamma_{exp} \approx 5 \times 10^{-11}$ eV $\approx 4 \times 10^{13} \Gamma$. Even if an experimental linewidth of $\Gamma_{exp} = 5 \times 10^{-11}$ eV could be reached (which would still be highly fascinating for many $\bar{\nu}_e$ experiments), the chances for the observation of Mössbauer $\bar{\nu}_e$ are drastically reduced because of two effects which are not present with conventional Mössbauer spectroscopy with photons: (a) an additional reduction of the recoilfree fraction because of *lattice expansion and contraction at the time of the nuclear transition*, and (b) the *direct influence of the binding energies* of ^{3}H and ^{3}He atoms in the metal matrix on the energy of the electron antineutrino. The variation of these binding energies in the inhomogeneous metal matrix can lead to a variation of lineshifts and thus to inhomogeneous line broadenings much larger than 10^{-9} eV (see Sect. 19.3.3.3). The Rare-Earth system ^{163}Ho–^{163}Dy offers several advantages, in particular a large probability of phononless emission and detection. However, magnetic relaxation processes and technological requirements still have to be investigated in detail to decide if the ^{163}Ho–^{163}Dy system is a promising alternative. Even if for Mössbauer $\bar{\nu}_e$ an experimental linewidth of not less than 10^{-7} eV (about 20 times the natural linewidth of the ^{57}Fe resonance) could be reached, highly interesting experiments concerning basic questions in physics could be performed.

Acknowledgements It is a pleasure to thank S. Roth, Physik-Department E15, Technische Universität München, for fruitful discussions. This work was supported by funds of the Deutsche Forschungsgemeinschaft DFG (Transregio 27: Neutrinos and Beyond), the Munich Cluster of Excellence (Origin and Structure of the Universe), and the Maier-Leibnitz-Laboratorium (Garching).

References

1. R.L.M. Mössbauer, Z. Physik **151**, 124 (1958)
2. W.M. Visscher, Phys. Rev. **116**, 1581 (1959)
3. W.P. Kells, J.P. Schiffer, Phys. Rev. C**28**, 2162 (1983)
4. R.S. Raghavan, hep-ph/0601079 v3, (2006)
5. W. Potzel, Phys. Scr. **T127**, 85 (2006)
6. R.S. Raghavan, arXiv: 0805.4155 [hep-ph] and 0806.0839 [hep-ph] (2008)
7. W. Potzel, J. Phys. Conf. Ser. **136**, 022010 (2008). arXiv: 0810.2170 [hep-ph]
8. W. Potzel, F.E. Wagner, Phys. Rev. Lett. **103**, 099101 (2009). arXiv: 0908.3985 [hep-ph]
9. R.S. Raghavan, Phys. Rev. Lett. **102**, 091804 (2009)
10. J.N. Bahcall, Phys. Rev. **124**, 495 (1961)
11. L.A. Mikaélyan, B.G. Tsinoev, A.A. Borovoi, Sov. J. Nucl. Phys. **6**, 254 (1968)
12. W. Potzel, Acta Physica Polonica **B40**, 3033 (2009). arXiv: 0912.2221 [hep-ph]
13. G. Breit, E. Wigner, Phys. Rev. **49**, 519 (1936)
14. M.J. Puska, R.M. Nieminen, Phys. Rev. B**10**, 5382 (1984)
15. D. Ceperley, University of Illinois at Urbana-Champaign, USA, private communication
16. B. Balko, I.W. Kay, J. Nicoll, J.D. Silk, Hyperfine Interact. **107**, 283 (1997)
17. R. Coussement, G. S'heeren, M. Van Den Bergh, P. Boolchand, Phys. Rev. B**45**, 9755 (1992)
18. W. Potzel et al., Hyperfine Interact. **72**, 197 (1992)
19. H.H. Wickman, G.K. Wertheim, in *Chemical Applications of Mössbauer Spectroscopy*, ed. by V.I. Goldanskii, R.H. Herber (Academic, New York, 1968), pp. 548; in particular Fig. 11.10

20. M.E. Stoll, T.J. Majors, Phys. Rev. B**24**, 2859 (1981) and references therein
21. G.K. Shenoy, F.E. Wagner (eds.), *Mössbauer Isomer Shifts* (North-Holland, Amsterdam, 1978)
22. A. Einstein, Ann. Physik **17**, 891 (1905)
23. A. Einstein, *The Meaning of Relativity*, 6th edn. (The Electric Book Company, London, 2001)
24. H. Wegener, *Der Mössbauereffekt und seine Anwendung in Physik und Chemie*, Hochschul-taschenbücher 2/2a (Bibliographisches Institut, Mannheim, 1966)
25. B.D. Josephson, Phys. Rev. Lett. **4**, 341 (1960)
26. J.P. Schiffer, Phys. Rev. Lett. **103**, 099102 (2009)
27. B. Kayser, Phys. Rev. D**24**, 110 (1981)
28. M. Beuthe, Phys. Rep. **375**, 105 (2003) and references therein
29. S.M. Bilenky, F. von Feilitzsch, W. Potzel, Phys. Part. Nucl. **38**, 117 (2007); and J. Phys. G: Nucl. Part. Phys. **34**, 987 (2007)
30. S.M. Bilenky, F. von Feilitzsch, W. Potzel, J. Phys. G: Nucl. Part. Phys. **35**, 095003 (2008). arXiv: 0803.0527 v2 [hep-ph]
31. E.Kh. Akhmedov, J. Kopp, M. Lindner, J. High Energy Phys. **0805**, 005 (2008). arXiv: 0802.2513 [hep-ph]
32. E.Kh. Akhmedov, J. Kopp, M. Lindner, J. Phys. G: Nucl. Part. Phys. **36**, 078001 (2009). arXiv: 0803.1424v2 [hep-ph]
33. S.M. Bilenky, F. von Feilitzsch, W. Potzel, J. Phys. G: Nucl. Part. Phys. **36**, 078002 (2009). arXiv: 0804.3409 [hep-ph]
34. S.M. Bilenky, F. von Feilitzsch, W. Potzel, Proceedings of the 13th International Workshop on Neutrino Telescopes, Venice 2009, edited by M. Baldo Ceolin, p. 315, arXiv:0903.5234 [hep-ph]
35. J. Kopp, J. High Energy Phys. **0906**, 049 (2009). arXiv: 0904.4346 [hep-ph]
36. W. Potzel, Phys. Part. Nucl. **42**, 1268 (2011), arXiv:1012.5000 [hep-ph]
37. S.M. Bilenky, Phys. Part. Nucl. **42**, 1012 (2011), arXiv:1012.4966 [hep-ph]
38. S.M. Bilenky, F. von Feilitzsch, W. Potzel, J. Phys. G: Nucl. Part. Phys. **38**, 115002 (2011), arXiv:1012.4966 [hep-ph]
39. DOUBLE CHOOZ collaboration, F. Ardellier et al., arXiv: hep-ex/0606025
40. T. Lasserre, Europhys. News **38**(N4), 20 (2007)
41. T. Kawasaki et al., AIP Conf. Proc. **981**, 202 (2008)
42. H. Minakata, S. Uchinami, New J. Phys. **8**, 143 (2006). hep-ph/0602046
43. H. Minakata, H. Nunokawa, S.J. Parke, R. Zukanovich Funchal, Phys. Rev. D**76**, 053004 (2007). arXiv: hep-ph/0701151
44. S.J. Parke, H. Minakata, H. Nunokawa, R. Zukanovich Funchal, Nucl. Phys. Proc. Suppl. **188**, 115 (2008). arXiv: 0812.1879 [hep-ph]
45. V. Kopeikin, L. Mikaelyan, V. Sinev. hep-ph/0310246v2 (2003)
46. C. Athanassopoulos et al., LSND collaboration. Phys. Rev. Lett. **81**, 1774 (1998)

Part IV
Epilogue

The final chapter starts with Mössbauer's "second love," the study of the mass structure of neutrinos by searching for neutrino flavor oscillations during their propagation in space and time, for which he initiated with Felix Boehm from CalTech the very first experiments using reactor antineutrinos. It concludes with what became known as "The second Mössbauer effect," a headline of a "Spiegel" article, announcing the creation of a Physics Department with 20 professors at the Technische Hochschule München in 1963 as a precursor of the university reform in Germany initiated by a student revolution in 1968 known by the catchphrase "Unter den Talaren der Muff von tausend Jahren".

Chapter 20
Neutrinos

F. von Feilitzsch and L. Oberauer

20.1 Neutrino Physics

Rudolph Mössbauer was one of the pioneers in modern experimental neutrino physics in Europe. Together with colleagues from USA and France (and later from Switzerland) he initiated and conducted neutrino oscillation experiments in Europe, which achieved a high sensitivity on very small neutrino mass differences, as we will show in the following chapters.

At that time (∼1978) finite neutrino masses and a possible mixing of neutrino states with the consequence of flavor oscillations were considered to be very exotic by many physicists. Today we know that neutrinos do oscillate and as a consequence the standard model of particle physics has to be extended. In addition the long standing solar neutrino puzzle could be solved by observing this phenomenon.

20.1.1 Phenomenology of Neutrino Oscillations

Neutrinos are elementary particles and exist in the form of three flavors which couple to the Z^0 exchange boson of weak interaction. This we know from the Z^0-decay width and from the observation of charged current neutrino interactions, in which the corresponding charged leptons (electron, muon, tau) appear in the end channel. Neutrinos are very light particles compared with charged leptons and quarks. Today an upper limit of 2.2 eV for the neutrino mass is coming from the tritium decay endpoint measurements performed at Mainz (Germany) [1] and Troitsk (Russia) [2].

F. von Feilitzsch (✉) · L. Oberauer
Physics Department, Technical University Munich, 85747 Garching, Germany
e-mail: franz.vfeilitzsch@ph.tum.de; lothar.oberauer@ph.tum.de

M. Kalvius and P. Kienle (eds.), *The Rudolf Mössbauer Story*,
DOI 10.1007/978-3-642-17952-5_20, © Springer-Verlag Berlin Heidelberg 2012

Bruno Pontecorvo suggested neutrino–antineutrino oscillations as explanation of the observed deficit of solar neutrinos as seen in the Homestake experiment by Davis [3], in close analogy to $K^0 - \bar{K}^0$-oscillations in the hadronic sector. Later he and others imposed neutrino flavor oscillations, a periodic transition in time of the probability to observe a distinct neutrino flavor. Precondition for neutrino oscillations are existing neutrino mass eigenstates which determine the propagation of neutrinos in vacuum. However neutrinos are created and detected in weak interactions. Flavor eigenstates v_α ($\alpha = e, \mu, \tau$) therefore have no fixed mass, but they are rather linear superpositions of the mass eigenstates v_i ($i = 1, 2, 3$):

$$\begin{pmatrix} v_e \\ v_\mu \\ v_\tau \end{pmatrix} = \begin{pmatrix} U_{e1} & U_{e2} & U_{e3} \\ U_{\mu 1} & U_{\mu 2} & U_{\mu 3} \\ U_{\tau 1} & U_{\tau 2} & U_{\tau 3} \end{pmatrix} \times \begin{pmatrix} v_1 \\ v_2 \\ v_3 \end{pmatrix}.$$

If no sterile neutrinos exist the matrix $U_{\alpha i}$ is unitary and

$$v_i = U_{\alpha i}^\dagger v_\alpha.$$

Here, we define the assignment v_α to v_i in such a way, that the absolute values of the diagonal elements of the mixing matrix are maximal. Hence, no mass hierarchy in the sense $m_j > m_i$ for $j > i$ is assessed a priori.

The mixing matrix $U_{\alpha i}$ (sometimes called Pontecorvo–Maki–Nakagawa–Sato matrix) has three real free parameter, which can be interpreted as rotation angles and one imaginary phase δ, which can cause CP-violation in the leptonic sector. In case the neutrino is its own anti-particle (a so-called Majorana particle) additional imaginary phases may occur. The matrix can be parameterized in the form

$$\begin{pmatrix} v_e \\ v_\mu \\ v_\tau \end{pmatrix} = \begin{pmatrix} 1 & 0 & 0 \\ 0 & c_{23} & s_{23} \\ 0 & -s_{23} & c_{23} \end{pmatrix} \begin{pmatrix} c_{13} & 0 & s_{13}e^{i\delta} \\ 0 & 1 & 0 \\ -s_{13}e^{i\delta} & 0 & c_{13} \end{pmatrix} \begin{pmatrix} c_{12} & s_{12} & 0 \\ -s_{12} & c_{12} & 0 \\ 0 & 0 & 1 \end{pmatrix} \begin{pmatrix} v_1 \\ v_2 \\ v_3 \end{pmatrix}.$$

Here, $s_{ij} = \sin \Theta_{ij}$ and $c_{ij} = \cos \Theta_{ij}$ with the rotation angles Θ_{ij}. The evolution of the mass eigenstates $v_i(t) = v_i(0) \exp(-i(E_i t - k_i x))$ will differ if E_i and k_i do not coincide and this will create interference effects leading to neutrino flavor oscillations. Here E_i and k_i are the energy and k-vector of the eigenstate v_i, respectively. Therefore a neutrino with a determined flavor at $t = 0$ (e.g. in the solar fusion reaction $p + p \rightarrow d + e^+ + v_e$) will rotate into another flavor, if the differences in the mass eigenvalues of v_i don't vanish and if the unitary matrix is not diagonal. In the simplified case of only two neutrino flavors the probability P to observe the same flavor at a distance L from the neutrino source is

$$P = 1 - \sin^2 2\Theta_{12} \cdot \sin^2 \left(1.267 \frac{\Delta m_{12}^2}{eV^2} \frac{L/m}{E/MeV} \right)$$

and the probability P to observe the other flavor is $P = 1 - P$. Here $\Delta m_{12}^2 = | m_2^2 - m_1^2 |$ is the quadratic mass splitting between the eigenvalues m_2 and m_1. The amplitude of these periodic functions is $\sin^2 2\Theta_{12}$ whereas the oscillation length is determined by the quotient $\Delta m_{12}^2 \cdot L / E$. Generally the oscillation of three flavors has to be considered. However, the two neutrino scenario is often a good approximation, as the oscillation phenomena in vacuum decouple due to the strong hierarchy of the measured Δm^2 values (see discussion of the oscillation results later).

In the standard model of particle physics neutrinos have no mass. Only left-handed neutrino states exist and helicity is a good quantum number. Furthermore the standard model predicts lepton flavor number conservation. Therefore the observation of neutrino oscillation is a clear evidence for physics beyond the standard model.

Basically there are two types of experiments searching for neutrino oscillations, appearance and disappearance experiments. In the former the appearance of neutrino interactions with a "strange" flavor is searched for. A typical example are accelerator experiments, where the neutrino energies are high enough to produce all flavors in charged current weak processes. The advantage of appearance experiments is their high sensitivity for the oscillation amplitude. Typical examples of disappearance experiments are reactor and solar neutrino experiments. In both cases one is looking for a deficit in the neutrino flux and for a distortion in the energy spectrum. In disappearance experiments the neutrino energies are not sufficient to produce charged muons or tau-leptons. However, with such experiments one can be very sensitive to small values of the mass splitting Δm^2, especially at long baselines L.

20.2 Oscillation Experiments at the ILL and Gösgen

20.2.1 The ILL Experiment

Rudolf Mössbauer proposed together with Felix Böhm from the California Institute of Technology in 1977 a first European–American experiment to search for neutrino oscillations at the nuclear research reactor (ILL) in Grenoble. Nuclear reactors are an intense source of electron antineutrinos, which are produced in the beta decays of fission isotopes. The proposal was based on publications of Bruno Pontecorvo and Samuel Bilenki proposing the possibility of neutrino oscillations. At that time this phenomenon was widely considered as a rather exotic model, even though the basic considerations used by Pontecorvo and Bilenki were based on general quantum mechanical principles.

First data from the Homestake solar neutrino experiment [3] indicated the possibility of oscillations as this experiment reported a neutrino flux deficit from the sun. However, for a reliable prediction of the solar neutrino signal in this experiment a rather good understanding of the complete solar fusion processes was required, and indeed an astrophysical explanation could not be excluded at that time. In a personal letter by Telegdi to Mössbauer and Böhm he expressed severe doubts that the proposed experiment would be justified.

The proposal sent by Mössbauer to the German Bundesministerium für Forschung und Technologie (BMFT) was as short as one page, but nevertheless accepted with a budget of about 1.5 million DM. It was certainly not very elaborate and even assumed rather optimistic detection efficiencies. The collaboration consisted of the French Institute de Science Nucleare in Grenoble (ISN), the CALTECH in California and the TUM in Munich. A first suggestion of the experimental set-up was proposed by Walther Mampe at that time at the Institut Laue Langevin (ILL) in Grenoble. This concept later was modified together with Cavignac and Vignon of the ISN. The detection of the electron antineutrinos emitted by the research reactor ILL with a nominal thermal power of 57 MW was based on the inverse beta decay reaction $\bar{\nu}_e + p \rightarrow e^+ + n$. The reaction threshold is at about 1.8 MeV. The positron was detected in an organic liquid scintillator together with both the 0.511 MeV gammas from the succeeding annihilation process.

Neutrino spectroscopy was possible by measuring the energy deposited and measured in the scintillator by the prompt positron signal, as the neutron carries away only a very small fraction of the energy. A delayed signal was detected from neutron capture in adjacent ^{3}He proportional counters. The search for these delayed coincidences allows an efficient rejection of background events coming from beta- and gamma radiation. Alpha events are suppressed due to the quenching effect in a liquid scintillator anyway. However, neutrons remained a dangerous source of background radiation.

The liquid scintillator was based on mineral oil to which 30% pseudocumene was added. It was produced by Nuclear Enterprice Ltd. under the product name NE235C and allowed for an efficient pulse shape discrimination to distinguish fast neutrons from gamma or beta induced signals. A total volume of 380 l of scintillator was used. The ^{3}He proportional chambers where developed together at the ISN and the TUM in Munich. The construction of the chambers was performed at the TUM physics department in Garching. The experimental preparations started in 1977, before funding was obtained in 1978. Data taking occurred from summer 1979 until the end of 1980. The detector was placed in the cavern B42 of the reactor containment at a distance of 8.75 m from the core. Figure 20.1 shows the detection principle as well as the schematic experimental set-up.

The data were compared to a calculated neutrino spectrum based on the weighted sum of neutrino spectra from the beta decays of fission products. For this a time period of more than one weak was taken to allow for contributions of long lived beta decaying isotopes. The expected neutrino spectrum was based partly on experimentally identified individual beta decaying fission isotopes and on theoretically evaluated rates performed by Petr Vogel from CALTECH, using nuclear physics parameters [4]. In addition independent direct measurements of the cumulative beta spectrum of fission isotopes from ^{235}U thermal neutron fission where taken using the BILL beta spectrometer at the ILL reactor. This high resolution spectrometer allowed for the placement of a fission target material close to the reactor core at a neutron flux of $\sim 10^{15}$ n/cm^2 s. The used ^{235}U sample was enclosed by two thin Ni-foils which stopped fission isotopes, whereas the subsequently emitted beta particles could be detected in the spectrometer. These measurements were

Detection principle   Detector assembly

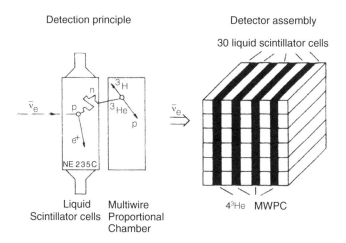

Fig. 20.1 Detection principle and schematic set-up of the ILL and Gösgen neutrino detector

extended to ^{239}U and ^{241}Pu fission beta decay spectra in a collaboration with Klaus Schreckenbach. The measured cumulative beta spectra were transformed into neutrino spectra taking into account an averaged Fermi function [5,6].

The data showed no clear evidence for neutrino oscillations, even though a $\sim$15% deficit of the neutrino flux was interpreted by different authors to provide an indication for neutrino oscillations. On the level of about 1.5 sigma an oscillation pattern in the detected neutrino spectrum could have been interpreted as neutrino oscillation with an neutrino mass squared difference of 2.2 eV^2 (or 1.1 eV^2 as an harmonic second minimum).

20.2.2 The Gösgen experiment

As there was no theoretical prediction on the expected neutrino oscillation wavelength and hence no restriction on the ratio L/E_ν, the next step in the search for neutrino oscillation required a change in the distance between the neutrino source (i.e. the reactor core) and the detector. Different options where considered, in particular the reactor station in Biblis (Germany) with two reactors each providing 2.5 GW thermal power and the reactor in Gösgen (Switzerland) providing one nuclear reactor with 2.8 GW thermal power.

The Swiss Institute for Nuclear Research (SIN), now named Paul-Scherrer-Institute, offered substantial experimental and administrative support and the experiment was decided to be performed at the Gösgen nuclear power plant. Three positions were used to allow for a relative measurement of the neutrino flux as a function of the distance between the reactor core and the detector. The positions were at 37.9, 45.9, and 64.7 m, respectively. The set-up of the ^{3}He-chambers was improved in order to allow for position determination in two dimensions. In addition the position dependence of pulse shape discrimination within each scintillator cell was determined by precise calibrations. All these improvements on the detector led to a further rejection of background events and made the neutrino measurement at a distance of 64.7 m possible. The signal to background ratio after all cuts was ~1 : 1 at this last distance. The background was determined experimentally during the annual reactor-off phases, which lasted about one month.

After the move from the ILL in Grenoble to Gösgen in 1980 data taking was started 1981 and ended in 1986 after the measurement of the neutrino spectra at all three positions. This allowed to reduce systematic uncertainties significantly, as it was the first measurement which was largely independent of the knowledge of the neutrino spectra emitted by the nuclear reactor core. In addition it allowed for an independent verification of the experimentally determined spectral shape emitted from the fission products of the Uranium- and Plutonium isotopes as determined in the experiments mentioned above. Figure 20.2 shows the positron energy spectra obtained at the three positions with respect to the expected flux and shape for the no-oscillation hypothesis.

No evidence for neutrino oscillations was found in the Gösgen experiments. As a consequence new limits on the mixing parameter were published [7]. For the first time the region $\sin^2(2\Theta) \sim 10^{-2}$ was investigated. Figure 20.3 shows the exclusion plots obtained from the Gösgen experiment. Today we know that the distance between a neutrino source with energies in the ~MeV range and the detector was to short to observe oscillations, as $\Delta m^2 \approx 8 \times 10^{-5}$ is to small for neutrino oscillation to develop in a measurable way.

20.2.3   The Neutrino Decay experiment at Bugey

Neutrino decay is an inevitable consequence of non-vanishing neutrino mixing and masses. For instance the radiative neutrino decay $\nu_j \rightarrow \nu_i + \gamma$ or even the decay $\nu_3 \rightarrow \nu_i e^+ e^-$ become possible. For the later process $m_3 - m_i > 2m_e \sim 1 \, \text{MeV}$ must hold which is excluded today, but was still open around 1990. At that time a most stringent upper limit $m_3 < 31 \, \text{MeV}$ was reported by the ARGUS group by investigating tau decays [8]. The decay in an electron–positron pair would occur on tree level and hence its probability p_{dec} scales like $p_{dec} \propto U_{e3}^2 \cdot (U_{e1}^2 + U_{e2}^2)$. Unitarity implies $U_{e1}^2 + U_{e2}^2 = 1 - U_{e3}^2 \sim 1$ as it was already known, that the admixture U_{e3} is small and hence $p_{dec} \propto U_{e3}^2$. An exact calculation yields for the transition rate $\Gamma_{e^+e^-}$ in the center of mass system

$$\Gamma_{e^+e^-} = \mid U_{e3} \mid^2 \Phi(m_3)(m_3/\text{MeV})^5 \times 3.5 \times 10^{-5} \, \text{s}^{-1},$$

with a phase space factor $\Phi(m_3) \sim 1$ for $m_3 \gg 1 \, \text{MeV}$.

The production of ν_3 at an electron (anti)neutrino source again scales with U_{e3}^2 and therefore the total probability to observe this decay mode $p_{obs} \propto U_{e3}^4$.

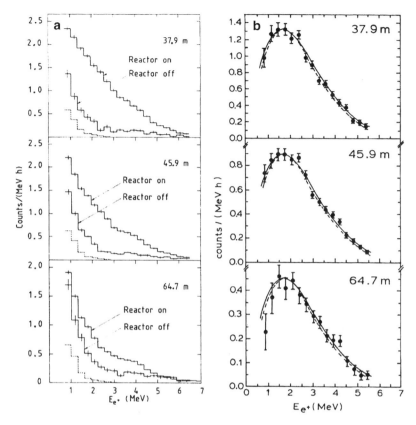

Fig. 20.2 Energy spectra of neutrino candidates (**a**) after cuts as measured in the Gösgen experiment for reactor on and off phases. Accidental background contribution is given by the *dashed lines*. (**b**) measured and predicted positron spectra for the no-oscillation hypothesis after background subtraction. *Solid line* for a fit on the reactor spectrum. *Dashed line* for a bin based analysis of the reactor spectrum

From 1991 until 1993 a search for radiative and $e^+ - e^-$ neutrino decay modes was performed at the nuclear power reactor at Bugey, France. Two x–y position sensitive multi-wire proportional chambers with dimensions $2m \cdot 2m$ each had been constructed to search for four-fold coincident signals from a $e^+ - e^-$-pair emitted by neutrino decay in a 8 m^3 large He-filled decay volume. Six additional counters acted as veto system against penetrating external background. The signature for radiative neutrino decay $\nu_j \rightarrow \nu_i + \gamma$ was the emission of photons inside the Helium bag and their detection in the proportional chambers at a threshold of about 2 keV. The detector was built up inside the reactor containment at an distance of 18.5 m to the core with a corresponding neutrino flux of $\sim 10^{13}$ cm^{-2} s^{-1}.

In order to determine background rates several reactor on–off sequences were used. No effect was observed and new limits on both decay modes could be set.

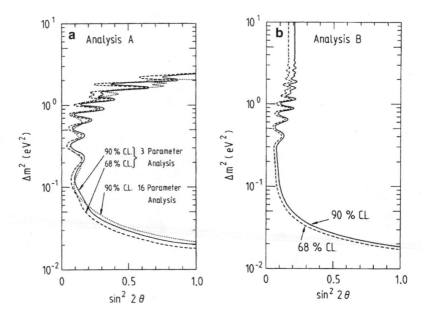

Fig. 20.3 Exclusion plots for the oscillation amplitude as function of the mass difference squared obtained in the Gösgen experiment. In analysis A only the comparison of data between the three different positions is used, whereas in B also the knowledge of the reactor fission spectra is included

For the $e^+ - e^-$ neutrino decay mode new limits of U_{e3}^2 in the mass range between 1 MeV and $\sim$5 MeV were obtained [9] as depicted in Fig. 20.4. A new limit on the radiative lifetime $\tau_\nu > 180\,\mathrm{s}(m_i/\mathrm{eV})$ (90% cl) was obtained [10]. Both constraints are still the most stringent limits obtained in laboratory experiments so far. In the standard model of particle physics with neutrino masses and mixing included the radiative decay mode appears on loop level and the corresponding transition rate $\Gamma_\gamma \sim |U_{e3}|^2 (m_3/\mathrm{eV})^5 \times 4 \times 10^{-37}\,y^{-1}$ is by far to small to deduce meaningful constraints on the mixing parameter $|U_{e3}|^2$.

20.3 Solar Neutrino Experiments

The energy released in the solar center is generated by thermonuclear fusion of hydrogen to helium. By the sum reaction $4\mathrm{H} + 2e^- \rightarrow\,^4\mathrm{He} + 2\nu_e$ a total energy of 26.73 MeV is released, of which the neutrinos carry away 0.59 MeV in average. From the well known solar luminosity $S = 8.5 \times 10^{11}\,\mathrm{MeV}/(\mathrm{cm}^2\,\mathrm{s})$ the total solar neutrino flux at Earth can be estimated to be $\Phi_\nu \sim 2 \cdot S/26.1\,\mathrm{MeV} \sim 6.5 \times 10^{10}\,\mathrm{cm}^{-2}\,\mathrm{s}^{-1}$.

Two cycles (pp and CNO) are responsible for the energy release inside of stars. However, for a relatively small star like the sun the pp-cycle is expected to dominate,

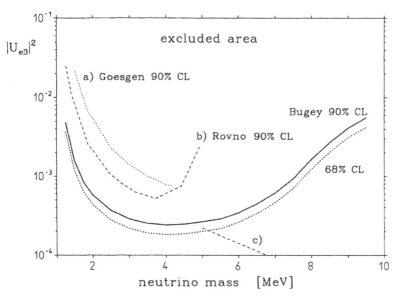

Fig. 20.4 Exclusion plot for $| U_{e3}^2 |$ (1-sigma and 90% cl) as obtained in the Bugey experiment in comparison with former reactor (Goesgen, Rovno) and accelerator (c) experiments

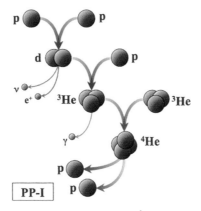

Fig. 20.5 Thermal nuclear fusion cycle ppI

as the high coulomb-barrier is suppressing the CNO-cycle. Indeed the results of all solar neutrino experiments today confirm, that in the sun the pp-cycle is the dominating process. In Figs. 20.5 and 20.6 the main thermal fusion reactions in the center of the sun are depicted.

The exact branching ratios of these fusion processes inside the sun depend on the nuclear cross sections at energies in the $\approx 10\,\text{keV}$ range, which corresponds to the central solar temperature. The branching between ppI and ppII/III is given by the cross section ratio of the (^{3}He,^{3}He) and (^{3}He,^{4}He) reactions and is crucial for the

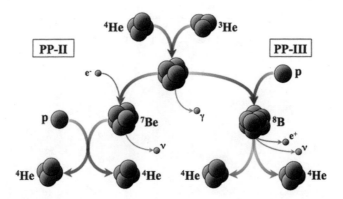

Fig. 20.6 Thermal nuclear fusion cycle ppII and ppIII

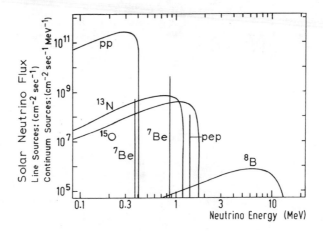

Fig. 20.7 Solar neutrino energy spectrum. Water Cherenkov experiments only can detect solar ^{8}B-neutrinos at high energies. The Gallium based radiochemical experiments GALLEX/GNO and SAGE provide integral information about all neutrino branches, including the most abundant pp-neutrinos as the energy threshold is 233 keV

strengths of the solar ^{8}B- and ^{7}Be-neutrino fluxes. These relevant cross sections have been measured at these low energies for the very first time by the LUNA experiment [11] and are important input parameter of solar models. In Fig. 20.7 the spectral distribution of solar neutrinos according to the solar model is shown.

With the pioneering Homestake neutrino experiment [3] Ray Davis proved the basic idea of thermal nuclear fusion processes as the mechanism of energy generation in the sun. Deep underground the production of Ar-atoms in a 615 t (metric tons) tank filled with perchlorethylen (C_2Cl_4) has been detected since 1970. They stem from the reaction $\nu_e + ^{37}\text{Cl} \rightarrow ^{37}\text{Ar} + e^-$ and were extracted from the target tank by He-flushing after an exposition of in average 60–70 days. The Ar-atoms decay back via electron capture with a lifetime of 50 days. This decay was detected in small proportional tubes. In average about 30 Ar-decays were counted after each extraction, proving the emission of neutrinos in the sun. This experiment opened the window for neutrino astronomy and for this pioneering work R. Davis was honored with the Nobel prize in 2002. However, there remained a puzzle. The measured neutrino rate was only ∼1/3 of the expected one. As the threshold for the reaction is rather high (814 keV), only a small part of the neutrinos emitted in the pp-cycle could be detected. It was argued, that the observed anomaly could be explained by changing parameters of the solar model.

The first direct detection of solar neutrinos with energies above ∼7 MeV succeeded in the Kamiokande experiment (in the underground facility at the Kamioka mine, Japan) via elastic neutrino electron scattering. The collaboration used a water Cherenkov detector with the possibility to determine the direction of the incident neutrino. Again a clear deficit in the neutrino flux was observed, but no direct proof for oscillations could be found. Today SuperKamiokande (SK) is the successor water Cherenkov experiment with a fiducial volume of 22.5 kt. The measured solar neutrino flux was in agreement with the Kamiokande result, but it improved the accuracy of the measurement significantly due to better statistics and lower background conditions. The Sun–Earth directionality of solar neutrinos was proven in an impressive manner, as depicted in Fig. 20.8. A detailed analysis of the ^{8}B-neutrino energy spectrum and a search for a day-night effect was performed (see [12] and references therein), as matter effects in the Sun and Earth predict spectral distortions in the energy regime above ∼5 MeV for small values of the mixing angle $\Theta_{12} \leq 1°$. However, no smoking gun for oscillations of solar neutrinos could be found and only large values for the mixing amplitude remained as a good solution. However, large mixing angles were not favored at that time as it was assumed, that

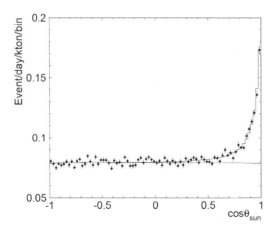

Fig. 20.8 $\cos\Theta_{sun}$ distribution of electron recoil tracks after elastic scattering of solar ^{8}B-neutrinos in SuperKamiokande. Relativistic electrons have tracks which show in the same direction as these of solar neutrinos. From [12]

lepton mixing should be like in the quark sector, where the off-diagonal elements in the unitary mixing matrix are small, indeed.

An important break-through in neutrino physics appeared in 1998. Besides solar neutrinos, also high energy ($E_\nu \sim 1$ GeV) atmospheric neutrinos are detected in SK. These neutrinos are part of the secondary cosmic rays and are emitted in charged π- and K-decays where only ν_μ and ν_e (plus their antiparticles), but no ν_τ are generated. In SK electron and muon like events can be separated at these energies and a clear disappearance of the ν_μ-signal depending on the azimuth angle was observed. The disappearance of ν_μ is maximal for up-going neutrinos [13]. In contrast the ν_e-signal is not disappearing at all and is in good agreement with the expected flux. This was the first clear evidence for neutrino oscillations that was published during the neutrino conference 1998 in Takayama, Japan. Detailed analysis shows, that $\nu_\mu \longleftrightarrow \nu_\tau$ oscillations had been revealed. The mixing amplitude is large, even close to the maximal value $\Theta_{23} \sim 45^0$ and the mass splitting in the range $\Delta m_{23}^2 \sim 10^{-3}$ eV^2. Now it became clear, that the leptonic and quark mixing matrix differ significantly and large mixing does occur with neutrinos.

20.4   The GALLEX and GNO experiment

In the last decade of the twentieth century two radiochemical experiments (GALLEX and SAGE) succeeded in measuring the integral electron neutrino flux at an energy threshold of 233 keV using the reaction $\nu_e +^{71} Ga \rightarrow^{71} Ge + e^-$. This allowed for the first time to comprehend all branches, including the most abundant pp-neutrinos (see Fig. 20.7) of the solar pp-chain.

The GALLEX project as well as the Gallium Neutrino Observatory (GNO), the successor experiment of GALLEX, was a radiochemical detector located in Hall A of the Gran Sasso underground laboratory. Cosmic rays are shielded effectively by an over-burden of $\sim$3600 m water equivalent. The experiment measured solar neutrinos for 12 years (1991–2003), thus monitoring one complete solar cycle ($\sim$11 years). GALLEX was performed within an international collaboration with Till Kirsten from the MPIK Heidelberg (Germany) as spokesman and under participation of the group of Rudolf Mössbauer at TUM.

About 101 t of a GaCl$_4$ solution in water and HCl acted as neutrino target. The GeCl$_4$ molecules were extracted by flushing N$_2$, then synthesis of GeH$_4$ and mixture with Xenon was performed in order to form a gas mixture which was filled in very small proportional tubes. There the back-decay of ^{71}Ge-atoms (electron capture, half-life 11.4d) in the energy regions of K- and L-capture (10.4 and 1.3 keV) was observed. Typically about ten atoms have been identified in one run, which corresponds to an exposition time of about one month.

In order to obtain the solar neutrino flux, the theoretical cross-section of the neutrino reaction on ^{71}Ga has to be used. This introduces systematic uncertainties, as there exist branches to excited states in ^{71}Ge and the total cross-section cannot be determined by the half-life of the back-decay alone. This and furthermore the

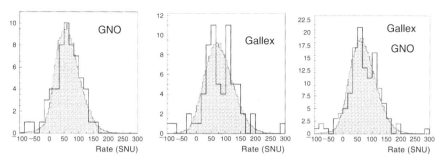

Fig. 20.9 Distribution of all single run results in GALLEX and GNO. The expected averaged value of solar models is at ~130 SNU. The averaged experimental value is (69.3 ± 5.5) SNU

total efficiency of the experiment was tested by an artificial ^{51}Cr neutrino source two times. The final GALLEX/GNO result on the neutrino flux is given in SNU (Solar Neutrino Unit), which corresponds to one neutrino capture per second in 10^{36} target atoms: (69.3 ± 5.5) SNU [14]. The distribution of single run results from GALLEX and GNO are shown in Fig. 20.9. This is in very good agreement with the SAGE experiment, which measured (66.9 ± 5.3) SNU [15]. Solar models predict a significant higher flux of about 130 SNU and the neutrino deficit is evident (5.5 sigma for GALLEX/GNO alone).

No astrophysical solution to the solar neutrino puzzle remained after GALLEX/GNO and SAGE as here the dominant neutrino pp-branch $p + p \rightarrow^2 H + e^+ + \nu_e$ is included. Of course, this was no proof for neutrino oscillations yet. The next steps in solar neutrino physics were the measurement of the energy distribution and flavor content in real time experiments.

20.5 Observation of Neutrino Flavor Transitions

The solar neutrino energy spectrum has its endpoint around 15 MeV. Hence, the appearance of ν_μ or ν_τ cannot be probed directly. However, flavor transition was found indirectly in the SNO (Sudbury Neutrino Observatory) heavy water Cherenkov experiment[16] by detecting solar neutrinos via these channels:

$$\nu_e + d \rightarrow e^- + 2p,$$
$$\nu_x + d \rightarrow \nu_x + p + n \quad (x = e, \mu, \tau),$$
$$\nu_x + e^- \rightarrow \nu_x + e^-.$$

The first reaction is a pure charge current interaction, whereas the second is neutral charge current and therefore flavor independent. The last reaction, elastic neutrino electron scattering, is also possible for all neutrino flavors, but the cross section for ν_e is higher with respect to $\nu_{\mu,\tau}$. About 1 kt of heavy water was used as target and the analysis threshold was around 5 MeV. Recently a new analysis with a lower threshold was presented [17]. In any case only solar ^{8}B-neutrinos

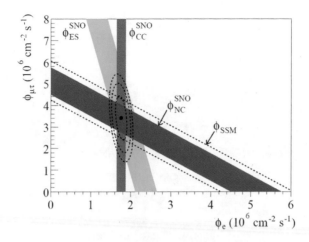

Fig. 20.10 Flux of solar ^{8}B-neutrinos which are μ or τ vs flux of ν_e deduced from the three reactions used in SNO [18]. Φ_{CC}, Φ_{NC}, and Φ_{ES} depict the measured neutrino fluxes vis charged current, neutral current, and electron scattering, respectively. The standard model value $\Phi_{\mu\tau} = 0$ is clearly excluded. The sum $\Phi_{\mu\tau} + \Phi_e \sim 5 \times 10^6$ cm^{-2} s^{-1} is in good agreement with the solar model (Φ_{SSM})

were observed in SNO. In the standard model, thus without flavor transition, the neutrino fluxes measured in the three reaction channels have to be equal. However, the observed ν_e flux was smaller with respect to the neutral current measurement by roughly a factor three (see Fig. 20.10).

This observation can only be explained by flavor transition of solar neutrinos. The measured rate of electron scattering events confirmed this scenario [18]. Therefore individual lepton number conservation is violated. The observed total neutrino flux is in good agreement with solar models and therefore the long standing solar neutrino puzzle was solved. Although neutrino oscillations was the most favored explanation for the SNO result, also other mechanism beyond the standard model were discussed.

However, the SNO result in conjunction with the Gallium experiments was very important for a second aspect. Neutrino propagation in matter and vacuum differs. Neutrinos undergo coherent forward scattering in matter, which leads in close analogy to photons to a refractive index or an effective neutrino mass. This effect depends on the matter density as well as on the neutrino energy. Furthermore this effect is different for ν_e and ν_μ or ν_τ, as ν_e scatter via charged and neutral current interaction with electrons in matter, but ν_μ and ν_τ solely via neutral current interaction. This introduces an additional phase difference between the mass states during propagation in matter and can influence the oscillation probability. For solar neutrinos the situations is even more complicated, as matter density varies dramatically during their flight through the Sun. It turns out, that solar neutrinos with energies above ~ 2 MeV are generated very close to the mass state ν_2 in the solar center, where matter density is very high, although they are born as electron neutrinos. With a large probability this state remains adiabatically as ν_2 during propagation from the center to the solar surface. However, ν_2 in vacuum couples mostly to ν_μ and ν_τ flavors. Numerical calculations yield a survival probability of about 30% for these high energy ν_e. Those neutrinos do not oscillate in the sense described above, they undergo flavor transition already inside the Sun. On the other hand vacuum oscillations are the dominating effect for low energy solar neutrinos, like the most abundant pp-neutrinos ($E_{max} = 430\,\text{keV}$) and their survival probability is around 60%. Therefore, one expects a transition in the spectrum between low energy and high energy solar neutrinos. Comparison between the Gallium and SNO/SK results gave a first indication of the matter effect. However, as long as no individual pp-branch below $\sim 2\,\text{MeV}$ could be measured the conclusion was quite indirect. It was one of the important aims of Borexino (see next section) to probe this effect by measuring the solar ^{7}Be-flux at $0.860\,\text{MeV}$.

20.6 The Proof of Neutrino Oscillations

A combined oscillation analysis of all solar neutrino results obtained so far indicated a common favored solution: the so called large mixing angle (LMA) scenario with $\Delta m_{12}^2 \sim 10^{-4}\,\text{eV}^2$ and rather large mixing angle $\Theta_{12} \sim 35°$. It became clear that this could be probed by neutrino experiments which are completely independent from solar physics.

Again in the Kamioka mine (Japan), the KamLAND experiment was measuring nuclear reactor neutrinos at an averaged distance of $\sim 180\,\text{km}$ from an array of Japanese and Korean power reactors. As target about 1 kt liquid scintillator is used. The detection reaction is, like it was in the former Gösgen experiment, the inverse beta decay on free protons in the liquid scintillator. In 2003 the first results were published [19] and a clear evidence for neutrino oscillations was found.

In Fig. 20.11 the experimental values of the ratio between measured and expected average $\bar{\nu}_e$-fluxes is depicted for all reactor experiments since 2003. Also shown is the expected curve for this ratio for oscillation parameters compatible with the LMA-scenario. It fits very well to the measured values. For such small values $\Delta m_{12}^2 \sim 10^{-4}\,\text{eV}^2$ neutrino oscillation is visible only at very long baselines. The second parameter set for neutrino oscillation was found.

In further analyses more and more details about neutrino oscillations have been detected. A clear signal for oscillations is the observation of the correct L/E_ν-dependence. This has been demonstrated in Superkamiokande for atmospheric as well as in KamLAND for reactor neutrino spectra. The spectral distortion due to neutrino oscillations and the L/E_ν-dependence in KamLAND have been published in [20] and are shown in Figs. 20.12 and 20.13. Due to the observed energy distortion KamLAND is very sensitive on Δm_{12}^2. On the other hand a common analysis on solar neutrino experiments delivers a high precision on the mixing angle Θ_{12}. Therefore a global analysis of both, reactor and solar results, is able to generate quite accurate values on neutrino oscillation parameters.

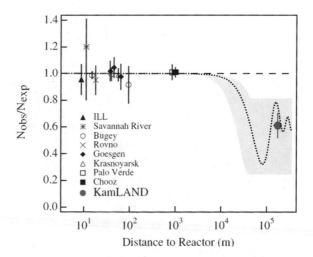

Fig. 20.11 Ratio between measured and expected averaged neutrino flux as function of the baseline for all reactor neutrino experiments. *Dashed line* shows the expected curve for $\Delta m_{12}^2 \sim 10^{-4}\,\mathrm{eV}^2$ and $E_\nu \sim 4\,\mathrm{MeV}$. From [19]

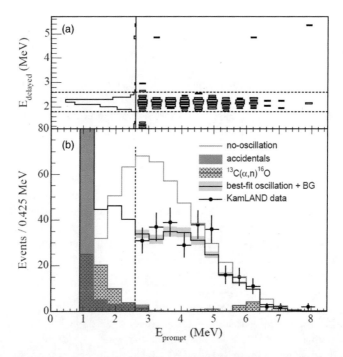

Fig. 20.12 Energy spectrum distortion of reactor neutrinos due to oscillation as seen in KamLAND (*lower figure*). Above the spectrum of the delayed event (i.e. from neutron capture on protons giving raise to 2.2 MeV gamma emission) is shown. The analysis threshold is 2.6 MeV and geo-neutrinos are no major background. The no-oscillation hypothesis is excluded significantly. From [20]

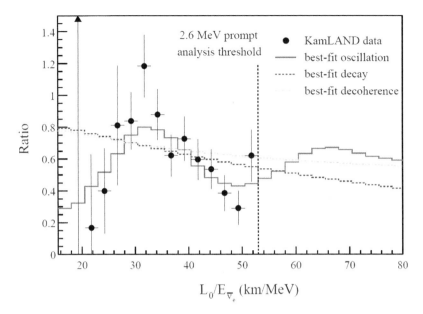

Fig. 20.13 Distribution of reactor neutrino data from KamLAND in L/E_ν and comparison with the expected curve from neutrino oscillations. Other, more exotic scenarios like neutrino decay and decoherence are excluded. From [20]

Neutrino oscillations have been seen in the meantime also by the high energy experiments K2K (Japan) and MINOS (USA) using accelerators as neutrino source. A first ν_τ-candidate appeared in the European OPERA experiment recently, confirming the Superkamiokande break-through result mentioned above. A review on new results in this field, albeit focusing on MINOS, can be found in [21].

The highlight in neutrino physics at low energies in the last years was Borexino. It is a solar neutrino experiment in the Gran Sasso underground laboratory (Italy), based on liquid scintillator technology. For the first time the low energy ^{7}Be-neutrinos ($E_\nu = 0.86$ MeV) could be observed directly in elastic neutrino electron scattering [22]. This was possible due to the larger light yield compared to a water Cherenkov detector, but the clue for the success was the development of an ultra-pure scintillator with mass concentrations of U and Th below 10^{-17}. The Compton-like recoil electron spectrum of ^{7}Be-ν as observed in Borexino is shown in Fig. 20.14. About 49 ± 5 counts per day in a fiducial volume of 100 t scintillator are observed, in very good agreement with the LMA-oscillation scenario. In case of no-oscillation Borexino should see $\sim 74 \pm 6$ events in 100 t and per day.

Recently the Borexino collaboration published a spectral analysis of solar ^{8}B-neutrinos with a threshold at 3 MeV [23, 24]. Again elastic scattering off electrons was used as detection reaction. In this energy range a successful cosmogenic background suppression is essential.

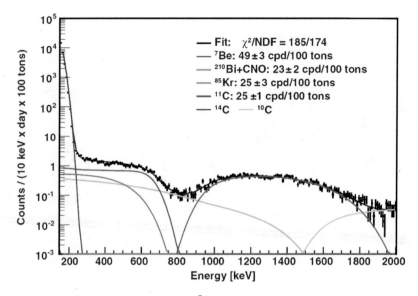

Fig. 20.14   Recoil energy spectrum of solar ^{7}Be-neutrinos after neutrino electron scattering and background contributions in Borexino. From [22]

The observed recoil energy spectrum is depicted in Fig. 20.15, compared with the LMA-solution and the no-oscillation hypothesis is clearly excluded again. Borexino is the first experiment measuring directly low and high energy solar neutrinos. In Fig. 20.16 the energy dependence of the ν_e-survival probability from vacuum to matter dominated flavor transition is shown. Here the Borexino and SNO results are used. The pp-flux is estimated by the Gallium results and subtraction of the ^{7}Be- and ^{8}B-contributions. Theoretical uncertainties of the solar neutrino fluxes are taken into account. As the matter effect only occurs for the case $m_2 > m_1$, we have evidence for normal hierarchy in this subsystem of neutrino mass eigenstates. It is notable, that this important result for particle physics could be achieved by solar neutrino experiments alone.

A global analysis of the results from all solar and reactor neutrino experiments has been performed recently [25]. The significance of neutrino oscillation is overwhelming. Neutrinos are massive and a new chapter in particle physics has been opened.

Up to now two mixing angles, Θ_{12} and Θ_{23} have been determined as well as the corresponding mass splittings. For the third angle, Θ_{13}, only upper limits exist, but global analysis may hint on a non-vanishing value. The limits come basically from a reactor neutrino experiment in Chooz, France, and new experiments (reactor as well as accelerator based) are currently under way to improve the sensitivity to Θ_{13} by at least one order of magnitude. Today the best values for the mixing angles (in degree) are [25]

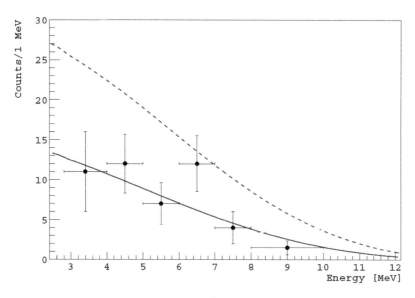

Fig. 20.15 Recoil energy spectrum of solar ^{8}B-neutrinos in Borexino. The no-oscillation hypothesis (*dashed line*) is excluded. Full line: LMA-scenario. From [23]

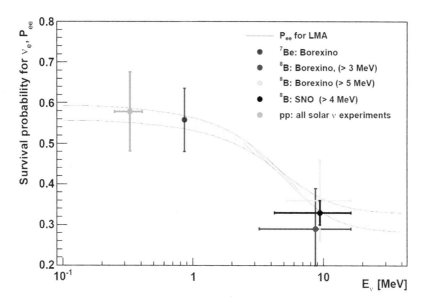

Fig. 20.16 Survival probability for solar neutrinos as function of energy and the expected shape according to the LMA-scenario. The error bars also include the theoretical uncertainties from the solar model. From [24]

$$\Theta_{12} = 34.4 \pm 1.0,$$
$$\Theta_{23} = 42.8^{+4.7}_{-2.9},$$
$$\Theta_{13} = 5.6^{+3.0}_{-2.7}.$$

Here normal mass hierarchy is assumed and the uncertainties quoted are at a one sigma level. Neutrino oscillation experiments only can deliver values on neutrino mass differences, but not on the absolute mass scale. The best experimental limit has been determined by ^{3}H-beta decay endpoint measurements: $m_i < 2.2\,\text{eV}$. The best values for the mass differences are

$$\Delta m^2_{21} = (7.6 \pm 0.2) \times 10^{-5}\,\text{eV}^2,$$
$$\Delta m^2_{31} = (2.46 \pm 0.12) \times 10^{-3}\,\text{eV}^2.$$

The square root of each Δm^2_{ij} is small against the upper limit of 2.2 eV. Therefore, neutrinos may be very hierarchical or degenerate, depending on the value of the lightest species. We have evidence for normal hierarchy in the ν_1, ν_2 subsystem. However, it is still open if $m_3 > m_1$ or whether the contrary case is realized.

Neutrino mass splitting is supposed to be the reason for the observed periodical time modulations of the K-shell electron capture (EC) decay rates of H-like heavy ions, measured at GSI in Darmstadt, Germany (see [26] and references therein). However, the interpretation of the observed time modulations as consequence of neutrino mass splitting is discussed in a controversial way (see e.g. [27]). If interpreted as consequence of the neutrino mass splitting, the value of Δm^2 is about two times larger compared to the KamLAND result. This discrepancy is interpreted in [28] as corrections to neutrino masses, caused by interaction of massive neutrinos with the strong Coulomb field of the daughter ions.

20.7   The Future of Neutrino Physics

Several important questions about intrinsic neutrino properties are still open.

- What is the absolute mass scale of neutrinos?
- Do we have normal or inverted mass hierarchy?
- What is the value of Θ_{13}?
- Are neutrinos Majorana or Dirac particles?
- Is there at least one CP-violating phase in the leptonic mixing matrix?
- Are there sterile neutrinos?

The absolute mass scale can be assessed experimentally by measuring beta spectra at the endpoint with great accuracy. An upper limit on $m_\nu = \sum m_i \cdot |U_{ei}|^2$ of 2.2 eV has been achieved in the Mainz [1] as well as in the Troitsk experiment [2] at 95% cl. This limit sets the scale on all neutrino masses m_i, as the oscillation results reveal that the mass differences are much smaller. In KATRIN, an tritium experiment in the starting phase at Karlsruhe, Germany, the sensitivity of the existing mass limit of 2.2 eV should be improved by roughly one order of magnitude. An overview about the state of the art of beta endpoint experiments and the potential of KATRIN can be found in [29].

Neutrinos are not massive enough to explain dark matter. However, they can have an important impact on structure formation in the universe on large scales. On the other hand astrophysical observations in conjunction with structure formation modeling can set limits on the sum of the masses of all neutrino species. Albeit model dependent these limits are interesting and in the range of about 1 eV (see e.g. [30]).

The question of mass hierarchy and Θ_{13} can be addressed by oscillation experiments which are sensitive to small mixing amplitudes. As $\Delta^2 m_{31}$ is known, the optimal baseline for experiments searching for Θ_{13} depends on the neutrino energy only: $L \approx (E/\text{MeV}) \cdot 0.5 \, \text{km}$. For accelerator appearance experiments ($E \sim \text{GeV}$) the appropriate distance between source and detector is a few hundred km, whereas reactor disappearance experiments operate at a range of 1–2 km. Both types of experiments are in the starting phase and they have in common the use of so called near detectors which should monitor the source and thereby minimize systematic uncertainties. The potential of reactor neutrino experiments searching for Θ_{13} are discussed in [31] and the corresponding information for future accelerator projects in e.g. [32, 33] and references therein.

If Θ_{13} is not too small, there is hope to discover CP-violation in future high energy experiments. Here one would search for differences between neutrino and antineutrino oscillations. In order to realize such a project, very powerful beams or even complete new concepts have to be accomplished. Recent information about the search for CP-violation can be found e.g. in [34, 35] and references therein.

The question whether the neutrino is a Dirac- or Majorana-type particle is investigated by searching for the neutrinoless double beta decay $A(Z) \rightarrow A(Z + 2) + 2e^-$. This ($\beta\beta 0\nu$)-decay would violate total lepton number conservation by $\Delta L = 2$ and can only occur when $\nu \equiv \bar{\nu}$ as illustrated in Fig. 20.17. The decay amplitude is sensitive to the so-called effective neutrino mass $m_{ee} = |\sum U_{ei}^2 \cdot m_i|$, which is a coherent sum over all mass eigenstates.

As discussed before the values U_{ei}^2 may contain complex phases, which could lead to partial cancellation of different terms in the sum and uncertainties in the nuclear matrix elements play an important role (see e.g. [36] or [37]). In addition other mechanism beyond the standard model, like right-handed weak charged currents, could contribute to the ($\beta\beta 0\nu$)-decay. In any case the observation of ($\beta\beta 0\nu$)-decay would be the proof, that at least one neutrino is a Majorana particle, as long as the corresponding physics is described by a gauge theory [38]. Hence, search for ($\beta\beta 0\nu$)-decay is the key experiment to probe the nature of neutrinos.

Double beta decay is only observable in even-even nuclei, like ^{76}Ge. Often source and detector are equivalent. In Ge-semiconductor detectors the ($\beta\beta 0\nu$)-decay would lead to a peak in the observed energy spectrum at the endpoint. Current limits on m_{ee} depend strongly on nuclear matrix elements and are in the range of $\sim 1 \text{eV}$. There is

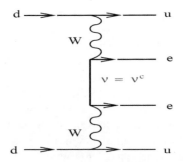

Fig. 20.17 Feynman-diagram of the neutrino-less double-beta decay. It can occur only, if the neutrino is a Majorana particle and effective neutrino masses m_{ee} (see text) may generate such transitions

a claim of a part of the Heidelberg-Moscow collaboration for a positive signal [39]. However, up to now it was not possible to confirm or to refuse this result.

Future experiments with substantial funding for construction, like CUORE, GERDA in Europe and EXO-200, SNO+ in North-America aim to reach sensitivities below $m_{ee} \sim 0.1$ eV. This would be an important step, as for an inverted mass hierarchy the expected value is between 0.01 and 0.1 eV, independent from CP-phase values. A recent overview about the experimental status and prospects in the search for $(\beta\beta0\nu)$-decay can be found in [40].

A further important aspect of neutrinos is their role as messengers from astrophysical objects. Solar neutrino physics and the detection of terrestrial neutrinos have been already mentioned. With a future large liquid scintillator detector (LENA, Low Energy Neutrino Astronomy in Europe, HANOHANO in USA) solar models as well as geophysical questions could be studied in more detail [41,42]. In addition. subtle questions arising for neutrino oscillations can be addressed with LENA, like the observation of neutrino decoherence with reactor neutrinos [43]. However, there are further scientific cases seen in conjunction with low energy neutrino astronomy:

- First detection of the diffuse supernova neutrino background in our universe.
- In case of a galactic supernova type II the study of a gravitational collapse via supernova neutrino detection.
- Search for proton decay, which is predicted in grand unified theories.
- Long baseline oscillation experiments with very high sensitivity on Θ_{13}, CP-violation, and mass hierarchy.

In Europe three types of detectors with masses of at least 50 kt are discussed: water Cherenkov (MEMPHIS), liquid argon (GLACIER), and liquid scintillator (LENA) [44–46]. Currently an European design study (LAGUNA, Large Apparatus for Grand Unification and Neutrino Astrophysics) is investigating possible underground facilities for such large experiments [47]. Also in the US at DUSEL (Deep Underground Science and Engineering Laboratory) and in Japan (Hyperkamiokande) very large water detectors are being planned.

References

1. Ch. Weinheimer, Nucl. Phys. B (Proc. Suppl.) **118**, 279 (2003)
2. V.M. Lobashev et al., Nucl. Phys. B (Proc. Suppl.) **118**, 282 (2003)
3. B.T. Cleveland et al., Astrophys. J. **496**, 505 (1998)
4. P. Vogel, Phys. Rev. **D29**, 1918 (1984)
5. F. von Feilitzsch et al., Phys. Lett. **118B**, 162 (1982)
6. K. Schreckenbach et al., Phys. Lett. **160B**, 325 (1985)
7. G. Zacek et al., Phys. Rev. **D34**, 2621 (1986)
8. Argus Collaboration, Phys. Lett. **B292**, 221 (1992)
9. C. Hagner et al., Phys. Rev. **D52**, 1343 (1995)
10. S. Schönert et al., Nucl. Phys. **B48**, 201 (1996)
11. R. Bonetti et al., Phys. Rev. Lett. **82**, 5205 (1999)
12. J.P. Cravens et al., SK Collaboration, Phys. Rev. **D78**, 032002 (2008)
13. Y. Fukuda et al., SK Collaboration, Phys. Rev. Lett. **81**, 1562 (1998)
14. M. Altmann et al., GNO Collaboration, Phys. Lett. **B616**, 174 (2005)
15. J.N. Abdurashitov et al., SAGE Collaboration, Phys. Rev. Lett. **83**, 4686 (1999)
16. J. Boger et al., SNO Collaboration, Nucl. Instrum. Meth. **A449** (2000)
17. B. Aharmim et al., SNO Collaboration, Phys. Rev. **C81**, 055504 (2010)
18. Q.R. Ahmad et al., SNO Collaboration, Phys. Rev. Lett. **89**, 011301 (2002)
19. K. Eguchi et al., KamLAND Collaboration, Phys. Rev. Lett. **90**, 021802 (2003)
20. T. Araki et al., KamLAND Collaboration, Phys. Rev. Lett. **94**, 081801 (2005)
21. A. Habig, Mod. Phys. Lett. **A25**, 1219 (2010)
22. C. Arpesella et al., Borexino Collaboration, Phys. Rev. Lett. **101**, 091302 (2008)
23. L. Oberauer et al., Borexino Collaboration, J. Phys. Conf. Ser. **203**, 012081 (2010)
24. G. Bellini et al., Borexino Collaboration, Phys. Rev. **D82**, 033006 (2010)
25. M.C. Gonzalez-Garcia, M. Maltoni, J. Salvado, J. High Energ. Phys. **04**, 056 (2010)
26. P. Kienle, Nucl. Phys. A **827**, 510c (2009)
27. C, Giunti, Phys. Lett. B **665**, 92 (2008)
28. R. Höllwieser et al., arXiv: 1102.2519 (2011)
29. E.W. Otten, C. Weinheimer, Rep. Progr. Phys. **71**, 086201 (2008)
30. V. Barger et al., Phys. Lett. **B595**, 55 (2004)
31. K. Anderson et al., arXiv:hep-ex/0402041 (2004)
32. V. Antonelli et al., Nucl. Phys. (Proc. Suppl.) **168**, 192 (2007)
33. M. Mezzetto, T. Schwetz, arXiv:1003.5800, J. Phys. G37, 103001 (2010)
34. H. Nunokawa, S.J. Parke, J. Valle, Progr. Part. Nucl. Phys. **60**, 338 (2008)
35. A. Bandyopadhyay et al., Rep. Progr. Phys. **72**, 106201 (2009)
36. A. Faessler, J. Phys. Conf. Ser. **203**, 012058 (2010)
37. F. Simkovich et al., Phys. Rev. **C77**, 45503 (2008)
38. J. Schechter, J. Valle, Phys. Rev. **D25**, 2951 (1982)
39. H.V. Klapdor-Kleingrothaus, J. Dietz, I.V. Krivosheina, O. Chkoverets, Nucl. Instrum. Meth. A **522**, 371 (2004)
40. S. Schönert, J. Phys. Conf. Ser. **203**, 012014 (2010)
41. M. Wurm et al., arXive:1004.3474 (2010)
42. K.A. Hochmuth et al., Earth Moon Planets **99**, 253 (2006)
43. B. Kayser, J. Kopp, arXive:hep-ph/1005.4081 (2010)
44. A. de Bellefon et al., arXiv:hep-ex/0607026 (2006)
45. A. Rubbia, J. Phys. Conf. Ser. **171**, 012020 (2009)
46. L. Oberauer, F. von Feilitzsch, W. Potzel, Nucl. Phys. (Proc. Suppl.) **138**, 105 (2005)
47. D. Autiero et al., Laguna cons. JCAP **0711**, 011 (2007)

Chapter 21
The Second Mössbauer Effect

Paul Kienle

"The Second Mössbauer Effect" was the headline of the "Spiegel" (a German weekly journal like the "Times") from May 13, 1964 for announcing the foundation of the "Physik-Department" at the Technische Hochschule München (THM), as it was still called at this time. Maier-Leibnitz was irritated by this headline because the article did not mention at all his contribution to the consolidation of the structure of the three physics institutes (Experimental, Technical, and Theoretical Physics) in the beginning of the sixties. Already in the late fifties ML's Institute for Technical Physics was overloaded with students working on their diploma or doctoral theses, because research in the new field of applied nuclear physics was very attractive and ML had for each student who applied an interesting research project. In the average, ML had to supervise between 150–200 diploma students, an impossible task. So, young postdoctoral students had to help him out by taking over the duties of professors. In a letter to the Bavarian Ministry of Education and Arts in 1957 he complained: "The directors of the institutes are hopelessly surcharged and the institutes are overcrowded, the resources for research projects are totally insufficient and lots of time and energy is wasted for finding additional resources."

In March 1962, Maier-Leibnitz sent a memoir for upgrading the physics at the THM, signed by himself and his colleagues Brenig, Riehl, and Wild to the Bavarian Ministry of Education and Arts and to the German Council of Science which had made already some recommendations to the Bavarian Government to improve the situation of physics at the THM. ML and his colleagues [1] demanded the foundation of a physics department at the THM with 20 professors in a single joint institute for the improved scientific and technical education of a large number of young talents, for not to give up Humboldt's ideal of the unity of research and education despite the large sizes of the universities, for competing with other countries having already successful institutions for research and education, for

P. Kienle (✉)
Physics Department, Technical University Munich, 85747 Garching, Germany
e-mail: paul.kienle@ph.tum.de

M. Kalvius and P. Kienle (eds.), *The Rudolf Mössbauer Story*,
DOI 10.1007/978-3-642-17952-5_21, © Springer-Verlag Berlin Heidelberg 2012

decreasing the load on universities with more and more institutes, for rejuvenating the faculty and attracting again emigrated German scientists, and finally for a successful continuation of the development started with the Munich Research Reactor. One specific request was a Department building near the "Atom-Ei" in Garching.

When Rudolf Mössbauer received the Nobel Prize in 1961, he was at the Californian Institute of Technology in Pasadena first as a research fellow on invitation by Felix Böhm (1960) and a short time later a faculty member and then, in context with the Nobel Prize, full professor. He enjoyed very much the lively scientific climate of the famous CalTech school with Richard Feynman, Murray Gell-Man, and his host Felix Böhm with whom he collaborated also later in the seventies in a search for neutrino oscillations. First the University of Bonn offered Mössbauer through the initiative of Prof. Wolfgang Paul, a later Nobel Laureate, a large Institute for Nuclear Physics which he declined after serious negotiations because he had something else in mind. After the Nobel Prize, the Bavarian government tried to call Rudolf Mössbauer on a chair back to the THM. In his negotiations, he requested as "sine qua non" condition the creation of a Physik-Department, as proposed by ML and his colleagues.

End of 1964, Mössbauer returned to the THM after main parts of his demands were indeed accepted by the Bavarian government. The three original physics institutes were merged into a single large institute, a department similar to the Anglo-Saxon system. The "Physik-Department" as it was called housed now 16 full chairs for physics, 10 experimental chairs and 6 for theory, in addition 4 so called "naked" full professor positions were installed, two of them attributed to experimental chairs and two to theory. So all together, the Department started with 20 full professors in accordance with the original plan of ML and his colleagues, 12 full professors for experimental physics, and 8 for theory. In addition, 256 permanent posts for scientific assistants and technical personal were authorized.

This development marked a revolutionary renewing of the German University system before the student revolution in 1968 which the Spiegel headlined "Unter den Talaren der Muff von tausend Jahren" (Beneath the gowns the musty smell of thousand years). The introduction of the department meant a break with the traditional German teaching and research system, which was characterized since the time of the "Geheimrats-Era" by one Professor with unlimited power holding solely the chair of a usually very large institute. In Munich, the German Science system experienced a change, which was noted in the "Kasten" a special listing of essential points from the above mentioned headline article on the student revolution by the Spiegel.

The main rules were:

- All professors of the Physik-Department do research and teaching as team work and cover in collaboration most of the interesting fields of research.
- All chair professors including the "naked" ones are equal and elect a Department Management Board consisting of three members, a Chairperson, a Deputy

responsible for research, and a Deputy responsible for education. They had to change in regular cycles.
- The heavy load of the many introductory courses for incoming students of natural sciences and engineering were to be distributed among all professors.
- Technical facilities, workshops, library, stock issue, etc. are governed by a central administration of the Department.

It took Mössbauer nearly two years of negotiations to achieve this "Second Mössbauer Effect," which was again proposed and carefully prepared by ML similarly to the first, but that one had been in contrast to the second one a true discovery by Mössbauer on his own. Actually somebody may wonder about so many theory chairs in physics at a Technical University. I think this was a pet idea of Maier-Leibnitz that experimentalists need many theorists to get balanced answers. The many new chairs could be occupied by experienced but also young scientists. The brain-drain to US Universities and Laboratories could be reversed. Besides Rudolf Mössbauer, Wolfgang Kaiser, Klaus Dransfeld, Edgar Lüscher, and G. Michael Kalvius returned from the Bell Laboratories, Illinois State University in Urbana, University of California Berkeley, and Argonne National Laboratory, respectively, to build up new disciplines such as laser physics, the structure and dynamics of condensed matter, interdisciplinary and applied research, and nuclear-solid-state applications at the Physik-Department. In the middle and late sixties, Haruhiko Morinaga from the University of Tokyo joined the Physik-Department and enforced together with Herbert Daniel from the Max-Planck Institute for Medical Research in Heidelberg, Achim Körner from Hamburg University, and myself from TH Darmstadt the nuclear physics and accelerator technique activities in Munich. The theory group with Wilhelm Brenig and Wolfgang Wild, who was later elected as President of the TUM, became Science Minister of Bavaria and President of the German Space Agency, was supplemented by young professors, such as Hansjörg Mang, Klaus Dietrich, Wolfgang Götze, Heinz Bilz, and Hartwig Schmidt enhancing forefront research and teaching in nuclear and condensed matter physics. In 1972 and 1974, the US-Americans Frederick Koch and Pierre Hohenberg joined the Physik-Department which thus developed a truly international flair. Frederick Koch was a pioneer of semiconductor physics and founded in 1987 together with the Faculty of Electrical Engineering and the support of the Siemens AG the "Walther Schottky Institute" for semiconductor research. Georg Alefeld, a scholar of Maier-Leibnitz, joined the Physik-Department and became famous in energy research and founded the Bavarian Center for applied energy research.

One demand was a new Department building near to the "Atom-Ei" in Garching, which was on the list of ML and colleagues. It was agreed on to be realized in two construction phases, because one section of the Physik-Department was planned to stay at the downtown site as long as teaching for the first two years students was done in a new downtown building. Finally, we moved in 1969 into the beautiful building with well-equipped laboratory space, central workshops, a spacious, well-bestowed library, three lecture halls of different sizes located around a spacious entrance hall and nice offices, designed by Professor Angerer, an architect of the

THM. We enjoyed these new facilities very much after many years of suffering in barracks and commuting between offices in town and laboratories in Garching. Close ties and collaborations developed among the various sub-institutes which covered a large area of research in experimental and theoretical physics. ML together with Mössbauer made me an offer to return from the TH Darmstadt to the THM with the attraction that I could build together with Prof. Ulrich Mayer-Berkhout an Accelerator-Laboratory in Garching commonly used by the LMU and THM nuclear physics groups. It was a 15 MV Tandem van de Graaf accelerator, ideally suited to accelerate heavy ions which was taken in operation in 1970 and became one of the two breeding places for heavy ion physics in Germany.

Mössbauer took in 1972 a leave of absence to follow Maier- Leibnitz as Director of the ILL in Grenoble, up till now a most powerful high-flux reactor with outstanding instrumentation for many applications of neutrons. With Mössbauer as director, the French-German collaboration was extended to include also the United Kingdom.

When Rudolf Mössbauer returned in 1977 to Munich, the Physik-Department was divided up in three only loosely connected institutes: one for theory, one for solid-state physics, and one for nuclear physics and nuclear-solid-state physics to which Mössbauer and I belonged. This was the consequence of a new Bavarian University Law of 1974. Its specifications resulted in the disintegration of the "Faculty for General Sciences" to which physics, chemistry, biology, and other science disciplines had belonged into smaller departments which had to be further subdivided into institutes. In the case of the already existing Physik-Department, three institutes had to be created in the new Faculty of Physics with a Dean, Deputy Dean, and an elected Faculty Board. According to my impression, there was only singular resistance in the Department against this step backward, documented by only two votes against this regulation in a Department meeting. Initially, the partition was considered by the Faculty Board as a legal necessity and thought to be a pure formality without affecting the Department idea in its roots. But, as is often the case, the three institutes developed more and more a life of their own and a trend to splitting emerged in the department in which all resources, positions, rooms, and laboratory space were divided up into three parts. This had the natural consequence that ill concealed envy crept up from time to time when the resources became scarcer and positions were reduced by some budget saving actions ("Personalabbau") of the Bavarian government. Especially, the fight about the space in the building became very serious. There was, however, a special cause for this dilemma, an extreme shortage of space. The plans for a second construction phase for the department building, again nicely designed by Professor Angerer, containing special meeting places, a faculty club, a cafeteria with ample space to meet had been completely finalized. Alarmed by the slowly decreasing number of incoming students, Maier-Leibniz and Georg Alefeld decided to reduce drastically the building program of the second construction phase. The first casualties were the "social rooms" were we could meet and discuss informally in a relaxed atmosphere. Time was wasted by redesigning a simple laboratory building. It was not recognized that this was disastrous for the collaborative spirit in the department which actually grows in easy accessible Cafeterias, such as at CERN and GSI in Darmstadt, where people come together informally and discuss their work and new ideas.

Mössbauer was furious about this development when he returned and became rather frustrated by governmental autocracy. He tried desperately to improve the conditions for research and teaching at the universities and to reduce the still prevalent bureaucracy with only slow success. He gave very carefully prepared excellent lectures and also focused, like everybody did, on his research, which in his case was neutrino physics. He was also an excellent student in this field completely new to him. When he became a corresponding member of the Russian Academy of Sciences, he invited many of the famous Russian scientists such as Goldanskii, Belaev, Kagan, and many others from whom all of us profited very much. In this way, we learned informally all about the progress of Russian science. From Belaev I learnt a lot of the progress of the Budger Institute in Novosibirsk, especially about electron cooling of proton beams which then was very successfully applied in the heavy ion storage ring in a construction project at GSI Darmstadt which I initiated and supervised.

When Mössbauer decided to change his field of physics, he felt that he and his group needed lectures in neutrino physics. So he invited via our close Japanese connection a young theory professor from the University of Tokyo, Masataka Fukugita, for a semester to give lectures on the theory of neutrinos. Recently, I met Fukugita again who actually wrote a famous book on basis of these lectures in Munich and he told me quite seriously, the best student he ever had was Rudolf Mössbauer. A few days after a lecture, Rudolf would come and correct all the mistakes he had made in the derivations during the lecture, because he had checked all formulae carefully.

From 1977, when Mössbauer returned to Munich until he became Emeritus in 1997, he took care of his teaching duties and focused on interesting research projects. He rejected an offer as Director of the Max-Plank Institute of Nuclear Physics in Heidelberg and did ground breaking research in the following three projects:

1. A solution of the phase problem in the structure determination of biological macromolecules and other coherent effects in resonance radiation.
2. Neutrino oscillations at the Gösgen nuclear power station in Switzerland.
3. Quasi particle trapping in a superconductive detector system exhibiting high-energy resolution.

Furthermore, he played a very important role in the realization of the GALLEX experiment in the Gran Sasso Laboratory (Italy). From 1991 to 1997, the GALLEX experiment measured the solar neutrino flux above an energy threshold of 233 keV. Its results contained for the first time neutrinos from the pp cycle and showed a remarkable deficiency of the expected flux.

In retrospect, one must concede that the "Fakultät für Physik", as it was called after splitting in three institutes, was scientifically further successful, despite the losses of coherence of the department system which was replaced by a sound rivalry of the three institutes which covered a large field in science and offered an excellent education for the students.

The Physik-Department is still playing an important role in contributing essentially to the special status of the Technische Universität München as one of the three first "Elite Universities" in Germany. Its research programs have a very wide spectrum ranging from particle-, nuclear-, and astrophysics, research with neutrons, biophysics, and studies of condensed matter properties to applications in energy research. The Physik-Department is entangled in the Campus Garching with Research Centers of the TUM, such as the Heinz-Maier-Leibnitz neutron source, FRM II, the Walter-Schottky-Institute for semiconductor research, WSI, the Institute for Advanced Study, IAS, and the Institute for Medical Engineering, IME.TUM. It is surrounded by a series of research facilities, such as the Maier-Leibnitz-Laboratory of the LMU and TUM, MLL, the Walther-Meissner-Institute for low temperature research, WMI, the Jülich-Center of Neutron Science, JCNS, the Helmholtz-Zentrum München, and the Bavarian-Center for applied Energy Research, ZAE. The Physik-Department collaborates with many Max-Planck-Institutes located in Garching and Munich, such as the MPI's for Astrophysics, Biochemistry, Extraterrestrial Physics, Physics, Plasma Physics, and Quantum Optics. The research is advanced by numerous Excellence Clusters, Sonderforschungsbereiche, and Transregios supported by the Deutsche Forschungs-Gemeinschaft:

21.1   Excellence Clusters

- Origin and structure of the Universe – the cluster of excellence for fundamental physics
- Nanosystems initiative Munich
- Munich-center for integrated protein science
- Munich-centre for advanced photonics

21.2   Sonderforschungsbereiche Transregios

- Solid-state based quantum information processing
- Kräfte in biomolekularen Systemen
- Neutrinos and beyond – weakly interacting particles in physics, astrophysics, and cosmology
- From electronic correlations to functionality

Within this network of these excellent research opportunities, the Physik-Department is looking in a bright future. Our field of nuclear-, particle-, and astrophysics received an enormous boost by the Excellence Cluster "Origin and

Structure of the Universe" and the Transregio "Neutrinos and beyond." Many junior research groups could be established, which push exciting fields to new frontiers.

References

1. Denkschrift zum Ausbau der Physik an der Technischen Hochschule München, 28. February 1962, TUM-Verwaltung, Registratur, Az

Index

Absolute ether theory, 269, 272
Absolute frame, 269
Absolute gravitational experiments, 367
Absolute inertial frame, 275
Absolute simultaneity, 275
Absolute space, 286
Absolute velocity, 269, 275, 276
(n, γ) absorption, 23
Accelerated red shift, 60
Accelerator-Laboratory, 420
Accelerator mass spectroscopy, 385
Accelerator neutrinos, 407
Acetyl CoA synthase, 249
Actinides, 124, 207
Advance, 29
After-effects, 116, 222
Allerton Conference, 78
Alpha-particle-X-ray spectrometer, 309
An axially symmetric electric field gradient,
    161
Angerer, 419
Anharmonic lattice vibrations, 115
Anisotropic f-factor, 120, 137
Anisotropy
    of the gravitational field, 369
    of inertia, 286
    of the inertial mass, 286
Anomalous transmission, 331
Antiferromagnetic exchange, 248
Antimony bonding properties, 117
Apollo program, 298
Applied energy research, 419
Applied nuclear physics, 417
APXS, 309
Aqueous processes, 307
Argonne National Laboratory, 70
Arrhenius law, 226, 233

Asymmetry parameter, 118, 284
A tensor, 245
Atmospheric-neutrino oscillation, 387
Atmospheric neutrinos, 404
Atom interferometer, 286, 287
Atomic Energy Commission, 107
197-Au, 186
Augmented plane wave method, 186
Average atomic displacement, 347

Backscattering, 302
Band mixing, 163
6s bands, 128
Band structure, 179
Band structure calculations, 184
Barium titanate transducer, 282
Bavarian University Law, 420
BCS quasi-particle operators, 155
BCS state, 237
Bell Laboratories, 104
β-decay, 374
Binaries, 287
Binding energies, 376, 378, 380, 388
Bio-inorganic chemistry, 351
Bio-molecules, 32
Biomimetic, 249
Biomimetic chemistry, 249, 253
Birch's law, 350
Black holes, 288
Blocking temperature, 235
Blueberries, 308
Blueshift, 284
Bound-state β-decay, 374
Bound-state tritium decay, 374
Brain-drain, 419
Brass, 136

M. Kalvius and P. Kienle (eds.), *The Rudolf Mössbauer Story*,
DOI 10.1007/978-3-642-17952-5, © Springer-Verlag Berlin Heidelberg 2012

Brilliance synchrotron radiation, 360
Broadening mechanisms, 369
Budger Institute, 421

California Institute of Technology, 30
CalTech Theses, 104
Carbonate minerals, 306
Catalysis, 353
Causality principle, 275
Centrifugal force, 286
Centripetal, 70
Change of the mean square radius, 154
Charge density at the nucleus, 180
Chart of nuclides, 147
Chemical binding, 7, 179
Chemical bonds, 380
Chemical effect, 21
Chemical elements where Mössbauer studies
    are possible, 147
Chemical environment of the resonating
    nucleus, 159
Chemical shift, 177
Classic energy spectroscopy, 363
Clock hypothesis, 265
Clock rates, 274, 279
Clocks, 271
Clocks on a rotating disk, 270
Coherent enhancement of the radiative channel,
    323
Coherent inelastic scattering, 326
Coherent properties, 32
Coherent scattering, 9
Coherent superposition of the mass eigenstates,
    385
Cold War, 106
Collective nuclear excitation, 276
Collinear laser spectroscopy, 160
Columbia hills, 304
Compounds, 30
Compton frequency, 286
Conservation of energy, 277
Constrained microscopic self-consistent mean
    field, 159
Contact field, 117, 122, 128
Contact plate, 303
Continuum-state β-decay, 374
Conversion electrons, 30, 217
Coriolis anti-pairing, 157
Cornell Meeting, 106
Coulomb excitation of the Mössbauer state,
    171
Covalency, 190
Covariant Ether Theories, 263, 274

CP-violation, 167, 413
Critical slowing down, 232
Cryogenic temperatures, 284
Cryoreduction, 252
Crystalline electric field interactions, 127
Cs lamellar compound, 120
Currents in inner-shell electrons, 167
Curvature of spacetime, 279
Curved spacetime, 271, 277–279, 287

Dark matter, 413
De-Broglie waves, 286, 287
Debye approximation, 268, 377
Debye integral, 381
Debye model, 264, 266, 381
Debye temperature, 112, 183, 264, 266, 285,
    377
Debye-Waller factor, 12, 83, 85, 296
Density functional theory, 186, 348
Department building, 418, 419
Deutsche Forschungs-gemeinschaft, 422
DFT calculations, 253
Diamagnetism, 203
Differential redshift measurements, 280
Diffractive and reflective optics, 341
Diffusion, 222, 237
Diffusional motion, 115
Digital time, 29
150–200 Diploma students, 417
Diploma thesis of Mössbauer, 20
Dirac-Fock calculations, 178
Dirac-Fock-Slater calculations, 184
Discovery of the M-effect, 37
Dispersion-like shape of resonance line, 329
Dispersion phenomena in Mössbauer spectra,
    168
Dispersion term, 128
Distant matter, 274, 288
Distinguished frame of reference, 274
Diurnal periodicity, 273
Doctor thesis of Mössbauer, 22
Doppler broadening, 375
Doppler effect, 6, 13, 16
Doppler shift, 5, 24
Double beta decay, 413
Double crystal monochromator, 348
Double exchange, 255
Double-loudspeaker drive, 29, 281
Double professorships, 33
Double pulsar, 287
Dragging of inertial frames, 286, 287
D-transition elements, 120
Dynamic shutter, 363

Dynamical beating, 324

Earth's rotation, 273
Effective field approximation, 130
Effective Hamiltonian, 245
Einstein Equivalence Principle, 277
Einstein frequency, 10
Elastic scattering of slow neutrons, 9
Electon-phonon coupling, 138
Electric field gradient, 284
Electric quadrupole splitting, 27, 175
Electromagnetic relaxation, 378
Electromagnetism, 287
Electron antineutrinos, 373, 374
Electron capture decays, 185
Electron density, 152
Electron hopping, 125, 226
Electron paramagnetic resonance, 243
Electron transfer, 247
Electronic fluorescence, 342
Elemental transformations, 383
Emission probability, 321
$E1$ Mössbauer transitions, 169
Energy and momentum scales, 85
Energy conservation, 279
Energy resolution, 15, 111
Enlarged phase conservation, 329
Entangled field modes, 370
Environmental effects, 71
Environmental science, 313
Equivalence of energy and mass, 265, 381
Equivalence principle, 263
Erbium oxide, 101
Escape velocity, 278
Ether, 269, 288
    drift, 269, 273
    theory, 269
Ether-drift experiments, 263, 272
Eu-compounds, 344
$EuFe_2.As_{1-x}P_x/_2$, 214
$Eu_3Fe_5O_{12}$, 207
Eu^{3+} ion, 207
Europium iron garnet, 127
Europium magnetism, 126
Europium monochalcogenides, 127
Evolution of the neutrino state, 386
Evolution of the neutrino wave function in
        space and time, 386
Excellence Clusters, 422
Exchange parameters (Eu), 127
Exchange splitting, 209
Excited atomic levels, 132
Exotic resonances, 367

Expansion and contraction of the lattice, 383
External high pressure, 207
External Magnetic Fields, 203
Extremely small hfi, 369

Faculty of Physics, 420
F-factor, 112
F-transition elements, 124
52-Fe, 185
57-Fe, 186
Fe^{IV}, 249
$Fe^{IV} = O$, 249
$Fe^{IV} Fe^{III}$ complex, 251
^{57}Fe nucleus, 280
Fe_3O_4, a ferrimagnet, 205
Fe protein, 257
Fermion creation and annihilation operators,
        155
Ferric compounds, 179
Ferromagnetic coupling, 250
Ferromagnetic relaxation, 129, 210
Ferrous compounds, 179
Fe_3S_4 clusters, 253
Fe_4S_4 clusters, 246, 257
Field equations, 263, 287
Field equations of gravitation, 277
Field gradient fluctuations, 228
Finger print applications, 195
First excited state, 147
Flat spacetime, 277
Flavor mixed neutrinos, 34
Flavour eigenstates, 385
Fluctuating hyperfine interactions, 378
Fluctuating magnetic fields, 378
Force constants, 348
Forefront research and teaching, 419
Forward scattering, 321
Frame-dragging forces, 286
Free ion field, 129
Frequency shift, 274
16 full chairs for physics, 418
20 full professors, 418
Fundamental constant variation, 367

Galactic center, 286
Galaxy, 286
GALLEX, 421
Gas proportional counter, 101
Gaseous γ-ray sources, 146
^{155}Gd, 211
Gd metal, 211
Geheimrats-Era, 418

Geiger counter, 106
General Relativity, 277, 279, 367
General Theory of Relativity, 276, 287, 288
Geodesy satellites, 288
Geometry of bond structure (Au), 123
Geophysical studies, 350
German Science system, 418
Giant Kohn anomaly, 137
Goethite, 301, 305
Gold coins (Celtic), 123
Gold compounds, 123, 192
Gold fine particles, 123
Gold ore, 123
Graphite lattice, 383
Gravitational effects, 57
Gravitational field, 271, 277–279
Gravitational force, 286
Gravitational potential, 271
Gravitational Red Shift(Ag), 139
Gravitational redshift, 57, 263, 276, 288, 367,
    382
    of antineutrinos, 374
    experiments, 280
    $\bar{\nu}_e$ measurements, 387
Gravitational shift, 281
Gravitational spin-orbit coupling, 288
Gravitational theories, 283
Gravitational time dilatation, 287
Gravitational-tensor interaction, 274
Gravitomagnetic effects, 288
Gravitomagnetism, 287
Gravity Probe B, 288
Gusev crater, 301
Gyroscope, 287

H-cluster, 249
^{3}H - ^{3}He system, 375, 377
Hamilton operator, 265
Harmonic crystal, 84, 88
Hartree-Fock calculations, 184
Harwell, 51
Heavy electron compound, 132
Heavy ion physics, 420
Heavy ion storage ring, 421
Hebel–Slichter anomaly, 237
Heidelberg, 22, 93
Heidelberg cyclotron, 26
Heisenberg uncertainty principle, 377
Heisenberg's uncertainty relation, 268
Hematite, 307
Hematite spherules, 309
Heme, 246
Heme-based protein, 351

Hermite polynomials, 92
High-brightness synchrotron radiation, 344
High-energy electron storage rings, 342
High energy transitions, 364
High pressure (Np), 134
High pressure (Sn), 116
High pressure (Zn metal), 137
High pressure (ZnO), 135
High-pressure studies, 346
High reflectivity, 326
High-resolution, 266
High-resolution isotope, 268
High-resolution Mössbauer spectroscopy, 282
High resolution resonances, 135
High-speed centrifuge, 263, 269, 274, 275, 288
High-speed rotors, 272
Hill Limit, 133
^{163}Ho - ^{163}Dy system, 383
Homogeneous broadening, 378, 388
Homogeneous line broadening, 378, 383
Hume-Rothery rules, 136
Hydrogen in Ta, 138
Hydrogenases, 249
Hydrothermal, 306
Hydrothermal processes, 301
Hydroxide sulfate, 307
Hyperfine anomaly, 123
Hyperfine field-^{57}Fe, 74
Hyperfine interactions, 101, 149, 287, 380
Hyperfine levels- ^{119}Sn, 78
Hyperfine levels-^{57}Fe, 74
Hyperfine splitting, 175
Hyperfine structure, 15
Hyperfine tensor, 130
Hyperfine-splitting, 379

127-I, 186
^{191}Ir, 22
Ilmenite, 299
Ilvait, 226
Imperfect lattice, 378
Implantation studies, 119, 120
Impurity, 89, 379
In-beam nuclear reactions, 171
Induced quadrupole interaction, 129
Inertial forces, 286
Inertial mass, 286
Inhomogeneous broadening, 378, 379
Inhomogeneous lattice, 380
Inhomogeneous line broadening, 380, 383, 388
Inhomogeneous metallic lattice, 380
Inner closed shell, 202
Institute Max von Laue-Paul Langevin, 33

Interatomic interactions, 347
Interference effect, 169
Interference of nuclear and electronic
        scattering, 329
Interference term, 163
Intermediate valence, 124, 126, 131, 227
Internal conversion, 184
Internal field, 16
Internal magnetic field, 199, 201, 203, 205,
        207, 209, 211, 213, 215, 217, 219,
        248
Internal stresses, 287
International flair, 419
Inverted mass hierarchy, 387
Iodine molecular, 118
193-Ir, 188
Iridium charge states, 121
Iron, 293
Iron meteorite, 309
Iron–sulfur cluster, 253
Isomer shift(s), 16, 76, 149, 152, 175, 268, 380
    calibration, 183
    and nuclear models, 154
Isotope shift, 152, 176
Isotropic exchange, 245
Itinerant magnetism, 134

Jahn–Teller, 228
Jarosite, 301

Kamacite, 310
KamLAND collaboration, 34
6.4 keV Fe X-rays, 303
14.4keV γ rays, 280
14.41 keV γ-rays, 303
93.31 keV transition in ^{67}Zn, 282
Kramers doublet, 247

Laboratorium für technische physik, 20
Laboratory system, 276
Lamb, 9
LAPW calculations, 136
Lathe-drive, 71
Lattice defects, 379
Lattice deformation energies, 377, 383
Lattice dynamical measurements, 135
Lattice dynamics, 287, 346
Lattice excitations, 377
Lattice expansion and contraction, 376, 384,
        388
Lattice vibrational spectrum, 8

Lattice vibrations, 112, 265, 288
Length contraction, 278
Lense-Thirring effect, 287
Life and habitability in the Solar System, 300
Ligand electron negativity, 135
Line broadening, 5, 6
Line overlap, 6
Line shift, 4, 5
Line width, natural, 4, 12
Liouville operator, 224
Liquid drop model, 181
Liquid scintillator, 409
Liquid water, 300
Lithium ion batteries, 194
Lithium tin intermetallic compounds, 194
Localized magnetism, 133
Local Lorentz frame, 277
Local Lorentz Invariance, 277
Local Position Invariance, 277
Local test experiment, 277
Lorentz frame, 278
Lorentzian line, 281
Lorentz invariance, 288
Lorentz transformations, 275
Loudspeaker drive, 105
Low energy Th resonance, 369
Low temperature facilities, 27
Lunar samples, 298

Mössbauer absorption, 272
Mössbauer antineutrinos, 373, 374
Mössbauer calendar, 69
Mössbauer effect, 263, 272, 288
Mössbauer elements, 112
Mössbauer measurement, 286
Mössbauer vs. μSR relaxation studies, 132
Mössbauer lines, 24
Mössbauer mineralogy, 310
Mössbauer periodic table, 112
Mössbauer physicists, 26
Mössbauer spectroscopy, 281, 287
Mössbauer time scale, 125
Mössbauer transition, 275
Mössbauer transitions observed by synchrotron
        radiation, 341
Mach's principle, 286
Magnetic frustration, 132
Magnetic hyperfine anomaly, 165
Magnetic hyperfine interaction(s), 72, 200
Magnetic hyperfine splitting, 150, 175
Magnetic lattice incommensurate, 207
Magnetic moment reversal, 74
Magnetic moments, 384

Magnetic phase transitions, 206
Magnetic relaxation, 388
    model, 378
    phenomena, 376, 384
Magnetite, 304
Magnetochemistry, 254
Magnetosrictive coefficients, 211
Maier-Leibnitz, 6
Main absorber, 282, 284
Main group elements, 192
5th main group elements, 114
Main rules, 418
Majorana vs. Dirac particle, 413
Malmfors, 6
Mars, 297
Mars climate, 306
Mars exploration rover (MER), 297
MarsExpress Beagle-2, 297
Martian meteorites, 298
Masataka Fukugita, 421
Mass anisotropy, 286
Mass eigenstates, 385–387
Mass hierarchy, 387
Mass-squared difference, 386
Materials Science, 351
Matrix isolated atoms, 188
Matrix isolation, 119
Max-Planck Institute for Medical Research in
    Heidelberg, 19
Max-Plank Institute of Nuclear Physics, 421
Max-Planck-Institutes, 422
Mean-square atomic displacement, 377
Mean square nuclear charge radius, 180
Memoir for upgrading the physics, 417
Meridiani Planum, 301, 307
Mesoscopic, 365
Metal–organic, 30
Meteorites, 294
Methane monooxygenase, 249
Method of Malmfors, 22
Metric, 277
    gravitational theories, 282
    tensor, 274
Michelson interferometer, 279
Micro-Brownian movement, 239
Microfocusing, 346
Micromachining, 365
Micrometer size samples, 346
Microscopic mean-field, 158
Microwave background radiation, 275
MIMOS II, 294
MIMOS II Mössbauer spectrometer, 303
MIMOS IIA, 311
Minimal width, 374

Minkowski geometry, 278
Minkowski space-time, 275
Mixed $M1–E2$ transition, 167
Mixed valence, 226
Mixing angle, 386, 410
Mixing matrix, 385, 386
Modern tests of the General Theory of
    Relativity, 286
Modulated spin structure, 130
MoFe protein, 256
Molecular crystals, 350
Moment, 16
Momentum transfer, 376
Monitor absorber, 282
Monitor system, 281
Monolayers, 353
Moon, 13
Motional narrowing, 226, 379
Multi-bend achromat storage-ring, 360
Multi-quasi particle isomers, 160
Multichannel analyzer, 29, 105
Munich Research Reactor, 418
Mylar windows, 281

NaI(Tl) scintillation detectors, 23
Nanoclusters, 353
Nanograins, 353
Nanophase ferric oxide, 304, 310
Nanoscopic, 365
Natural abundance, 111
Natural linewidth, 263, 282, 284, 376, 378, 387
Natural width, 23, 266, 374
Nature of neutrino oscillations, 386
Nature of the hyperfine fields, 165
Nb lattice, 377
Nb metal, 376
Néel–Brown model, 233
Negative internal field-^{57}Fe, 75
Neptunium monochalcogenides, 133
Neptunium valence states, 132
Neptunyl bond, 134
Neptunyls, 133
Neutral current weak interactions, 406
Neutrino decay, 398
Neutrino detection, 92
Neutrino masses, 387, 393
Neutrino mass hierarchy, 413
Neutrino mass splitting, 412
Neutrino mixing, 393
Neutrino oscillation, 33, 374, 386, 395, 421
Neutrino physics, 37, 421
Neutrinos and beyond, 423
Neutrino spectroscopy, 407

Neutron absorption in crystals, 83
Neutron resonances, 23
Neutron stars, 287
New frontiers, 423
Newtonian gravitational potential, 271, 279
61-Ni, 188
Nickel alloys, 121
Nilsson model, 101
Nilsson single-particle wave functions, 156
Nitrogenase, 243, 244, 256
Nobel Prize, 287
Non-Doppler broadened Gamma rays, 26
Non-equilibrium systems, 362
Non-metric theories of gravitation, 283
Non-stationary phenomenon, 386
Non-white noise relaxation, 235
Normal mass hierarchy, 387
Note of Maier-Leibnitz, 27
^{237}Np, 208
NpAl$_2$, 208
NRIXS, 359
NSF, 359
Nuclear charge radius, 180
Nuclear decay, 276
Nuclear decay signal, 345
Nuclear exciton, 319, 370
Nuclear exciton polariton, 319, 324
Nuclear forward scattering, 359
Nuclear hyperfine anomaly between excited
        and ground state, 150
Nuclear lifetimes, 30, 151
Nuclear lighthouse effect, 276
Nuclear magnetic resonance technique, 286
Nuclear polarization, 72
Nuclear quantum optics, 370
Nuclear resonance absorption, 20
Nuclear resonance fluorescence, 145
Nuclear resonance inelastic X-ray scattering,
        340
Nuclear resonant inelastic scattering, 359
Nuclear resonant scattering cross-section, 342
Nuclear shapes, 150
Nuclear spin-spin interaction, 379
Nuclear techniques, 171
Nuclear transformations, 380
Nuclear Zeeman effect, 116
Nukleare festkörperphysik, 20
Null experiment, 274
Null gravitational experiments, 367

Occupation number, 9
Olivine, 304
Olivine-basalt, 305

Optical isotope shift, 176
Optical nuclear clocks, 370
Optical theory, 320
Orbital and spin contributions to the nuclear
        moment, 165
Orbital precession, 287
Organometallic tin compounds, 116
Origin and Structure of the Universe, 423
Origin of inertia, 286
189-Os, 188
Oscillation length, 387
Oscillation parameters, 387
Overlap effect in iodine, 118
Oxidation state, 295

Pair breaking via Coriolis effects, 153
Paramagnetic compounds, 203
Paramagnetic hyperfine patterns, 229
Paramagnetic hyperfine splitting, 129
Particle physics, 385
Pendellösung in the energy and time domains,
        333
Perihel precession, 287
Persistent fluctuations, 132
Phase coherence, 85, 318
Phase difference, 287
Phase problem, 32, 421
Phobos Grunt, 313
Phonon density of states, 135, 349
Phonon dispersion, 347
Phonon excitations, 349
Phonon-less emission and detection, 383
Phononless transitions, 376
Phonon spectrum, 264
Physics department, 417
Physik-Department, 32, 418
Piezo drive for Zn, 135
Piezoelectric drive, 283
Piezoelectric transducer, 273, 284
Planetary science, 313
Planetary surfaces, 297
Planet formation, 300
Point charge model, 128
Polarization, 202
Polarization effects, 285
Polarized spectra, 73
Position sensitive detector, 276
Precession, 288
Precision energy-shift measurements, 282
Preferred-frame effects, 277
Preferred-location effects, 277
Pressure dependence, 208
Principle of uniqueness of free fall, 277

Probability density, 90
Probability for recoilless process, 10
Probability of nuclear transitions, 9
Probability of phononless emission and
        detection, 388
Probability of phononless transitions, 383
Propagation beating, 324
Propagation mode of nuclear polariton, 323
Proportional counters, 20
Protein, 238
Proton decay, 414
195-Pt, 188
Pulsars, 287
Pulsed source, 341
Pump-probe studies, 366

QNFS, 360
Quadrupole interaction, 102
Quadrupole shift, 76
Quadrupole splitting, 150, 284
Quadrupole tensor, 284
Quantized energy state, 8
Quantum beat, 276
Quantum beating, 324
Quantum interference, 320
Quantum-mechanical matter waves, 286
Quantum phase coherence, 335
Quantum phase transitions, 159
Quasi-elastic, 238, 239
Quasielastic line, 238
Quasi-elastic nuclear forward scattering, 360

Radiation damage, 101, 365
Radiation detectors, 98
Radical SAM superfamily, 256
Radio pulsar, 287
Radiochemical neutrino experiments, 404
Raman process, 132
Random frequency changes, 379
Rare earth, 100, 124
Rare-earth system, 374
Reactor neutrinos, 395
Recoil energy, 4, 9, 377
Recoil-energy loss, 6, 25
Recoil free fraction, 99, 112, 282, 377
Recoil free line, 24
Recoilfree resonant antineutrino emission and
        absorption, 375
Recoilless emission and absorption, 14, 15
Recoilless γ-ray emission, 147
Recoilless line, 13

Recoilless nuclear resonance absorption of
        γ-radiation, 37
Recoilless transition, 9
Redistribution of bond structure, 118
Redshift, 284
Reference absorber, 284
Reference frame, 278
Reference standard for isomer shift, 182
Relativistic effects, 124, 380, 384
Relativistic phenomena, 263
Relativistic temperature shift, 266, 268
Relaxation, 378
    broadening, 126
    operator, 224
    spectra, 129, 221
Renewing of the German University system,
        418
Research facilities, 422
Research programs, 422
Resonance absorption, 3, 14, 269
    cross section (Ge), 138
    effect, 104
    line, 23
Resonance broadening, 328
Resonance cross section, 375, 385
Resonance fluorescence, 3
Resonant absorption, 49
Resonant and non-resonant scattering, 32
Resonant self absorption, 139
Resonant spectral density, 375
Resonant time spectra, 63
Rest mass, 387
Rest-mass energy, 286
Robotic space missions, 293
Rotating star, 287
Rotation of the coordinate axes, 275
Rotational excitations, 30
Rotational nuclear excitation, 129
Rotational transitions, 156
Rotor, 269, 273
Rotor disk, 270
Rotor experiment, 269, 270
^{99}Ru, 213
97-Ru, 185
99-Ru, 185, 188
Ruby glass, 123
Russian Academy of Sciences, 421
Ruthenium charge states, 121
Ruthenium magneto-superconducting, 213

Satellite, 271
121-Sb, 186
Schwarzschild radius, 279

Second Mössbauer conference, 31
Second Mössbauer effect, 32, 419
Second-order Doppler shift, 176, 183, 264, 276, 381
Seeding scheme, 361
Selective absorption, 27
S-electron density at the Mössbauer nucleus, 287
S-electron polarization, 117, 122
Self-consistent cranking model, 156
Self-gravitation, 277
Self-seeding, 362
Shapiro delay, 287
Shim plates, 26
Short-lived radioactive beams, 171
Short range crystallographic order, 136
Si-drift detectors, 311
Signal-to-noise ratio, 345
Single-line absorber, 273
Single-line systems, 285
Siroheme, 246
Six line iron spectrum, 76
Slow paramagnetic relaxation, 229, 230
Small particles, 233
119-Sn, 186
SNO, 405
Solar-neutrino oscillation, 387
Solar neutrinos, 402
Solar system, 293
Solid-state effects, 284, 287
Solid state physics, 27
Sonderforschungsbereiche, 422
Sound velocity, 349
Source and absorber, 6, 7
Source experiments, 115
Space, 420
Space-time homogeneity, 275
Space-time isotropy, 275
Spatial anisotropy, 286
Spatial coherence, 318
Special Relativity, 277
Special Theory of Relativity, 263, 265, 272, 288, 380
Speed of light, 279
Speed-up effect, 334
S-p hybridization, 116
Spiegel, 417
Spin-crossover, 228
Spin density wave, 214
Spinel, 285
Spin expectation value, 245
Spin Hamiltonian, 244, 255
Spin–lattice, 223
Spin-lattice relaxation, 132

Spinning charge, 287
Spinning mass, 287
Spin reorientations, 210
Spin-spin interactions, 378
Spin–spin relaxation, 131, 223
Spin transition, 190, 252
Standing wave mode, 326
Sterile neutrinos, 387
Sternheimer factor, 103
Stochastic fluctuations, 223
Stochastic processes, 378
Stochastic relaxation, 379
Stokes, 3
Sulfides, 295
Sulfite reductase, 246
Super- and subradiant states, 92
superconducting state, 216
Superconductive detector, 421
Supernova neutrinos, 414
Superparamagnetic fluctuations, 223
Superparamagnetic relaxation, 233
Superposition of mass eigenstates, 386
Suppression effect, 327
Surfaces, 217, 351
SXFEL, 361
Synchrotron Mössbauer Source, 335
Synchrotron Mössbauer spectroscopy, 340
Synchrotron radiation, 275, 276, 288

181-Ta, 188
Ta as dilute impurity, 137
Technische Hochschule München, 19
Temperature control, 281
Temperature decrease, 7
Temperature dependence of hyperfine field, 78
Temperature dependence of the electronic absorption, 21
Temperature shift, 266
Temporal coherence, 318
Terrestrial planets, 300
Test body, 277
Test body trajectories, 277
Tests of time reversal invariance, 167
Theoretical calculations, 378
Thermal excitation, 103
Thermal motion, 5
Thermonuclear fusion, 400
Third Generation Facilities, 360
Thomson cross-section, 342
Three-body process, 374
Three-level system, 378
Thulium 169, 100
Thulium metal, 102

Thulium oxide, 101
Time dilatation, 263, 264, 278
Time-dilatation effect, 380
Time discrimination, 342
Time filtering, 75
Time reversal, 123
Time scanning, 105
Time-energy uncertainty principle, 378, 379
Tin bonding, 116
Tin compounds, 193
Total reflection, 27
Transferred field, 116
Transferred hyperfine field, 216
Transferred hyperfine interaction, 78
Transmission integral, 120
Transregios, 422
Transuranic elements, 124
Tritium trick, 384
Twin paradox, 265
Two-body process, 374

Ultracentrifuge rotor, 273
Ultra-low divergence, 362
Ultra-short base lines, 387
Undulators, 342, 360
Universe, 286

Vacancy formation, 120
Valence state, 190
Variations of the binding energies, 380, 383
Vector interaction, 274
Velocity drive, 28
Velocity relative to distant matter, 274
Venus, 294

Vibrational spectra, 346
Victoria crater, 309
Volcanic activity, 306

Walther Schottky Institute, 419
Wave-particle duality, 84
Weak Equivalence Principle, 277
Weak-field limit, 279
Weak interaction, 376
Weathering processes, 301
White noise approximation, 131, 223
Whole cells, 259
Wien2k, 186
Wolfgang Paul, 418
World lines, 278

XFEL, 361
X-ray Free Electron Laser, 361
X-ray sources, 359

Yttrium-iron garnet (Sn), 117

Zero-field splitting, 244
Zero-point energy, 379, 382
Zero-point motion, 268, 287, 288, 377
Zero-point vibrations, 381
Zn chalcogenides, 135
Zn-Cu alloy, 136
Zn metal, superconducting, 137
Zn nanocrystalline, 136
^{67}Zn resonance, 282
ZnO, 284